Discrete Mathematics and Combinatorics

Discrete Mathematics and Combinatorics

T. Sengadir
Associate Professor
Department of Mathematics
SSN College of Engineering
Chennai, India.

PEARSON

Chennai • Delhi • Chandigarh

ISBN 978-81-317-1405-8

First Impression

Published by Dorling Kindersley (India) Pvt. Ltd., licensees of Pearson Education in South Asia.

Head Office: 7th Floor, Knowledge Boulevard, A-8(A), Sector-62, Noida.
Registered Office: 14 Local Shopping Centre, Panchsheel Park, New Delhi 110 017, India.

Laser typeset by Sigma Business Process Pvt. Ltd., Chennai.

Printed in India at Rahul Print-o-Pack.

Dedicated to

Visalakshy (my mother), Kamatchy and Rajagopalan who taught in vernacular Government schools and taught me well.

Contents

Foreword

Topics in discrete mathematics play an important role in many modern engineering and technological fields like computer science, communication engineering and network security. Number theory, which did not have any application during the times of Ramanujam and Hardy, has now turned out to be a highly useful topic in fields like cryptography. There is a strong need for a text book which would give an idea of the applications of mathematical ideas in real life.

I am glad to note that Dr T. Sengadir, Department of Mathematics, SSN College of Engineering, has written such a text book and it gives me immense pleasure to note that Pearson Education is bringing it out. The material in this book can be comprehended on self-study. I am sure that the student community would be benefited by this text book. I recommend that the copies of this book be made available in the libraries of all institutions of higher learning where subjects like computer science, information and communication are taught.

Professor S. Salivahanan,
Principal,
SSN College of Engineering,
Chennai

Preface

The teaching of Mathematics to engineering and science students is a real challenge. Their mathematical skills may surpass those of students of mathematics per se. However, there is a need for the instructor to make students feel that the topics are relevant to them. A typical student of mathematics is quite comfortable when the instructor starts a topic with a definition, gives some examples and goes on to prove theorems, whereas an engineering or a science student often wonders why a lot of fuss is made about simple things in the name of mathematical rigor. While it is true that students should eventually become rigorous in their approach to problem solving, undue stress on mathematical rigor right from the start may distract many potentially good students. When teaching mathematics to non-mathematics students, it is useful to start with a real-life example before giving formal definitions. Examples may be chosen in such a way as to make students realize the importance and the relevance of the definition. In this book I have provided such examples, wherever feasible. I have used this approach in the classroom and have found that it sets the students thinking and gets them interested in the topics under discussion.

The chief aim of this book is to provide a useful self-study material for the student. There are no pre-requisites needed, and the appendices present a useful overview of the essential basics. The use of logic and mastery of the language of sets and functions are essential parts of understanding discrete mathematics. The first two chapters are written in such a way that learners who have no previous exposure to these ideas can master them on their own.

This book can be used as a text book of discrete mathematics for graduate and postgraduate students of Computer Science and Engineering. To the extent possible, I have made the discussions resemble the transcription of an actual lecture. Example problems given in the book may be worked out by the instructor in the class room, while the Practice problems may be assigned to the students. Students are expected to solve these problems before proceeding further. This will not only enhance students' understanding of the concepts discussed but will also prepare them for the next topic.

Acknowledgements

I was fortunate to have been exposed to excellent teaching of mathematics teaching right from my school days in the semi-rural atmosphere of Lawspet, Pudhucherry. I acknowledge each and every teacher of mathematics at the Department of Mathematics, Regional Institute of Education, Mysore and the teachers at the Department of Mathematics, Indian Institute of Technology, Mumbai.

I thank the Management, SSN College of Engineering, for providing the latest computing facilities and an excellent atmosphere conducive to academic pursuits that has helped me prepare the manuscripts easily. I am obliged to the Head of my Department, Professor S. Narasimman for his encouragement and support in writing this book. It is a pleasure to acknowledge all my colleagues, especially Dr R. Sujatha and Professor B. Praba who were a source of inspiration when I was working this project.

I am grateful to my wife S. Chitra, a dedicated scientist, who takes care of the house without allowing the demands of her vocation to weigh her down. Perhaps I should also add that in the guise of restoring order in the house, she has, at times, misplaced the manuscripts! My daughter C. S. Yalini could have helped me to complete this book a bit faster. Nevertheless, she was a pleasurable distraction.

Finally, I thank the editorial team at Pearson whose patience and co-operation greatly helped this book in taking its present form.

T. Sengadir

About the Author

Dr T. Sengadir received his Ph. D in Mathematics from the Department of Mathematics, Indian Institute of Technology, Mumbai. After holding a post-doctoral Fellowship at the Institute of Mathematical Sciences, Chennai, he joined the Department of Mathematics, Indian Institute of Science as a lecturer. Currently, he is Assistant Professor at the Department of Mathematics, SSN College of Engineering, Chennai. He is an active researcher committed to the teaching of Mathematics at all levels. He has been a resident faculty in the Mathematics Training and Talent Search programmes organized by the National Board of Higher Mathematics. He is an advisor of Pavalareru Thamizh Vazhith Thodakkappalli (Pavalareru Tamil Medium Primary School), Kalpakkam, which provides quality education to children of all economic strata at a low cost. He is also a member of Reward Trust, a trust which aims to promote the quality of education in rural areas.

CHAPTER 1

Equations, Inequalities and Basic Logic

Learning Objectives

On completing this chapter, you should be able to:

- recall some basic ideas related to linear equations, quadratic equations, linear inequalities and quadratic inequalities
- solve simple equations and inequalities by the method of drawing graphs
- make logically precise statements in some simple and familiar situations

1.1 EQUATIONS AND THEIR SOLUTIONS

$$ax + b = 0 \tag{1.1}$$

is probably the simplest of all mathematical equations. We assume that the constants a and b are known and try to find the value of x, which would satisfy equation (1.1).

Quantities whose values have to be found out from the given equations are called *unknowns*. Equation (1.1) is an example of a linear equation in one unknown. The solution to equation (1.1) is given by

$$x = \frac{-b}{a}.$$

Next, let us consider the equation

$$px + q + m - nx + k(sx + u) = t(hx - e) + jx + v. \tag{1.2}$$

This is slightly more complicated than equation (1.1) but it can be easily simplified to take the form of equation (1.1). That is, we rewrite equation (1.2) as

$$(p - n + ks - th - j)x + (q + m + ku + te - v) = 0 \tag{1.3}$$

and its solution is

$$x = \frac{-q - m - ku - te + v}{p - n + ks - th - j}. \tag{1.4}$$

Let us now try to solve the following equation

$$2x + 3 + 1 - 5x + (-2)(x + 3) = (2x - 4) - 7x. \tag{1.5}$$

We can directly solve this equation, but if we want to apply the formula (1.4), it is clear that $p = 2$, $q = 3$, $m = 1$, $n = 5$, $k = -2$, $s = 1$, $u = 3$, $t = 1$, $h = 2$, $e = 4$, $j = -7$ and $v = 0$. The numerator is -2 and the denominator $= (2 - 5) + (-2)(1) - 2 + 7 = 0$. Hence, equation (1.5) has no solution.

This simple example tells us that, even in simple situations, we have to be careful with theoretical considerations. The solution given for equation (1.1) is valid when $a \neq 0$, and equation (1.4) is valid when $p - n + ks - th - j \neq 0$.

So, we have observed that the equation

$$0x = -2$$

has no solution.

From this discussion we understand that:

A linear equation in one unknown may not have a solution.

Can you think of a linear equation in one unknown with infinitely many solutions?

Consider the equation

$$0x = 0.$$

Then, it is clear that any number x is a solution of this equation.

It is interesting to study linear equations with the help of graphs. We know that the graph corresponding to the equation $y = ax + b$ is a straight line (Fig. 1.1).

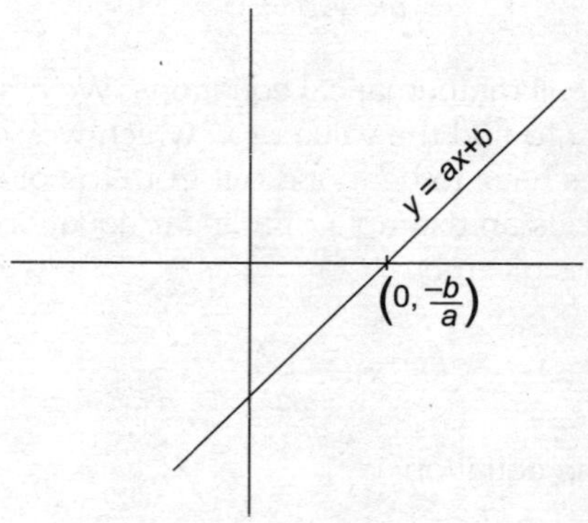

Figure 1.1 $y = ax + b$

If this line is not parallel to the x axis (that is, if $a \neq 0$), it meets the x axis at the point $[(-b/a), 0]$. Note that the x co-ordinate of the point of intersection of the line $y = ax + b$ and the x axis is the solution to equation (1.1).

In fact, for any expression $E(x)$, the solutions of $E(x) = 0$ are the x co-ordinates of the points of intersection of the graph of $y = E(x)$ and the x axis.

PRACTICE 1.1

How would you describe the solution of $ax + b = c$ graphically?
Hint: Think of the line $y = c$. This line is parallel to the x axis.

Next, we consider the quadratic equation

$$ax^2 + bx + c = 0.$$

The key fact in understanding the solution of a quadratic equation is as follows:

The square root of a negative number is not a real number.

The simplest quadratic equation is

$$x^2 + c = 0.$$

Assume that $c \leq 0$. Then the above equation has two real solutions, which are given by $x = \pm\sqrt{-c}$. If $c > 0$, then $-c$ is negative and hence this equation has solutions $\pm\, i\sqrt{c}$, which are imaginary numbers. Here i is the imaginary square root of the negative number -1.

PRACTICE 1.2

Draw the graphs of $y = x^2 - 4$ and $y = x^2 + 4$. Note that the second graph does not meet the x axis anywhere. This illustrates the fact that a negative number does not have a real square root.

Now, we shall learn about the nature of solutions of a general quadratic equation

$$ax^2 + bx + c = 0,$$

with the help of graphs. Let us assume that $a > 0$. If we draw the graphs of $y = ax^2 + bx + c$ (Fig. 1.2) for various values of a, b and c, we find that the graphs are curves known as parabolas. We also find that relative to the x axis, there are three possibilities.

(a) The graph is completely above the x axis.
(b) The graph touches the x axis.
(c) The graph cuts the x axis at two points.

The above graphical possibilities correspond to the following algebraic possibilities:

(a) There is no real solution to the equation $ax^2 + bx + c = 0$.
(b) There is exactly one real solution to the equation $ax^2 + bx + c = 0$.
(c) There are two real solutions to the equation $ax^2 + bx + c = 0$.

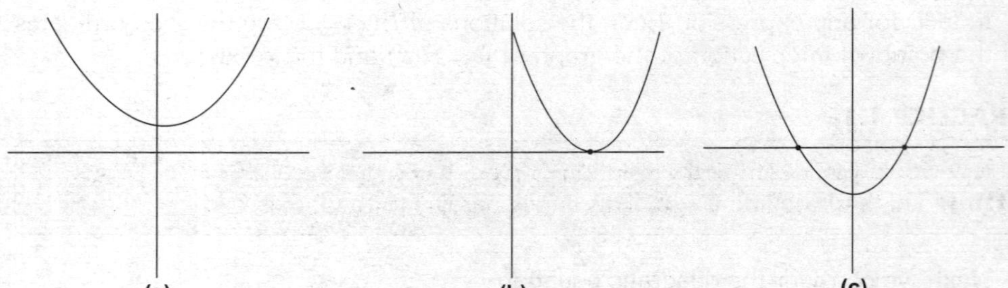

Figure 1.2 $y = ax^2 + bx + c$

When $a < 0$, all the above three graphs get inverted, but the statements regarding the nature of solutions of the quadratic equation remain the same.

Next, we give the explicit algebraic solution to the quadratic equation. The solutions of the quadratic equation $ax^2 + bx + c = 0$, $a \neq 0$, are given by the formula

$$\frac{-b \pm \sqrt{b^2 - 4ac}}{2a}.$$

When we use the above formula to solve the equation $x^2 + 14x + 50 = 0$, we obtain

$$\frac{-14 \pm \sqrt{-4}}{2}.$$

Since a negative number is present inside a square root, the equation $x^2 + 14x + 50 = 0$ does not have any real solution. But using the property $i^2 = -1$, we get the complex solutions as

$$-7 \pm i.$$

The quantity $b^2 - 4ac$, which is in the square root symbol of the solution formula, is known as the discriminant of the equation.

Table 1.1 summarizes the nature of the solutions of $ax^2 + bx + c = 0$ in terms of the discriminant.

TABLE 1.1
NATURE OF SOLUTIONS OF $ax^2+bx+c=0$

The value of $b^2 - 4ac$	Nature of the solutions
+	Real and distinct
0	Real and equal
–	Complex roots

1.2 LOGICAL IMPLICATIONS AND EQUIVALENCES

No doubt calculations are very important in mathematics, but in the absence of correct logical reasoning, blind calculations may lead to errors. We shall illustrate this by a well-known example of an extraneous root.

Consider the problem of solving the equation

$$\sqrt{x+5}+\sqrt{x+21}=\sqrt{6x+40}. \tag{1.6}$$

This equation involves square roots, and hence the natural step is to square both the sides.

By doing this we obtain

$$(x+5)+(x+21)+2\sqrt{(x+5)(x+21)}=6x+40. \tag{1.7}$$

We find that we have not yet eliminated the square root and so we must square again. But a little thinking tells us that squaring both the sides of equation (1.7) is not going to eliminate the square root. So, we push the square root symbol to one side as follows:

$$(x+5)+(x+21)-(6x+40)=-2\sqrt{(x+5)(x+21)}. \tag{1.8}$$

Note that at the above stage, we were guided by logic and not by a blind method. Now, simplification and squaring of equation (1.8) gives us

$$3x^2+2x-56=0 \tag{1.9}$$

which can be solved and we obtain $x=4$ and $x=(-14)/3$.

Standard text books tell us that we can substitute these values in equation (1.6) to verify which of them is a *genuine* solution and which is an *extraneous* solution.

We find that $x=4$ satisfies equation (1.6) and $x=(-14)/3$ does not. Thus, the only solution is $x=4$.

Now, let us consider another example.

Solve

$$x=(x^2+12x)^{\frac{1}{3}}. \tag{1.10}$$

The natural thing for us to do is to cube both the sides. After cubing and simplification, we get the cubic equation

$$x^3-x^2-12x=0 \tag{1.11}$$

which can be easily factorized into

$$x(x+3)(x-4)=0. \tag{1.12}$$

Its solutions are given by $x = 0$, $x = -3$ and $x = 4$. If we now substitute these values in equation (1.10), we find that there are no extraneous solutions. All these values are solutions of the equation we started with.

Now, let us see when we may get extraneous solutions and when we may not. Let us make the following observations:

(i) Every solution to equation (1.6) is a solution to equation (1.9).
(ii) Every solution to equation (1.10) is a solution to equation (1.12).
(iii) Every solution to equation (1.12) is a solution to equation (1.10).

But the following statement is **false**.

(iv) Every solution to equation (1.9) is a solution to equation (1.6).

We now take up a real-life situation to understand what is actually happening here. This would help us to focus purely on logical considerations without getting confused by equations. Let us consider the following statement :

If today is a Sunday, then today is a holiday.

Now, if today is a holiday, can we assert that today is a Sunday? Obviously not. Because, there are many holidays which do not fall on Sundays. Not many persons make this kind of a logical error in day-to-day life. But while doing mathematics, this kind of error is very common. If we think that all solutions to equation (1.9) are also solutions to equation (1.6), it is like concluding that every holiday is a Sunday.

Consider the following statements which are true

If $A = B$, then $A^2 = B^2$. (S_1)
If $A = B$, then $A^3 = B^3$. (S_2)

Next, consider the statements

If $A^2 = B^2$, then $A = B$. (S_3)
If $A^3 = B^3$, then $A = B$. (S_4)

Both the above statements are false because $(-1)^2 = 1^2$ but $1 \neq -1$. If ω is a complex cube root of 1, then $\omega^3 = 1^3$ but $1 \neq \omega$.

But if A and B are real numbers, then S_4 is true and that is why we did not get any extraneous roots for equation (1.10) by cubing it.

Based on these observations, it can be said that every solution to

$$f(x) = g(x) \tag{1.13}$$

is also a solution of

$$[f(x)]^2 = [g(x)]^2 \tag{1.14}$$

but not the other way round. That is why some solutions of equation (1.14) are not solutions of equation (1.13), which are called extraneous solutions.

Definition 1.1: If P is a statement and Q is a conclusion that follows from P, in logical notation, we write this as

$$P \rightarrow Q. \tag{1.15}$$

The converse of equation (1.15) is defined as the statement

$$Q \rightarrow P \tag{1.16}$$

which may or may not be true. When both equations (1.15) and (1.16) are valid, we write

$$P \leftrightarrow Q$$

and say that P and Q are logically equivalent.

A statement of the form (1.15) is called a *logical implication* and a statement of the form (1.16) is known as a *logical equivalence*.

Using this notation, now we can write

$$\sqrt{x+5}+\sqrt{x+21}=\sqrt{6x+40} \longrightarrow 3x^2+2x-56=0 \tag{1.17}$$

and the converse of equation (1.17) is false.

Next, under the assumption that x is a real number,

$$x=(x^2+12x)^{\frac{1}{3}} \longrightarrow x(x+3)(x-4)=0 \tag{1.18}$$

is valid and further, its converse

$$x(x+3)(x-4)=0 \longrightarrow x=(x^2+12x)^{\frac{1}{3}} \tag{1.19}$$

is also valid. We write this as

$$x=(x^2+12x)^{\frac{1}{3}} \longleftrightarrow x(x+3)(x-4)=0. \tag{1.20}$$

Example 1.1

Examine whether the following implications are true or not:

(i) $x \leq 3 \rightarrow 2x \leq 6$.
(ii) $x > 5 \rightarrow xy > 5y$.
(iii) $x = 1 \rightarrow x^3 = 27$.
(iv) $x = 10 \rightarrow 4x - 3 = 37$.
(v) $x = -2 \rightarrow x^2 = 4$.
(vi) $x^2 = 100 \rightarrow x = 10$.

SOLUTION

We can easily see that (i), (iv) and (v) are all true implications. Clearly, (ii) is false.

If we take $y = -1$, and $x = 10$, then (ii) becomes

$$10 > 5 \rightarrow -10 > -5$$

which is absurd. Hence, (ii) is false.

(vi) is also false because $x^2 = 100$ includes the two possibilities $x = 10$ and $x = -10$. However, let us also note that the converse of (vi) is true. ■

Example 1.2

The following are all true:

(i) It rains $\rightarrow$ Roads get wet.
(ii) $x = 3 \rightarrow x + 5 = 8$.
(iii) $x = 3 \rightarrow x^2 = 9$.
(iv) $\triangle ABC$ is an equilateral triangle $\rightarrow \angle A = 60^0$.
(v) In $\triangle ABC$, $AB = AC \rightarrow \angle B = \angle C$.

Let us observe that converses of (i), (iii) and (iv) are not true. But converses of (ii) and (v) are true. Therefore, we can write

(i) $x = 3 \leftrightarrow x + 5 = 8$.
(ii) In $\triangle ABC$, $AB = AC \leftrightarrow \angle B = \angle C$. ■

PRACTICE 1.3

Identify the true implications in the following. Also, identify implications whose converses are true.

(i) In $\triangle ABC$, $AB \geq AC \geq BC \rightarrow \angle C \geq \angle B \geq \angle A$.
(ii) $x < 3 \rightarrow x < 5$.
(iii) $x^2 - 10x + 25 = 0 \rightarrow x = 5$.
(iv) $x^2 + 8x + 15 = 0 \rightarrow x = -3$.
(v) $x > 2 \rightarrow x = 5$.

1.3 SOLVING LINEAR INEQUALITIES

Example 1.3

Imagine an institution that has allotted a maximum of Rs 12,10,000 to procure computers for 15 of its staff members. The estimated transport and installation charge is Rs 25,000. We want to find the maximum price for a single computer. If x is the possible price of a computer, then to spend within the allotted budget, we must have

$$15x + 25{,}000 \leq 12{,}10{,}000. \tag{1.21}$$

We can easily calculate the maximum price for a computer as Rs 79,000. Equation (1.21) is an example of a Linear inequality and any x satisfying $x \leq 79{,}000$ is a solution to equation (1.21). Equation (1.21) is an example of a Linear inequality and any x satisfying $x \leq 79{,}000$ is a solution to equation (1.21). ■

Example 1.4

Solve the following Linear inequality

$$3 \leq 4x + 6 \leq 9. \tag{1.22}$$

SOLUTION

If we substitute $x = 0$ in equation (1.22), we get

$$3 \leq 6 \leq 9$$

which is true. If we substitute $x = 1$ in equation (1.22) we get

$$3 \leq 10 \leq 9$$

which is false. If we substitute $x = -2$ in equation (1.22), we get

$$3 \leq -2 \leq 9$$

which is also false. So, we say that $x = 0$ is a solution to the inequality (1.22) and $x = 1$ and $x = -2$ are not solutions to equation (1.22). We would like to know all the solutions to equation (1.22). We know that, when $a \neq 0$, the Linear equation $ax + b = 0$ has only one solution. But, in general an inequality such as equation (1.22) has many solutions. The procedure to solve this inequality is not very much different from that of solving a linear equation. Let us now solve equation (1.22).

$$3 \leq 4x + 6 \leq 9. \tag{1.22}$$

$$\leftrightarrow \quad 3 - 6 \leq 4x \leq 9 - 6. \tag{1.23}$$

$$\leftrightarrow \quad -3 \leq 4x \leq 3. \tag{1.24}$$

$$\leftrightarrow \quad \frac{-3}{4} \leq x \leq \frac{3}{4}. \tag{1.25}$$

So, any x that satisfies equation (1.22) also satisfies equation (1.25). That is, x is between $(-3)/4$ and $3/4$. Further, if x is between $(-3)/4$ and $3/4$, then x will satisfy equation (1.22). Hence, the solutions of equation (1.22) are precisely the numbers between $(-3)/4$ and $3/4$. Now, let us consider the inequality $33 < -10x + 3 < 43$. If we try to solve this inequality as we did in the previous case, we get the result $-3 < x < -4$, which is meaningless. But, common sense tells us that the solutions are numbers between -4 and -3. Now, let us carefully examine the steps involved in the solution. It is easy to see that for any c,

$$(a < b) \quad \leftrightarrow \quad (a + c < b + c).$$

In fact, this is what we use for getting equation (1.23) from equation (1.22) with $c = -6$. Now, the question is the following:

$$\text{Is } (a < b) \leftrightarrow (ac < bc)?$$

In general, the answer is *no*. But, in getting equation (1.25) from equation (1.24), we multiply the inequalities by $c = 1/4$, which is positive. In this case, we have reached the correct conclusion. But, if c is negative, from $a < b$, we cannot conclude that $ac < bc$. In fact, we must conclude that $ac > bc$. Now, we correct the rules in the form of a proposition. The proof is given in the appendix. ■

Proposition 1.1

The following rules for inequalities are true:

(i) For any c, $a < b \leftrightarrow a + c < b + c$.
(ii) For any c, $a \le b \leftrightarrow a + c \le b + c$.
(iii) For any $c > 0$, $a < b \leftrightarrow ac < bc$.
(iv) For any $c > 0$, $a \le b \leftrightarrow ac \le bc$.
(v) For any $c < 0$, $a < b \leftrightarrow ac > bc$.
(vi) For any $c < 0$, $a \le b \leftrightarrow ac \ge bc$. ■

Example 1.5

Solve the following inequalities:

(i) $-2x + 9 \ge 0$.
(ii) $5x - 4 < 0$.
(iii) $3 \le 4x + 2 < 9$.

SOLUTION

(*i*)

$$\begin{aligned} -2x + 9 &\ge 0 \\ \Rightarrow -2x &\ge -9 \\ \Rightarrow x &\le \frac{-9}{-2} \\ \Rightarrow x &\le \frac{9}{2} \end{aligned}$$

(*ii*)

$$\begin{aligned} 5x - 4 &< 0 \\ \Rightarrow 5x &< 4 \\ \Rightarrow x &< \frac{4}{5} \end{aligned}$$

(*iii*)

$$\begin{aligned} 3 \le\ & 4x + 2 \le 9 \\ \Rightarrow 2 \le\ & 4x \le 7 \\ \Rightarrow \frac{2}{4} \le\ & x \le \frac{7}{4}. \end{aligned}$$

■

1.4 RANGES FOR LINEAR EXPRESSION

Example 1.6

Imagine that a mathematician wants to buy 3 kg of sugar. He does not own a car or a scooter, and his bicycle is also not in a good condition. He cannot get an auto-rikshaw near his house and so he must walk to the shop. There is an auto stand near the shop. So, he can get an auto-rikshaw to come back home. This would cost him Rs 30. He does not remember the exact price of 1 kg of sugar. He knows that it is somewhere between Rs 14 and Rs 16. He wants to estimate the total expense for buying the sugar and coming back by an auto. ■

This problem can be posed mathematically as follows: If $14 \leq x \leq 16$, what are the possible values for $y = 3x + 30$? By following the rules for inequalities we can show that the mathematician's total expense would be between Rs 72 and Rs 78.

The rule we have used in the above solution is stated as a proposition below:

Proposition 1.2

(1) If $p > 0$, then

$$a < x < b \leftrightarrow pa + q < px + q < pb + q.$$

(2) If $p < 0$, then

$$a < x < b \longrightarrow pa + q > px + q > pb + q.$$

These rules are obtained by combining the rules of Proposition 1.1 in Section 1.3. ■

PRACTICE 1.4

Give a graphical interpretation to the rules of Proposition 1.1.

Example 1.7

There are 15 staff members in the department of Electrical Engineering of Gandhi Institute of Technology. Each of the staff members would be given a PC costing between Rs 50,000 and Rs 70,000. If the total cost of transportation and installation is expected to be Rs 30,000, estimate the expenditure.

SOLUTION

Mathematically, this problem is to find all possible values of $y = 15x + 30{,}000$ when all the possible values of x lie between 50,000 and 70,000. We solve it as follows:

$$50{,}000 \le x \le 70{,}000.$$
$$\leftrightarrow \quad 15 \times 50{,}000 + 30{,}000 \le 15x + 30{,}000 \le 15 \times 70{,}000 + 30{,}000.$$
$$\leftrightarrow \quad 7{,}80{,}000 \le 15x + 30{,}000 \le 10{,}80{,}000.$$

Thus, the estimated cost is between 7,80,000 and 10,80,000. ■

1.5 QUADRATIC INEQUALITIES

In Section 1.3 we solved inequalities of the type

$$p \le ax + b \le q.$$

In this section we shall learn about procedures for solving

$$p \le ax^2 + bx + c \le q.$$

First, let us consider the simple quadratic inequality $x^2 \leqslant b$. Since $x^2 \ge 0$, if $b < 0$, there is no solution for this inequality. Here, we must point out that the statement $p < q$ has no meaning if p and q are complex numbers.

So, let us assume that $b \ge 0$. By *taking square roots on both the sides,* we obtain $x \le \sqrt{b}$. Have we solved the inequality? No. Take $b = 25$. Let us consider

$$x^2 \le 25. \tag{1.26}$$

The above procedure tells us that

$$x \le 5. \tag{1.27}$$

Now, $x = -6$ satisfies equation (1.27), but $x = -6$ does not satisfy equation (1.26) since $(-6)^2 > 25$. In other words, the logical implication

$$x^2 < a^2 \rightarrow x < a$$

is **false**.

The actual solution to equation (1.26) is given by $-5 \le x \le 5$. This is illustrated in Figure 1.3.

Generalizing from the above example, we obtain the first rule for a quadratic inequality as follows. Assuming $b \ge 0$,

$$x^2 < b \leftrightarrow -\sqrt{b} < x < \sqrt{b}.$$

Now, let us consider the set of all solutions of

$$a < x^2 < b. \tag{1.28}$$

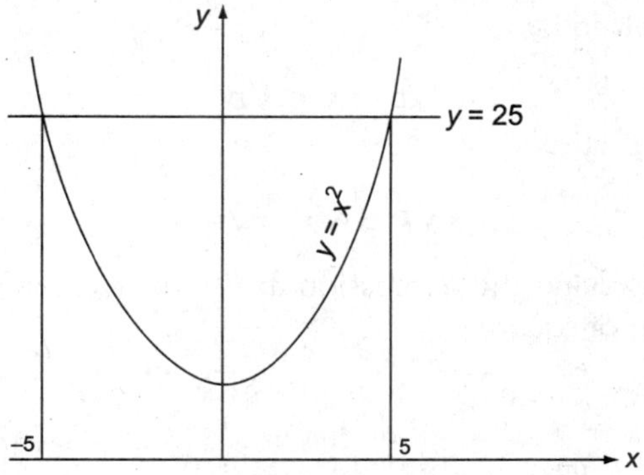

Figure 1.3 Solution of $x^2 \leq 25$

If $a < 0$, the condition $a < x^2$ always holds and equation (1.28) simply reduces to that of $x^2 < b$ for which we already know the solutions. If b is negative. then clearly there is no solution to this inequality. So, let us now consider equation (1.28) with the conditions that $a \geq 0$ and $b \geq 0$ and naturally with $a < b$. A little bit of thought tells us that the solutions of equation (1.28) are given by

$$\sqrt{a} < x < \sqrt{b}$$

and

$$-\sqrt{b} < x < -\sqrt{a}.$$

Figure 1.4 solves this problem graphically.

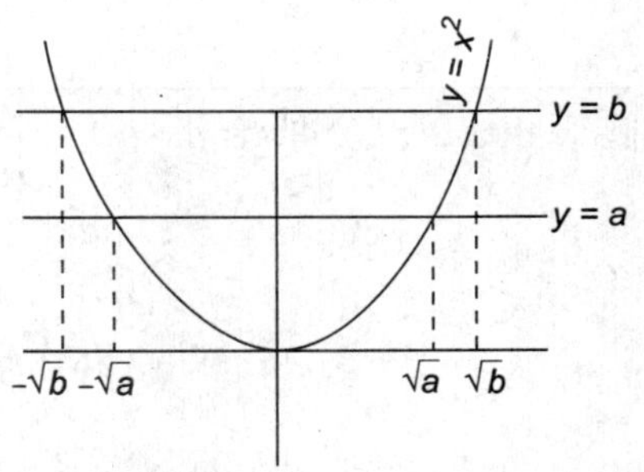

Figure 1.4 Solutions of $a < x^2 < b$

From the graph, we find that the solutions of equation (1.28) are given by

$$\sqrt{a} < x < \sqrt{b}$$

and

$$-\sqrt{b} < x < -\sqrt{a}$$

We can easily see how one has to modify the solutions for inequalities like

$$a < x^2 \leq b.$$

The solutions are given by

$$\sqrt{a} < x \leq \sqrt{b}$$

and

$$-\sqrt{b} \leq x < -\sqrt{a}.$$

The inequalities involving the expression $ax^2 + bx + c$ can be simplified by the method of completion of square.

Example 1.8

Solve $x^2 + 6x - 9 < 17$.

SOLUTION

$$\begin{aligned} & x^2 + 6x - 8 < 17. \\ \Leftrightarrow\ & x^2 + 2(3)(x) + 9 - 9 - 8 < 17. \\ \Leftrightarrow\ & (x+3)^2 - 17 < 17. \\ \Leftrightarrow\ & (x+3)^2 < 34. \\ \Leftrightarrow\ & -\sqrt{34} < (x+3) < \sqrt{34}. \\ \leftrightarrow\ & -\sqrt{34} - 3 < x < \sqrt{34} - 3. \end{aligned}$$

In this example, we have rewritten $x^2 + 6x - 8$ as $(x+3)^2 - 17$. This is the procedure known as completion of square. ■

PRACTICE 1.5

Let $a \geq 0$. Show graphically that the solutions of

$$x^2 > a$$

are given by

$$x > \sqrt{a}$$

and

$$x < -\sqrt{a}.$$

Remark 1.1

If $a < 0$, any real number x is a solution to the inequality $x^2 > a$. ■

Example 1.9

Solve $x^2 - 9x + 15 > 0$.

SOLUTION

$$x^2 - 9x + 15 > 0.$$

$$\leftrightarrow \left(x - \frac{9}{2}\right)^2 - \frac{81}{4} + 15 > 0.$$

$$\leftrightarrow \left(x - \frac{9}{2}\right)^2 > \frac{21}{4}.$$

$$\leftrightarrow \left(x - \frac{9}{2}\right) \leq -\sqrt{\frac{21}{2}}$$

or

$$\left(x - \frac{9}{2}\right) \geq \sqrt{\frac{21}{2}}.$$

$$\leftrightarrow x \leq \frac{9}{2} - \sqrt{\frac{21}{2}}$$

or

$$x \geq \frac{9}{2} + \sqrt{\frac{21}{2}}.$$

■

1.6 RANGES FOR QUADRATIC EXPRESSION

In this section, we find all the possible values for $ax^2 + bx + c$ if $p < x < q$.

Example 1.10

If $-3 < x < 5$, what are the possible values of x^2? Blindly squaring leads to the solution $9 < x^2 < 25$, which is false. For example, take $x = -2$. The inequality $-3 < -2 < 5$ is true, but $9 < 4 < 25$ is **not true.** By looking at the graph of $y = x^2$, we find that $-3 < x < 5 \to 0 < x^2 < 25$ (Fig. 1.5).

In general, the conclusion

$$p < x < q \quad \to \quad p^2 < x^2 < q^2$$

is **false.**

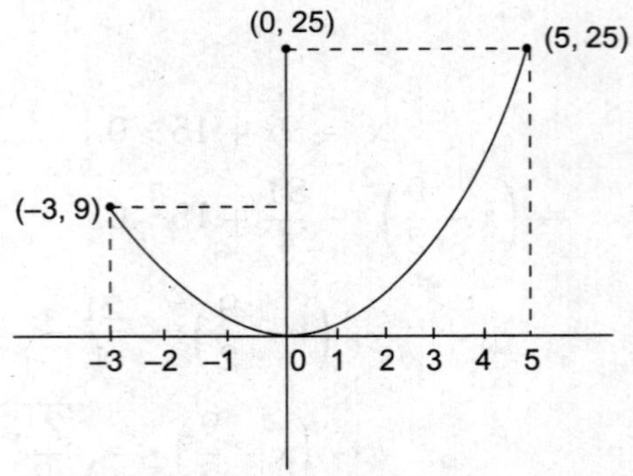

Figure 1.5 Solution of a quadratic inequality. ■

Now, let us list the correct statements:

(a) If $0 \le p < q$, then $p < x < q \leftrightarrow p^2 < x^2 < q^2$.
(b) If $p < q \le 0$, then $p < x < q \leftrightarrow p^2 > x^2 > q^2$.

Example 1.11

If $-3 < x < 5$, what are all the possible values of $3x^2 - 4x + 3$?

SOLUTION

We break the given condition into two parts. First, consider the values of x satisfying $0 < x < 5$.

$$0 < x < 5.$$
$$\leftrightarrow \quad 0 < x^2 < 25.$$
$$\leftrightarrow \quad 0 < 3x^2 < 75. \tag{1.29}$$
$$0 < x < 5.$$
$$\leftrightarrow \quad 0 > -4x > -20.$$
$$\leftrightarrow \quad -20 < -4x < 0. \tag{1.30}$$

By adding equations (1.29) and (1.30), we obtain

$$-20 < 3x^2 - 4x < 75$$

and finally

$$0 < x < 5 \leftrightarrow -17 < 3x^2 - 4x + 3 < 78. \tag{1.31}$$

Next, consider the values of x satisfying $-3 < x < 0$.

$$-3 < x < 0$$
$$\leftrightarrow \quad 9 > x^2 > 0$$
$$\leftrightarrow \quad 27 > 3x^2 > 0$$
$$\leftrightarrow \quad 0 < 3x^2 < 27. \tag{1.32}$$
$$-3 < x < 0$$

$$\leftrightarrow \quad 0 < -4x < 12. \tag{1.33}$$

By adding equations (1.32) and (1.33), we obtain

$$0 < 3x^2 - 4x < 39$$

and finally

$$-3 < x < 0 \leftrightarrow 3 < 3x^2 - 4x + 3 < 42. \tag{1.34}$$

The case $x = 0$ is not covered by equation (1.31) or (1.34). But if $x = 0$, then the given expression has the value 3. If $y = x^2 - 4x + 3$, then all possible values of y are given by either $-17 < y < 78$ or $30 < y < 42$ or $y = 3$. Thus, the required answer is simply $-17 < y < 78$. (The other possibilities are absorbed in this inequality.) ■

SUMMARY

- The equation $ax + b = c$ does not have any solution when $a = 0$ and $c \neq b$.
- Every number x is a solution of the equation $ax + b = c$ when $a = 0$ and $c = b$.
- Negative numbers do not have real square roots.
- The square root of -1 is the imaginary number i.
- The quadratic equation $ax^2 + bx + c = 0$ has either two distinct solutions or equal solutions or two complex solutions depending on whether the discriminant is either positive or zero or negative, respectively.
- If the truth of the statement Q follows from the truth of the statement P, then we write $P \to Q$.
- $Q \to P$ is called the converse of $P \to Q$.
- The truth of the converse of a statement does not follow from the truth of the statement.
- When both $P \to Q$ and $Q \to P$ are true, we say that P and Q are logically equivalent. This fact is denoted by $P \leftrightarrow Q$.
- Let $a > 0$. Then the solutions of the inequality $p < ax + b < q$ are given by

$$\frac{p-b}{a} < x < \frac{q-b}{a}.$$

- Let $a < 0$. Then the solutions of the inequality $p < ax + b < q$ are given by

$$\frac{q-b}{a} < x < \frac{p-b}{a}.$$

- If $p < x < q$, then $ap + b < ax + b < aq + b$, provided $a > 0$.
- If $p < x < q$, then $aq + b < ax + b < ap + b$, provided $a < 0$.
- If $b > 0$, then $x^2 < b \to -\sqrt{b} < x < \sqrt{b}$.
- If $a < x < b$, $a > 0$ and $b > 0$, then $a^2 < x^2 < b^2$.
- If $a < x < b$, $a < 0$ and $b > 0$, then $0 \leq x^2 < p^2$ where p is the bigger of $-a$ and b.

PROBLEMS

1. Find the condition that the equation $3(mx + n) - 4x = px + 5$ does not have any solution.
2. Find the condition that the above equation has infinitely many solutions.
3. Solve

   (i) $9 < -4x - 16 \leq 28$.
   (ii) $56 > 7x + 49 > 4$.

4. Imagine a long straight road on which your house, your friend's house, a park and a vacant house are located. Your friend's house, the park and the vacant house are, respectively, located at distances of 15 km, 40 km and 85 km from your house. If your friend starts driving his car at time $t = 0$ from his house at a uniform speed of 50 km per hour, find the time duration during which he would be between the park and the vacant house.
5. If $y = 108x - 15$, find the possible values of y when x takes the values between -23 and 45.
6. Solve

$$x^2 + 12x - 28 < 0.$$

7. Show that there is no solution for the quadratic inequality

$$x^2 - 18x + 85 \leq 5.$$

8. Show that every x satisfies

$$x^2 - 16x + 67 \geq 0.$$

9. Solve

$$-x^2 + 57x + 103 > 0.$$

10. Find the possible values of $y = 4x^2 - 7x + 9$ when x is between 7 and 17.
11. Draw the graph of $y = x^3$. Using graphs show that

$$a < x^3 < b \leftrightarrow a^{\frac{1}{3}} < x < b^{\frac{1}{3}}.$$

Here $p^{1/3}$ is the unique real cube root of p.

FURTHER READING

www.coolmath.com.
www.purplemath.com.
www.classzone.com/books/algebra1.

Hall, H. A and Knight, S.R. (2001), *Higher Algebra*, Surjeet Publications, Delhi.
Barnard, S. and Child, J.M. (2002),*Higher Algebra*, A.I.T.B.S. Publishers and Distributors, Delhi.

CHAPTER

2

Sets, Functions and Relations

Learning Objectives

On completing this chapter, you should able to:

- work with the concepts of a set, its elements and subsets
- define the sets of natural numbers, integers, rational numbers, real numbers, complex numbers and their various subsets
- find unions and intersections of given sets
- define the concepts of a relation and a function
- solve problems on reflexive, symmetric and transitive relations
- solve problems on one-to-one and onto functions
- find the images of subsets of domains and inverse images of subsets of the co-domain for some simple functions
- find left inverses, right inverses and the inverse of some simple functions whenever they exist
- prove basic results on one-one and onto functions
- define the notions of finite and infinite sets
- precisely define the number of elements of a finite set

2.1 SET THEORY

The best way to understand what a set is, is to look at some examples:

(i) $\{01111111, 10111111, 11011111, 11101111, 11110111, 11111011, 11111101, 11111110\}$.

(ii) $\{a, m, l, k\}$.

(iii) {Muthu, Thames, Jupiter}.

(iv) {Palaru, Kaveri, Thenpennai, Vaigai, Bhavani, Manimuthaaru, Thamirabarani...}.
(v) {Mercury, Mars, Earth, Venus, Jupiter, Uranus, Neptune, Pluto}.
(vi) $\{1, 4, 9, 16, 25, \ldots\}$.
(vii) $\{23.890, 123465438\}$.
(viii) {34, Saddam Hussein, Everest, Tagore}.
(ix) {S. Chandrasekar, C.V. Raman, R.N. Tagore, Mother Teresa, Amarthya Sen}.
(x) {Albert Einstein, Thiru Mular, Premchand, Kandekar, Krishnadevaraya}.
(xi) The set of all human beings who eat peanuts.
(xii) The set of all animals in Borivili National Park, Mumbai.
(xiii) $\mathbb{N} = \{1, 2, 3, \ldots\}$ = The set of Natural numbers.
(xiv) $\{1, 8, 27, 64, 125, \ldots\}$.
(xv) $\left\{n^3 : n \in N\right\}$.
(xvi) The set of all positive perfect cubes.
(xvii) $\left\{\begin{pmatrix} 3 & 9 & 7 \\ 4 & -7 & 6 \end{pmatrix}\right\}$.
(xix) $\{1100, 0110, 0011, 1010, 0101, 1001\}$.

Roughly speaking, a set is a collection of objects. Given a set A and an object x, it should be clear whether x belongs to the collection or not. In mathematics, we take 'set' as a concept which is not defined. The objects of the collection are called the *elements* of the set.

Note that there need not be any kind of a 'special relationship' between the elements of a set. Examples (viii) and (x) illustrate this. These kind of sets can be described only by listing all their elements. If the elements of a set have some common property, it becomes easier to describe the set.

For example, (i) is a set of all **bytes** containing only one 0, (iv) is the set of rivers flowing in Tamil Nadu and (ix) is the set of all Indian Nobel laureates. Note that the sets (xiv), (xv) and (xvi) are the same. In (xiv) the elements are listed. (xv) is the mathematical description of the set and (xvi) is the verbal description. The set (xvii) contains just one object, namely the 2×3 matrix $\begin{pmatrix} 3 & 9 & 7 \\ 4 & -7 & 6 \end{pmatrix}$. The set (xix) is a set of all strings of length four having two zeroes and two ones.

Definition 2.1: A set containing only one element is called a *singleton*.

Example 2.1

The set of all female prime ministers of India is a singleton. Let A be the set of all rivers flowing in India. Then, Ganga, Godavari and Vaigai are all elements of A. Nile, being an African river, is not an element of A. These facts are denoted as follows:

$$\text{Ganga} \in A$$

$$\text{Nile} \notin A.$$

Let C be the set of all rivers originating in Himachal Pradesh and flowing through Tamil Nadu. It is obvious that the set C does not contain any element. Such a set is defined as the empty set and it is denoted by ϕ. ■

Example 2.2

The set of all bytes containing nine ones is an empty set. ■

Example 2.3

$\left\{x \in \mathbb{R} : x^2 + 1 = 0\right\} = \phi$, $\left\{x \in \mathbb{N} : 3x = 5\right\} = \phi$, ■

Subsets: Let I be the set of all rivers flowing in India and let T be the set of all rivers flowing into Tamil Nadu. How are these two sets related to each other? Informally speaking, T is a part of I. Since every river flowing into Tamil Nadu is also flowing in India, every element of T is also an element of I. We say that T is a subset of I and I is a superset of T. Note that an element of I need not be an element of T.

If S is the set of all students of a class, G is the set of all girl students of the same class and B is the set of all boy students, then both G and B are subsets of S.

Definition 2.2: Let A and B be sets. If every element $b \in B$ is also an element of A, then we say that B is a *subset* of A. This fact is denoted by $B \subseteq A$ and is read as B is contained in A. A is called a *superset* of B. We write this as $A \supseteq B$ and read it as A contains B.

Example 2.4

Consider the sets $A = \{2, -23, 7, 8, 0, 5\}$, $B = \{-30, -29, -28, \ldots, 0, 1, 2, \ldots, 8, 9, 10\}$, $C = \{2, -23, 10\}$ and $D = \{1, 2, 3, \ldots\}$. It is clear that the following hold:

(i) $A \not\subseteq D$.
(ii) $B \not\subseteq D$.
(iii) $C \not\subseteq D$.
(iv) $A \subseteq B$.
(v) $C \subseteq B$.
(vi) Neither A is contained in C nor C is contained in A. ■

Example 2.5

From the definition of *subset* it is clear that $A \subseteq A$. Further, the empty set is a subset of every set. We shall prove this statement logically in Chapter 3. ■

Example 2.6

Let $A = \{1, 2, 3, a, b\}$, $B = \{2, 5, b, c\}$, $C = \{a, 3\}$, $D = \{2, 5, c\}$. Then we have,

(a) Neither $A \subseteq B$, nor $B \subseteq A$.
That is, neither A contains B nor B contains A.
(b) $C \subseteq A$.
(d) $D \subseteq B$. ■

Definition 2.3: If $B \subseteq A$ and $A \neq B$, then B is called a *proper subset* of A. This is denoted as $B \subset A$.

Example 2.7

Let $A = \{x, y, z\}$. Then we have
(i) $\{x, z\} \subset A$.
(ii) $\{y\} \subset A$.

Now, let A be an arbitrary set and x be an arbitrary object. $x \in A$ is read as x is an element of A or x belongs to A. $x \notin A$ is read as x is not an element of A or x does not belong to A.

In mathematics, we come across situations where we have to prove that two sets A and B got by different definitions are the same. This is proved by showing that each is a subset of the other. ■

Example 2.8

Let A be a set of all letters of the English alphabet, B be the set of all English letters appearing in the sentence **I am a good girl** and C be the set of all letters appearing in the sentence **Pack my box with five dozen liquor jugs** and $D = \{a, d, g, i, l, m, o, r\}$. Show that $A = C$ and $B = D$.

SOLUTION

It is clear that B, C and D are all subsets of A. First, we show that $A = C$. Since $C \subseteq A$, we have to show that $A \subseteq C$. By picking a letter from A and examining the sentence 'Pack my box with five dozen liquor jugs', we find that every element of A occurs in the sentence. Thus, $A \subseteq C$ and hence $A = C$. Next, every letter occurring in the sentence $B = D$. ■

2.1.1 Sets of Numbers

$N = \{1, 2, 3, \ldots\}$ is the set of natural numbers. $\mathbf{Z} = \{\pm 1, \pm 2, \pm 3, \ldots\}$ is the set of integers. Now, -3 is *not* a natural number and $\frac{-2}{3}, \frac{5}{6}$ are *not* integers. $\frac{-2}{3}, \frac{5}{6}$ are examples of rational numbers. A rational number is nothing but a number that can be expressed as a **ratio** of two **integers**. The set of rational numbers $\mathbf{Q}$ is defined as

$$\mathbf{Q} = \left\{\frac{p}{q} : p \in \mathbf{Z}, q \in \mathbf{Z}, q \neq 0\right\}.$$

Since $p = \frac{p}{1}$ for any integer p, every integer is a rational number. In other words, $Z \subset \mathbf{Q}$.

We are all familiar with the decimal representation of fractions. For example $\frac{4}{10} = 0.4$, $\frac{1}{8} = 0.125$, $\frac{1}{3} = .33333\ldots$ $\frac{2}{7} = 0.285714285714285714\ldots$ Decimal expansions of the rational numbers $\frac{4}{10}$ and $\frac{1}{8}$ stop after a finite number of decimals. Such decimals are called *terminating* decimals. Decimal expansions of $\frac{1}{3}$ and $\frac{2}{7}$ do not stop, but a pattern of some digits gets repeated. These decimal expansions are called *periodic*. **A rational number has a decimal expansion, which is either terminating or periodic, and vice-versa.**

There are numbers that are not rational. These numbers have a decimal expansion, which is **neither terminating nor periodic**. Examples of such numbers are $\sqrt{2}$, π, e, etc.

$$\sqrt{2} = 1.4142135623730950488016887240 97\ldots$$
$$\pi = 3.14159265358979323846264338327 95\ldots$$
$$e = 2.71828182845904523536028747135 27\ldots$$

We can easily give further examples of numbers whose decimal expansions neither terminate nor have a repeating pattern. For example

$$3.010010001000010000001\ldots$$

For this number, the number of zeros between successive 1s increases by one. Next,

$$1.525525555552555555555555555555555552\ldots$$

In this number, the number of 5s between the decimal point and the first 2 is 1! Similarly the number of 5s between the second and the third 2 is 2!, the number of 5s between the third and the fourth 2 is 3! and so on.

Such numbers are called **irrational numbers**. Rational and irrational numbers together form the set of real numbers, which is denoted by the symbol R.

Definition 2.4: A real number x is said to be positive if $x > 0$ and it is said to be negative if $x < 0$. A real number x is said to be non-negative if $x \geq 0$

We are familiar with the rules of multiplication for real numbers given below:

$$\text{Positive} \times \text{Positive} = \text{Positive}$$
$$\text{Negative} \times \text{Negative} = \text{Positive}$$
$$\text{Positive} \times \text{Negative} = \text{Negative.}$$

From these rules, it is clear that if y is negative, then there is no real number x such that $x^2 = y$. We shall prove this fact. If $x = 0$, then $x^2 = 0$ and hence y being negative we cannot have $x^2 = y$. Then, x has to be either positive or negative. In both these cases, x^2 has to be positive. Hence, $x^2 = y$ is not possible. In other words, negative reals do not have real square roots. This fact leads us to consider complex numbers.

Complex Numbers: Complex numbers have been created to solve equations like $x^2 + 1 = 0$, which do not have any solution in the set of real numbers.

Definition 2.5: $\pm i$ are the two solutions of the equation $x^2 + 1 = 0$. A complex number is a number of the type $x + iy$ where x and y are real numbers.

Example 2.9

(i) The set $\{1, 3, 5, 7, \ldots\}$ is a subset of $\mathbb{N}$. We can also write this set as

$$\{x \in \mathbb{N} : x \quad \text{is odd}\}.$$

(ii) The set $\{-3, -2, -1, 0, 1, 2, 3\}$ can be described as

$$\{x \in \mathbf{Z} : -3 \leq x \leq 3\}.$$

(iii) Is the above set same as

$$\{x \in \mathbb{R} : -3 \leq x \leq 3\}?$$

Certainly not. For example, $2.9 \in \{x \in \mathbb{R} : -3 \leq x \leq 3\}$, but $2.9 \notin \{x \in \mathbf{Z} : -3 \leq x \leq 3\}$. While $\{x \in \mathbf{Z} : -3 \leq x \leq 3\}$ is a finite set, $\{x \in \mathbb{R} : -3 \leq x \leq 3\}$ is not a finite set.

(iv) Consider the set

$$\{\ldots, -48, -36, -24, -12, 0, 12, 24, 48, 36, \ldots\}.$$

This set is nothing but the set of integral multiples of 12. We can also describe it as

$$\{n \in \mathbf{Z} : n \text{ is a multiple of } 12\}.$$

An integer n is a multiple of 12 if there is $k \in Z$ such that $n = 12k$. Thus, the above set can also be written as

$$\{12k : k \in \mathbf{Z}\}.$$

(v) Consider the set

$$\{12, 24, 48, 36, \ldots\}.$$

This set is nothing but the set of positive integral multiples of 12. We can describe it as

$$\{12k : k \in \mathbb{N}\}.$$

(vi) Let $a, b \in \mathbb{R}$. The following subsets of $\mathbb{R}$ are known as *intervals*:

$$\begin{aligned}
[a, b] &= \{r \in \mathbb{R} : \quad a \leq r \leq b\} \\
(a, b] &= \{r \in \mathbb{R} : \quad a < r \leq b\} \\
[a, b) &= \{r \in \mathbb{R} : \quad a \leq r < b\} \\
(a, b) &= \{r \in \mathbb{R} : \quad a < r < b\} \\
[a, \infty) &= \{r \in \mathbb{R} : \quad a \leq r\} \\
(a, \infty) &= \{r \in \mathbb{R} : \quad a < r\} \\
(\infty, a] &= \{r \in \mathbb{R} : \quad r \leq a\} \\
(\infty, a) &= \{r \in \mathbb{R} : \quad r < a\} \\
\mathbb{R}^+ &= [0, \infty).
\end{aligned}$$

Note that the symbol ∞ plays only a symbolic role and *we do not treat it as a number in any sense.* Note also that the first four intervals will be empty sets if $b < a$.

$[a, b]$ is called a *closed interval* and (a, b) an *open interval*. The intervals $(a, b]$ and $[a, b)$ are called *semi-open intervals*. ■

The following example helps us understand the intervals better.

Example 2.10

(a) $3.4 \in (3, 7]$.
(b) $7 \notin [5, 7)$.
(c) $-9 \in (-\infty, -8)$.
(d) $7 \notin [8, \infty)$.
(e) $3.000001, 7.9999999 \in (3, 8)$.
(f) $e \in \{x \in [2, 3] : x \text{ is irrational}\}$.
(g) $\pi \notin \{x \in (3, 4) : x \text{ is rational}\}$.
(h) $\frac{22}{7} \in \{x \in (3, 4) : x \text{ is rational}\}$. ■

Example 2.11

Figure 2.1 illustrates the intervals $(-\infty, -2)$, $(4, \infty)$ and $[-1, 6])$.

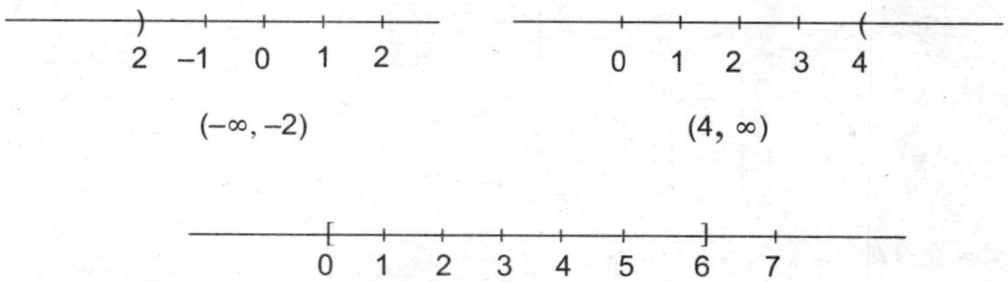

Figure 2.1 Intervals ■

PRACTICE 2.1

Illustrate the intervals $[0, \infty)$, $[7, 15]$ and $(-\infty, -3)$ using figures.

Example 2.12

Show that $[2, 9] = \{x \in R : 5 \le 4x - 3 \le 33\}$

SOLUTION

Let $A = [2, 9]$ and $B = \{x \in R : 5 \le 4x - 3 \le 33\}$.
We have

$$2 \le x \le 9 \rightarrow 5 \le 4x - 3 \le 33.$$

That is,

$$x \in A \rightarrow x \in B.$$

Thus, $A \subseteq B$.
Further,

$$5 \le 4x - 3 \le 33 \rightarrow 2 \le x \le 9.$$

That is,

$$x \in B \rightarrow x \in A.$$

Thus, $B \subseteq A$.
Therefore,

$$A = B.$$ ■

Below, we describe some sets whose elements are solutions of equations:

Example 2.13

(a) $\left\{x \in \mathbb{R} : x^2 - 7x - 30 = 0\right\} = \{10, -3\}$.
(b) $\left\{x \in \mathbf{C} : x^2 - 7x - 30 = 0\right\} = \{10, -3\}$.
(c) $\left\{x \in \mathbb{R} : x^2 + 14x + 50 = 0\right\} = \phi$.
(d) $\left\{x \in \mathbf{C} : x^2 + 14x + 50 = 0\right\} = \{-7 + i, -7 - i\}$.
(e) $\{x \in \mathbb{R} : \sin x = 0\} = \{n\pi : n \in \mathbf{Z}\}$.
(f) $\left\{x \in \mathbb{R} : \tan x = \sqrt{3}\right\} = \{n\pi + \frac{\pi}{3} : n \in \mathbf{Z}\}$.
(g) $\left\{x \in \mathbb{R} : \cos x = \frac{1}{\sqrt{2}}\right\} = \{2n\pi \pm \frac{\pi}{4} : n \in \mathbf{Z}\}$.
(h) $\left\{x \in \mathbb{R} : \cos x = \frac{1}{\sqrt{2}},\ 0 \le x \le \pi\right\} = \{\pi/4\}$.
(i) $\left\{x \in \mathbb{N} : x^2 - 5 = 0\right\} = \phi$.
(j) $\left\{x \in \mathbb{R} : x^2 - 5 = 0\right\} = \left\{\sqrt{5}, -\sqrt{5}\right\}$. ■

Example 2.14

Let $A = \{3n + 1 : n \in \mathbf{Z}\}$. A is nothing but the set of all integers, which give a remainder of 1 when divided by 3. In other words,

$$A = \{\ldots, -5, -2, 1, 4, 7, \ldots\}.$$

Now, we can easily check that $300 \notin A$, $301 \in A$ and $302 \notin A$. ■

PRACTICE 2.2

(1) Rewrite the following sets by listing the elements:

(i) $\{x \in \mathbb{N} : x \text{ is a divisor of } 120\}$.
(ii) $\{x \in \mathbf{Z} : x \text{ is a multiple of } 7\}$.
(iii) $\{x \in \mathbb{N} : x \text{ is a prime and } x \le 17\}$.
(iv) $\{x \in \mathbb{N} : x \text{ is a perfect square}\}$.
(v) $\{x \in \mathbb{N} : 32 \le x \le 36 \text{ is a perfect square}\}$.
(vi) $\{x \in \mathbb{N} : x \le 14\ x \text{ is relatively prime to } 14\}$.

(2) Rewrite the following sets using mathematical/verbal description:

(i) $\{6, 18, 54, 162, \ldots\}$.
(ii) {Nehru, Lal Bahadhur Sastri, Indira Gandhi, Morarji Desai, Charan Singh, Chandrashekhar, Rajiv Gandhi, Narasimha Rao, I.K. Gujral, Deve Gowda, V.P. Singh, Vajpai, Manmohan Singh}.
(iii) $\{6, 11, 16, 21, \ldots\}$.
(iv) $\{101, 102, 103, \ldots\}$.
(v) {Dharma, Bheema, Arjuna, Nakula, Sakadeva}.
(vi) {*Julius Caesar*, *Antony*}.

Next, we consider an example related to matrices.

Example 2.15

Let

$$A = \left\{ \begin{pmatrix} a & b \\ c & a \end{pmatrix} : a, b, c \in R \right\}$$

and

$$B = \left\{ \begin{pmatrix} 1 & a \\ b & c \end{pmatrix} : a, b, c \in R \right\}.$$

A is the set of all 2×2 matrices with real entries such that the diagonal elements are equal.

B is the set of all 2×2 matrices with real entries such that the element in the first row, first column is 1. So, we have,

$$\begin{pmatrix} -3 & 7 \\ 6 & -3 \end{pmatrix} \in A, \begin{pmatrix} 3i & 7 \\ 6 & 3i \end{pmatrix} \notin A, \begin{pmatrix} 7 & 8 \\ 67 & 9 \end{pmatrix} \notin A$$

and

$$\begin{pmatrix} 10 & 7 \\ 6 & 1 \end{pmatrix} \notin B, \begin{pmatrix} 1 & 7i \\ 6 & 65 \end{pmatrix} \notin B, \text{ but } \begin{pmatrix} 1 & 87.7 \\ 34 & 0.92 \end{pmatrix} \in B.$$ ■

Definition 2.6: An $n \times n$ matrix is said to be singular if $|A| = 0$. It is said to be non-singular if $|A| \neq 0$. $GL_n(R)$ is the set of all $n \times n$ non-singular matrices.

$$SL_2(R) = \{A \in GL_2(R) : |A| = 1\}.$$

Cartesian Product of Sets: Let us recall that points in the xy plane are represented by pairs of numbers such as $(3, 4)$, $(-2, 1)$. Figure 2.2 illustrates this

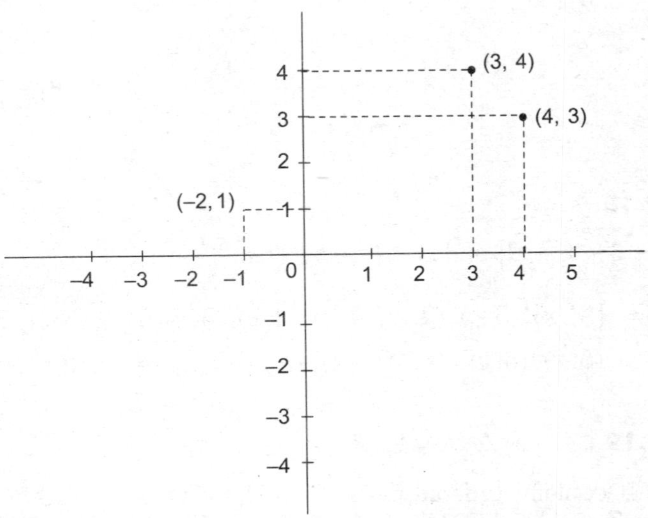

Figure 2.2 Cartesian coordinates

Note that the points (4, 3) and (3, 4) are different. Hence, they are called *ordered pairs of numbers*. A pair is a collection of two objects. Often, we would like to arrange the objects of a pair in some order. That is, we would like to specify which one is the first object and which is the second. Such a pair is called *an ordered pair*. If a pair consists of two objects x and y and if we take x as the first and y as the second, then the ordered pair is written as (x, y).

Definition 2.7: Let A and B be sets. Their Cartesian product $A \times B$ is the set of all ordered pairs (x, y) with $x \in A$ and $y \in B$. In other words

$$A \times B = \{(x, y) : x \in A, y \in B\}.$$

Example 2.16

Let $A = \{7, 9\}$, $B = \{K, T, G\}$. Then,

$$A \times B = \{(7, K), (7, T), (7, G)(9, K), (9, T), (9, G)\}.$$

Note that $(T, 9)$ is not an element of $A \times B$. But $(T, 9) \in B \times A$. ■

Example 2.17

Next, we consider $\mathbf{Z} \times \mathbb{N}$. We cannot list down all the elements of $\mathbf{Z} \times \mathbb{N}$, but we can indicate the elements as follows:

$$\begin{aligned}\mathbf{Z} \times \mathbb{N} = \{&(0, 1), (0, 2), (0, 3) \ldots \\ &(1, 1), (1, 2), (1, 3) \ldots \\ &(-1, 1), (-1, 2), (-1, 3) \ldots \\ &\ldots\ldots\ldots\ldots\ldots\ldots \\ &\ldots\ldots\ldots\ldots\ldots\ldots \\ &\ldots\ldots\ldots\ldots\ldots\ldots\}.\end{aligned}$$ ■

Example 2.18

$A = \{n \in \mathbb{N} : 3 \leq n < 9\}$ and $B = \{x, y, z\}$ Then,

$$\begin{aligned}A \times B = \{&(3, x), (3, y), (3, z), (4, x), (4, y), (4, z)(5, x), (5, y), (5, z) \\ &(6, x), (6, y), (6, z), (7, x), (7, y), (7, z), (8, x), (8, y), (8, z)\}.\end{aligned}$$ ■

Example 2.19

The set $\mathbb{R} \times \mathbb{R}$ contains ordered pairs like $(-2, 4)$, $(1.3, \pi)$, $(23.567, \sqrt{5})$, etc. For both sets $\mathbf{Z}$ and $\mathbb{N}$, their elements can be arranged as first element, second element, etc. (see Section 2.9). On the contrary for real numbers, the elements of

$\mathbb{R}$ cannot be arranged in such a fashion. So, there is no way we can even indicate a possible list for $\mathbb{R} \times \mathbb{R}$ as in the case of $\mathbf{Z} \times \mathbb{N}$. But we can think of these ordered pairs as points in the xy plane and this helps us in visualizing $\mathbb{R} \times \mathbb{R}$ geometrically. $\mathbb{R} \times \mathbb{R}$ is denoted by $\mathbb{R}^2$. We think of $\mathbb{R}^2$ as the xy plane. Refer to Figure 2.2. ■

Example 2.20

An element of $\mathbb{R}^2 \times \mathbb{R}$ looks like $((x, y), z))$. But we denote $\mathbb{R}^2 \times \mathbb{R}$ as $\mathbb{R}^3$ and the elements as (x, y, z). We think of (x, y, z) as a point in the xyz space and $\mathbb{R}^3$ as the space. ■

Example 2.21

Let $A = \{x \in \mathbb{R} : 1 \leq x \leq 3\}$ and $B = \{x \in \mathbb{R} : 2 \leq x \leq 5\}$. $A \times B$ is the rectangular region in the xy plane whose vertices are the points $(1, 2)$, $(1, 5)$, $(3, 2)$ and $(3, 5)$. ■

Example 2.22

Let $A = \left\{(x, y) \in \mathbb{R}^2 : y = 2x + 1\right\}$. Then, $A \subseteq R \times R$ and further

$$(0, 1) \in A, \quad \left(\frac{-1}{2}, 0\right) \in A, \quad (3, 7) \in A.$$

Now, if we plot all these three points in the xy plane, we find that they are co-linear. In fact, geometrically, A is nothing but the line in the xy plane whose equation is $y = 2x + 1$. $(0, 0) \notin A$, $(-3, 8) \notin A$. In other words, this line does not pass through the origin or through the point $(-3, 8)$. ■

Example 2.23

$$P = \left\{(x, y, z) \in \mathbb{R}^3 : x + 2y - 3z = 10\right\}$$

is a plane that is a subset of the space $\mathbb{R}^3$. ■

Example 2.24

Consider the equations

$$\frac{x - x_0}{l} = \frac{y - y_0}{m} = \frac{z - z_0}{n}$$

representing a line in space passing through the point (x_0, y_0, z_0) whose direction ratios are (l, m, n) The para-matric equation of a straight line in space is given by

$$x = x_0 + lt$$

$$y = y_0 + mt$$
$$z = z_0 + nt.$$

In other words,

$$L = \{(x_0 + lt, y_0 + mt, z_0 + nt) : t \in \mathbb{R}\}$$

is a line in space. ■

PRACTICE 2.3

(i) Can you think of an element of $\{x \in R : -3 \leq x \leq 3\}$, which is not in $\{x \in Q : -3 \leq x \leq 3\}$?

(ii) Let $A = \{29n : n \in Z\}$ and $B = \{5n + 2 : n \in N\}$. Which of the following is true?

(a) $0 \in A$.
(b) $57 \in A$.
(c) $9193 \in A$.
(d) $7 \in B$.
(e) $-52 \in B$.
(f) $102 \in B$.
(g) $(502, 290) \in B \times A$.
(h) $(0, 27) \in A \times B$.
(i) $(41, 3) \in A \times B$.

Example 2.25

Let $P = \left\{(x, y, z) \in R^3 : x + 2y - 2z + 17 = 0\right\}$ and $L = \{(2t + 1, 3t - 6, 4t + 3) : t \in R\}$. Show that L is a proper subset of P.

SOLUTION

An element in L is of the form $(2t+1, 3t-6, 4t+3)$ for some $t \in R$. To show that $L \subseteq P$, we have to show that every such element belongs to P. Letting $x = 2t+1$, $y = 3t - 6$, $z = 4t + 3$, we find that

$$\begin{aligned} x + 2y - 2z + 17 &= 2t - 1 + 2(3t - 6) - 2(4t + 3) + 17 \\ &= 2t - 1 + 6t - 12 - 8t - 6 + 17 \\ &= -2. \end{aligned}$$

Thus, $(2t + 1, 3t - 6, 4t + 3) \in P$. So, $L \subseteq P$. Now, to show that L is a proper subset of P, we have to produce an element $(x, y, z) \in P$, which is not in L. Take $(17, 0, 0) \in P$. We show that $(17, 0, 0) \notin L$. Suppose that there is a $t \in \mathbb{R}$ such that $2t + 1 = 17, 3t - 6 = 0$ and $4t + 3 = 0$, then we get $t = 8$, $t = 2$ and $t = -0.75$, which is impossible. Hence, $(17, 0, 0) \notin L$ Therefore, $L \subset P$. ■

2.1.2 Union and Intersection of Sets

Example 2.26

Consider the inequality

$$x^2 - 17x + 56 > 0. \tag{2.1}$$

By using the techniques of Section 1.5, we find that an x, which satisfies equation (2.1), has to satisfy one of the following inequalities:

$$x < 7 \tag{2.2}$$
$$x > 8 \tag{2.3}$$

and vice-versa. Using the idea of sets, we can write this fact as

$$a \in \left\{x : x^2 - 17x + 56 > 0\right\} \leftrightarrow a \in (-\infty, 7) \text{ or } a \in (8, \infty)$$

Consider the inequalities

$$4 > x + 3 \tag{2.4}$$
$$x + 3 > 1 \tag{2.5}$$
$$4 > x + 3 > 1. \tag{2.5}$$

The solution set of equation (2.4) is $(-\infty, 1)$ and the solution set of equation (2.5) is $(-2, \infty)$. A solution of equation (2.6) must satisfy both equations (2.4) and (2.5). Hence, we have

$$\{x : 4 > x + 3\}$$
$$a \in \{x : 4 > x + 3 > 1\} \leftrightarrow a \in \{4 > x + 3\} \text{ and } a \in \{x : x + 3 > 1\}.$$ ■

This example motivates the definitions of union and intersection of sets.

Definition 2.8: If A and B are sets $A \cup B$ and $A \cap B$ are defined as

$$\begin{aligned} A \cup B &= \{x : x \in A \text{ or } x \in B\} \\ A \cap B &= \{x : x \in A \text{ and } x \in B\}. \end{aligned}$$

Example 2.27

Let $A = \{a, b, c\}$ and $B = \mathbf{Z}$, the set of integers. Then

$$A \cup B = \{a, b, c, 0, \pm 1, \pm 2, \pm 3, \ldots\}.$$ ■

Example 2.28

Let $A = N$, $B = \{5, 6, 7, 8\}$, $C = \{p, q, r\}$, $D = \{7, 8, 9, 10, 11, 12\}$, E be the set of all alphabets of English and $F = \{l, m, n, o, p, q\}$.

(i) $A \cap B = \{5, 6, 7, 8\} = B$.
(ii) $C \cap F = \{p, q\}$.
(iii) $B \cap D = \{7, 8\}$.
(iv) $E \cap F = F$.
Now, the sets A and E do not have any common elements. Hence, the set $A \cap E$ does not have any element. Hence $A \cap E = \phi$. Also, Clearly, $B \cap F = \phi$. ■

Example 2.29

$$[1, 3] \cup [2, 4] = [1, 4] \; [1, 3] \cap [2, 4] = [2, 3].$$

We can talk about union of more than two sets. For example, if we have five sets A_1, A_2, A_3, A_4 and A_5 their union $A_1 \cup A_2 \cup A_3 \cup A_4 \cup A_5$ is the set containing all the elements of the five given sets, this union is also written as

$$\bigcup_{i=1}^{5} A_i.$$

In this example, there are only 5 sets and the index (or the label) i is an element of the set $\{1, 2, 3, 4, 5\}$. Consider n sets A_1, A_2,...,A_n. Let us denote the set $\{1, 2, \ldots, n\}$ by Λ. The union of the n sets can now be written as

$$\bigcup_{i=1}^{n} A_i = \bigcup_{i \in \Lambda} A_i.$$

Next, we are going to allow the index set Λ to be any arbitrary set. If for each $\lambda \in \Lambda$ there corresponds a set A_λ, we have a set of sets, which can be denoted by

$$\{A_\lambda : \lambda \in \Lambda\}.$$

The union of all these sets is nothing but

$$\{x : x \in A_\lambda \quad \text{for some} \quad \lambda \in \Lambda\}$$

which is denoted by

$$\bigcup_{\lambda \in \Lambda} A_\lambda.$$

The intersection of all these sets is nothing but

$$\{x : x \in A_\lambda \quad \text{for all} \quad \lambda \in \Lambda\}$$

which is denoted by

$$\bigcap_{\lambda \in \Lambda} A_\lambda.$$

■

Example 2.30

(a) Let $A = (3, 7.8]$, $B = \mathbb{N}$. Then $A \cap B = \{4, 5, 6, 7\}$.
(b) Let $A = [9, 17]$ and $B = [8, 15]$ Then $A \cup B = [8, 17]$ and $A \cap B = [9, 15]$.
(c) Let $A_\lambda = [0, \lambda]$, $\lambda \in \mathbb{R}^+$. Then

$$\bigcup_{\lambda \in R^+} A_\lambda = [0, \infty)$$

and

$$\bigcap_{\lambda \in R^+} A_\lambda = \{0\}.$$

Venn Diagrams: Venn Diagrams are symbolic representations of sets. A set is represented by an oval-shaped figure or a portion of the oval shape. We illustrate union and intersection of two sets in Figure 2.3.

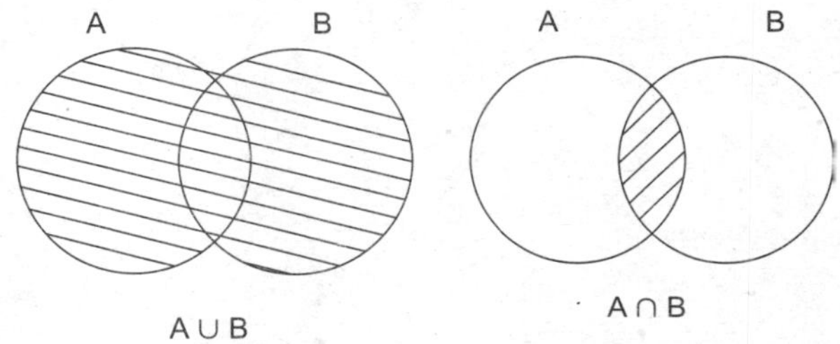

Figure 2.3 Union and intersection of sets

The following example illustrates the use of Venn Diagrams in understanding relationships between sets. The formal logical proof is presented in Section 3.5.

Example 2.31

Prove

$$A \cap (B \cup C) = (A \cap B) \cup (A \cap C)$$

by using Venn diagram.

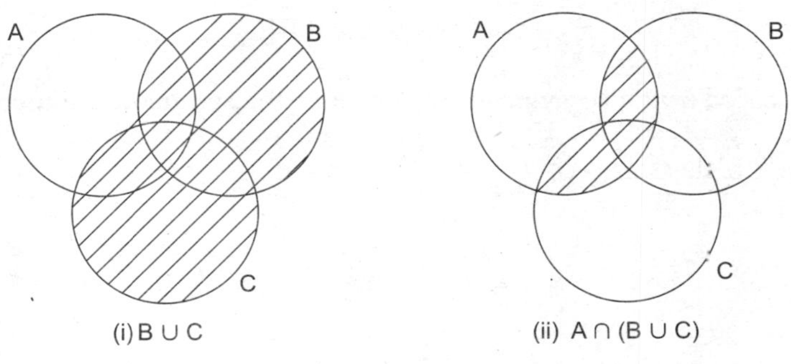

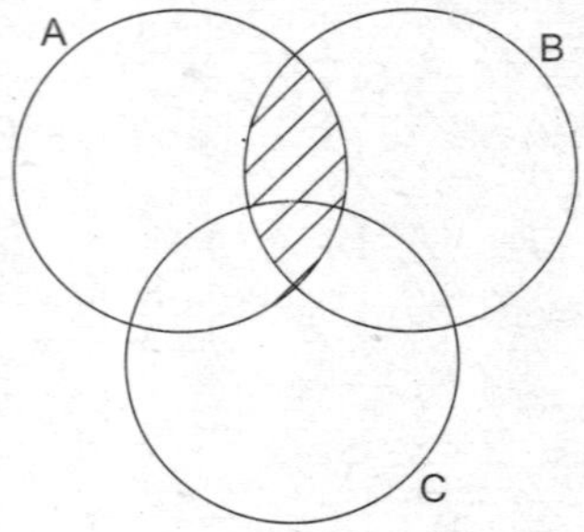

(iii) $A \cap B$

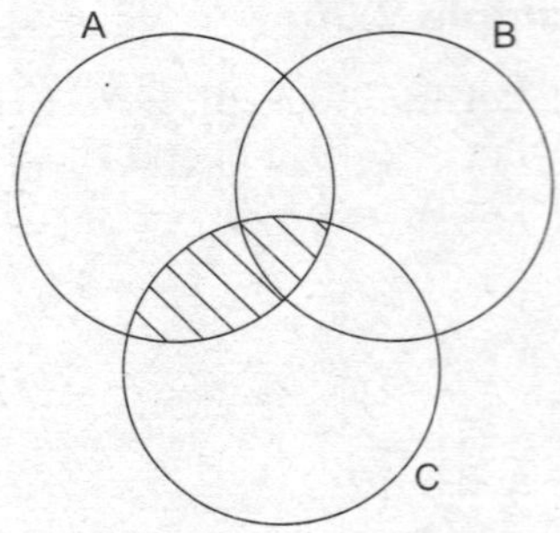

(iv) $A \cap C$

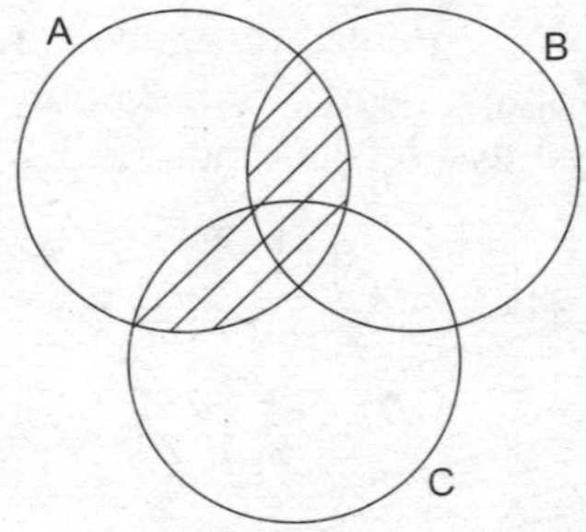

(v) $(A \cap B) \cup (A \cap C)$

Figure 2.4

The proof follows from the observation of that in Figure 2.4, the diagrams for (ii) and (v) are identical. ■

PRACTICE 2.4

Prove that

$$A \cup (B \cap C) = (A \cup B) \cap (A \cup C)$$

using Venn Diagrams.

Definition 2.9: Let A and B be sets. The set $A - B$ is defined as

$$\{x \in A : x \notin B\}.$$

$A - B$ is called as the compliment of B in A or the set theoretic difference between A and B.

Figure 2.5 illustrates $A - B$.

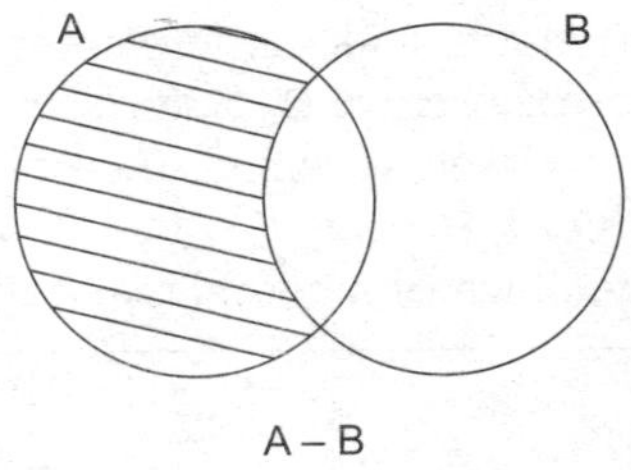

Figure 2.5 $A - B$

Example 2.32

Let $A = \{1,2,3,4,5,6,7,8,9,10\}$, $B\{3,8,10\}$, $C = \{5,7\}$, $D = \{2,3,7,10\}$
Then we have

(i) $A - B = \{1,2,4,5,6,7,9\}$.
(ii) $B - A = \phi$.
(iii) $B - C = B$.
(iv) $C - B = C$.
(v) $D - C = \{2,3,10\}$.
(vi) $C{-}D = \{5\}$. ■

Example 2.33

Show that $A - (A - B) = A \cap B$ using Venn diagrams (Fig. 2.6).

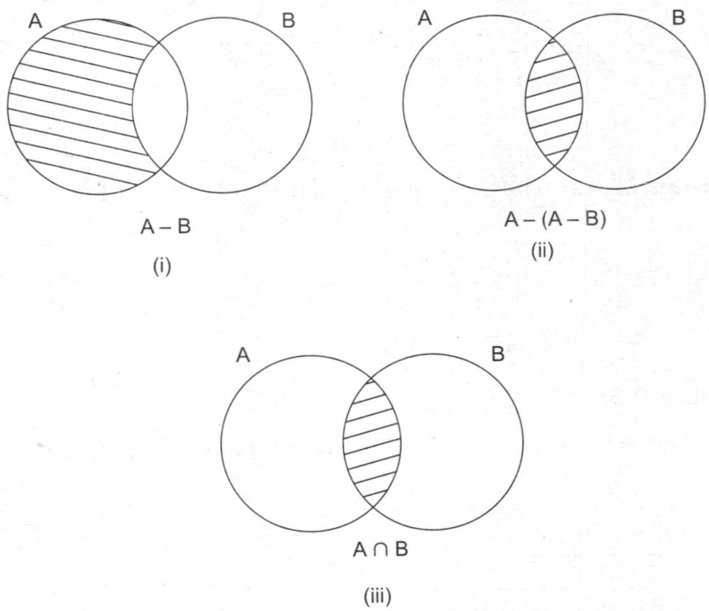

Figure 2.6 $A - (A - B) = A \cap B$

The proof follows from the observation that the diagrams for (ii) and (iii) are identical.

There are situations in mathematics where some set U plays a prominent role and all sets we deal with are subsets of U. Such a set U is described as the universal set. If a universal set U has been specified, then $U - A$ is simply denoted by A^c and is called *the compliment of A*.

Figure 2.7 illustrates the idea of a compliment:

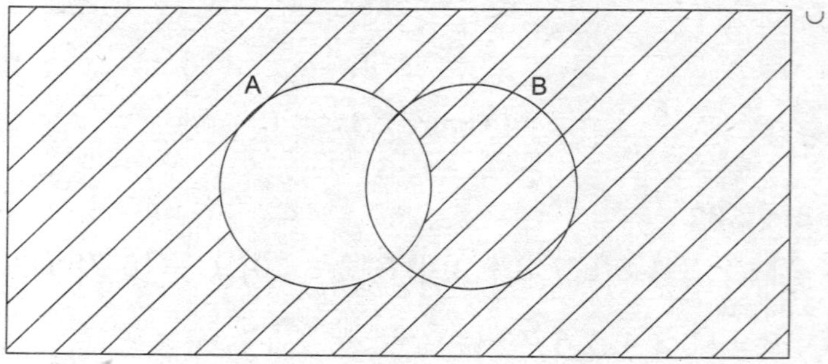

Figure 2.7 Compliment of a set

Example 2.34

Let $U = \{a, b, c, d, e, f, g, h\}$ be the universal set and let $A = \{c, f, g\}$, $B = \{a, b, c\}$. Then we have,

(i) $U^c = \phi$.
(ii) $A^c = \{a, b, d, e, h\}$.
(iii) $B^c = \{d, e, f, g, h\}$.
(iv) $\phi^c = U$.

De Morgan Laws: The following equalities for sets are known as De Morgan Laws:

(i) $(A \cup B)^c = A^c \cap B^c$.
(ii) $(A \cap B)^c = A^c \cup B^c$.

Example 2.35

Show that $(A \cup B)^c = A^c \cap B^c$ using Venn diagrams (Fig. 2.8).

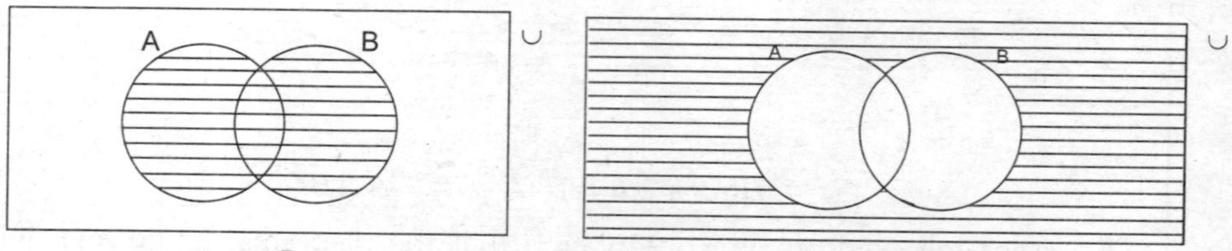

(i) A ∪ B (ii) (A ∪ B)^C

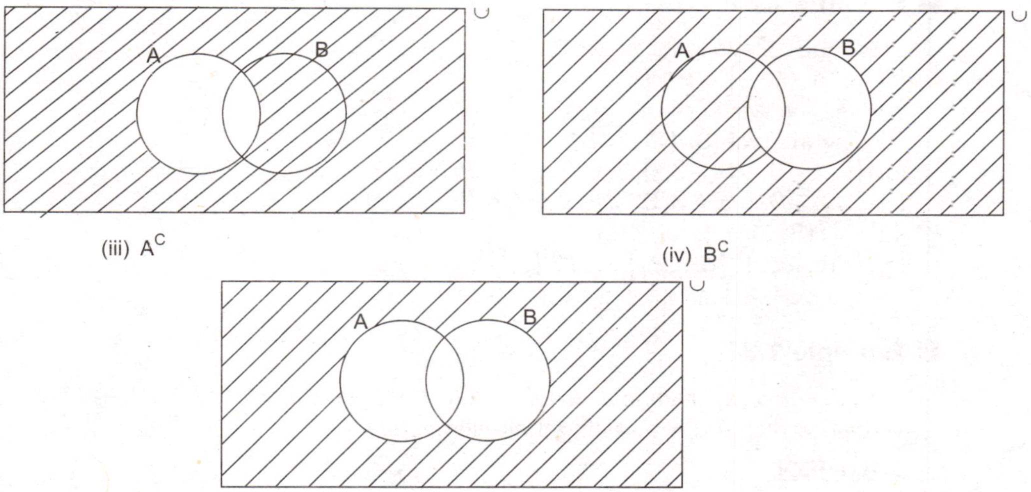

(iii) A^C

(iv) B^C

(v) $A^C \cap B^C$

Figure 2.8 De Morgan's law ■

PRACTICE 2.5

Show that $(A \cap B)^c = A^c \cup B^c$ using Venn diagrams.

Definition 2.10: Let A be a set. The power set of a set $P(A)$ is defined as the set of all subsets of A. That is, $B \subseteq A$ if and only if $B \in P(A)$.

Example 2.36

Let $A = \{4, 9\}$ Then, ϕ, $\{4\}$ $\{9\}$, $\{4, 9\}$ are all subsets of A. Hence,

$$P(A) = \{\phi, \{4\}, \{9\}, \{4, 9\}\}.$$ ■

PRACTICE 2.6

If $S = \{p, q, r\}$, write down $P(S)$.
Hint: $P(S)$ contains eight elements.

Proposition 2.1

If A is a finite set that contains n elements, then $P(A)$ contains 2^n elements.

Remark 2.1

The proof of the above statement would be given in Chapter 4. In view of this result, $P(A)$ is also denoted by 2^A. This notation is valid even if A is not a finite set.

Example 2.37

Study the following facts:
(a) $\{2, 4, 6, \ldots\} \in P(N)$.
(b) The interval $(3, 4) \notin P(Q)$.
(c) The interval $(3, 4) \in P(R)$.
(d) $\{\ldots, -9, -6, -3, 0, 3, 6, 9, \ldots\} \notin P(N)$.
(e) $\{\ldots, -9, -6, -3, 0, 3, 6, 9, \ldots\} \in P(Z)$.
(f) $Q \cap (3, 4) = \{x \in Q : 3 < x < 4\} \in P(Q)$.
(g) $\{\{a\}, \{a, c\}, \{a, b, c\}\}. \subset P(\{a, b, c\})$. ■

Example 2.38

Let $A = \{x, y, z\}$. Define $J = \{X \in P(A) : x \in X\}$ and $K = \{X \in P(A) : y \notin X\}$. Write down Q and R by listing their elements.

SOLUTION

Q is the set of all subsets of A containing x. Thus, $Q = \{\{x\}, \{x, y\}, \{x, z\}, \{x, y, z\}\}$. R is the set of all subsets of A not containing y. Thus, $R = \{\phi, \{x\}, \{x, z\}\}$. ■

2.2 RELATIONS

Example 2.39

Consider the problem of arranging a given set of words in the alphabetical order. The first step in this problem is to group the words having the same first letter. This idea of '*having the same first letter*' is an example of a relation on the given set of the words. Under this *relation*, the words *Zebra*, *Zorn*, *Zoo* are related to each other. The words **Shakespeare** and **Kamban** are *not related* but **Kamban** and **Kalidasa** are related.

The second step in the alphabetical arrangement of the words involves the relation 'having the same first two letters'. Under this new relation, **John** and **Joshua** are related but **Mary** and **Mecca** are not related. ■

Example 2.40

Imagine that an invisible extra-terrestrial being is visiting one of our villages. Let us also imagine that he somehow finds the names of all the people there. Now, he wants to learn the meaning of the word 'Uncle'. He makes the following observations:

Kannan addresses Veeran as Uncle. Mythili addresses Veeran as Uncle. Kumar addresses Ramu as Uncle. Ramu addresses Muthu as Uncle, etc. As far as he is concerned, the relationship Uncle is associated with the set

$$\{(\text{Kannan, Veeran}), (\text{Mythili, Veeran}), (\text{Kumar, Ramu}), (\text{Ramu, Muthu})\}.$$ ■

Example 2.41

The following statements describe some relationships between numbers:
(a) 7 is a factor of 56.
(b) 9 is less than 999.
(c) 169 is the square of -13.
(d) 16 is the cube root of 4096. ■

Example 2.42

Consider the relationship expressed by the description 'a is a factor b' for elements of the set $\{1, 2, 3, 4\}$. Imagine someone who does not understand this factor business. How to tell him which elements are related to which under this relationship? We can simply give him the following list:
(a) 1 is a factor of 1.
(b) 1 is a factor of 2.
(c) 1 is a factor of 3.
(d) 1 is a factor of 4.
(e) 2 is a factor of 2.
(f) 2 is a factor of 4.
(g) 3 is a factor of 3.
(h) 4 is a factor of 4.

The ordered pairs $(1,1)$, $(1,2)$, $(1,3)$, $(1,4)$, $(2,2)$, $(2,4)$ $(3,3)$, $(4,4)$ completely describe this relation as far the set $\{1, 2, 3, 4\}$ is concerned.

Example 2.41 lists some possible relationships between numbers. Let us now consider two sets X and Y. What are the possible relationships one can think of between two sets? We list three possibilities:
(a) The number of elements of X is more than that of Y.
(b) X is the compliment of Y.
(c) X is a subset of Y.
In Example 2.43, we consider a set of sets and list the pairs whose relationship is given by (c) in this list. ■

Example 2.43

Let $A = \{0\}$, $B = \{1\}$, $C = \{1, 2\}$, $D = \{1, 2, 3\}$. Let $S = \{A, B, C, D\}$. Note that S is a set of sets. We consider the relation 'X is a subset of Y' on elements of S. The list of related elements is (B, C), (D, C).

Now, consider two general sets A and B. We want to consider a relationship between the elements of these two sets. Some elements of A are related to some elements of B in some manner. Let us collect all the pairs (a, b) such that a is related to b. Recall that $(a, b) \in A \times B$, where $A \times B$ is the cartesian product of the sets A and B. With these ideas in mind, now we define the concept of a relation as follows. ■

Definition 2.11: Let A and B be two sets. A relation R from A into B is a subset of $A \times B$. If $(a, b) \in R$, we say that a is related to b and denote this fact by aRb. A relation from A into A is called a relation on A.

Example 2.44

In the following, a relation is described using a figure. $A = \{6, 9, 87\}$, $B = \{1, 5, 9, 86\}$. R is defined as follows: aRb if and only if $a \leq b$. Then,

$$R = \{(6, 9), (6, 86), (9, 9), (9, 86)\}.$$

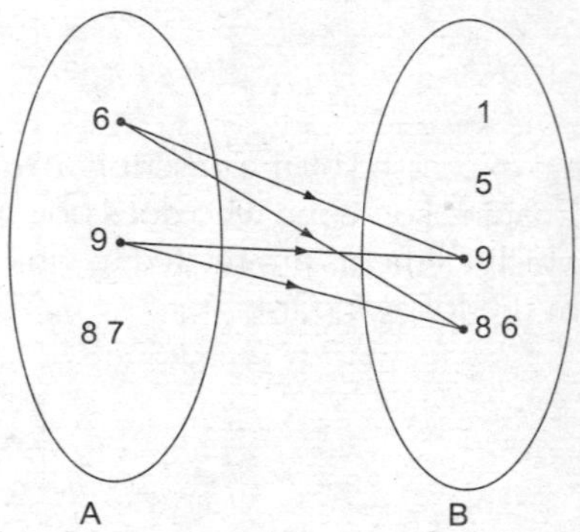

Figure 2.9 Arrow diagram of a relation

In general, after drawing the sets A and B, with A on the left and B on the right, an arrow is drawn from a to b if aRb (Fig. 2.9). ■

PRACTICE 2.7

Draw figures for the following relations:

(i) Let $A = \{a, b, c, d\}$ and $B = \{34, 2, -3\}$.
Define $R = \{(a, -3), (a, 2), (b, 2), (c, 2), (d, 34), (d, 2), (d, -3)\}$.
(ii) Let $A = \{1, 2, 3, 4, 5, 6, 7, 8, 9, 10\}$ and $B = \{14, 15, 17, 64, 100\}$. The relation R is defined as aRb if a is a divisor of b.

PRACTICE 2.8

Consider the following relation on R:

$$R = \left\{(x, y) \in R^2 : y^2 = \cos x\right\}.$$

Which of the following are true?
(i) $0 \; R \; \frac{\pi}{2}$.
(ii) $\frac{\pi}{2} \; R \; 0$.
(iii) $0 \; R \; 1$.
(iv) $0 \; R \; -1$.
(v) $0 \; R \; \pi$.
(vi) $\frac{\pi}{4} \; R \; \frac{1}{\sqrt{2}}$.
(vi) $\frac{\pi}{4} \; R \; \frac{1}{2^{\frac{1}{4}}}$.

2.2.1 Relations With Special Properties

Example 2.45

Let B be the set of all bytes. That is, the set of all strings of length 8 consisting of zeros and ones. We define the following two relations on B.

$a\ R_1\ b$ if the number of zeros in a is strictly less than the number of zeros in b.

$a\ R_2\ b$ if both a and b have same first bit.

Let $a = 00010100$, $b = 00000000$, $c = 00110011$, $d = 01111111$, $e = 10100100$ and $f = 11111111$. Then, we have the following:

(i) $a\ R_1\ b$, $d\ R_1\ e$ and $e\ R_1\ a$ hold but $a\ R_1\ c$ is false.

(ii) $c\ R_2 c$, $a\ R_2\ b$ and $e\ R_2\ f$ hold but $d\ R_2\ e$ is false.

Further, we note that $x\ R_2\ x$ is true for all $x \in B$. We say that R_2 is a reflexive relation, whose definition is given below. ■

Definition 2.12: A relation R on A is said to be reflexive if $a\ R\ a$ for all $a \in A$.

Example 2.46

Let $A = N$. Define $a\ R\ b$ if a divides b. This relation is reflexive.

Proof

As every $a \in N$ divides itself, $a\ R\ a$ and hence R is reflexive. ■

Example 2.47

Let L be the set of all lines in a plane. We define R as $l_1 R\, l_2$ if l_1 is perpendicular to l_2.

Clearly, this relation is not reflexive. We argue as follows: If this relation is reflexive, then it implies that $l_1 R l_1$. That is, l_1 is perpendicular to l_1, which is not true. ■

PRACTICE 2.9

Consider the following relations defined on $\mathbb{R}$:

(i) $a\ R_1\ b$ if $a < b$.

(ii) $a\ R_2\ b$ if $a \leq b$.

Which one of the above relations is reflexive?

Example 2.48

Let B be the set of all bytes given in Example 2.45. We have already observed that R_2 is reflexive and R_1 is not reflexive. In fact, $x\ R_1\ x$ is *not* true for *any* of $x \in B$. Now, we give an example of a relation R_3 on B such that $a\ R_3\ a$ hold for some a but not for all $a \in B$.

We define R_3 as follows: $a\ R_3\ a$ if the first bit of a is the same as the last bit of b. For this relation, we have the following:

(i) 00100010 R_3 00100010 is true.
(ii) 10000000 R_3 10000000 is false.
(iii) 00100011 R_3 11111110 is true.

Clearly, R_3 is not reflexive because of (ii). ■

Example 2.49

Let $A = \{x, y, z, u, v\}$ and $R = \{(x, x), (x, y), (x, v), (y, z), (z, z), (z, u), (u, u), (v, v)\}$. This relation is not reflexive since $y\ R\ y$ does not hold. ■

Example 2.50

Let us define the term sister as follows: A female y is called a sister of x if y is a daughter of the father of x. With this definition consider the following two relations:

(i) Let H be the set of all human beings and define R_1 on H as aR_1b if b is a sister of a.
(ii) Let F be the set of all females and define R_2 on F as aR_1b if b is a sister of a.

R_1 is not reflexive. Because, if x is a male, xR_1x is *false*. But R_2 is reflexive because as per our definition of the term sister, a female is her own sister. So, for every $x \in F$, xR_2x is true.

The relations R_1 and R_2 are described in an identical manner. But as the underlying sets are different, their natures differ. ■

Definition 2.13: A relation R on a set A is said to be symmetric if whenever $x\ R\ y$ then $y\ R\ x$.

Example 2.51

Consider Example 2.45. R_2 is symmetric but not R_1. Further, R_3 of Example 2.48 is also not symmetric. ■

Example 2.52

Let L be the set of all lines in a plane. Define R as $l_1 R\, l_2$ if l_1 is parallel to l_2. Clearly, this relation is both reflexive and symmetric. ■

Example 2.53

Let L be the set of all lines in a plane. Define R as $l_1 R\, l_2$ if l_1 is perpendicular to l_2. Clearly, this relation is symmetric. (Recall that this relation is not reflexive.) ■

Example 2.54

Let $A = \{2, 6, 5, 10\}$ and define R on A as $a\,R\,b$ if b is a multiple of a. We can easily see that

$$R = \{(2,2), (2,6), (2,10), (6,6), (5,5), (5,10), (10,10)\}.$$

This relation is reflexive but not symmetric. Note that $(2,10) \in R$ but $(10,2) \notin R$. That is, $2\ R\ 10$ but not $10\ R\ 2$. ■

Example 2.55

Let R on R be defined as

$$R = \left\{(x, y) \in R^2 : x^2 + y^2 = 1\right\}$$

This relation is symmetric. For, if xRy, then $x^2+y^2 = 1$. But then, $y^2+x^2 = 1$ and hence $y\ R\ x$. ■

PRACTICE 2.10

Consider the relation $\left\{(x, y) \in R^2 : x^2 + 4y^2 = 1\right\}$. Is this symmetric?

Definition 2.14: A relation R on A is said to be *anti-symmetric* if whenever $a\ R\ b$ and $b\ R\ a$, we have $a = b$.

Example 2.56

Let $A = \{2, 6, 5, 10\}$ and define R on A as $a R b$ if b is a multiple of a. It is clear that if a is a multiple of b and b is a multiple of a, then $a = \pm b$. But as there are no negative integers in A, we must have $a = b$. Thus, this relation is anti-symmetric. ■

Example 2.57

Let $A = \{-2, 2, 6, 5, 10\}$ and define R on A as $a R b$ if b is a multiple of a. Note that 2 is a multiple of -2 and -2 is a multiple of 2. So, $-2\ R\ 2$ and $2\ R\ -2$. But $2 \neq -2$. Hence, this relation is not anti-symmetric. ■

PRACTICE 2.11

(i) Show that the following relation R on $\mathbb{R}$, the set of real numbers, is anti-symmetric.

$$a\ R\ b \Leftrightarrow a \leq b.$$

(ii) Let R be defined on N as aRb if b is a factor (or divisor) of a. Show that R is anti-symmetric.

(ii) Let R be defined on Z as aRb if b is a factor (or divisor) of a. Show that R is **not** anti-symmetric.

Definition 2.15: A relation R on A is said to be transitive, if $a\ R\ b$ and $b\ R\ c$ imply $a\ R\ c$.

Example 2.58

Let B be the set of all bytes. Define R_4 on B as $a\ R_4\ b$ if a and b have the same number of zeros. It is obvious that R_4 is transitive. ■

Example 2.59

Let S be any set and let $P(S)$ be the set of all subsets of S. Define R on $P(S)$ as $A\ R\ B$ if $A \subseteq B$. Show that R is reflexive, anti-symmetric and transitive.

SOLUTION

Let $A \in P(S)$. This means that $A \subseteq S$. Now, clearly, $A \subseteq A$. Thus, $A\ R\ A$ and hence R is reflexive.

Now, let $A, B \in P(S)$. Now, if $A\ R\ B$ and $B\ R\ A$, we have $A \subseteq B$ and $B \subseteq A$. So, we get $A = B$. Thus, R is anti-symmetric.

Next, $A\ R\ B$ and $B\ R\ C$ imply that $A \subseteq B$ and $B \subseteq C$. This gives us $A \subseteq C$ and so $A\ R\ C$. Hence, R is transitive. ■

Example 2.60

Let $S = \{p, q, r\}$. Define R as follows:

$$R = \{(p, p), (q, q), (r, r), (p, q), (q, p), (q, r), (r, q)\}.$$

Clearly, this relation is reflexive and symmetric. Note that pRq, qRr but p is not related to r. This relation is not transitive. ■

Example 2.61

Give an example of a relation that is neither reflexive nor symmetric but transitive.

SOLUTION

Let $A = \{a, b, c\}$ and $R = \{(a, b), (b, c), (a, c)\}$. ■

Example 2.62

Let $A = \{\alpha, 3, \beta\}$ and define R as $\{(\alpha, 3), (3, \beta)\}$. This relation is not transitive. Because we have $\alpha\ R\ 3$ and $3\ R\ \beta$ but we do not have $\alpha\ R\ \beta$. ■

Example 2.63

Give an example of a relation that is symmetric and transitive but not reflexive.

SOLUTION

Let $A = \{a, b, c\}$ and $R = \{(b, b), (c, c), (b, c), (c, b)\}$. This relation is not reflexive since $(a, a) \notin R$. We can easily check that it is both symmetric and transitive. ■

PRACTICE 2.12

Give an example of a relation that is reflexive and transitive but not symmetric.

2.2.2 Equivalence Relations and Equivalence Classes

Consider the set L of all lines in a plane and the relation R defined as $l_1 R l_2$ if l_1 is parallel to l_2. We have seen that this relation is both reflexive and symmetric. Further, if l_1 is parallel to l_2 and l_2 is parallel to l_3, then we know that l_1 is parallel to l_3. Thus, R is also transitive. This relation is an example of an *equivalence relation*. Its definition is given below.

Definition 2.16: A relation R on a set S is called an equivalence relation if

(i) R is reflexive.
(ii) R is symmetric.
(iii) R is transitive.

Example 2.64

The easiest example of an equivalence relation is an equality. Let A be any subset of $\mathbb{R}$. The relation R defined as $a\ R\ b$ if $a = b$ is obviously an equivalence relation. ■

Example 2.65

Let A be a finite set and $P(A)$ be its power set. A relation R on $P(A)$ is defined as $X\ R\ Y$ if X and Y have the same number of elements. Now, we show that R is an equivalence relation. $X\ R\ X$ means that X and X have the same number of elements, hence it is true. Next, let $X\ R\ Y$. This means that X and Y have the same number of elements. This can also be stated as Y and X have the same number of elements. So, $Y\ R\ X$. Hence R is symmetric. Further, $X\ R\ Y$, and $Y\ R\ Z$ imply that X and Y have the same number of elements and Y and Z have the same number of elements. It is clear that X and Z will have the same number of elements. So, $X\ R\ Z$ and R is transitive. We have now proved that R is an equivalence relation.

In particular, if $A = \{0, 1\}$ the above relation on $P(A)$ is expressed as a subset of $P(A) \times P(A)$ as follows:

$$R = \{(\phi, \phi), (\{0\}, \{0\}), (\{1\}, \{1\}) (\{0\}, \{1\}), (\{1\}, \{0\}), (\{0, 1\}, \{0, 1\})\}.$$ ■

Example 2.66

Define R on $\mathbb{R}$ as $x\ R\ y$ if $x - y$ is an integer. Now, we show that this is an equivalence relation. $x - x = 0$ and 0 is an integer. Hence $x\ R\ x$. So, this relation is reflexive.

Now let $x\ R\ y$. This implies that $x - y$ is an integer. Then, $y - x$ is also an integer. So, $y\ R\ x$ and R is symmetric.

Now let $x\ R\ y$ and $y\ R\ z$. These imply that $x - y$ is an integer and $y - z$ is an integer. Since $x - z = (x - y) + (y - z)$, $x - z$ is also an integer. So, $x\ R\ z$ and R is transitive. Hence R is an equivalence relation. ■

Example 2.67

Define R on **Z** as $x\ R\ y$ if $x - y$ is even. Show that this is an equivalence relation.

(a) R is reflexive: $x - x = 0$ and 0 is even. Hence xRx.

(b) R is symmetric: Let $x\ R\ y$. Then $x - y$ is even. Now, $(y - x) = -(x - y)$ and hence is also even. Thus, $y\ R\ x$.

(c) R is transitive: Let $x\ R\ y$ and $y\ R\ z$. Then $x - y$ and $y - z$ are even. Now, $(x - z) = (x - y) + (y - z)$ and hence is also even. Thus, $x - z$ is even. So, $x\ R\ z$.

Hence, R is an equivalence relation. ■

PRACTICE 2.13

Define R on **Z** as $x\ R\ y$ if $x - y$ is a multiple of 3. Show that this is an equivalence relation.

2.3 PARTITIONS

In this section, we discuss the partitions of a set. Before describing the partition of a set we first present a set of examples to understand the ideas behind the partitions of a set.

Example 2.68

(a) Let S be the set of students of a class. Let A and B be, respectively, the set of girl students and the set of boy students of the same class. We have

(i) $A \cup B = S$.
(ii) $A \cap B = \phi$.

(b) Let S be the set of all human beings. Let M be the set of all males, F be the set of all females and N be the set of all noble laureates.

(i) $S = M \cup F \cup N$.
(ii) $M \cap F = \phi$, $M \cap N \neq \phi$ and $F \cap N \neq \phi$.

(c) Let $S = N$. $A = \{3, 6, 9, \ldots\}$, $B = \{4, 7, 10, \ldots\}$ and $C = \{5, 8, 11, \ldots\}$.

(i) $S = A \cup B \cup C$.
(ii) $A \cap B = \phi$, $B \cap C = \phi$ and $C \cap A = \phi$.

(d) Let $S = \{3, 4, 5, 6, 7, 8, 9\}$. Let A be the set of all even numbers in S and B be the set of all prime numbers in S.

(i) $S \neq A \cup B$.
(ii) $A \cap B = \phi$.

In all the above examples, we have a set S and some subsets of it. In (a), (b) and (c), the union of the given subsets is S. In (a), (c) and (d) the subsets are disjoint. Examples (a) and (c) illustrate the idea of the *partition* of a set. ■

Definition 2.17: Let S be a set and $\{A_\lambda : \lambda \in \Lambda\}$ be a collection of non-empty subsets of S. We say that this collection defines a partition of S if

(i) $S = \bigcup_{\lambda \in \Lambda} A_\lambda$.
(ii) $E_{\lambda_1} \cap E_{\lambda_2} = \phi$ if $E_{\lambda_1} \neq E_{\lambda_2}$.

Remark 2.2

Note that we do not demand that $E_{\lambda_1} \neq E_{\lambda_2}$ for $\lambda_1 \neq \lambda_2$.

Example 2.69

For $k \in N$, let $A_k = [k-1, k]$ and $B_k = [k-1, k)$. We have the following observations:
(a) $\mathbb{R}^+ = \bigcup_{k=1}^{\infty} A_k$.
(b) $\mathbb{R}^+ = \bigcup_{k=1}^{\infty} B_k$.
(c) $A_k \cap A_{k+1} = \{k\}$ and $A_k \cap A_l = \phi$ if $l > k+1$.
(d) $B_k \cap B_l = \phi$ if $B_k \neq B_l$.
So, $\{A_k : k \in N\}$ is not a partition of R^+, whereas $\{B_k : k \in N\}$ is a partition of R^+. ■

Example 2.70

Let $y \in \mathbb{R}$. Define $A_y = \{(x, y) : x \in \mathbb{R}\}$. Show that $\{A_y : y \in \mathbb{R}\}$ is a partition of $\mathbb{R}^2$.

SOLUTION

Each A_y is a subset of $\mathbb{R}^2$ and hence their union would also be a subset of $\mathbb{R}^2$. We now show that $\mathbb{R}^2 \subseteq \bigcup_{y \in R} A_y$. Let $(x, y) \in \mathbb{R}^2$. Clearly $(x, y) \in A_y$. Hence, $(x, y) \in \bigcup_{y \in R} A_y$. Hence, $\mathbb{R}^2 \subseteq \bigcup_{y \in R} A_y$. Finally $\mathbb{R}^2 = \bigcup_{y \in R} A_y$. Next, we must show that $A_y \cap A_z = \phi$ if $A_y \neq A_z$. Let us note that $y = z \leftrightarrow A_y = A_z$. Now, $A_y \neq A_z \rightarrow y \neq z$ and hence $(x, y) \neq (x, z)$ for any x. So, $A_y \cap A_z = \phi$. ■

Remark 2.3

Each A_y is nothing but a horizontal line in the xy plane. In particular, A_0 is the x axis and $A_{5.1}$ is the line $y = 5.1$.

Example 2.71

Define R on Z as aRb if $a - b$ is a multiple of 4. Under this equivalence relation, 0 is related to 0, 4, 8, etc. Similarly, 1 is related to 1, 5, 9, etc. So, now let us define the following subsets of Z as $[0] = \{0, \pm 4, \pm 8, \pm 12, \ldots\}$.
$[1] = \{\ldots, -11, -7, -3, 1, 5, 9, \ldots\}$.
$[2] = \{\ldots, -10, -6, -2, 2, 6, 10, \ldots\}$.
$[3] = \{\ldots, -9, -5, -1, 3, 7, 10, \ldots\}$.

For $r = 0, 1, 2, 3$, $[r]$ is nothing but the set of all integers that leave a remainder r when divided by 4. Now, any integer when divided by 4 would leave a remainder that is 0, 1, 2 or 3. From these observations, it is clear that these four subsets of $\mathbf{Z}$ define a partition of $\mathbf{Z}$. ■

Theorem 2.1

Let S be a set and R be an equivalence relation. For $x \in S$, define the equivalence class of x as

$$[x] = \{y \in S : x\,R\,y\}.$$

$\{[x] : x \in S\}$ is a partition of S.

Proof

First, we show that if $[x] \neq [y]$, then $[x] \cap [y] = \phi$. We prove this by contradiction. Let $[x] \neq [y]$. If $[x] \cap [y] \neq \phi$, then there is $z \in [x] \cap [y]$. So, xRz and yRz. Using symmetry, we get zRx. By transitivity, xRy. Using this, we shall show that $[x] \subset [y]$. Let $w \in [x]$. Then xRw or wRx Combining this with transitivity and the fact that xRy, we get wRy and hence $w \in [y]$. This shows that $[x] \subseteq [y]$. In a similar way, we can show that $[y] \subset [x]$ and finally $[x] = [y]$, a contradiction.

Next, we have to show that

$$S = \bigcup_{x \in S} [x].$$

It is clear that $\bigcup_{x \in S} [x] \subseteq S$. So, we have to show that

$$S \subseteq= \bigcup_{x \in S} [x].$$

Now, by reflexivity, $x \in [x]$ and hence any $x \in S$ is element of $\bigcup_{x \in S} [x]$. Thus, $S = \bigcup_{x \in S} [x]$. Thus, $\{[x] : x \in S\}$ is a partition of S. ■

Example 2.72

Let C be the set of all strings of length 5 consisting of zeros and ones. Let R be defined as follows: aRb if a and b have the same number of zeros. Then, R is an equivalence relation and we have

$$[11111] = \{111111\}$$
$$[01111] = \{01111, 10111, 11011, 11101, 11110\}$$
$$[00111] = \{00111, 01011, 01101, 01110, 10011, 101011, 10110, 11001, 11010, 11100\}$$
$$[00011] = \{11000, 10100, 10010, 10001, 01100, 010100, 01001, 00110, 00101, 00011\}$$
$$[00001] = \{00001, 10000, 01000, 00100, 00010\}$$
$$[11111] = \{11111\}.$$

The above 6 sets form a partition of C. ■

Example 2.73

Consider the set of all lines L in a plane and the equivalence relation on L defined as $l_1 R l_2$ if l_1 is parallel to l_2. For a given line l, the equivalence class $[l]$ is nothing but the set of all lines parallel to l. ■

Example 2.74

Imagine a district affected by floods. The 7 important towns in the district are denoted by A, B, C, D, E, F and G. The following table gives us the information regarding whether the road connecting a pair of cities is usable.

	A	B	C	D	E	F	G
A	√	×	×	×	√	×	×
B	×	√	×	√	×	×	×
C	×	×	√	×	√	×	×
D	×	√	×	√	×	×	×
E	√	×	√	×	√	×	×
F	×	×	×	×	×	√	√
G	×	×	×	×	×	√	√

If √ is marked in the row for X and the column for Y, it means that the road between the towns X and Y is usable. If × is marked, it either means that there is actually no direct road between X and Y or that the direct road between them has been affected by the floods.

Now on the set of these towns, we define a relation as follows: We say that $X\ R\ Y$ if we can travel from X to Y. It is clear that this relation is an equivalence relation. The equivalence class $[X]$ is nothing but all towns that can be reached from X. The following diagram will be useful to find the equivalence class of each town:

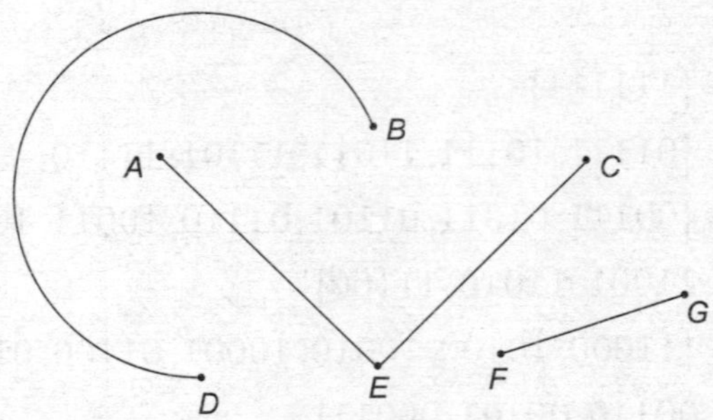

Figure 2.10

$$[A] = \{A, C, E\}$$
$$[B] = \{B, D\}$$
$$[C] = \{A, C, E\}$$
$$[D] = \{B, D\}$$
$$[E] = \{A, C, E\}$$
$$[F] = \{F, G\}$$
$$[G] = \{F, G\}$$

Note that there are only 3 equivalence classes since $[A] = [C] = [E]$, $[B] = [D]$, $[F] = [G]$.

Figure 2.10 gives an example of a graph. We shall learn more about graphs in the chapter on graph theory. ■

PRACTICE 2.14

Let R be defined on Z as mRn if $m - n$ is a multiple of k. Show that R is an equivalence relation and find all the equivalence classes.

2.3.1 Partial Order

Definition 2.18: A relation R on a set S is called a partial order if it is

(i) R is reflexive.
(ii) R is anti-symmetric.
(iii) R is transitive.

■ Example 2.75

Let $S = \{1, 2, 3, \ldots, 10\}$ and R is defined as aRb if $a \leq b$. Then R is a partial order.

Proof

(i) For any $a, b \in S$, we have $a \leq a$. Hence R is reflexive.
(ii) If $a \leq b$ and $b \leq a$, then $a = b$. Hence R is anti-symmetric.
(iii) If $a \leq b$ and $b \leq c$, then $a \leq c$. So, R is transitive.

Thus, R is a partial order. ■

PRACTICE 2.15

Let S be any set and let $P(S)$ be the set of all subsets of S. ARB if $A \subseteq B$. Show that R is a partial order.

Counter Example: Let $S = \{\phi, \{1\}, \{2\}, \{1,2\}\}$. Let $n(A)$ be the number of elements of A. Define ARB if $n(A) \leq n(B)$. Then R is not a partial order.

(i) $n(A) = n(A)$ and hence ARA for all $A \in S$.
(ii) Now, consider $A = \{1\}$ and $B = \{2\}$. We have $n(A) \leq n(B)$ and $n(B) \leq n(A)$, since $n(A) = n(B) = 1$. Thus ARB and BRA. But, $A \neq B$. Hence this relation is not anti-symmetric.
(iii) If $n(A) \leq n(B)$ and $n(B) \leq n(C)$, then $n(A) \leq n(C)$. So, R is transitive.

Thus, R is only reflexive and transitive but not anti-symmetric. Thus R is not a partial order.

Example 2.76

Let R on $\mathbf{Z}$ be defined as mRn if n is a multiple of m. Then, R is not a partial order.
(i) Every integer is a multiple of itself. Hence nRn for all $n \in \mathbf{Z}$. Thus R is reflexive.
(ii) Let mRn and nRm. That is m is a multiple of n and m is a multiple of n. This is possible when $m = \pm n$; hence, we cannot conclude that $m = n$. So, R is not anti-symmeric.
(iii) If b is a multiple of a and c is a multiple of b, then c is a multiple of a. So, R is transitive.

The given relation is reflexive, transitive but not anti-symmetric. So, it is not a partial order. ■

Example 2.77

Let R be defined as N as mRn if n is a multiple of m. Then, R is a partial order.
(i) Every natural number is a multiple of itself. Hence nRn for all $n \in N$. Thus R is reflexive.
(ii) Let mRn and nRm. That is, m is a multiple of n and m is a multiple of n. Since both m and n are positive, $m = n$. So, R is anti-symmetric.
(iii) If b is a multiple of a and c is a multiple of b, then c is a multiple of a. So, R is transitive.

The given relation is reflexive, anti-symmetric and transitive So, it is a partial order. ■

PRACTICE 2.16

Which of the following relations are partial orders?

(i) R on $\mathbf{Z}$ is defined as mRn if n is a divisor of m.
(ii) R on N is defined as mRn if n is a divisor of m.

Remark 2.4

It is customary to denote a partial order by $\leq$ even if the relation is not the usual 'less than or equal to' relation.

Example 2.78

Consider the following figures that are known as directed graphs (Fig. 2.11):

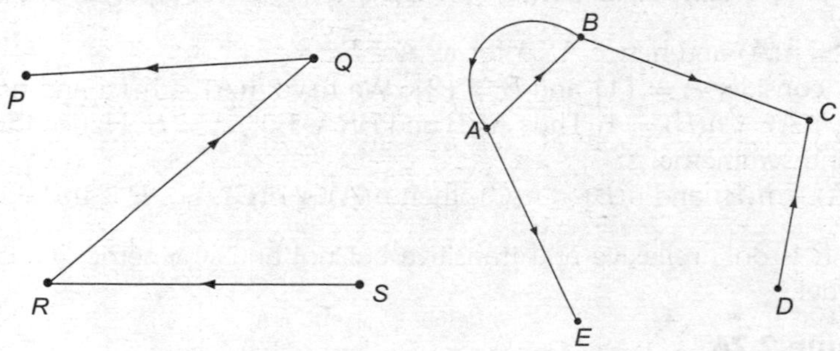

Figure 2.11 A directed graph

We say that $X\ R\ Y$ if either $X = Y$ or if there are edges leading from X to Y.

This relation is reflexive and transitive. It is clear that for the first graph, R is a partial order, because there are no two distinct vertices X and Y such that we can go from X to Y and also from Y to X.

But for the second graph the distinct vertices A and B are such that ARB and BRA. So, R is not anti-symmetric and is hence not a partial order. ■

2.4 FUNCTIONS

We know that the period of oscillation of a simple pendulum *depends* on the length of the string. In fact, the precise relationship is given by

$$T = 2\pi\sqrt{\frac{l}{g}}. \tag{2.7}$$

Here l is the length of the pendulum and g is the acceleration due to gravity. If we know the value of the length of the pendulum, we can calculate the period of oscillation T. In the language of mathematics, we say that T **is a function of** l. To be precise, T depends on g too. Thus, T is a function of both l and g. But if we remain on the surface of our planet Earth, g is a constant and so we treat T as a function of the single variable l.

Now, imagine that we fix the length of our pendulum and undertake a space journey to various planets in the universe. T is now a function of g, which changes from planet to planet. If l increases T increases. If g increases, T actually decreases. So, we see that an expression like (2.7) can be used to define at least three different functions viz, (i) fix l, vary g (ii) fix g, vary l, and (iii) vary both. This tells us that a function is much more than a formula. Further, what happens to T if g is zero? $l = 0$ gives us $T = 0$. But $l = -5$ is meaningless both mathematically and physically. Thus, it is not enough to have a formula; we must know the set of data for which the formula is meaningful. This is the basis for the idea of a mathematical function.

Definition 2.19: A function f from a set A into a set B is a rule that assigns precisely one element of B to each element of A. A is called the domain of the function and B is called the co-domain of the function. If $b \in B$ is the element that is assigned to $a \in A$, then we write $f(a) = b$. b is called the image of a under f and a is called a pre-image of b under f.

Example 2.79

Let A be the set of all students of an institute and B be the set of all positive integers. We can think of a function, which we shall call as the age function, as the one that assigns the age of a student (number of whole years completed after birth) to every student. Note that in our description of the term 'function', we did not demand that every element of B to be associated with some element of A. No student of the institute would be of age 3 or 96 but this does not prevent these ages to be an element of B. Note also that we did not demand that two different elements of A to be associated with two different elements of B. Students of the same class are very likely to have the same age and again this is perfectly all right.

Of course, no student can have have two ages and this is an illustration to the requirement that a function cannot assign two different elements to an element of A.

The notation $f : A \longrightarrow B$ tells us that f is a function from the set A into the set B. ■

Example 2.80

Let A and B both be sets of all living human beings. Can we define $f : A \longrightarrow B$ as

$$f(x) = \text{brother of } x?$$

No, we cannot, as
(i) x may not have a brother.
(ii) x may have more than one brother. ■

Example 2.81

Why can't we think of a function as just an expression or a method to calculate $f(x)$ for a given value of x? Why do we bother about domains, etc.? Let us consider the expression $f(x) = \frac{1}{x+2}$. Imagine that you have written a program that calculates $f(x)$ for various values of x during the course of its implementation. Imagine a situation, where x becomes -2. The computer will obviously give an error message as this value gives rise to the meaningless fraction $\frac{1}{0}$. It is better that we take this fact into account and design the program such that $f(x)$ is calculated only when $x \neq -2$.

Mathematically speaking, we permit A to be *any subset* of R *not* containing -2 and then define $f : A \longrightarrow R$ as $f(x) = \frac{1}{x+2}$.

The largest subset D of R such that a proper function $f : D \longrightarrow R$ with $f(x) = \frac{1}{x+2}$ can be defined is $D = R - \{-2\}$. ■

Example 2.82

Consider the expression log(log(log log(x))). Try and see what happens if we calculate log(log(log log(5))) using a calculator. When we press the button for the third time the calculator gives an error message. Thus, the expression log(log(log log(x))) cannot be computed for $x = 5$.

We shall now find the 'largest domain' for this expression. For $\log r$ to be defined we must have $r > 0$. Hence, if log(log(log log(x))) has to be defined, then we must have $(\log(\log\log(x))) > 0$

$$\begin{aligned} &\rightarrow \quad \log(\log\log(x)) > 0 \\ &\rightarrow \quad \log(\log x) > 1 \\ &\rightarrow \quad \log x > e \\ &\rightarrow \quad x \geq e^e. \end{aligned}$$

Hence the required set is $[e^e, \infty)$.

Statements like 'Let f be the function defined as $f(x) = \log(\log(\log\log(x)))$' are ambiguous. We analyse the expression log(log(log log(x))) and find that it is meaningful only when $x \geq e^e$. Now we can define $f : [e^e, \infty)$ as $f(x) = \log(\log(\log\log(x)))$. ■

Example 2.83

Define $f : \mathbb{R} \longrightarrow \mathbb{R}$ as $f(x) = x^3 - x + 3$. ■

Example 2.84

Let $A = R - \{2, -3\}$. Define $g : A \longrightarrow \mathbb{R}$ as $g(x) = \frac{1}{x^2+x-6}$. In Examples 2.83 and 2.84 algebraic expressions are used to define the function. But this need not be the case always.

A legitimate function $f : A \longrightarrow \mathbb{R}$ may be defined using different algebraic expressions for different subsets of A and there are functions such that there may not be a 'formula' to 'calculate' $f(x)$ using the 'value' of x. ■

Example 2.85

Let A be the set of all living human beings whose mothers are alive. Let B be the set of all living females. For $x \in A$, $f(x)$ is defined as the mother of x. ■

Example 2.86

Let $A = [0, \infty)$ and $B = \mathbb{R}$. Define $f : A \longrightarrow B$ as $f(x) = 1$ for all $x \in [0, 1]$ and $f(x) = x - 2$ for all $x \in (1, \infty)$.

The graph of this function is given in Figure 2.12.

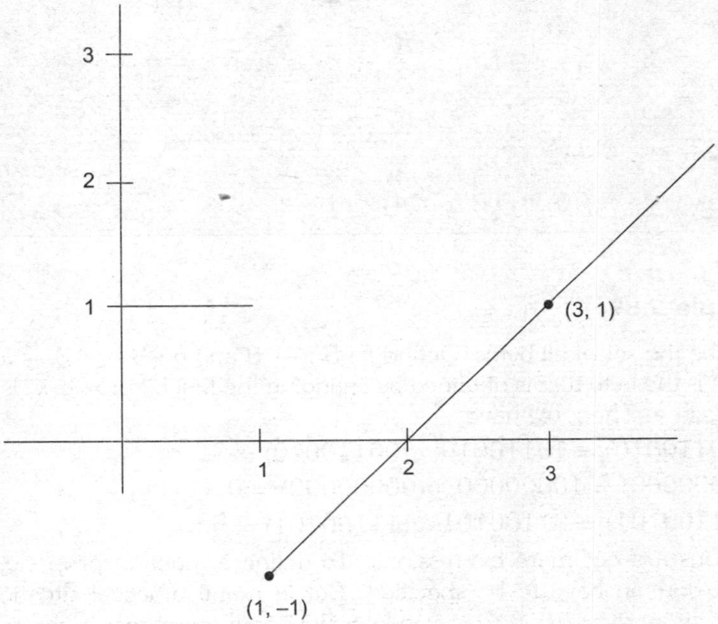

Figure 2.12 A directed graph ■

Example 2.87

Let $f : \mathbb{R} \longrightarrow \mathbb{R}$ be defined as

$$\begin{aligned} f(x) &= 2x - 3, \ \ x \le 9 \\ &= x^2 - 4x + 7, \ \ 9 < x < 100 \\ &= cos\pi x, \ \ 100 \le x. \end{aligned}$$

Find $f(7)$, $f(8)$ $f(9)$ $f(10)$ and $f(100)$.

SOLUTION

Since $7, 8, 9 \in (-\infty, 9]$, $f(7) = 2 \times 7 - 3 = 11$, $f(8) = 2 \times 8 - 3 = 13$ and $f(9) = 2 \times 9 - 3 = 15$
Since $10 \in (9, 100)$, $f(10) = 10^2 - 4 \times 10 + 7 = 67$.
Since $100 \in [100, \infty)$, $f(100) = \cos 100\pi = (-1)^{100} = 1$. ■

Example 2.88

Let A be the set of all M.P's. $f : A \longrightarrow R$ is defined as $f(x)$ = Height of x in cm. When x is a woman and $f(x)$ = Weight of x in kg. Clearly, $f(SoniaGandhi)$ is the height of Mrs Sonia Gandhi in cm and $f(ManmohanSingh)$ is the weight of Mr Manmohan Singh in kg. ■

PRACTICE 2.17

Let $f : Z \longrightarrow Z$ be defined as

$$f(n) = \frac{n}{2} \quad \text{if } n \text{ is even}$$
$$= \frac{n-1}{2} \quad \text{if } n \text{ is odd.}$$

Find the values of $f(0), f(7), f(-14), f(9)$.

Example 2.89

Let B be the set of all bytes. Define $f : B \longrightarrow B$ and $g : B \longrightarrow Z$ as follows:

$f(x)$ is the byte that is obtained by changing the first bit of x. $g(x)$ is the number of ones in x. Then, we have

(i) $f(00110010) = 10110010$, $g(00110010) = 3$.

(ii) $f(0000000) = 10000000$, $g(00000000) = 0$.

(iii) $f(11100101) = 01100101$, $g(11100101) = 5$.

Functions are not mere expressions. To define a function properly, its domain and co-domain have to be specified. But in many practical situations, we first encounter an expression $E(x)$ and later have to find out the values of x for which the expression $E(x)$ makes sense. The following definition is useful in this context. ■

Definition 2.20: The *natural domain* for an expression $E(x)$ is the set of all x for which $E(x)$ is meaningful.

Example 2.90

Find the natural domain for the expression $\sqrt{x+5}$.

SOLUTION

$\sqrt{u}$ is meaningful when $u \geq 0$. Thus, we have to find out all values of x, for which $x+5$ is non-negative. In other words, we have to solve the inequality

$$x + 5 \geq 0$$

i.e.

$$x \geq -5.$$

Hence, the natural domain for the function defined by the expression $\sqrt{x+5}$ is $[-5, \infty)$. ■

Example 2.91

Find the natural domain for the expression $\sqrt{x^2 - 9x - 10}$.

SOLUTION

$\sqrt{u}$ is meaningful when $u \geq 0$. Thus, we have to find out all values of x, for which $x^2 - 9x + 10$ is non-negative. In other words, we have to solve the inequality

$$x^2 - 9x + 10 \geq 0.$$

By using the ideas from Section 1.5 of Chapter 1, we find that $x \leq -1$ or $x \geq 10$. ■

Example 2.92

Consider Figure 2.13.

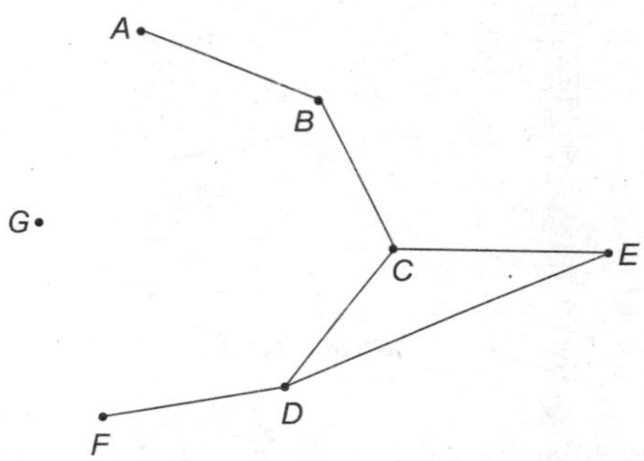

Figure 2.13

Let $V = \{A, B, C, D, E, F\}$ be the set of vertices of this graph. Recall that 2^V is the set of all subsets of V. We say that a vertex X is a neighbour of Y if there is an edge connecting them. For example, B is a neighbour of A. F is not a neighbour of A.

$f : V \longrightarrow 2^V$ is defined as follows $f(X)$ is the set of all neighbours of X. $f(A) = \{B\}$. $f(B) = \{A, C\}$, $f(C) = \{B, D, E\}$, $f(D) = \{C, E, F\}$, $f(E) = \{C, D\}$ and $f(F) = \{D\}$ and finally $f(G) = \phi$, the empty set. ■

2.4.1 Connection Between Functions and Relations

Let $A = \{1, 2, 3, 4\}$ and $B = \{a, b, c\}$. Define $f : A \longrightarrow B$ as $f(1) = c$, $f(2) = a$, $f(3) = f(4) = b$. We can draw Figure 2.14 for this function:

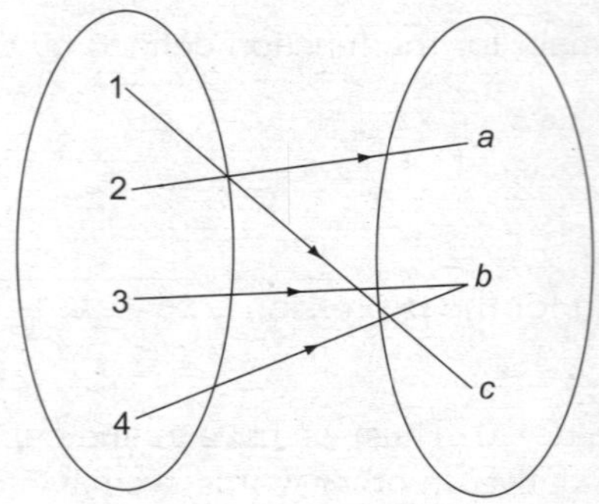

Figure 2.14

Next, consider Figure 2.15.

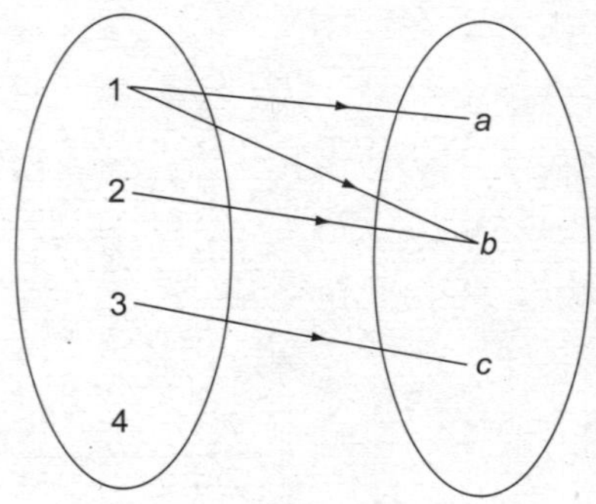

Figure 2.15

This diagram represents the relation $R = \{(1, a), (1, b), (2, b)(3, c)\}$. Does this diagram represent a function? No. Roughly speaking, the reasons are

(i) $f(1)$ would have 'two values'.
(ii) $f(4)$ 'has no value'.

To be precise, we have the following :

Proposition 2.2

Let $R \subseteq A \times B$ be a relation. Then, there exists a function $f : A \longrightarrow B$ such that

$$(x, y) \in R \leftrightarrow (x, f(x)) \in R$$

if and only if

(i) for all $x \in A$ there exists $y \in B$ such that $(x, y) \in R$.
(ii) $(x, y_1), (x, y_2) \in R \leftrightarrow y_1 = y_2$.

Pictorially speaking, a relation would give rise to a function provided exactly *one* line starts from *each* element of A and reaches an element of B.

Example 2.93

Let $R = \left\{(x, y) : y^2 = x\right\}$. Does R define a function?

SOLUTION

Note that $(25, 5)$ and $(25, -5)$ are elements of R. Hence the condition $(x, y_1), (x, y_2) \in R \leftrightarrow y_1 = y_2$ is not true. So, this relation does not define a function. ■

2.5 FUNCTIONS WITH SPECIAL PROPERTIES

Many equations in mathematics can be formulated as

$$f(x) = y_0$$

where f is a function, y_0 is given and x is the required solution. On encountering such an equation, we would like to ask the following questions:

(i) is there at least one solution ?
(ii) if the answer to the above question is yes, how many solutions are there?

Note that the answer to these questions depend on the function and also the value of y_0. If f is a real valued function of a real variable, then these questions can be answered graphically. Draw the graph of $y = f(x)$ and the line $y = y_0$. If the equation $f(x) = y_0$ has a solution, then the graph and the above-mentioned line will intersect. If (x_1, y_0) (x_2, y_0), etc. are the points of intersection, then x_1, x_2, etc. are solutions of $f(x) = y_0$ (Fig. 2.16).

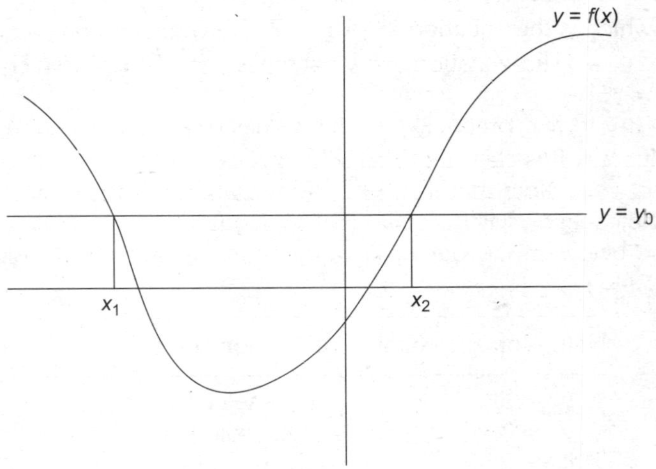

Figure 2.16 Solving $f(x) = y_0$ graphically

In the following examples we analyse these questions for some functions.

Example 2.94

Consider the equations

$$\cos x = y_0. \tag{2.8}$$

$$\tan x = y_0. \tag{2.9}$$

First, consider equation (2.8). If $y_0 = -1$, then $x = \pi$ is a solution to equation (2.8). If $y_0 = \frac{\sqrt{3}}{2}$, $x = \frac{\pi}{6}$ is a solution. If $y_0 = 2$, as there is no value of x for which $\cos x = 2$ and equation (2.8) has no solution. In fact, from trigonometry we know that equation (2.8) has a solution only if $|y_0| \leq 1$.

Next, consider equation (2.9). It has at least one solution for any $y_0 \in R$. This is because for any y_0 we can find an x with $\tan x = y_0$. ■

Example 2.95

Let $f : \mathbf{C} \longrightarrow \mathbf{C}$ be defined as $f(z) = z^2 + 3$. Can we solve the equation

$$f(z) = y_0$$

for all values of y_0? The above equation is nothing but

$$z^2 + 3 = y_0.$$

It is clear that the equation has two solutions, namely, the complex square roots of $y_0 - 3$. ■

Example 2.96

Let $f : R \longrightarrow R$ be defined as $f(x) = x^2 + 3$. Let $y_0 = 2$. Consider the equation $f(x) = y_0$, which is the equation $x^2 + 3 = 2$. This equation has no real solution. Next, let $y_0 = 4$. The equation, now becomes $x^2 + 3 = 4$, which has the two solutions $x = \pm 1$.

Next, we try to find precisely for what values of y_0, the equation $x^2 + 3 = y_0$ has real solutions. It is clear that real solutions exist if $\sqrt{y_0 - 3}$ is real. Hence we must have $y_0 \geq 3$. Note that if $y_0 = 3$, the equation has the only solution 0 and if $y_0 > 3$, the equation will have two real solutions.

The table below shows our observations for the number of solutions of the equation $f(x) = y_0$.

Nature of y_0	Number of solutions of $x^2 + 3 = y_0$
$y_0 < 3$	No solution
$y_0 = 3$	One solution
$y_0 > 3$	Two solutions

■

Example 2.97

Let $\mathbf{M}_2$ be the set of all 2×2 matrices with real entries. Define $f : \mathbf{M}_2 \longrightarrow R$ as $f(A) = |A|$. Let $y \in R$ be given. Consider the equation $f(A) = y$, which is nothing but $|A| = y$. There is no procedure to solve this kind of an equation. We should see if we can find a matrix whose determinant is y. This can easily be constructed. Take $A = \begin{pmatrix} 1 & 0 \\ 0 & y \end{pmatrix}$. $|A| = y$ and hence for any given $y \in R$, we have found $A \in \mathbf{M}_2$ such that $f(A) = y$. ■

Example 2.98

Let

$$X = \left\{ \begin{pmatrix} \cos\theta & -\sin\theta \\ \sin\theta & \cos\theta \end{pmatrix} : \theta \in R \right\}.$$

Define $f : X \longrightarrow R$ as $f(A) = |A|$. Note that this is a constant function. It is easy to show that $f(A) = 1$ for all $A \in X$. Hence, obviously the equation $f(A) = 8$ has no solution such that $A \in X$. Thus, we have found $y = 8$, which is an element of the co-domain R such that there is no $A \in X$ with $f(A) = y$.

In all these above examples, for a given y, we examined whether there is a solution to $f(x) = y$ and if solutions exist, we tried to find the number of solutions. Informally, a function is onto if $f(x) = y$ has at least one solution for every y in the co-domain. A function is one-to-one if there is at most one solution to $f(x) = y$. The formal definitions are given below: ■

Definition 2.21:

(a) A function $f : A \longrightarrow B$ is said to be onto if for every $y \in B$ there is some $x \in A$ such that $f(x) = y$.
(b) A function $f : A \longrightarrow B$ is said to be one-to-one if $f(x_1) = f(x_2)$ implies that $x_1 = x_2$.

The next proposition relates the properties of functions to solutions of equations.

Proposition 2.3

Let $f : A \longrightarrow B$ be a function. Consider the equation

$$f(x) = y. \tag{2.10}$$

(i) The function f is *onto* if equation (2.10) has *at least* one solution $x \in A$ for every $y \in B$.
(ii) The function f is *one-to-one* if equation (2.10) has *at most* one solution $x \in A$ for every $y \in B$.

Example 2.99

$f_1 : \mathbb{R} \longrightarrow \mathbb{R}^+$ is defined as $f_1(x) = x^2$.

Clearly for every non-negative real number b, there is an element $a \in \mathbb{R}$ such that $f_1(a) = b$, that is, $a^2 = b$. In fact, there are two values of a, namely, $\pm\sqrt{b}$. Hence this function is onto, but not one-to-one.

If we draw the graph of the function defined in Example 2.99, we observe that for any positive b, the line $y = b$ intersects the graph at two places (Fig. 2.17).

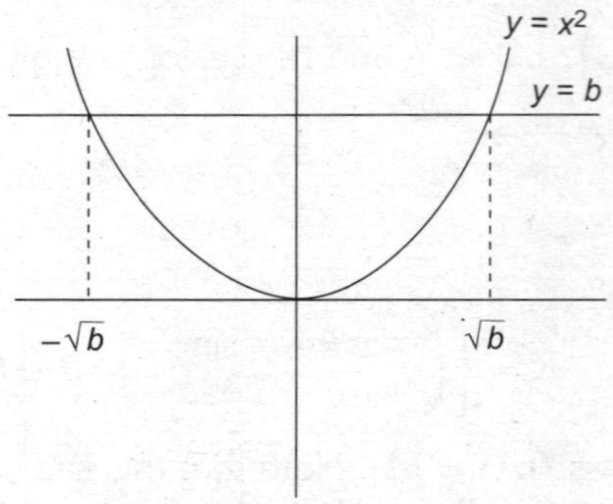

Figure 2.17

■

Example 2.100

$f_2 : \mathbb{R} \longrightarrow \mathbb{R}$ is defined as $f_2(x) = x^3$.

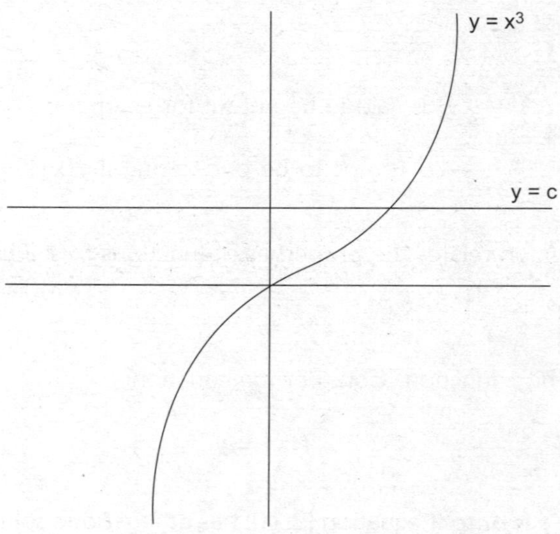

Figure 2. 18

Now any line $y = c$ meets the graph of this function precisely at one point (Fig. 2.18). Thus, f_2 is an example of a *one-to-one* and onto function.

Let us prove these conclusions analytically.

Let $y \in \mathbb{R}$. Consider $f(x) = y$, which is the equation $x^3 = y$. Clearly, this equation has the only real solution $x = y^{\frac{1}{3}}$. Hence this function is one-to-one and onto.

Notice that f_1 of Example 2.99 is onto but not one-to-one and f_2 of Example 2.100 is both one-to-one and onto. The following function is one-to-one but not onto. ■

Example 2.101

$f_3 : R^+ \longrightarrow R$ is defined as $f_3(x) = \frac{x}{1+x}$.

Obviously $f_3(x) < 1$ for all $x \in \mathbb{R}$ and hence for $b = 1$ there is no $a \in \mathbb{R}$ such that $f(a) = 1$. Therefore f is not onto.

To show the function to be one-to-one, consider $f(x_1) = f(x_2)$, which is nothing but $\frac{x_1}{1+x_1} = \frac{x_2}{1+x_2}$. By cross multiplication, we get, $x_1 + x_1x_2 = x_2 + x_1x_2$ from which it follows that $x_1 = x_2$. Hence this function is one-to-one. ■

Example 2.102

Consider the following relations defined pictorially (Fig. 2.19). Which of them are functions? Which of them are one-to-one and which of them are onto?

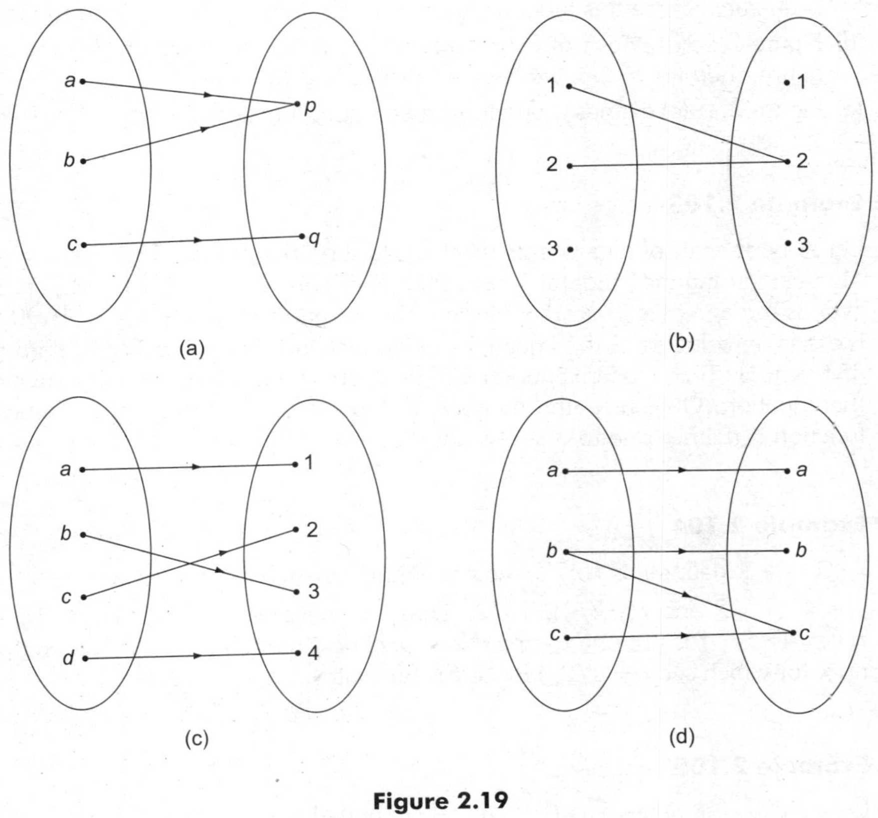

Figure 2.19

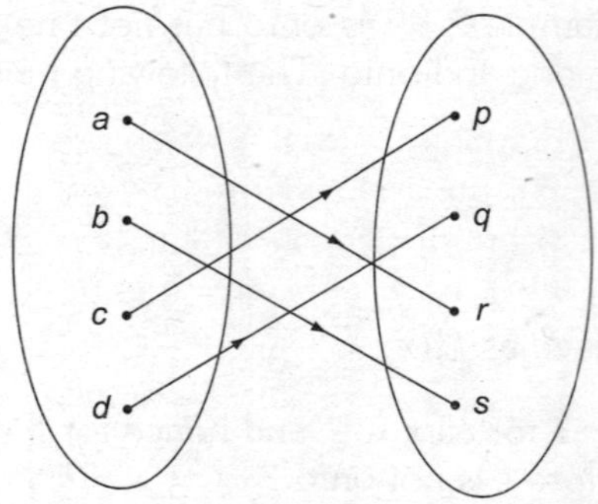

(e)

Figure 2.19 *Contd.*

SOLUTION

(a) Figure 2.19(a) defines a function. Since every element of the co-domain has a pre-image, it is onto. Since $f(a) = f(b) = p$, f is not one-to-one.

(b) Figure 2.19(b) does not define a function. If R is the relation defined by this figure, then for $3 \in A$, and there is no $y \in B$ with $(3, y) \in \mathbb{R}$.

(c) Figure 2.19(c) defines a function. The pre-image of every element in the range is unique. Hence this function is one-to-one. Note that 5 is not in the range.

(d) Figure 2.19(d) does not define a function. If $\mathbb{R}$ is the relation defined by this figure, then for $b \in A$, we have $(b, b) \in \mathbb{R}$ and $(b, c) \in \mathbb{R}$.

(e) Figure 2.19(e) defines a one-to-one and onto function. ■

Example 2.103

Let A be the set of all students who joined the first Year B. Tech Course at IIT, Mumbai, during the academic year 1999–2000 and B be $\mathbb{N}$. Let for $x \in A$, define $f(x)$ as the age of x in years. Consider the equation $f(x) = y$. If $y = 1000$, then there is no solution to the equation. Hence f is not onto. If $y = 18$, then many first year B. Tech students during 1999–2000 of age 18 would be a solution to the equation. Obviously, the number of solutions is more than one. Hence this function is neither one-to-one nor onto. ■

Example 2.104

$f : \mathbb{R} \longrightarrow \mathbb{R}$ defined as $f(x) = \sec x$ is neither one-to-one nor onto.

$f(\pi/4) = \sqrt{2}$ and $f(-\pi/4) = \sqrt{2}$. Thus, we have $\pi/4 \neq -\pi/4$ but $f(\pi/4) = f(-\pi/4)$. So, f is not one-to-one. Now, we know that $|\sec x| \geq 1$. Hence there is no x for which $\sec x = 1/2$. Hence f is not onto. ■

Example 2.105

Let $f : \mathbf{C} \longrightarrow \mathbb{R}$ defined as $f(x + iy) = x$. Show that it is onto but not one-to-one.

SOLUTION

If $z = x + iy$, $f(z)$ is nothing but the real part of z. This function is not one-to-one. Consider the complex numbers $z_1 = 3 + 4i$ and $z_2 = 3 - 7i$. Clearly, $z_1 \neq z_2$ but $f(z_1) = f(z_2) = 3$. This function is onto. Let $a \in \mathbb{R}$. We must find $z \in \mathbf{C}$ such that $f(z) = a$. We take $z = a + i$. It is clear that $f(z) = a$ and hence the function is onto. ■

Example 2.106

Let $f : [8, \infty) \longrightarrow \mathbb{R}$ defined as $f(x) = x^3 - 8x^2 + 5x + 14$ is not onto.

SOLUTION

$f(x)$ can be factorized into $x^3 - 8x^2 + 5x + 14 = (x+1)(x-2)(x-7)$. Thus, the solutions of $x^3 - 8x^2 + 5x + 14 = 0$ are $x = -1$, $x = 2$ and $x = 7$. But we find that $-1, 2, 7 \notin [8, \infty)$. Thus 0 does not have a pre-image. Hence f is not onto. ■

Example 2.107

Let $f : \mathbb{R} \longrightarrow \mathbb{R}$ be defined as $f(x) = a_0x^n + a_1x^{n-1} + \cdots + a_{n-1}x + a_n$ be a polynomial function of degree n. Here a_0, a_1,...,a_n are real numbers such that $a_0 \neq 0$. Show that, If n is odd, then f is onto.

SOLUTION

Let $y \in \mathbb{R}$ be given. We must find $x \in \mathbb{R}$ such that $a_0x^n + a_1x^{n-1} + \cdots + a_n = y$. But such an x is a real solution to the equation

$$a_0x^n + a_1x^{n-1} + \cdots + a_n - y = 0.$$

The result follows from the fact that any odd degree polynomial equation has *at least* one real root. ■

Example 2.108

Let $f : \mathbb{R} \longrightarrow \mathbb{R}$ be defined as $f(x) = ax^2 + bx + c$ with $a \neq 0$. Show that f is not onto.

SOLUTION

To show that the function is not onto we should find $y \in \mathbb{R}$ such that the equation $ax^2 + bx + c = y$ has no real solution. Let us assume that $a > 0$.

$$ax^2 + bx + c = a\left(x + \frac{b}{2a}\right)^2 - \frac{b^2}{4a} + c.$$

Clearly,

$$f(x) \geq -\frac{b^2}{4a} + c$$

since $\left(x + \frac{b}{2a}\right)^2 \geq 0$. Choosing $y < -\frac{b^2}{4a} + c$, we find that the equation $f(x) = y$ has no solution. Thus f is not onto. The case $a < 0$ is left as an exercise. ■

Graphical solution:

If $a > 0$, the graph of $y = ax^2 + bx + c$ is a parabola, which faces upwards and it is above a certain horizontal line. Hence, a horizontal line below this line will not intersect the parabola.

If $a < 0$, the graph of $y = ax^2 + bx + c$ is a parabola, which faces downwards and it is below a certain horizontal line. Hence, a horizontal line above this line will not intersect the parabola (Fig. 2.20).

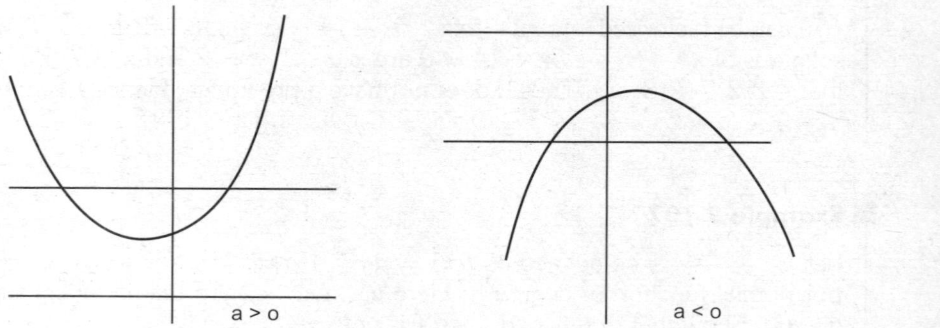

Figure 2.20

PRACTICE 2.18

Examine which of the following functions are onto/one-to-one:

(a) $f : \mathbb{N} \longrightarrow \mathbb{N}$ defined as $f(n) = 2n$.
(b) $f : \mathbb{R} \longrightarrow \mathbb{R}$ defined as $f(x) = 2x$.
(c) $f : \mathbb{R} \longrightarrow \mathbb{R}$ defined as $f(x) = x^2$.
(d) $f : \mathbf{C} \longrightarrow \mathbb{R}$ defined as $f(z) = f(x + iy) = y$.
(e) $f : [-25, -6] \longrightarrow R$ defined as $f(x) = x^2 + 10x + 31$.
(f) $f : \mathbb{R} \longrightarrow \mathbb{R}$ defined as $f(x) = x^2 - 7x + 9$.

Example 2.109

Let $f : \mathbb{R} \longrightarrow \mathbb{R}$ be defined as $f(x) = 7 - |2x + 3|$. Show that this function is neither one-to-one nor onto.

SOLUTION

It is clear that $f(x) = 0$ when $2x + 3 = 7$ or when $2x + 3 = -7$. Solving these two equations, we get $x_1 = 2$, $x_2 = -5$. Clearly, $x_1 \neq x_2$ but $f(x_1) = f(x_2) = 0$. So, f is not one-to-one. Now, to show that f is not onto, note that $|2x + 3| \geq 0$ for all

$x \in R$. Hence $7 - |2x + 3| \leq 7$ for all x. By choosing $y = 8$, we find that there is no solution to $f(x) = 8$. Therefore, f is not onto. ■

Example 2.110

Show that $f : [2, 7] \longrightarrow R$ defined as $f(x) = 4x^2 - 5x - 9$ is one-to-one.

SOLUTION

Consider the graph of $y = 4x^2 - 5x - 9 = 4\left(x - \frac{5}{8}\right)^2 - \frac{676}{64}$. It is symmetric about the line $x = \frac{5}{8}$. Hence, if the line $y = y_0$ meets the parabola at the two points (x_1, y_0) and (x_2, y_0), with $x_1 < x_2$, then $x_1 < \frac{5}{8} < x_2$. Hence $x_1 \notin [2, 7]$. So, we cannot find both x_1 and x_2 in the domain $[2, 7]$ with $f(x_1) = f(x_2)$. Thus, the function is one-to-one (Fig. 2.21).

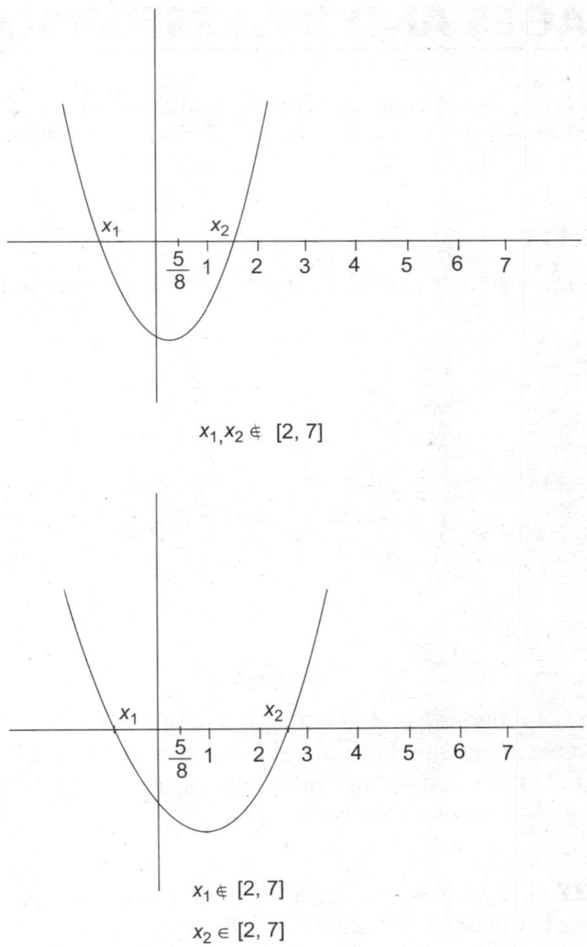

Figure 2.21

The same thing can be made precise by the following argument. Let $x_1, x_2 \in [2,7]$ be such that $f(x_1) = f(x_2)$. Denote their common values by y_0. Then x_1 and x_2 are the real solutions to $4x^2 - 5x - (9 + y_0) = 0$. Clearly, the discrimnant $25 - 4(4)(-1)(9+y_0) = 25 + 16(9+y_0) \geq 0$. Further, x_1 and x_2 must be given by the solutions

$$\frac{-(-5) \pm \sqrt{25 + (9 + y_0)}}{8}.$$

We find that one solution must be greater than $\frac{5}{8}$ and the other solution must be less than $\frac{5}{8}$. So, at most one of x_1 or x_2 would be in $[2,7]$. Hence, the function is one-to-one. ■

2.6 IMAGES AND INVERSE IMAGES

For a function $f : X \longrightarrow Y$ and $x \in X$, the element $f(x)$ is called the *image* of x under the function f. If $f(x) = y$, then x is called an *inverse image* or *pre-image* of y.

Example 2.111

Let $f : \{a, b, c\} \longrightarrow \{a, b, c, d, e\}$ be the function described by Figure 2.22.

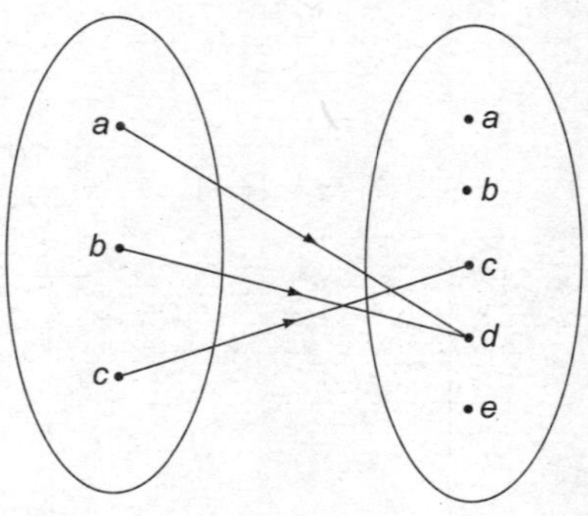

Figure 2.22

The image of a under f is d, the image of c under f is c. Note that b does not have any inverse image and both a and b are inverse images of d.
In the next definition, we define the image of a subset of the domain and the inverse image of a subset of the co-domain. ■

Definition 2.22: Let $f : X \longrightarrow Y$ be a function and let $A \subseteq X$ and $B \subseteq Y$. Then, $f(A)$, the image of A under f, is defined as follows:

$$f(A) = \{f(x) : x \in A\}.$$

$f^{-1}(B)$, the inverse image of B under f, is defined as follows:

$$f^{-1}(B) = \{x \in X : f(x) \in B\}.$$

Remark 2.5

Note that $f(A) \subseteq Y$ and $f^{-1}(B) \subseteq X$.

Example 2.112

Let $X = \{a, b, c, d, e\}$, $Y = \{1, 2, 3, 4, 5, 6, 7\}$ and $f : X \longrightarrow Y$ be given by Figure 2.23.

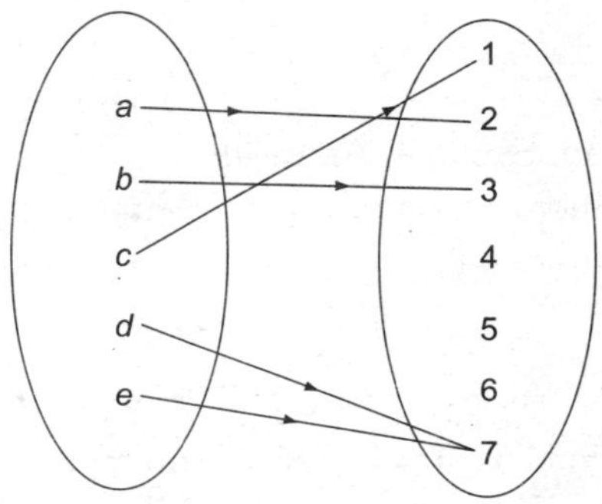

Figure 2.23

Let $A=\{d, e\}$, $B=\{a, b, c\}$, $C=\{e\}$ and $P = \{7\}$, $Q=\{4, 5, 6\}$, $R=\{2, 3, 6\}$. Find $f(A)$, $f(B)$, $f(C)$ and $f^{-1}(P)$, $f^{-1}(Q)$ and $f^{-1}(R)$.

SOLUTION

$f(A) = \{7\}$, $f(B) = \{1, 2, 3\}$, $f(C) = \{7\}$.
$f^{-1}(P) = \{d, e\}$, $f^{-1}(Q) = \phi$, $f^{-1}(R) = \{a, b\}$. ■

Example 2.113

Consider the following list of a group of students of the same class and their respective ages:

Name	Age
Murugan	23
Selvi	22
Subramanyam	21
James	22
Fathima	23
Ramasamy	22
Mary	21
Aparajitha	23

Let X be the set of all students listed above. Define $f : X \longrightarrow N$ as $f(x) =$ Age of x. Let

$A = \{$ Fathima, Selvi, Murugan, James$\}$.
$B = \{$ Murugam, Aparajitha, Subramanyam$\}$.
$C = \{21, 22, 23, 24\}$.
$D = \{23\}$.
$E = \{21, 22\}$.
$F = \{26, 27, 28\}$.
Find the images of A and B. Also, find the inverse images of D, E and F.

SOLUTION

Note that A and B are subsets of the domain X and C, D, E, F are subsets of the co-domain N. We have,
(a) $f(A) = \{22, 23\}$.
(b) $f(B) = \{21, 23\}$.
(c) $f^{-1}(C) = X$.
(d) $f^{-1}(D) = \{$Murugan, Fathima, Aparajitha$\}$.
(e) $f^{-1}(E) = \{$Selvi, Subramanyam, James, Ramasamy, Mary$\}$.
(f) $f^{-1}(F) = \phi$, the empty set. ∎

Example 2.114

Let $f : \mathbf{C} \to \mathbb{R}$ be defined as $f(z) = Re(z)$. Find $f(s)$.
where $S = \{-4 + 3i, 7 - 4i, -i, 8 + 6i, 8 - 9i\}$.

SOLUTION

$f(S) = \{-4, 7, 0, 8\}$. ∎

PRACTICE 2.19

In the previous example, find $f(\{-4 + 3i, 8 + 6i, 8 - 9i\})$ and $f^{-1}(\{0, 8\})$.

PRACTICE 2.20

Consider a group of M.E. students along with their age as given below:

Aandaal	22
Kannan	21
Naseem	21
John	25
Thamizhchelvi	21
Jyostna	21

Let X be the set of all students listed above and $Y = N$. Define $f : X \longrightarrow Y$ as $f(x) =$ Age of x. Let B_1=N, B_2= $\{21\}$, B_3= $\{n \in N : n \geq 24\}$, B_4=$\{21, 25\}$. Find the inverse images of the above sets.

Example 2.115

Let $f : \mathbb{R} \to \mathbb{R}$ be defined as $f(x) = 3x - 5$. Find the images of $\{3, 7, -15\}$, $\{0\}$, $(0, 6]$, $\mathbb{N}$ and the inverse images of $\{9, -24, 16\}$, $\{6\}$, $(8, 10)$ and $\mathbb{N}$.

SOLUTION

$f(\{3, 7, -15\}) = \{f(3), f(7), f(-15)\} = \{3 \times 3 - 5, 3 \times 7 - 5, 3 \times (-15) - 5\} = \{4, 16, -50\}$.
$f(\{0\}) = \{-5\}$.

By drawing the graph of $y = 3x - 5$, we can conclude that $f((0, 6]) = (-5, 13]$ (Fig. 2.24).

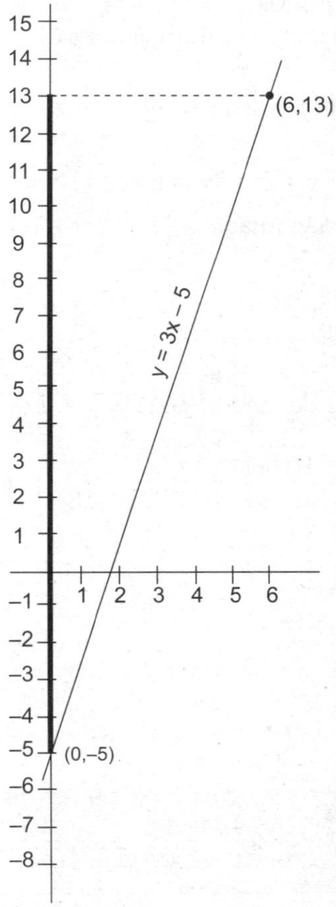

Figure 2.24

But, let us look at a formal proof for this:

$$\begin{aligned} f((0, 6]) &= \{f(x) : x \in (0, 6]\} \\ &= \{3x - 5 : 0 < x \leq 6\}. \end{aligned}$$

Using Proposition 1.3 of Chapter 1,

$$0 < x \leq 6 \leftrightarrow -6 < 3x - 5 \leq 13.$$

Hence

$$\{3x - 5 : 0 < x \leq 6\} = \{y : -6 < y \leq 13\}.$$

So, $f((0, 6]) = (-6, 13]$.
Next, $f(\mathbb{N}) = \{f(1), f(2)), f(3), \ldots\} = \{-3, 0, 3, 6, \ldots\}$.
Now, we shall find $f^{-1}(\{9, -24, 16\})$.
By definition, $f^{-1}(\{9, -24, 16\}) = \{x \in \mathbb{R} : 3x - 6 \in \{9, -24, 16\}\}$. Thus, the required inverse image is the set of the solutions of the three equations: $3x-6 = 9$, $3x - 6 = -24$ and $3x - 6 = 16$.
Hence, $f^{-1}(\{9, -24, 16\}) = \left\{5, -6, \frac{22}{3}\right\}$.
To find $f^{-1}(\mathbb{N})$:
By definition, $f^{-1}(\mathbb{N}) = \{x \in \mathbb{R} : 3x - 6 \in \mathbb{N}\}$. Now, if $3x - 6 = n$, then $x = \frac{n+6}{3}$.
Hence, the required inverse image is $\left\{\frac{n+6}{3} : n \in \mathbb{N}\right\} = \left\{\frac{7}{3}, \frac{8}{3}, 3, \ldots\right\}$. ■

Example 2.116

Let $f : GL_2(\mathbb{R}) \longrightarrow \mathbb{R}$ be defined as $f(A) = |A|$. Let

$$S = \left\{\begin{pmatrix} 10 & 7 \\ 6 & 1 \end{pmatrix}, \begin{pmatrix} 7 & 5 \\ 21 & 12 \end{pmatrix}, \begin{pmatrix} 1 & 0 \\ 0 & 1 \end{pmatrix}\right\}.$$

Find $f(S)$.

SOLUTION

$f(S) = \{-32, -21, 1\}$. ■

PRACTICE 2.21

Define $f : \mathbb{R}^2 \longrightarrow \mathbb{R}$ as $f(x, y) = 4x - 3y$.
Let $B = \{(0, 0), (-3, 6), (7, 8), (6, -4), (-12, -34)\}$. Find $f(B)$.

Example 2.117

Define $f : R \longrightarrow R^2$ as $f(x) = (6x, -x)$. Find the image of $\{3, -6, 10, 24\}$.

SOLUTION

Let $A = \{3, -6, 10, 24\}$. Then,

$$f(A) = \{(18, -3), (-36, 6), (60, -10), (144, -24)\}.$$ ■

Example 2.118

Define $f : \mathbb{R}^2 \longrightarrow \mathbb{R}$ as $f(x, y) = 4x - 3y$. Let $B = \{8\}$. Find $f^{-1}(B)$.

SOLUTION

$$\begin{aligned} f^{-1}(B) &= \{(x, y) \in R^2 : f(x, y) \in B\} \\ &= \{(x, y) \in R^2 : 4x - 3y = 8\}. \end{aligned}$$

Hence, the required inverse image is nothing but the straight line whose equation is $4x - 3y = 8$. ■

PRACTICE 2.22

Define $f : \mathbb{R}^2 \longrightarrow \mathbb{R}$ as $f(x, y) = 6x + y$. Let $B = \{0, 3\}$. Find $f^{-1}(B)$.
Hint: The answer is a pair of parallel lines.

Example 2.119

Let $f : GL_2(\mathbb{R}) \longrightarrow GL_2\mathbb{R}$ be defined as $f(A) = A + 3I$, where I is the identity matrix. Let

$$S = \left\{ \begin{pmatrix} 10 & 7 \\ 6 & 1 \end{pmatrix}, \begin{pmatrix} 7 & 4 \\ 21 & 12 \end{pmatrix}, \begin{pmatrix} 1 & 0 \\ 0 & 1 \end{pmatrix} \right\}.$$

Find $f(S)$.

SOLUTION

$$f(S) = \left\{ \begin{pmatrix} 13 & 7 \\ 6 & 4 \end{pmatrix}, \begin{pmatrix} 10 & 4 \\ 21 & 15 \end{pmatrix}, \begin{pmatrix} 4 & 0 \\ 0 & 4 \end{pmatrix} \right\}.$$

Let $f : X \longrightarrow Y$ and $A = \{x_1, x_2, \ldots, x_m\}$ be a finite subset of X containing m elements (see Section 2.8). $f(A)$ is simply $\{f(x_1), f(x_2), \ldots, f(x_m)\}$. Now, can we say that the number of elements of $f(A)$ is also 4? The answer is ***no***. $f(A)$ need not have m elements, because for some i and j with $i \neq j$, it may so happen that $f(x_i) = f(x_j)$. We give one such example below. ■

Example 2.120

Let $f : \mathbb{R} \rightarrow \mathbb{R}$ be defined as $f(x) = \sec x$. $S = \{0, \pi, 2\pi\}$. $f(S) = \{\sec 0, \sec \pi, \sec 2\pi\} = \{1, -1, 1\} = \{1, -1\}$. Here, we have an example of a set and a function f such that S has 3 elements but $f(S)$ has only two elements. ■

Example 2.121

Let $\mathbf{M}_2$ be the set of all 2×2 matrices with real entries. Define $f : \mathbf{M}_2 \longrightarrow R$ as $f(A) = |A|$.

Let

$$S = \left\{ \begin{pmatrix} 1 & 0 \\ 0 & 1 \end{pmatrix}, \begin{pmatrix} \sqrt{2} & 1 \\ 1 & \sqrt{2} \end{pmatrix}, \begin{pmatrix} 5 & 3 \\ 4 & 1 \end{pmatrix}, \begin{pmatrix} -7 & 0 \\ 0 & 1 \end{pmatrix} \begin{pmatrix} 0 & 7 \\ 1 & 0 \end{pmatrix} \right\}.$$

It is clear that $f(S) = \{1, -7\}$. Here, S has 5 elements but $f(S)$ has only two elements. ■

Example 2.122

Consider the following three functions:

(i) $f : \mathbf{Z} \longrightarrow \mathbb{R}$ defined as $f(n) = n^2$.
(ii) $g : \mathbb{R} \longrightarrow \mathbb{R}$ defined as $g(x) = x^2$.
(iii) $h : \mathbb{N} \longrightarrow \mathbb{R}$ defined as $f(n) = n^2$.

Let $S = \{1, 2, 3, 4, 5, 6, 7, 8, 9\}$.

$$\begin{aligned} f^{-1}(S) &= \{1, -1, 2, -2, 3, -3\} \\ g^{-1}(S) &= \Big\{1, -1, \sqrt{2}, -\sqrt{2}, \sqrt{3}, -\sqrt{3}, \\ &\qquad 2, -2, \sqrt{5}, -\sqrt{5}, \sqrt{6}, -\sqrt{6}, \sqrt{7}, -\sqrt{7}, 2\sqrt{2}, -2\sqrt{2}, 3, -3\Big\} \\ h^{-1}(S) &= \{1, 2, 3\}. \end{aligned}$$

■

Example 2.123

Let $h : \mathbb{R} \longrightarrow \mathbb{R}$ be defined as $h(x) = 2x - 3$. Find the inverse image of the interval $[-3, 8]$.

SOLUTION

$$\begin{aligned} f^{-1}([-3, 8]) &= \{x \in \mathbb{R} : f(x) \in [-3, 8]\} \\ &= \{x \in \mathbb{R} : 2x - 3 \in [-3, 8]\} \\ &= \{x \in \mathbb{R} : -3 \leq 2x - 3 \leq 8\}. \end{aligned}$$

We note that the required inverse image is nothing but the solution set of the inequality $-3 \leq 2x - 3 \leq 8$. We solve it as in Section 1.3:

$$\begin{array}{rcccl} -3 & \leq & 2x - 3 & \leq & 8 \\ -3 + 3 & \leq & 2x & \leq & 8 + 3 \\ 0 & \leq & 2x & \leq & 11 \\ 0 & \leq & x & \leq & \frac{11}{2}. \end{array}$$

Thus,

$$f^{-1}[-3, 8] = \left\{ x \in \mathbb{R} : 0 \leq x \leq \frac{11}{2} \right\}$$

which is nothing but the interval $[0, \frac{11}{2}]$. ■

Example 2.124

Find the image of $[-3,7]$ under the function $f:\mathbb{R}\longrightarrow\mathbb{R}$ defined as $f(x)=x^2$.

SOLUTION

We use the techniques from Section 1.6 of Chapter 1.

$$\begin{aligned} f([-3,7]) &= \{f(x) : x\in[-3,7]\} \\ &= \left\{x^2 : -3\le x\le 7\right\} \\ &= \left\{x^2 : -3\le x<0\right\}\cup\left\{x^2 : 0\le x\le 7\right\} \\ &= \left\{x^2 : 0<x^2\le 9\right\}\cup\left\{x^2 : 0\le x^2\le 49\right\} \\ &= (0,9]\cup[0,49] \\ &= [0,49]. \end{aligned}$$

The above example motivates us to examine sets like $f(A\cup B)$. ■

Proposition 2.4

Let $f:X\longrightarrow Y$ be a function. Let A and B be subsets of X and C and D be subsets of Y. Then, we have

(i) $f(A\cup B)=f(A)\cup f(B)$.
(ii) $f(A\cap B)\subseteq f(A)\cap f(B)$.
(iii) $f^{-1}(C\cup D)=f^{-1}(C)\cup f^{-1}(D)$.

Remark 2.6

To show that two sets are equal, we show that each one is a subset of the other. To show that S_1 is a subset of S_2, we must show that an arbitrary element $x\in S_1$ is an element of S_2.

Proof of Theorem:

(i) First we show that $f(A\cup B)\subseteq f(A)\cup f(B)$:

$$\begin{aligned} & y\in f(A\cup B) \\ \Rightarrow\ & y=f(x) \text{ for some } x\in A\cup B \\ \Rightarrow\ & y=f(x) \text{ for some } x\in A \text{ or } y=f(x) \text{ for some } x\in B \\ \Rightarrow\ & y\in f(A) \text{ or } y\in f(B) \\ \Rightarrow\ & y\in f(A)\cup f(B) \\ \Rightarrow\ & f(A\cup B)\subseteq f(A)\cup f(B). \end{aligned}$$

Next, we show that $f(A) \cup f(B) \subseteq f(A \cup B)$:

$$\begin{aligned}
& y \in f(A) \cup f(B) \\
\Rightarrow\ & y \in f(A) \text{ or } y \in f(B) \\
\Rightarrow\ & y = f(x) \text{ for some } x \in A \text{ or } y = f(x) \text{ for some } x \in B \\
\Rightarrow\ & y = f(x) \text{ for some } x \in A \cup B \\
\Rightarrow\ & y \in f(A \cup B) \\
\Rightarrow\ & f(A) \cup f(B) \subseteq f(A \cup B).
\end{aligned}$$

So, $f(A \cup B) = f(A) \cup f(B)$.

(ii)

$$\begin{aligned}
& y \in f(A \cap B) \\
\Rightarrow\ & y = f(x) \text{ for some } x \in A \cap B \\
\Rightarrow\ & y = f(x) \text{ for some } x \in A \text{ and } y = f(x) \text{ for some } x \in B \\
\Rightarrow\ & y \in f(A) \text{ and } y \in f(B) \\
\Rightarrow\ & y \in f(A) \cap f(B) \\
\Rightarrow\ & f(A \cap B) \subseteq f(A) \cap f(B).
\end{aligned}$$[1]

(iii) We show that $f^{-1}(C \cup D) \subseteq f^{-1}(C) \cup f^{-1}(D)$:

$$\begin{aligned}
& x \in f^{-1}(C \cup D) \\
\Rightarrow\ & f(x) \in C \cup D \\
\Rightarrow\ & f(x) \in C \text{ or } f(x) \in D \\
& x \in f^{-1}(C) \text{ or } x \in f^{-1}(D) \\
\Rightarrow\ & x \in f^{-1}(C) \cup f^{-1}(D) \\
\Rightarrow\ & f^{-1}(C \cup D) \subseteq f^{-1}(C) \cup f^{-1}(D)
\end{aligned}$$

Next, we can show that $f^{-1}(C) \cup f^{-1}(D) \subseteq f^{-1}(C \cup D)$ just by reversing the arguments above. Hence (iii) follows.

PRACTICE 2.23

Let $f : X \longrightarrow Y$ be a function and C and D be subsets of Y. Show that

$$f^{-1}(C \cap D) = f^{-1}(C) \cap f^{-1}(D).$$

---

[1] In general, $f(A \cap B) = f(A) \cap f(B)$. Suppose $y \in f(A) \cap f(B)$. This implies that $y = f(a)$ for some $a \in A$ and $y = f(b)$ for some $b \in B$. From this, we cannot conclude that $y = f(x)$ for some $x \in A \cap B$. Hence the equality is not valid in (ii).

Example 2.125

Show by an example that $f(A \cap B) \neq f(A) \cap f(B)$.

SOLUTION

Let $f : \mathbf{Z} \longrightarrow$ be defined as $f(n) = n^2$. Let $A = \{2, 1\}$ and $B = \{2, -1\}$. We have $A \cap B = \{2\}$. $f(A \cap B) = \{4\}$. But $f(A) = \{1, 4\}$ and $f(B) = \{1, 4\}$, $f(A) \cap f(B) = \{1, 4\} \neq \{4\}$. ■

Definition 2.23: The range $R(f)$ of a function $f : X \longrightarrow Y$ is defined as the image of the whole domain X; that is, $R(f) = f(X)$.

Example 2.126

Let $f : \mathbb{R} \longrightarrow \mathbb{R}^+$ be defined as $f(x) = x^2$ and $g : \mathbb{R} \longrightarrow \mathbb{R}$ be defined as $g(x) = x^2$. Note that $f(\mathbb{R}) = g(\mathbb{R}) = \mathbb{R}^+$. The range and co-domain of f are the same but for g its range and co-domain are different.

Let us observe that in the previous example f is onto but g is not. In fact, the following proposition is easy to prove: ■

Proposition 2.5

Show that $f : X \longrightarrow Y$ is onto if and only if $f(X) = R(f) = Y$.

Proof

Let f be onto. We shall show that $f(X) = Y$. It is clear that $f(X) \subseteq Y$. We need only to show that $Y \subseteq f(X)$.

Let $y \in Y$. The function is onto. Hence, there is $x \in X$, such that $y = f(x)$. But this implies that $y \in f(X)$ and hence $Y \subseteq f(X)$. So, $Y = f(X)$.

Next, let us assume that $f(X) = Y$ and show that f is onto. Let $y \in Y$. Since $Y = f(X)$, $y \in f(X)$. From the definition of the image of a set, we see that there is $x \in X$ such that $y = f(x)$. Hence f is onto.

In the next example, we show how the concept of inverse image may be useful. ■

Example 2.127

Consider the problem of finding the natural domain for the expression $\sec^{-1}(6x + 8)$. We know that $\sec^{-1}(u)$ is meaningful when $|u| \geq 1$. Thus, the required domain D is nothing but

$$\{x \in \mathbb{R} : |6x + 8| \geq 1\} = \{x \in R : |6x + 8| \in [1, \infty)\}.$$

Now defining $g : \mathbb{R} \longrightarrow \mathbb{R}$ as $g(x) = |6x + 8|$, the above set is simply the inverse image of $[1, \infty)$ under g. To find this set, we have to consider the two inequalities $6x + 8 \geq 1$ and $6x + 8 \leq -1$.

$$6x + 8 \geq 1$$

$$6x \geq 1 - 8$$

$$6x \geq -7$$
$$x \geq \frac{-7}{6}.$$

$$6x + 8 \leq -1$$
$$6x \leq -1 - 8$$
$$6x \leq -9$$
$$x \leq \frac{-9}{6}.$$

Thus,

$$\begin{aligned} D &= \left\{ x \in R : x \geq \frac{-7}{6} \text{ or } x \leq \frac{-9}{6} \right\} \\ &= \left[\frac{-7}{6}, \infty\right) \cup \left(-\infty, \frac{-9}{6}\right]. \end{aligned}$$

Let us now generalize this example. Let $f : U \longrightarrow V$ and $g : X \longrightarrow Y$ be given functions. We would like to consider the expression $f(g(x))$. If this expression should be defined, then we must have $x \in X$ and $g(x) \in U$. Since the co-domain for g is Y, $g(x) \in U \cap Y$.

Thus, the largest subset of X, for which $f(g(x))$ is meaningful is

$$\{x \in X : g(x) \in U \cap Y\}.$$

But the above set is the inverse image $g^{-1}(U \cap Y)$.

The proper mathematical notion for studying expressions like $f(g(x))$ is the notion of *composition of functions*, which is defined in the next section. ■

PRACTICE 2.24

Find the inverse images of B in the following:

(1) $f : \mathbb{N} \longrightarrow N$ defined as $f(n) = n^3$. $B = \{1, 2, 3, 4, 5, 6, 7, 8, 9\}$.
(2) $f : \mathbb{R} \longrightarrow \mathbb{R}$ defined as $f(x) = x^3$. $B = \{1, 2, 3, 4, 5, 6, 7, 8, 9\}$.
(3) $f : \mathbf{Z} \longrightarrow N$ defined as $f(n) = n^2$. $B = \{1, 2, 3, 4, 5, 6, 7, 8, 9\}$.
(4) $f : \mathbb{R}^+ \longrightarrow \mathbb{R}^+$ $f(x) = \sqrt{x}$. $B = [3, 4]$.

PRACTICE 2.25

Prove the following:

(i) $f^{-1}(y)$ is non-empty if and only if $y \in \mathbb{R}(f)$.
(ii) $f : A \longrightarrow B$ is onto if and only if $f^{-1}(y)$ is non-empty for every $y \in B$.
(iii) For any subset $D \subset B$, $f^{-1}(D) = f^{-1}(D \cap R(f))$.

PRACTICE 2.26

Find the range of the following functions:

(a) $f : \mathbb{R} \longrightarrow \mathbb{R}$ defined as $f(x) = 2x + 5$.
(b) $f : \mathbb{R} \longrightarrow C$ defined as $f(x) = \cos x + i \sin x$.
(c) $f : [-3, 7] \longrightarrow \mathbb{R}$ defined as $f(x) = x^2$.
(d) $f : \mathbb{R} \longrightarrow \mathbb{R}$ defined as $f(x) = -x^2 + 12x - 35$.
(e) $f : [5, 7] \longrightarrow \mathbb{R}$ defined as $\sqrt{-x^2 + 12x - 35}$.
(f) $f : \mathbb{R} \longrightarrow \mathbb{R}$ as $f(x) = 2\cos x - 9$.

2.6.1 Composition of functions

The idea of composition of functions is useful to study mathematical expressions such as $\log(\sin x)$, e^{3x^2-4x+3} $\sin(\sqrt{x+5})$, etc.

Consider the expression $\sin^{-1}(7x + 2)$. This expression is meaningful only when $-1 \leq 7x + 2 \leq 1$. For the expression $\sin^{-1}(\log x)$, we must have $x > 0$ and $-1 \leq \log x \leq 1$.

Let us now think of an expression $f(g(x))$. It is clear that x should be in the domain of g and $g(x)$ should be in the domain of f. Let x be in the domain of g. $g(x)$ is in the co-domain of g. For $f(g(x))$ to be meaningful $g(x)$ should be in the domain of f. So, if we let the co-domain of g to be the domain of f, then $f(g(x))$ is well defined.

Definition 2.24: Let $g : A \longrightarrow B$ and $f : B \longrightarrow C$ be given functions. We define a function $h : A \longrightarrow C$ as follows:

$$h(x) = f(g(x)).$$

The function h defined above is denoted by $f \circ g$. Figure 2.25 illustrates h:

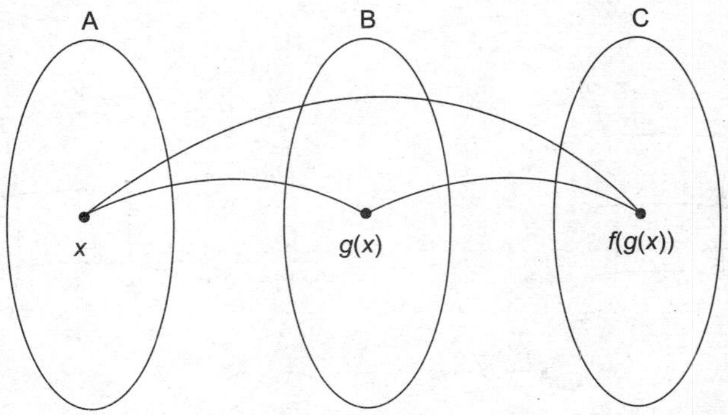

Figure 2.25 Composition of functions

Let A be any set. What is the simplest function $f : A \longrightarrow A$ one can think of? Since $f(x)$ must be in A, the most natural choice is $f(x) = x$. This function is called the identity function on A.

Definition 2.25: For any set A the *identity function* I_A is defined as $I_A(x) = x$ for all $x \in A$.

Let $f, g : A \longrightarrow B$ be two functions. When are they equal? It is clear that f and g are equal when $f(x) = g(x)$ for all $x \in A$.

Proposition 2.6

For any function $f : A \longrightarrow B$, we have $I_B \circ f = f$ and $f \circ I_A = f$.

Proof

Let $x \in A$. Then $(I_B \circ f)(x) = I_B(f(x)) = f(x)$. Therefore, $I_B \circ f = f$. Next, $(f \circ I_A)(x) = f(I_A(x)) = f(x)$. So, $f \circ I_A = f$. ■

Example 2.128

Show that the compositions of two onto functions is onto.

SOLUTION

Let $f : X \longrightarrow Y$ and $g : Y \longrightarrow Z$ be onto. Let $h = f \circ g$. We need to show that for an arbitrary $z \in Z$, there is $x \in X$, such that $h(x) = z$. Now since g is onto, there exists $y \in Y$ such that $g(y) = z$. Since f is onto there exists $x \in X$ such that $f(x) = y$. But $y = g(z)$. Thus $f(g(t) = z$ and the result follows. ■

PRACTICE 2.27

Show that the composition of two one-to-one functions is one-to-one.

Illustrative Example: Let $A = \{1, 2, 3\}$ and $B = \{a, b, c, d\}$ Define $f : A \longrightarrow B$ as $f(1) = a$, $f(2) = b$ and $f(3) = c$. Now, we would like to define a function from B to A, which is some kind of an inverse to f. We can easily define $g(a) = 1$, $g(b) = 2$ and $g(c) = 3$. But, how to define $g(d)$? We can simply define $g(d)$ to be any arbitrary element of A. Let us define $g(d) = 3$ (Fig. 2.26).

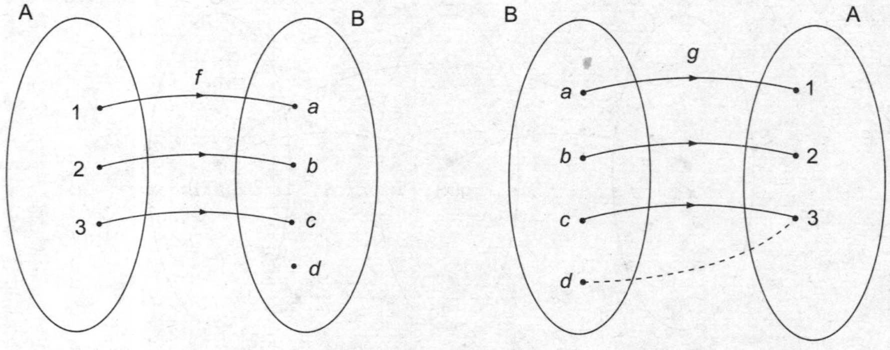

Figure 2.26

We can easily check that $(g(f(x)) = x$ for all x. Hence $g \circ f = I_A$. Let us note that $f \circ g(d) = f(g(d)) = f(2) = b$ and hence $f \circ g \neq I_B$.

PRACTICE 2.28

Define $f : \{1, 2, 3\} \longrightarrow \{a, b\}$ as an onto function. Construct $g : \{a, b\} \longrightarrow \{1, 2, 3\}$ such that $f \circ g = I_B$ and examine if $g \circ f = I_A$.

Definition 2.26: Let $f : A \longrightarrow B$ be a function. A function $g : B \longrightarrow A$ such that $f \circ g = I_B$ is called a *right inverse* of f.

A function $g : B \longrightarrow A$ such that $g \circ f = I_A$ is called a *left inverse* of f.

Proposition 2.7

If $f : A \longrightarrow B$ is an onto function, then f has a right inverse $g : B \longrightarrow A$.

Proof

For any $y \in B$, we have already seen that $f^{-1}(y)$ is non-empty. Choose an element, say, x in $f^{-1}(y)$ and call it $g(y)$. As $f(x) = y$, we have $f(g(y)) = y$. g is a right inverse of f.

Note that a function f can have many right inverses. ■

PRACTICE 2.29

Give examples to illustrate the fact that a function may have many right inverses.

It is interesting to note that the existence of a right inverse tells us that the function is onto.

Proposition 2.8

If f has a right inverse, then the function is onto.

Proof

Let $g : B \longrightarrow A$ be a right inverse for f. We have $f \circ g = I_B$. That is, for any $y \in B$, $f(g(y)) = y$. Now, to show that f is onto, for $y \in B$, we must find $x \in A$ with $f(x) = y$. Let $y \in B$ and take $x = g(y)$. Now, $f(x) = f(g(y)) = y$ and hence we have got the required $x \in A$. Thus, f is onto. ■

PRACTICE 2.30

Show that a function $f : A \longrightarrow B$ has a left inverse $g : A \longrightarrow B$, if and only if f is one-to-one.

Definition 2.27: We say that a function $f : A \longrightarrow B$ is *invertible* if there is a function $g : B \longrightarrow A$ such that $f(g(y)) = y$ for all $y \in B$ and $g(f(x)) = x$ for all $x \in A$. This function g is called the *inverse* of f and it is denoted by f^{-1}.

The following proposition follows from the definitions easily.

Proposition 2.9

A function $f : A \longrightarrow B$ is invertible if and only if

(a) f has a left inverse $g_L : B \longrightarrow A$.
(b) f has left inverse $g_R : B \longrightarrow A$.
(c) $g_R = g_L$.

Theorem 2.2

A function $f : A \longrightarrow B$ is invertible if and only if f is one-to-one and onto.

Before proving this theorem, let us understand what we mean by saying that two functions $g : A \longrightarrow B$ and $h : E \longrightarrow F$ are equal. We must have $A = E$ and $E = F$. That is, both functions should have *the same domains* and *same co-domains*. Further, $g(x) = h(x)$ must hold for *all* $x \in A$.

I Proof of Theorem 2.2:

Let f be invertible. Then, there is $g : B \longrightarrow A$ such that $f \circ g = I_B$ and $g \circ f = I_A$. Hence, f has a right inverse and a left inverse. Thus, by the previous results f is one-to-one and onto.

Now, let f be one-to-one and onto. Then f has a left inverse g_L and a right inverse g_R. For a function to be invertible, there must be a single function, which is its left as well as its right inverse. So, if we can show that $g_R = g_L$, then the same function is both a left and a right inverse for f, and hence f would be invertible. We shall now show that $g_R = g_L$.

Note that both g_L and g_R have the same domain B and co-domain A. Thus, now we only have to show that

$$g_R(y) = g_L(y) \quad \text{for all} \quad y \in B.$$

Let $y \in B$. Since the function f is one-to-one and onto, there is a unique $x \in A$ such that $y = f(x)$. From $g_L \circ f = I_x$, we get $g_L(y) = g_L(f(x)) = x$ and hence $g_L(y) = x$. Now, we must show that $g_R(y) = x$. Suppose $g_R(y) \neq x$. By the one-to-one property of f, we must have $f(g_R(y) \neq f(x)$. But $f(g_R(y)) = y$ and hence we get $f(x) \neq y$ as a contradiction. So, $g_R(y) = x$ and the proof is complete.

II Proof of Theorem 2.2:

We shall now prove Theorem 2.2 without using the left and right inverses.

Let f be one-to-one and onto. We should define $g : B \longrightarrow A$ having the desired properties. For every $y \in B$, there is a unique $x \in A$ such that $f(x) = y$. So, we define $g(y)$ as x.

Let $y \in B$. $g(y) = x$ where $f(x) = y$. Hence $f(g(y)) = f(x) = y$.

Let $x \in A$. From the definition of g, it is clear that $g(f(x)) = x$.

Next, let f be invertible. We should show that f is one-to-one and onto. Let $g = f^{-1}$ be its inverse. To show that f is one-to-one, consider $f(x_1) = f(x_2)$. Now, $g(f(x_1)) = g(f(x_2))$ and hence $x_1 = x_2$. Thus, f is one-to-one.

To prove that f is onto, consider $y \in B$. We should find $x \in A$ such that $f(x) = y$. Let $x = g(y)$. Now, $f(x) = f(g(y)) = y$ and hence f is onto. ■

2.7 SOME BASIC FUNCTIONS AND THEIR PROPERTIES

What do we mean when we say that 3.14 is an approximation of π? This means that on the real line these two numbers are 'close' to each other. The numbers -7.1 and -7 are closer to each other than the pair 10 and -1000. The modulus function is useful to measure the distance between two numbers. $|x|$ is just the distance between origin and the point x. We formally define the modulus function $|| : \mathbb{R} \longrightarrow \mathbb{R}^+$ as follows.

2.7.1 The Modulus Function

$$|x| = x, \; x \geq 0 \tag{2.11}$$

$$= -x, \; x < 0. \tag{2.12}$$

The distance between two points x and y is nothing but $|x - y|$. The statement $|x| \leq a$ is equivalent to the statement $-a \leq x \leq x$.

PRACTICE 2.31

Show that the function $f : \mathbb{R} \longrightarrow \mathbb{R}$ defined as $f(x) = |x|$ is neither one-to-one nor onto.

2.7.2 Trigonometric Functions and their Inverses

In this section, we assume that the reader is familiar with trigonometric functions. The interested reader may refer to the Appendix on Trigonometric functions.

Example 2.129

Let $f : R \to R$ be defined as $f(x) = \sin x$. This function is neither one-to-one nor onto. To show that a function is not one-to-one, we need two values x_1 and x_2 such that $x_1 \neq x_2$ but $f(x_1) = f(x_2)$. We take $x_1 = 0$ and $x_2 = 2\pi$. Clearly, $0 \neq 2\pi$ but $f(x_1) = \sin 0 = 0$ and $f(x_2) = \sin 2\pi = 0$. Hence, this function is not one-to-one. To show that a function is not onto, we should find $y \in R$ such that there is no solution $x \in R$ for $\sin x = y$. Recall that $|\sin x| \leq 1$. So, taking $y = 2$, we know that there is **no** solution to $\sin x = 2$. ■

Example 2.130

Define $f : R \to [-1, 1]$ as $f(x) = \sin x$. This function is not one-to-one but onto. The proof of the fact that this function is not one-to-one is similar to the previous example. Now, we show that this function is onto.

Let $y = 0$. We take $x = 0$ and get $\sin x = y$.

Let $y = 1$. We take $x = \pi/2$ and get $\sin x = y$.

Let $0 < y < 1$. Construct a right-angled triangle whose sides are y, $\sqrt{1-y^2}$ and 1. If x is the measure of the angles that are opposite to the side of length y, then $\sin x = y$. If $-1 \le y \le 0$, then by the previous arguments, we can find x_1 such that $\sin x_1 = -y$. Hence, by taking $x = -x_1$, we get $\sin x = y$.

Thus f is onto. ■

Example 2.131

Define $f : [0, \pi/2] \to [-1, 1]$ as one-to-one but not onto. If x and y are angles less than or equal to 90^0 and $\sin x = \sin y$, then $x = y$. Thus, this function is one-to-one. If $x \in [0, \pi/2]$, then $0 \le \sin x \le 1$. So, for $y = -1 \in [-1, 1]$, there is no $x \in [0, \pi/2]$ for which $\sin x = -1$. Thus, f is not onto. ■

PRACTICE 2.32

(i) Show that $f : R \to R$ defined as $f(x) = \cos x$ is neither one-to-one nor onto.
(ii) Show that $f : R \to [-1, 1]$ defined as $f(x) = \cos x$ is not one-to-one but onto.

Example 2.132

Let $S = \{x \in R : x \neq (2n+1)(\pi/2), n = 0, \pm1, \pm2, \ldots\}$. Let $f : S \longrightarrow R$ be defined as $f(x) = \tan x$. Then, show that f is not one-to-one.

SOLUTION

As before, $f(0) = f(\pi) = 0$ and hence f is not one-to-one. ■

PRACTICE 2.33

Show that the function of the previous example is onto.

PRACTICE 2.34

Analyse the functions associated with the expressions $\csc x$, $\sec x$ and $\cot x$.

As we have already noted, the function $f : \mathbb{R} \longrightarrow \mathbb{R}$ defined as $f(x) = \sin x$ is not invertible. So, what do we mean by the expression $\sin^{-1}(x)$? Consider the function $g : [-\pi/2, \pi/2] \longrightarrow [-1, 1]$ defined as $g(x) = \sin x$. g is one-to-one onto and hence is invertible. This inverse is denoted by $\sin^{-1} x$.

Consider the function $h : [0, \pi] \longrightarrow [-1, 1]$ defined as $h(x) = \cos x$. h is one-to-one onto and hence is invertible. This inverse is denoted by $\cos^{-1} x$.

PRACTICE 2.35

Define $tan^{-1}(x)$.

The Square Root Function: We say that a is a square root of b if $a^2 = b$. We know that there is no real number whose square is -25 and so -25 does not have any *real* square root. Further, 25 has two square roots namely, 5 and -5. For a non-negative real number x, the symbol $\sqrt{x}$ denotes its *positive* square root.

So, we define $\surd : \mathbb{R}^+ \longrightarrow \mathbb{R}^+$ as the inverse of the function $h : \mathbb{R}^+ \longrightarrow \mathbb{R}^+$ defined as $h(x) = x^2$.

It is clear that $(\sqrt{x})^2 = x$. Of course, this statement makes sense only when $x \geq 0$.

Now , what is $\sqrt{x^2}$? If x is real, $x^2 \geq 0$ and hence it is meaningful to talk about $\sqrt{x^2}$. But what is its value ? Recalling the fact that $\surd$ denotes the positive square root of the number inside, we find that $\sqrt{5^2} = \sqrt{(-5)^2} = 5$. Hence $\sqrt{x^2} = |x|$.

2.7.3 Quadratic Functions

Example 2.133

Define $f : \mathbb{R} \rightarrow \mathbb{R}$ as $f(x) = x^2 - 7x + 9$. Find the range of f.

SOLUTION

Recall that range of f is the set

$$\left\{x^2 - 7x + 9 : x \in R\right\}.$$

Let us re-write f as $f(x) = (x - \frac{7}{2})^2 - \frac{49}{4} + 9 = \left(x - \frac{7}{2}\right)^2 - \frac{13}{4}$. Hence $f(x) \geq \frac{-13}{4}$ for all $x \in \mathbb{R}$. This implies that the range is contained in $(-\frac{13}{4}, \infty)$. Next, we show that the range is actually $(-\frac{13}{4}, \infty)$. To show this, we must show that $(-\frac{13}{4}, \infty)$ is contained in the range of f. Let $y \in (-\frac{13}{4}, \infty)$. This means that $\frac{-13}{4} \leq y$. To find $x \in \mathbb{R}$ such that $f(x) = y$, we have to solve $x^2 - 7x + 9 = y$. Consider

$$\begin{aligned}
x^2 - 7x + 9 &= y \\
\left(x - \frac{7}{2}\right)^2 - \frac{13}{4} &= y \\
\left(x - \frac{7}{2}\right)^2 &= \frac{13}{4} + y \\
\left(x - \frac{7}{2}\right) &= \pm\sqrt{\frac{13}{4} + y}.
\end{aligned}$$

Since $\frac{-13}{4} \leq y$, $0 \geq y + \frac{13}{4}$. Thus, we can take the real square root of $\frac{13}{4} + y$. Finally, we get two values of x, namely $x = \frac{7}{2} \pm \sqrt{\frac{13}{4} + y}$. Therefore, $\left(\frac{-13}{4}, \infty\right)$ is in fact the range of f. ■

PRACTICE 2.36

Find the range of $g : \mathbb{R} \longrightarrow \mathbb{R}$ defined as $g(x) = -3x^2 + 7x - 9$.

Example 2.134

Find the range of $g : [-3, 5] \longrightarrow R$ defined as $g(x) = 3x^2 - 4x + 3$.

SOLUTION

This problem is nothing but a rephrasing of Example 1.11 of Chapter 1. Hence, the required answer is $[-17, 78]$. ■

2.8 INCREASING AND DECREASING FUNCTIONS

Look at the graphs of the three functions given below:

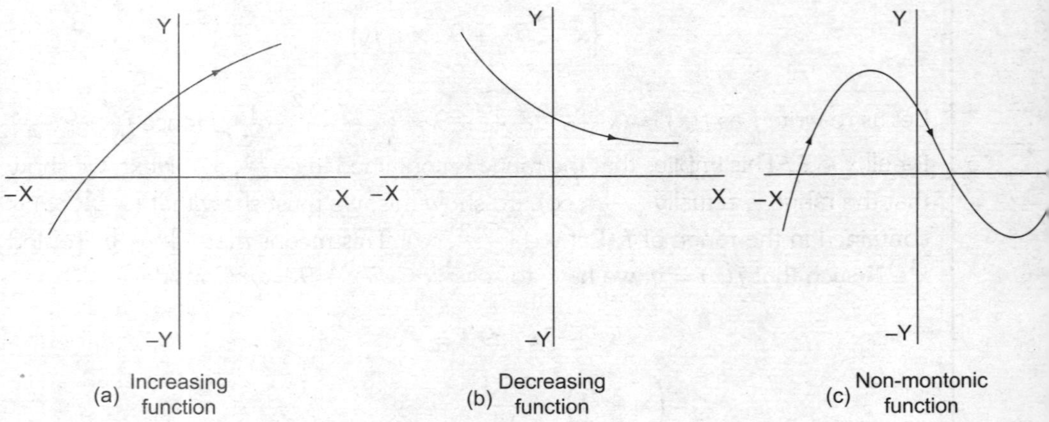

Figure 2.27

In Figure 2.27(a), as the value of x moves from left to right, the value of $f(x)$ moves upwards. In Figure 2.27(b), as the value of x moves from left to right, the value of $f(x)$ moves downwards. But in Figure 2.27(c), as the value of x moves from left to right, the value of $f(x)$ moves up and down. The first function is called an increasing function and the second function is called a decreasing function. Increasing or decreasing functions are called monotonic functions. The third function is neither increasing nor decreasing. Such functions are called non-monotonic.

Definition 2.28: Let $X, Y \subset \mathbb{R}$ and $f : X \to Y$ be a function. Then
(i) f is said to be *increasing* if $x_1 \leq x_2 \to f(x_1) \leq f(x_2)$.
(ii) f is said to be *decreasing* if $x_1 \leq x_2 \to f(x_1) \geq f(x_2)$.
(iii) f is said to be *strictly increasing* if $x_1 < x_2 \to f(x_1) < f(x_2)$.
(iv) f is said to be *strictly decreasing* if $x_1 < x_2 \to f(x_1) > f(x_2)$.
(v) f is said to be *monotonic* if f is either decreasing or increasing.
(vi) f is said to be *strictly monotonic* if f is either strictly decreasing or strictly increasing.

Example 2.135

(1) $f : \mathbb{R} \longrightarrow \mathbb{R}$ defined as $f(x) = x^3$ is an increasing function.
(2) $f : \mathbb{R} \longrightarrow \mathbb{R}$ defined as $f(x) = e^{-x}$ is a strictly decreasing function. ■

2.8.1 A Technique to Find the Range of Some Nice Functions

In this section, the reader is required to be familiar with the notion of a derivative of a function.

Example 2.136

Consider the following examples
(i) Let $f : [a, b] \to \mathbb{R}$ be defined as $f(x) = px + q$ and $p > 0$. The range of f is clearly $[pa + q, pb + q]$.
(ii) Let $f : [-10, 15] \to \mathbb{R}$ be defined as $f(x) = x^3$. Its range is $[-1000, 3375]$. Based on the above two examples one is tempted to conclude that the range of $f : [a, b] \longrightarrow \mathbb{R}$ is $[f(a), f(b)]$. But this conclusion is **not** valid. Consider the next example.
(iii) Let $f[-3, 5] \longrightarrow \mathbb{R}$ be defined as $f(x) = x^2$. Its range is not $[9, 25]$. For example, if 2 is taken as the domain then $f(2) = 4$ must be in the above range. But $4 \notin [9, 25]$. By looking at Figure 2.28, we find that its range is $[0, 25]$.

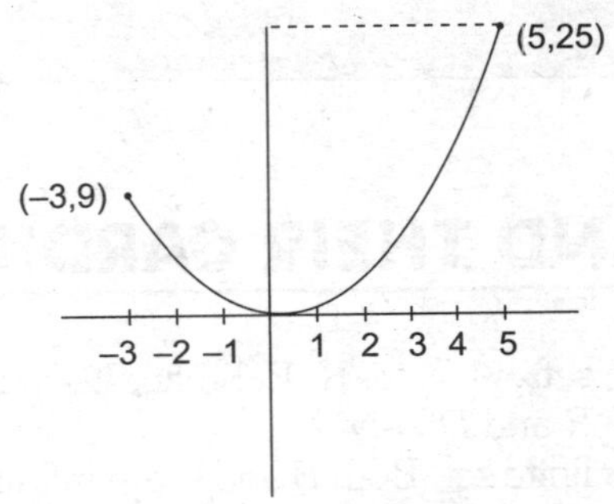

Figure 2.28 ■

The following proposition tells us how to find the image of intervals if the functions are continuous and monotonic.

Proposition 2.10

Let $f : (a, b) \to R$ be an increasing *continuous* function. Then $f[(a, b)] = (f(a), f(b))$. Let $f : [a, b]$ be a decreasing *continuous* function. Then $f((a, b)) = (f(b), f(a))$.

Proposition 2.11

Let $f : (a, b) \to R$ be a function such that $f'(x) \geq 0$ for all $x \in (a, b)$. Then, f is an increasing function in (a, b). Let $f : (a, b) \to \mathbb{R}$ be a function such that $f'(x) \leq 0$ for all $x \in (a, b)$. Then, f is a decreasing function in (a, b).

Example 2.137

Find the range of $f : [-4, 5] \longrightarrow \mathbb{R}$ defined as $f(x) = 2x^3 + 3x^2 - 12x + 60$. $f'(x) = 6x^2+6x-12 = 3(x^2+2x-4)$. We shall find the intervals where f' has the same sign. Applying the methods from Section 1.5, we find that $(x^2+2x-4) \geq 0$ in $(-\infty, -2) \cup [1, \infty)$ and $x^2 + 2x - 4 < 0$ in $(-2, 1)$. ■

Example 2.138

Find the range of the following functions:
(c) $f : [-3, 7] \longrightarrow R$ defined as $f(x) = x^3 - x^2$.
(d) $f : R \longrightarrow R$ defined as $f(x) = -x^4 + 12x - 35$.
(e) $f : [5, 7] \longrightarrow R$ defined as $\sqrt{-x^2 + 12x - 35}$.
(f) $f : R \longrightarrow R$ as $f(x) = 2\cos x - 9$. ■

PRACTICE 2.37

(1) Find the the natural domains for the following expressions:

(i) $f(x) = \frac{x^2-7x+9}{x^2+16x+69}$.
(ii) $f(x) = \frac{1}{x^2+10x-160}$.
(iii) $f(x) = \frac{1}{\sqrt{x^2+10x-160}}$.
(iv) $f(x) = \sqrt{\cos x}$.

2.9 SETS AND THEIR CARDINALITY

Consider the following sets: $A = \{$A.R. Rahman, Ilayaraaja, R.D. Burman$\}$, $B = \{2, 4, 6, \ldots\}$, $C = R$ and $D = N$.

A is an example of a finite set. Both B and C are infinite sets, but somehow we feel that C has 'more' number of elements than B. B is a proper subset of D and hence D

has more elements than B. But we intuitively feel that both B and D have the same 'size'. We may also feel that the 'size' of $\mathbb{R}^2$ is much more than that of $\mathbb{R}$, but we shall see that it is not so. Comparing the 'sizes' of infinite sets is a tricky business. To compare the sizes of *finite sets*, we simply count the number of elements. But we cannot do this with infinite sets.

We count the elements of a finite set as 'one, two, three,...', etc. and assign the last uttered number as the number of elements in that set. It is clear that for a set that is not finite, there is no last element. If a finite set has, say, five elements, then when we count its elements we would have called the names of all elements in the set $\{1, 2, 3, 4, 5\}$. In general, if a finite set has n elements, it is related to the set $\{1, 2, 3, \ldots n\}$. This set is denoted by N_n. Having said so much, let us now start with precise definitions and rigorous proofs.

Definition 2.29: An empty set is finite. A non-empty set S is said to be finite, if there exists $n \in \mathbb{N}$ such that there is a one-to-one and onto function $f : S \longrightarrow N_n$. n is called the cardinality of the set and we say that the set S has n elements. This is denoted as $|S| = n$.

Proposition 2.12

The set $\{1, 2, \ldots, n\}$ is finite and has n elements.

Proof

The identity function $I_S : S \longrightarrow S$ is one-to-one and onto.[2] ■

Example 2.139

Show that the set $\{3, 4, 5, 6, 7, 8\}$ is finite and has six elements.

Define $f : \{3, 4, 5, 6, 7, 8\} \longrightarrow N_6$ as $f(3) = 1$, $f(4) = 2$, $f(5) = 3$, $f(6) = 4$, $f(7) = 5$ and $f(8) = 6$. By drawing the arrow diagram for this function, we find that it is one-to-one and onto (Fig. 2.29).

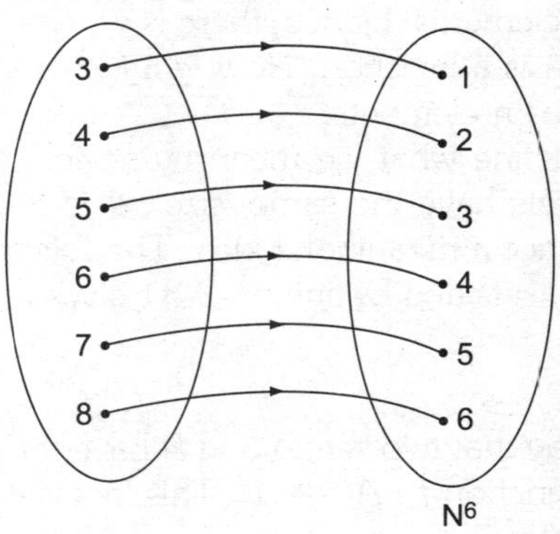

Figure 2.29

Hence the given set has six elements. ■

[2]In this section, the reader might feel that certain obvious results are proved in a roundabout manner. But the point is that although the results are obvious and intuitively clear, their proofs need to be rigorous.

Example 2.140

Show that the set $S = \{\text{Amitabh, Rajini, Mommutty, Rajkumar}\}$ has four elements.

SOLUTION

Define $f : S \to N_4$ as $f(\text{Amitabh}) = 1$, $f(\text{Rajini}) = 2$, $f(\text{Mommutty}) = 3$ and $f(\text{Rajkumar}) = 4$.

The above two examples illustrate how such a simple thing as 'counting' looks complicated under a rigorous definition.

A one-one onto map is called a **bijection**. An important question in this regard is the following:

If a finite set A **has** n **elements, can it have** m **elements**?

In other words,

Does there exist a bijection $f : N_n \longrightarrow N_m$?

The answer to the above question is **no** if $m \neq n$. But we have to prove it in a logical manner. A complete proof will be given in Chapter 5 after we learn the method of Mathematical Induction. In the meantime, we shall assume the following result. ■

Proposition 2.13

Let A be a finite set and let B be a non-empty proper subset of A. Then B is finite and $0 < |B| < |A|$.

Proposition 2.14

Let $f : A \longrightarrow B$ be a one-to-one and onto function. If A has n elements, then B also has n elements.

Proof

A is a finite set having n elements. Hence, there is $g : A \longrightarrow N_n$, which is one-to-one and onto. $f^{-1} : B \longrightarrow A$ is a bijection. Now, $g \circ f^{-1}$ is a bijection from B into N_n. Thus, B is finite and it has n elements.

Next, our aim is to define what we mean by an infinite set and what we mean when we say that two sets have the same 'size'. It is obvious that for infinite sets, 'number of elements' is not a meaningful idea. The following definition tells us how to compare two arbitrary sets and Definition 2.31 gives us the idea of an infinite set. ■

Definition 2.30: We say that two sets A and B have the same cardinality if there is a one-to-one and onto function $f : A \longrightarrow B$. This fact is denoted by $A \sim B$ and they are also called *similar* sets.

The proof of the following proposition easily follows from Definition 2.30 and Proposition 2.14.

Proposition 2.15

Two finite sets A and B have the same cardinality if and only if they have the same number of elements.

PRACTICE 2.38

Prove Proposition 2.15.

Definition 2.31: We say that a set A is an infinite set if there is a non-empty proper subset B such that A and B have the same cardinality.

Proposition 2.16

The sets N and R are infinite.

Let $A = \{2, 3, 4, \ldots\}$, which is a proper subset of N. Define $f : A \longrightarrow N$ as $f(n) = n - 1$. f is one-to-one and onto. Hence N is infinite.

Let $B = (-1, 1)$. Define $f : R \longrightarrow B$ as $f(x) = \frac{x}{1+|x|}$. Thus, R is an infinite set.

Proposition 2.17

A set is infinite if and only if it is not finite.

Sketch of the Proof

Let A be an infinite set. Then there is a non-empty proper subset B of A such that there is a bijection $f : B \longrightarrow A$. If A is finite, then by Proposition 2.13, B is also finite and $|B| < |A|$. Thus, we have bijections $h : B \longrightarrow N_m$ and $g : A \longrightarrow N_n$ with $m < n$.

Now, $g_0 f_0 h^{-1}$ is a bijection of N_n into N_m. But such a bijection is not possible. Thus, A is not finite. Now, let A be not finite. We have to prove that A is infinite. That is, we have to get a non-empty proper subset of B, which is objective to A. $A \neq \phi$. Hence there is $x_1 \in A$. Now, consider $A - \{x_1\}$. If $A - \{x_1\}$ is empty, then A contains only one element, which is a contradiction.

Continuing, we get distinct elements $x_1, x_2, \ldots \in A$. Let $B = A - \{x_1\}$. Define $f : B \longrightarrow A$ as follows: $f(x_i) = x_{i-1}$ and $f(x) = x$ for $x \neq x_i$. It is easy to show that f is one-to-one and onto. Hence A is infinite.

Figure 2.30 illustrates $f : B \longrightarrow A$.

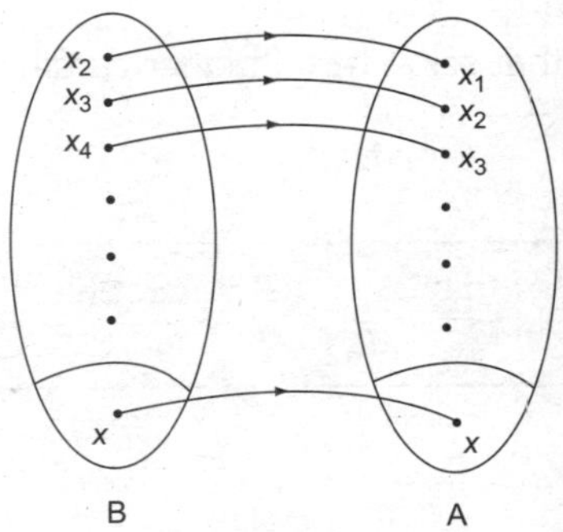

Figure 2.30

■

Proposition 2.18

Let A and B be two disjoint sets such that $|A| = n$ and $|B| = m$. Then $A \cup B$ has $m + n$ elements.

Proof

Let $f : A \longrightarrow N_n$ and $g : B \longrightarrow N_m$ be one-to-one and onto functions. We must produce a one-to-one and onto function $h : A \cup B \longrightarrow N_{m+n}$. We define h as follows:

$$\begin{aligned} h(x) &= f(x) \text{ for } x \in A \\ &= g(x) + n \text{ for } x \in B. \end{aligned}$$ ■

Proposition 2.19

Show that **Z** and $\mathbb{N}$ have the same cardinality.

SOLUTION

Define $f(\mathbf{Z}) \longrightarrow \mathbb{N}$ as $f(0) = 1$, $f(2) = 3$, $f(3) = 5$, etc. That is $f(n) = 2n + 1$ if $n \geq 0$.

Further, $f(-1) = 2$, $f(-2) = 4$, etc. That is, $f(n) = 2|n|$, if $n < 0$. We can easily show that f is one-to-one and onto. The following diagram illustrates this.

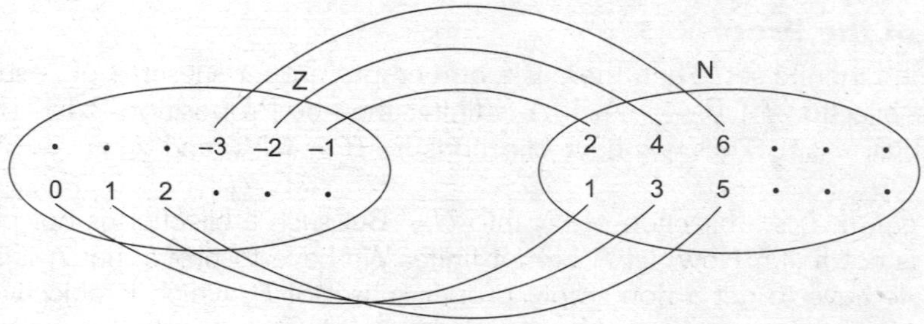

Figure 2.31

We have learnt that **Z**, $\mathbb{N}$, $\mathbb{R}$ all are infinite sets and **Z** and N have the same cardinality. Though Z contains infinite elements that are not in $\mathbb{N}$, **Z** and N have the same 'size'. But $\mathbb{R}$ is much bigger than $\mathbb{N}$. This statement is proved by Proposition 2.20 (Fig. 2.31).

Definition 2.32: We say that set A has a higher cardinality than B if there cannot exist an onto function $f : B \longrightarrow A$.

PRACTICE 2.39

For finite sets A and B show that $|A| > |B|$ if and only if there cannot be an onto function $f : B \longrightarrow A$.

Theorem 2.3

For any set X, the power set $P(X)$ has higher cardinality than X.

Proof

If the theorem were to be false, then there is an onto function $f : X \longrightarrow 2^X$. Let $S = \{x : x \notin f(x)\}$. Let $x \in S$ be such that $f(x) = S$. If $x \in S$, then $x \notin f(x) = S$. If $x \in S = f(x)$, then $x \notin S$, a contradiction.

The proof of Proposition 2.20 is beyond the scope of this book as some knowledge of real analysis is required. The reader may refer to *Mathematical Analysis* by Tom Apostle. ■

Proposition 2.20

There is a bijection $f : \mathbb{R} \longrightarrow P(\mathbb{N})$.

Now combining Proposition 2.20 and Theorem 2.3 gives the following result.

Proposition 2.21

$\mathbb{R}$ has higher cardinality than $\mathbb{N}$.

Definition 2.33: We say that S is *countably infinite* if there exists $f : \mathbb{N} \to S$, which is one-to-one and onto. We say that a set S is *countable*, if it is either finite or countably infinite. A set that is not countable is called uncountable.

In view of Proposition 2.21 and Definition 2.33, R is an uncountable set.

SUMMARY

- A set is a collection of objects not necessarily related to each other.
- The objects are called elements of the set.
- If every element of A is an element of B, then we say that A is a subset of B and denote this by $A \subseteq B$.
- Real numbers can be represented by points on a straight line called the real line.
- The union of a collection of sets is the set whose elements belong to at least one of the sets of the collection.
- The intersection of a collection of sets is the set whose elements belong to each of the sets of the collection.
- A relation R from A into B is a subset of $A \times B$.
- A relation R on A is said to be reflexive if $a\,R\,a$ for all $a \in R$.
- A relation R is said to be symmetric if $a\,R\,b \to b\,R\,a$.
- A relation R on A is said to be transitive if $a\,R\,b$ and $b\,R\,c \to a\,R\,c$.
- A relation R is said to be anti-symmetric if $a\,R\,b \to$ and $b\,R\,a \to a = b$.
- A relation is said to be an equivalence relation if it is reflexive, symmetric and transitive.
- A relation is said to be a partial order if it is reflexive, anti-symmetric and transitive.
- The equivalence classes of an equivalence relation on a set A partition the set A and a partition gives rise to an equivalence relation on A.
- A function $f : A \to B$ associates an element of B to each of an element of A. A is called the domain of the function and B is called the co-domain of the function.
- A function $f \quad A \longrightarrow B$ is said to be one-to-one if $f(x) = f(y) \to x = y$.
- A function $f : A \longrightarrow B$ is said to be onto if for every $y \in B$, there is $x \in A$ such that $f(x) = y$.
- A function $f : A \longrightarrow B$ has an inverse function $g : B \longrightarrow A$ if and only if it is one-to-one and onto.
- A set A is said to be finite if there is a one-to-one and onto function $f : \{1, 2, 3, \ldots, n\} \longrightarrow A$ and the number of elements $|A|$ is n.

SUMMARY (Continued)

- An infinite set is a set that is not finite.
- The sets $\mathbb{N}$ and $\mathbb{R}$ are infinite sets.
- A set A is said to be countably infinite if there is a one-to-one and onto function $f : \mathbb{N} \longrightarrow A$.
- The sets $\mathbb{N}$ and $\mathbf{Z}$ are examples of countably infinite sets.
- The set $\mathbb{R}$ is uncountable.

PROBLEMS

1. Rewrite the following sets by listing the elements:

   (i) $\{x \in \mathbb{N} : x \text{ is a power of } 2\}$.
   (ii) $\{x \in \mathbf{Z} : x \text{ is a factor of } 360\}$.
   (iii) $\{x \in \mathbb{N} : x \text{ is even and } x \leq 23\}$.
   (iv) $\{x \in \mathbb{N} : x \text{ gives a remainder of 1 when divided by } 10\}$.
   (v) $\{x \in \mathbb{N} : 125 \leq x \leq 1000 \text{ is a perfect cube}\}$.
   (vi) $\{x \in \mathbb{N} : 23 < x < 29 \quad x \text{ is a prime }\}$.
   (vii) $\left\{x \in R : x^2 + 12x - 325 = 0\right\}$.
   (viii) $\left\{x \in \mathbf{Z} : 2x^2 - 3x - 2 = 0\right\}$.

2. Give an example of a set of real values of a, b and c such that

$$\left\{x \in Q : ax^2 + bx + c = 0\right\} \neq \left\{x \in R : ax^2 + bx + c = 0\right\}.$$

3. Rewrite the following sets using mathematical/verbal description:

   (i) $\{5, 25, 125, 625, \ldots\}$.
   (ii) {George Washington, John Adams,..., Abraham Lincoln,..., John F. Kennedy, ..., Ronald Regan, H.W. George Bush, Bill Clinton, H. George Bush}.
   (iii) $\{6, 11, 16, 21, \ldots\}$.
   (iv) $\{101, 102, 103, \ldots\}$.

4. Write down the power set of $A = \{a, b, 23, \pi\}$.
5. Write down the power set of the power set of $\{3\}$.
6. Find $P \times Q$ where $P = \{a\}$ and $Q = \{1, 2, 3, 4, 5\}$.
7. Illustrate the rectangle of Example 2.21 by a figure.
8. In Example 2.38, find $K \cup J$ and $K \cap J$.
9. Show that the set

$$\left\{\begin{pmatrix} a & b \\ c & d \end{pmatrix} : a, b, c, d \in R, ad - bc = 1\right\}$$

is a subset of $GL_2(R)$, which is the set of all non-singular 2×2 matrices.

10. Let

$$\mathbf{O} = \left\{ \begin{pmatrix} \cos\theta & -\sin\theta \\ \sin\theta & \cos\theta \end{pmatrix} : \theta \in R \right\}$$

$$SL_2(R) = \left\{ \begin{pmatrix} a & b \\ c & d \end{pmatrix} : a, b, c, d \in R, ad - bc = 1 \right\}.$$

Show that $\mathbf{O} \subseteq SL_2(R)$. Find an element in $SL_2(R)$, which is not in $\mathbf{O}$

11. Let $L_1 = \left\{(x, y) \in R^2 : 3x + 4y = 7\right\}$ and $L_2 = \left\{(x, y) \in R^2 : x - 4y = 5\right\}$. Find $L_1 \cap L_2$.
12. Find the union and intersection of the intervals $[-3, 7]$ and $[-4, 5]$.
13. Show that $\bigcup_{n=1}^{\infty}[-n, n] = R$.
14. Show that $\bigcap_{n=1}^{\infty}\left[\frac{-1}{n}, \frac{1}{n}\right] = \{0\}$.
15. Show that the line $\{(t, 2 - 4t, 3t + 5) : t \in R\}$ is contained in the plane $\left\{(x, y, z) \in R^3 : x + y + z = 7\right\}$.
16. Show that the line $\left\{(t, t, \sqrt{2}t) : t \in R\right\}$ is contained in the cone $\left\{(x, y, z) \in R^3 : x^2 + y^2 = z^2\right\}$.
17. Find $A \cap B$ where $A = \{(t, 2t - 2, 3t - 2) : t \in R\}$ and $B = \left\{(x, y, z) \in R^3 : x + y + z = 3\right\}$.
18. Show that $A - B = A - (A \cap B)$ using Venn Diagrams.
19. Let $\mathbb{R}$ be the universal set. Find the compliments of $[1, 2]$, $(1, 2)$, $(1, 2]$, $[1, 2)$, $(1, \infty)$, $(-\infty, 2]$.
20. In Figure 2.32, we have given the arrow diagrams for some relations. Examine which of the following relations are reflexive, symmetric and transitive:

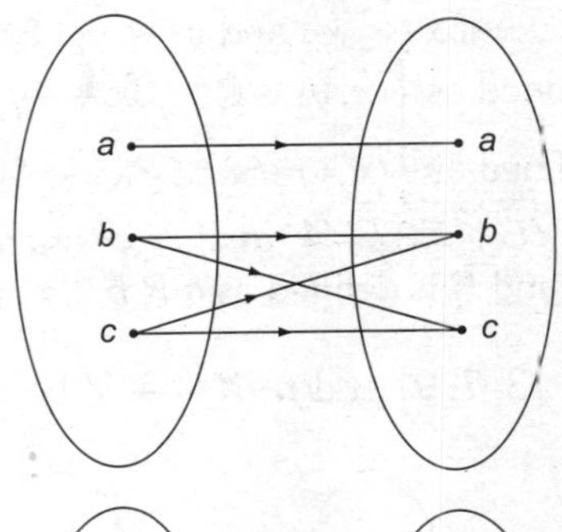

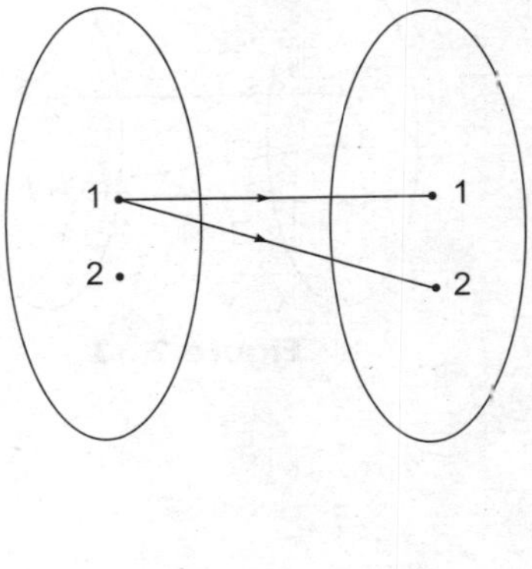

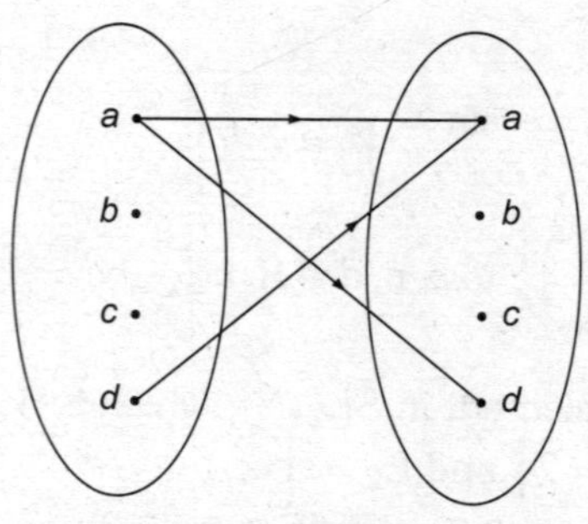

Figure 2.32

21. Let $A = \{1, 2, 3, 4, 5, 6, 7, 8, 9, 10\}$. Define $\mathbb{R}$ as
$$x\,R\,y \;\leftrightarrow\; x^2 \le y.$$
Draw the arrow diagram for this relation. Express $\mathbb{R}$ as a subset of $A \times A$. Examine whether this relation is reflexive, symmetric or transitive.
22. Give an example of a set A and a relation on A, which is symmetric, transitive but not reflexive.
23. Define $\le$ on $\mathbb{R}^2$ as $(x_1, y_1)\ R\ (x_2, y_2)$ if $x_1 \le x_2$. Show that $\le$ is reflexive and transitive but neither symmetric nor anti-symmetric.
24. Examine the properties of the relation on $S = \{1, 2, 3, \ldots, 10\}$ defined as $\mathbb{R} = \{(x, y) \in S \times S : x + y = 10\}$.
25. Consider the relation on R^3 defined as $(x_1, y_1, z_1)\ R\ (x_2, y_2, z_3)$ if $(x_2 - x_1) + (y_2 - y_1) + (z_2 - z_1) = 0$. Show that R is an equivalence relation.
26. Let B be the set of all bytes. R is defined as follows: $a\ R\ b$ if both a and b have the same first bit and the same last bit. Show that R is an equivalence relation and identify the equivalence classes.
27. Define $\le$ on $\mathbb{R}^2$ as $(x_1, y_1)\ R\ (x_2, y_2)$ if $x_1 \le x_2$ and $y_1 \le y_2$. Show that $\le$ is a partial order.
28. Show that the relation R on $\mathbb{R}$ defined as $\left\{(x, y) \in \mathbb{R}^2 : 3x + 4y = 9\right\}$ is a function.
29. Show that the relation R on $\mathbb{R}$ defined as $\left\{(x, y) \in \mathbb{R}^2 : 3x^2 + 4y^2 = 7\right\}$ is a not a function.
30. Let $f, g, h : \mathbb{R} \longrightarrow \mathbb{R}$ be defined as $f(x) = 3x - 4$, $g(x) = x^2$ and $h(x) = e^x$. Find $f \circ f$, $g \circ h$, $h \circ g$ and $g \circ f$.
31. Let $S = \{2, 3, 4, 5, 6, 7, 8, 9, 10\}$ and R is defined as $a\ R\ b$ if a is relatively prime to b. Write R as a subset of $S \times S$.
32. Let $X = \{m, n, l, h, g, e\}$ and $Y = \{3, 7, 9\}$ and $f : X \longrightarrow Y$ be defined by Figure 2.33.

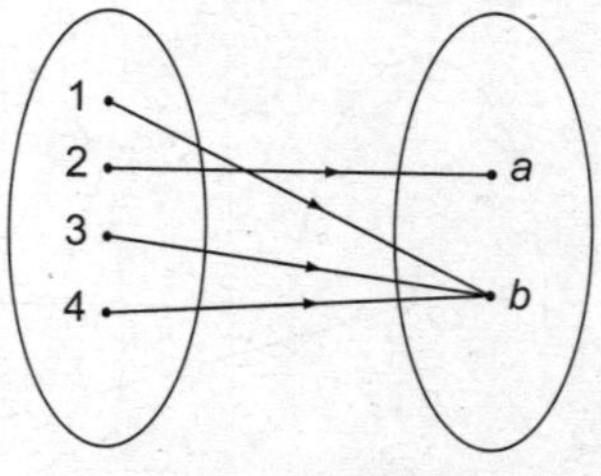

Figure 2.33

Find a right inverse $g : Y \longrightarrow X$.

33. Let $X = \{p, q, r\}$ and $Y = \{1, 2, 3, 4, 5\}$ and f be given by Figure 2.34.

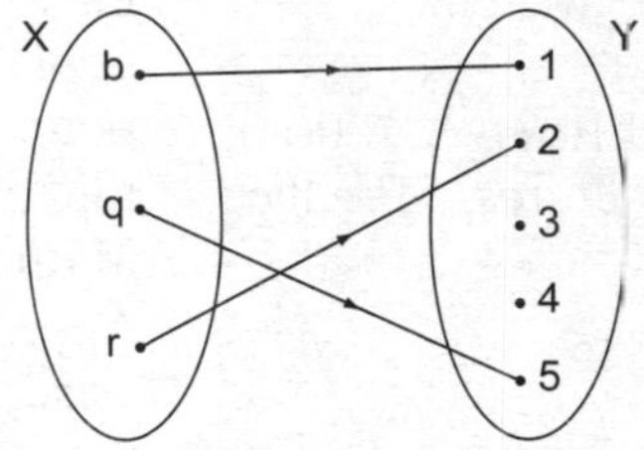

Figure 2.34

Find two different left inverses $g : Y \longrightarrow X$.

34. Show that $f : \mathbb{R} \to \mathbb{R}$ defined as $f(x) = 3\tan x + 9$ is onto.
35. Show that $f : \mathbb{N} \to \mathbb{N}$ defined as $f(n) = n^3$ is not onto.
36. Show that $f : \mathbb{R}^3 \longrightarrow \mathbb{R}$ defined as $f(x, y, z) = x - 2y + z$ is onto but not one-to-one. Find a right inverse to f.
37. Show that $f : \mathbb{R} \longrightarrow \mathbb{R}^2$ defined as $f(x) = (x, 8x)$ is one-one but not onto. Find a left inverse to f
38. Show that $f : X \longrightarrow Y$ is onto if and only if $f^{-1}(B) \neq \phi$ for every non-empty $B \subseteq Y$.
39. Show that $f : X \longrightarrow Y$ is one-to-one if and only if $f^{-1}(y)$ is either a singleton or empty for every $y \in y$.
40. Show that $f : X \longrightarrow Y$ is one-to-one and onto if and only if $f^{-1}(y)$ is a singleton for every $y \in y$.
41. Let X be the set of all 2×2 matrices with integer entries and $f : X \longrightarrow Z$ be defined as $f(A) = |A|$. Show that f is onto but not one-to-one.
42. Show that $f : [0, \infty) \longrightarrow [0, 1)$ defined as $f(x) = \frac{x}{1+x}$ is one-to-one and onto. Also, find the inverse function.
43. Show that $f : [1, \infty) \longrightarrow [-1, \infty)$ defined as $f(x) = x^2 - 2x - 3$ is one-to-one and onto. Also, find the inverse.
44. Let $X = \left\{ \begin{pmatrix} n & n \\ n & 4n \end{pmatrix} : n \in N \right\}$.
45. Show that $f : B \longrightarrow B$ of Example 2.89 is invertible by finding the inverse of f.
46. Show that $g : B \longrightarrow \mathbf{Z}$ of Example 89 is neither one-to-one nor onto. How will you modify the co-domain of g to obtain an onto function?
47. Let $f : \{u, v, w, x, y\} \longrightarrow \{p, q, r\}$ be defined as $f(u) = r$, $f(v) = q$, $f(w) = r$, $f(x) = q$, $f(y) = r$. Draw an arrow diagram for this function. Find the images of $\{v, y\}$, $\{u, v, x\}$.
48. Let $f : [0, \infty) \longrightarrow \mathbb{R}$ be defined as $f(x) = x^2 - 4x + 64$. Find the images of $\{-2, 4, 67\}$ and $[4, 7)$. Find the inverse images of $\{45, 6\}$ and $[-1, 1]$.
49. Let $f : \mathbb{N} \longrightarrow \mathbb{N}$ and $g : \mathbb{N} \longrightarrow \mathbb{N}$ be defined as $f(n) = n^3$ and $g(n) = n^2 + 1$. Find $f \circ g$ and $g \circ f$.
50. Find the natural domains for the following expressions:

(i) $f(x) = \sec^{-1}(4x - 9)$.

(ii) $\dfrac{x^2 - 6x + 9}{x^2 + 10x - 45}$.

(iii) $\sin^{-1}(x^2 - 7x + 12)$.

(iv) $\sqrt{x^2 + 12x - 35}$.

(v) $\dfrac{1}{\sqrt{x^2 + 12x - 35}}$.

FURTHER READING

http://www.maths.mq.edu.au/ wchen/ lndmfolder/lndm.html.

Kolman, B., Busby, R.C. and Ross, S. (2001), *Discrete Mathematical Structures*, Prentice Hall of India, New Delhi.

Lipschutz, S. (1981), *Set Theory and Related Topics*, McGraw-Hill Book Company, Singapore.

Lipschutz, S. and Lipson, M. (1999), *Discrete Mathematics*, Tata McGraw-Hill, New Delhi.

Apostol, T.M. (2002), The *Mathematical Analysis*., Narosa Publishing House, New Delhi.

CHAPTER

3

Logic

Learning Objectives

On completing this chapter, you should be able to:

- identify statements
- assign truth values to complex statements obtained by combining atomic statements with the help of logical connectives
- identify tautologies and contradictions
- write down the normal forms of given statements
- prove or disprove the validity of logical implications by the methods of truth table, inference theory or automatic theorem proving
- identify statement functions
- formulate statements using quantifiers in addition to logical connectives
- prove or disprove the validity of logical implications of predicate calculus by the method of inference theory

Imagine a child returning home from school. She finds a pair of slippers on the door step, which does not belong to any of her family members. She immediately concludes that a guest has arrived. This is an example of logical inference. With the help of her observation, she deduces the presence of a guest without actually seeing the guest.

Imagine that a murder has been committed in Mumbai at about 3 p.m. and the police suspects Mr Kuppan. But suppose that someone has met and talked with Mr Kuppan at 2.15 p.m in Chennai. A plane, which is the fastest mode of transport between Mumbai and Chennai, takes around an hour to cover the distance. So, we conclude that Mr Kuppan could not have committed the murder. In real life we depend on *common sense to make such logical deductions*. But mathematical logic is somewhat different. We shall now try to understand this.

Consider the following statement:

$$\textbf{Every Thursday, Parthasarathy goes for swimming.} \tag{3.1}$$

In real life, the above statement could actually mean that he goes for swimming almost on all Thursdays. For example, if he has a fever on a Thursday he may not go to swim. But in formal logic, we do not permit such vagueness.

The above statement could also be written as

$$\textbf{If today is a Thursday, then Parathasarathy goes for swimming.} \tag{3.2}$$

Let us now consider the following statements:
P : Today is a Thursday Q : Parthasarathy goes for swimming.
Thus, equation (3.2) can be written as

$$\textbf{If P then Q.} \tag{3.3}$$

As we have learnt in Section 1.2, this is symbolically written as

$$P \longrightarrow Q.$$

Now, can we assert that if today is not a Thursday, then Parthasarathy does not go for swimming? No. The statement (3.2) does not say anything about what Parthasarathy does on a day other than a Thursday. But if Parthasarathy does not go for swimming on a particular day, that day cannot be a Thursday. To summarize:

(a) We cannot deduce: **If P is not true, then Q is not true.**
(b) We can deduce: **If Q is not true, then P is not true.**

Formal logic is the study of the process of these kind of deductions. To understand logic, we have to understand what a *statement* is.

Statements Consider the following sentences:

(i) Mr George Washington was the first President of the United States.
(ii) Mr Karunanidhi is the Collector of Kanchipuram District.
(iii) Ms Jayalalitha is the Prime Minister of Tamil Nadu.
(iv) Computers are dangerous.
(v) $x + 3 = 9$.
(vi) $(x + 1)^2 = x^2 + 2x + 1$.
(vii) $2 + 10 = 210$.
(viii) $5^3 = 125$.

We make the following observations:

(a) (i), (vi) and (viii) are true.
(b) (ii) and (vii) are false.
(c) (iii) is absurd and as far as (v) is concerned, unless we know what x is, we cannot say whether (v) is true or not.

We say that (i), (vi), (viii), (ii) and (vii) are statements. (iii) and (v) are not statements. That is, **a statement is an assertion that is either true or false.**

3.1 NEW STATEMENTS OUT OF OLD

In this section, we learn how to obtain new statements by combining old statements. Also, we learn how to get the truth value of the new statement depending on the truth values of the old statement. The discussion of this section formally known as statement calculus.

Consider the following statements:

(a) Muthu is earning Rs 25,000 per month.
(b) Muthu is interested in classical music.

Now, we can use the above statements to create new statements as follcws:

(c) Muthu is earning Rs 25,000 per month *or* Muthu is interested in classical music.
(d) Muthu is earning Rs 25,000 per month *and* Muthu is interested in classical music.

(i) (a) could be true and (b) could be true.
(ii) (a) could be true and (b) could be false.
(iii) (a) could be false and (b) could be true.
(iv) (a) could be false and (b) could be false.

Let us make the following *common sense* observations regarding the truths of statements (c) and (d):

(I) When (i) holds, both (c) and (d) are true.
(II) When (ii) holds (c) is true but (d) is false.
(III) When (iii) holds, (c) is true but (d) is false.
(IV) When (iv) holds, both (c) and (d) are false.

Consider the statements:

(a) If it rains, then crops grow well.
(b) If n is a multiple of 10, then n is a multiple of 5.

These statements illustrate one more way of obtaining a new statement using two given statements p and q. That is, we can formulate the statement

$$\text{If } p \text{ then } q.$$

Let us now make a list of simple new statements that can be formed by using the given statements p and q.

(a) p or q,
(b) p and q,
(c) If p, then q,
(d) Not p.

The mathematical notations for these new statements are as follows:

p or q	$p \vee q$
p and q	$p \wedge a$.

If p, then q	$p \to q$
not p	$\sim p$.

The symbols $\vee$, $\wedge$, $\rightarrow$, $\sim$ are called *connectives*. Symbols like p and q are called statement variables or atomic statements. $p \vee q$ is called the *disjunction* of p, q. $p \wedge q$ is called the *conjunction* of p, q.

$p \rightarrow q$ is called an *implication*.

It is clear that we can build complex statements using the connectives and statement variables. These are called statement formulas, For example, $\sim p \vee (q \rightarrow (r \wedge s))$ is a statement formula. Now,

$p(\vee(q \wedge r)$ is not a properly formed statement formula. We shall define the idea of a *well-formed formula (wff).*

Definition 3.1: A well-formed formula is defined using the following rules:

(a) A single statement variable is a wff.
(b) If P is a wff, then $\sim P$ is a wff.
(c) If P and Q are wffs, then $(P \vee Q)$, $(P \wedge Q)$, $(P \rightarrow Q)$, $(P \leftrightarrow Q)$ are wffs.
(d) A string made up of statement variables, connectives and brackets is a wff, if and only if it is obtained by using finitely many applications of the above rules.

A truth table for a wff is a summary of observations such as we have made above for wffs. We now present the truth tables for statements involving these basic connectives. These tables are based on the following observations:

(i) $p \vee q$ is true when at least one of p, q is true. Otherwise it is false
(ii) $p \wedge q$ is true only when both p and q are true
(iii) P is true when $\sim P$ is false and vice versa
(iv) Only a true conclusion can follow from a true hypothesis. Any conclusion can follow from a false hypothesis.

p	q	$p \to q$
T	T	T
T	F	F
F	T	T
F	F	T

p	q	$p \vee q$
T	T	T
T	F	T
F	T	T
F	F	F

p	q	$p \wedge q$
T	T	T
T	F	F
F	T	F
F	F	F

p	$\sim p$
T	F
F	T

Example 3.1

Write the following in symbolic forms:
(a) Ramasamy is gentle and firm.
(b) If Mr Ram makes coffee in the morning, then Mrs Ram teaches well.
(c) John is neither weak nor jealous.
(d) Abdul is intelligent but not proud.
(e) If Mr Arul prepares breakfast, then Mrs Arul does not sing well.
(f) Sridhar is either kind or innocent.

SOLUTION

First, we have to break the above statements into basic statements. It is better that the basic statements do not contain words like *not*, *neither*, etc. Then, we use the connectives to get the symbolic forms. The answers are given below:

(a) p – Ramasamy is gentle. q – Ramasamy is firm.
The given statement in symbolic form is:

$$p \wedge q.$$

(b) p – Mr Ram makes coffee in the morning. q – Mrs Ram teaches well.
The given statement in symbolic form is:

$$p \rightarrow q.$$

(c) p – John is weak. q – John is jealous.
The given statement in symbolic form is:

$$\sim p \wedge \sim q.$$

(d) p – Abdul is intelligent. q – Abdul is proud.
The given statement in symbolic form is:

$$p \wedge \sim q.$$

(e) p – Mr Arul prepares breakfast. q – Mrs Arul sings well.
The given statement in symbolic form is:

$$p \rightarrow \sim q.$$

(f) p – Sridhar is kind. q – Sridhar is innocent.
The given statement in symbolic form is:

$$p \vee q.$$

■

PRACTICE 3.1

(1) Kannan is not rich and Kannan is happy.
(2) Razia is either hardworking or intelligent.
(3) If it does not rain, Mangai eats 'pani puri'.
(4) If there are mosquitoes, then Mathews does not sleep.
(5) If Ram sings or Rahim tells a story, Robert is happy.
(6) If Ram sings, then Rahim tells a story and Robert is happy.
(7) Thirumavalavan does not watch TV, when Murugan goes for a concert.
(8) It is not true that when it rains water problems get solved.

Example 3.2

Draw truth tables for:

(i) $p \wedge \sim q$.
(ii) $\sim q \rightarrow p$.
(iii) $\sim (\sim p \wedge q)$.
(iv) $p \vee (q \wedge r)$.

SOLUTION

p	q	$\sim q$	$p \wedge \sim q$
T	T	F	F
T	F	T	T
F	T	F	F
F	F	T	F

p	q	$\sim q$	$\sim q \rightarrow p$
T	T	F	T
T	F	T	T
F	T	F	T
F	F	T	F

$\sim p$	q	$\sim p \wedge q$	$\sim (\sim p \wedge q$
F	T	F	T
F	F	F	T
T	T	T	F
T	F	F	T

p	q	r	$q \wedge r$	$p \vee (q \wedge r)$
T	T	T	T	T
T	T	F	F	T
T	F	T	F	T
T	F	F	F	T
F	T	T	T	T
F	T	F	F	F
F	F	T	F	F
F	F	F	F	F

PRACTICE 3.2

Draw truth tables for the following:

(i) $\sim p \vee q$.
(ii) $q \wedge \sim p$.
(iii) $\sim (p \vee q)$.
(iv) $p \wedge (q \vee r)$.
(v) $(p \vee q) \wedge (q \vee r)$.

We know that the truth values $P \rightarrow Q$ and $Q \rightarrow P$ are independent of each other. But there are also interesting situations when both these statements hold. We recall the following examples from Chapter 1:

(1a) If n is even, then n^2 is even.
(1b) If n^2 is even, then n is even.
(2a) If all the three sides of a triangle are equal, then all the three angles are also equal.
(2b) If all the three angles of a triangle are equal, then all the three sides are also equal.

Let us also recall that when both the implications $P \to Q$ and $Q \to P$ are valid, we write $P \leftrightarrow Q$. In this situation, Q is true when P is true and further Q is true only when P is true. Thus, $P \leftrightarrow Q$ is just the statement P if and only if Q.

PRACTICE 3.3

Write down the truth table for $P \leftrightarrow Q$.

3.2 LOGICAL EQUIVALENCES AND IMPLICATIONS

The statement

'Abdul Kalam is the President of India or Abdul Kalam is the President of India'

does not contain any more information than the simple statement

'Abdul Kalam is the President of India'.

Two statements that may appear different, but have exactly the same meaning can be called equivalent statements. In mathematical logic, we look at the truth values to define equivalence.

Definition 3.2: Two statement formulas, which have the same truth value irrespective of the the truth values of the atomic statements, are called *equivalent statements*. If A and B are equivalent, we denote it by $A \Leftrightarrow B$.

Example 3.3

Consider the two statements P and $P \vee P$. Both are true when P is true and both are false when P is false. Hence

$$P \Leftrightarrow P \vee P.$$

Similarly,

$$P \Leftrightarrow P \wedge P.$$

How to prove that two statement formulas are equivalent? Draw a truth table having a column for each of these formulas. If they have identical entries, they are equivalent. In the following, we illustrate this method for one of the De Morgan Laws. ■

De Morgan Laws: The following equivalences are called De Morgan laws:

(i) $\sim (P \vee Q) \Leftrightarrow \sim P \wedge \sim Q$.

(ii) $\sim (P \wedge Q) \Leftrightarrow \sim P \vee \sim Q$.

Example 3.4

Prove the first De Morgan Law given above.

SOLUTION

We find that the columns for $\sim (P \vee Q)$ and $\sim P \wedge \sim Q$ have identical entries. Hence

$$\sim (P \vee Q) \Leftrightarrow \sim P \wedge \sim Q).$$

p	q	$\sim P$	$\sim Q$	$P \vee Q$	$\sim (P \vee Q)$	$\sim P \wedge \sim Q$
T	T	F	F	T	F	F
T	F	F	T	T	F	F
F	T	T	F	T	F	F
F	F	T	T	F	T	T

■

PRACTICE 3.4

Prove that $\sim (P \wedge Q) \Leftrightarrow \sim P \vee \sim Q$.

Remark 3.1

Note the similarity between the above laws and the set theoretic De Morgan Laws given by:

(i) $(A \cup B)^c = A^c \cap B^c$.
(ii) $(A \cap B)^c = A^c \cup B^c$.

Example 3.5

Show that $P \longrightarrow Q \Leftrightarrow \sim P \vee Q$.

As the columns for $P \longrightarrow Q$ and $\sim P \vee Q$ are identical, we conclude that

$$P \longrightarrow Q \Leftrightarrow \sim P \vee Q.$$

P	Q	$\sim P$	$P \longrightarrow Q$	$\sim P \vee Q$
T	T	F	T	T
T	F	F	F	F
F	T	T	T	T
F	F	T	T	T

■

Example 3.6

Check whether $(P \vee Q) \wedge R$ and $P \vee (Q \wedge R)$ are equivalent.

From the last two columns, we find that the given statements are not equivalent.

P	Q	R	$P \vee Q$	$Q \wedge R$	$(P \vee Q) \wedge R$	$P \vee (Q \wedge R)$
T	T	T	T	T	T	T
T	T	F	T	F	F	T
T	F	T	T	F	T	T
T	F	F	T	F	F	T
F	T	T	T	F	T	F
F	T	F	T	F	F	F
F	F	T	F	F	F	F
F	F	F	F	F	F	F

■

PRACTICE 3.5

Show the following equivalences:

(i) $P \Leftrightarrow (P \wedge Q) \vee (P \wedge \sim Q)$.
(ii) $(P \wedge Q) \to R \Leftrightarrow (R \vee \sim P) \wedge (R \vee \sim Q)$.
(iii) $(P \to Q) \wedge R \Leftrightarrow (\sim P \wedge R) \vee (R \wedge Q)$.

Distributive laws

(i) $(X \vee Y) \wedge Z \Leftrightarrow (X \wedge Z) \vee (Y \wedge Z)$.
(ii) $(X \wedge Y) \vee Z \Leftrightarrow (X \vee Z) \wedge (Y \vee Z)$.

The proof of the distributive laws is left as an exercise for the reader in the following problem.

PRACTICE 3.6

Show that the distributive laws hold by using the truth tables.

The following example illustrates the use of logic in set theory. In Section 2.1, the distributive law

$$A \cap (B \cup C) = (A \cap B) \cup (A \cap C)$$

of set theory has been proved using Venn diagrams. But the following is the formal logical proof. Naturally, the key step in the proof involves the use of the distributive law for statements.

Example 3.7

Prove

$$A \cap (B \cup C) = (A \cap B) \cup (A \cap C).$$

SOLUTION

We show that

$$x \in A \cap (B \cup C) \leftrightarrow x \in (A \cap B) \cup (A \cap C).$$

$$
\begin{aligned}
& x \in A \cap (B \cup C) \\
\leftrightarrow\ & x \in A \wedge (x \in B \vee x \in C) \\
\leftrightarrow\ & (x \in A \wedge x \in B) \vee (x \in A \wedge x \in C) \quad \text{(distributive law for statements)} \\
\leftrightarrow\ & (x \in A \cap B) \vee x \in A \cap C) \\
\leftrightarrow\ & x \in (A \cap B) \cup (A \cap C).
\end{aligned}
$$
■

PRACTICE 3.7

Check the validity of the following equivalences using truth tables:

(i) $P \vee \sim Q \Leftrightarrow \sim P \vee Q$.
(ii) $P \vee \sim (Q \vee R) \Leftrightarrow (P \vee \sim Q) \wedge (P \vee \sim R)$.
(iii) $(P \vee Q) \longrightarrow R \Leftrightarrow \sim ((\sim P \to Q) \wedge \sim R)$.
(iv) $A \longrightarrow (P \wedge Q) \Leftrightarrow (A \wedge \sim P) \longrightarrow Q$.
(v) $((P \wedge Q \wedge R) \to C) \wedge (R \to (P \vee Q \vee C)) \Leftrightarrow (R \wedge (P \leftrightarrow Q)) \to C$.
(vi) $P \wedge \sim (Q \to R) \Leftrightarrow (P \wedge Q) \vee (P \wedge \sim R)$.
(vii) $(P \wedge Q) \to R \Leftrightarrow P \to (Q \vee R)$.

Example 3.8

This example shows that we can prove some equivalences without using truth tables. We prove (ii) of the previous exercise using De Morgan and distributive laws:

$$
\begin{aligned}
P \vee \sim (Q \vee R) &\Leftrightarrow P \vee (\sim Q \wedge \sim R) \quad \textit{(DeMorgan Law)} \\
&\Leftrightarrow (P \vee \sim Q) \wedge (P \vee \sim R) \quad \textit{(Distributive Law)}.
\end{aligned}
$$
■

3.3 TAUTOLOGIES AND CONTRADICTIONS

In this chapter, we noted that the converse of the statement $P \longrightarrow Q$ is just $Q \longrightarrow P$. Consider the following statements:

(i) If today is a Sunday, then today is holiday
(ii) If today is not a Sunday, then today is not a holiday
(iii) If today is a holiday, then today is a Sunday
(iv) If today is not a holiday, then today is not a Sunday.

We note that (iii) is the converse of (i). We say that (ii) is the inverse of (i), and (iv) is the contrapositive of (ii). We find that the statement and its contrapositive are equivalent, and that converse and inverse statements are equivalent to each other.

Example 3.9

Draw the truth tables for the following:
(i) $(P \to Q) \vee (P \wedge \sim Q)$.
(ii) $(P \to (Q \vee R)) \wedge P \wedge \sim Q \wedge \sim R$.
(iii) $(P \vee (Q \to R)) \wedge P$.

SOLUTION

(i)

P	Q	$\sim Q$	$P \longrightarrow Q$	$P \wedge \sim Q$	$(P \longrightarrow Q) \vee (P \wedge \sim Q)$
T	T	F	T	F	T
T	F	T	F	T	T
F	T	F	T	F	T
F	F	T	T	F	T

(ii)

P	Q	R	$Q \vee R$	$\sim Q$	$\sim R$	$P \longrightarrow (Q \vee R)$	$P \longrightarrow (Q \vee (Q \vee R)) \wedge (P \wedge \sim Q \wedge \sim R)$
T	T	T	T	F	F	T	F
T	T	F	T	F	T	T	F
T	F	T	T	T	F	T	F
T	F	F	F	T	T	F	F
F	T	T	T	F	F	T	F
F	T	F	T	F	T	T	F
F	F	T	T	T	F	T	F
F	F	F	F	T	T	T	F

(iii)

P	Q	R	$Q \vee R$	$P \longrightarrow (Q \vee R)$	$(P \vee (Q \to R)) \wedge P$
T	T	T	T	T	T
T	T	F	T	T	T
T	F	T	T	T	T
T	F	F	F	F	F
F	T	T	T	T	F
F	T	F	T	T	F
F	F	T	T	T	F
F	F	F	F	T	F

■

We find that the first statement formula is *always true* irrespective of the truth values of the atomic statements. The second one is *always false* irrespective of the truth values of the atomic statements. The statement formula (i) is called a tautology and (ii) is called a contradiction. Note that (iii) is neither a tautology nor a contradiction.

Definition 3.3: A statement formula whose truth value is true irrespective of the truth values of the atomic statements is called a *tautology*. It is denoted by **T**.

A statement formula whose truth table is false irrespective of the truth value of the atomic statements is called a contradiction. It is denoted by **F**.

Example 3.10

It is obvious that $P \vee \sim P$ is a tautology. ■

PRACTICE 3.8

(i) Define a *contradiction*.
(ii) Construct a simple contradiction involving a single statement and its negation.

Example 3.11

If **T** is a tautology, show that $P \Leftrightarrow P \wedge \mathbf{T}$.

SOLUTION

Let P be true. Then since **T** is always true, $P \wedge \mathbf{T}$ is true.

Let P be false. Clearly, $P \wedge \mathbf{T}$ is false. Thus, P and $P \wedge \mathbf{T}$ have the same truth values always. So, $P \Leftrightarrow P \wedge \mathbf{T}$. ■

PRACTICE 3.9

Show that $P \Leftrightarrow P \vee \mathbf{F}$.

The simplest method to show whether a statement formula is a tautology or a contradiction is to construct its truth table. But for some simple cases, we can avoid the truth table.

Example 3.12

Show that $(P \longrightarrow (Q \vee R)) \wedge P \wedge \sim Q \wedge \sim R$ is a contradiction *without using a truth table.*

SOLUTION

$$\begin{aligned}
&(P \longrightarrow (Q \vee R)) \wedge P \wedge \sim Q \wedge \sim R \\
&\quad \Leftrightarrow \quad (\sim P \vee (Q \vee R)) \wedge P \wedge \sim Q \wedge \sim R \\
&\quad \Leftrightarrow \quad (\sim P \vee Q \vee R) \wedge P \wedge \sim Q \wedge \sim R \\
&\quad \Leftrightarrow \quad \sim (P \wedge \sim Q \wedge \sim R) \\
&\qquad\qquad \wedge P(\wedge \sim Q \wedge \sim R).
\end{aligned}$$

The last formula is of the form $\sim A \wedge A$ and hence this is a contradiction. ■

Example 3.13

Show that $Q \vee (P \wedge \sim Q) \vee (\sim P \wedge \sim Q)$ is a tautology without using a truth table.

SOLUTION

$$
\begin{aligned}
Q \vee (P \wedge \sim Q) \vee (\sim P \wedge \sim Q) &\Leftrightarrow Q \vee ((P \vee \sim P) \wedge \sim Q) && \text{(Distributive Law)} \\
&\Leftrightarrow Q \vee (\mathbf{T} \wedge \sim Q) && (P \vee \sim P \text{ is a tautology}) \\
&\Leftrightarrow Q \vee \sim Q && (\mathbf{T} \wedge \sim \mathbf{Q} \Leftrightarrow \sim Q) \\
&\Leftrightarrow \mathbf{T}.
\end{aligned}
$$

■

3.3.1 Logical Implication

It is easy to see that $(P \wedge Q) \rightarrow Q$ is a tautology. In other words, whatever may be the statements P and Q, Q is true provided $P \wedge Q$ is true. This is an example of a *logical implication* and we write this as $(P \wedge Q) \Rightarrow Q$.

Definition 3.4: $A \Rightarrow B$ if and only if $A \rightarrow B$ is a tautology.

Example 3.14

Let us understand the difference between $\Rightarrow$ and $\rightarrow$. We can easily check that $P \vee Q \rightarrow P$ is *not* a tautology. So, we *cannot* write

$$P \vee Q \Rightarrow P.$$

But in some special situations

$$P \vee Q \rightarrow P \tag{3.4}$$

could be valid. For example, let P be the statement that Ram is happy. Let Q be the statement that today is a holiday. If it so happens that Ram would be happy on a holiday, then clearly equation (3.4) is valid. ■

Example 3.15

Show that

$$(P \vee Q) \wedge (Q \rightarrow P) \Rightarrow P.$$

SOLUTION

Going by the definition, we must show that

$$((P \vee Q) \wedge (Q \rightarrow P)) \rightarrow P$$

is a tautology.

P	Q	$P \vee Q$	$Q \rightarrow P$	$(P \vee Q) \wedge (Q \rightarrow P)$	$((P \vee Q) \wedge (Q \rightarrow P)) \rightarrow P$
T	T	T	T	T	T
T	F	T	T	T	T
F	T	T	F	F	T
F	F	F	T	F	T

SOLUTION

Instead of making the full truth table, we can also prove this implication as demonstrated below:

P	Q	$P \vee Q$	$Q \to P$	$(P \vee Q) \wedge (Q \to P)$
T	T	T	T	T
T	F	T	T	T
F	T	T	F	F
F	F	F	T	F

From the truth table, we find that whenever $(P \vee Q) \wedge (Q \to P)$ is true P is also true. This proves the implication. ■

PRACTICE 3.10

Show that

$$P \wedge (Q \vee R) \wedge \sim R \Rightarrow Q.$$

Example 3.16

Show that $\phi \subset A$ for any set A.

Proof.
Let us recall the definition of a subset $X \subseteq Y$ if every element of X is an element of Y. That is $X \subseteq Y$ if $x \in X \to x \in Y$.
Now we have to show that $x \in \phi \to x \in A$. But $x \in \phi$ is false since ϕ does not have any element. We know that $P \to Q$ is a valid implication if P is false. Thus, the result is proved. ■

3.4 NORMAL FORMS

A single statement formula can have many logically equivalent formulas. If $\mathbf{T}$ is a tautology, then the truth value of $P \wedge \mathbf{T}$ depends only on the truth value of P. Thus,

$$P \Leftrightarrow P \wedge \mathbf{T} \tag{3.5}$$

Similarly,

$$P \Leftrightarrow P \vee \mathbf{F} \tag{3.6}$$

where $\mathbf{F}$ is a contradiction. So, we have the following equivalences:

(i) $P \Leftrightarrow P \vee (Q \wedge \sim Q)$.
(ii) $Q \Leftrightarrow Q \wedge (P \vee \sim P)$.
(iii) $P \Leftrightarrow P \vee (Q \wedge \sim Q) \vee ((Q \wedge R) \wedge (\sim Q \vee \sim R))$.

As there are infinitely many equivalent formulas, it is desirable to write a given statement formula in some *standard* equivalent form. What is the use of this? For example, if two statement formulas have the same standard form, then they are equivalent to each other. Now, we shall learn about some standard forms. There are two standard forms known as Disjunctive Normal Forms and Conjunctive Normal forms. Together, these are known as *normal forms*.

Remark 3.2

We can think of $\vee$ as similar to $+$ and $\wedge$ as similar to $\times$. Now, **T** is like 1 and **F** is like 0. Thus, equation (3.5) is similar to $a \times 1 = a$ and equation (3.6) is similar to $a + 0 = a$.

Hereafter, conjunction will be called *product* and disjunction will be called *sum*.

Definition 3.5: A sum of atomic statements and their negations is an elementary sum. Single atomic statements and their negations are also elementary sums.

A product of atomic statements and their negations is an elementary product. Single atomic statements and their negations are also elementary products.

Example 3.17

$\sim P \vee Q$, $P \vee \sim Q \vee R$, $P \wedge P$, $P \vee \sim P \vee Q$, $\sim P \vee Q \vee R$ are all examples of elementary sums. Single atomic statements like Q and negations like $\sim R$ are also elementary sums.

$(P \to Q) \vee R$, $\sim (P \vee R) \vee R$ are **not** elementary sums. ■

PRACTICE 3.11

Give examples of statement formulas that are elementary products and formulas that are not elementary products.

Disjunctive Normal Forms: Disjunctive Normal Form (**DNF**) of a statement formula is a logically equivalent formula, which is a *sum of elementary products*.

Conjunctive Normal Forms: Conjunctive Normal Form (**CNF**) of a statement formula is a logically equivalent formula, which is a *product of elementary sums*.

The following example illustrates how to apply an appropriate distributive law to obtain normal forms.

Example 3.18

Obtain the disjunctive and conjunctive normal forms of $(P \vee (Q \wedge R)) \wedge R$.

SOLUTION

First we obtain disjunctive normal form:

$$\begin{aligned}(P \vee (Q \wedge R)) \wedge R &\Leftrightarrow (P \wedge R) \vee ((Q \wedge R) \wedge R)\\ &\Leftrightarrow (P \wedge R) \vee (Q \wedge R \wedge R)\\ &\Leftrightarrow (P \wedge R) \vee (Q \wedge R).\end{aligned}$$

Next, we obtain the conjunctive normal form:

$$\begin{aligned}(P \vee (Q \wedge R)) \wedge R &\Leftrightarrow ((P \vee Q) \wedge (P \vee R)) \wedge R \\ &\Leftrightarrow (P \vee Q) \wedge (P \vee R) \wedge R.\end{aligned}$$

We must pay careful attention to the first step in the above solutions. For DNF, the symbol $\vee$ links elementary products. So our aim is to pull $\vee$ out of the brackets. So, the distributive law that is used is

$$(A \vee B) \wedge C = (A \wedge C) \vee (B \wedge C).$$

For CNF, the symbol $\wedge$ links elementary sums. So our aim is to pull $\wedge$ out of the brackets. So, the distributive law that is used is

$$(A \wedge B) \vee C = (A \vee C) \wedge (B \vee C).$$

General steps to get DNF and CNF:

(i) Use $A \to B \leftrightarrow \sim A \vee B$ to get rid of $\to$.
(ii) Using De Morgan laws, apply $\sim$ to any non-atomic part of the formula.
(iii) Then apply appropriate distributive laws. ■

Example 3.19

Obtain the normal forms of $(P \vee Q) \to \sim (Q \wedge R)$.

SOLUTION

$$\begin{aligned}(P \vee Q) \to \sim (Q \wedge R) &\Leftrightarrow \sim (P \vee Q) \vee (\sim (Q \wedge R)) \\ &\Leftrightarrow (\sim P \wedge \sim Q) \vee (\sim Q \vee \sim R).\end{aligned}$$

$(\sim P \wedge \sim Q) \vee (\sim Q \vee \sim R)$ is the DNF.
Now, to get CNF, we have to pull $\wedge$ out of the bracket. So, we do

$$\begin{aligned}(\sim P \wedge \sim Q) \vee (\sim Q \vee \sim R) &\Leftrightarrow (\sim P \vee (\sim Q \vee \sim R)) \wedge (\sim Q \vee (\sim Q \vee \sim R)) \\ &\Leftrightarrow (\sim P \vee \sim Q \vee \sim R) \wedge (\sim Q \vee \sim Q \vee \sim R) \\ &\Leftrightarrow (\sim P \vee \sim Q \vee \sim R) \wedge (\sim Q \vee \sim R).\end{aligned}$$ ■

Definition of a Max Term: A Max term is an elementary sum in which precisely one of an atomic statement or its negation is present.

PRACTICE 3.12

Define a Min Term.

In a situation where the number of atomic statements is two, there are 4 min terms and four max terms. For example, with P and Q, all the max terms are given by

$$P \vee Q, \ \ P \vee \sim Q, \ \ \sim P \vee Q, \ \ \sim P \vee \sim Q.$$

All the min terms are given by

$$P \vee Q, \ \ P \vee \sim Q, \ \ \sim P \vee Q, \ \ \sim P \vee \sim Q.$$

$P \vee \sim P$ is not a max term since both P and $\sim P$ are present.
$P \wedge Q \wedge \sim Q$ is not a min term since both Q and $\sim Q$ are present.

PRACTICE 3.13

Write down all possible max and min terms involving P, Q and R. How many max/min terms are possible with n atomic statements?

Remark 3.3

A single atomic statement is not a max or a min term.

3.4.1 Principal Normal Forms

It is obvious that

$$P \vee (Q \wedge R) \Leftrightarrow (P \wedge P) \vee (Q \wedge R)$$

and both these equivalent statements are in disjunctive normal form. We shall now learn about the principal disjunctive normal form, which gives a *unique* representation. Similarly, principal conjunctive normal form gives a unique representation of the conjunctive normal forms.

Example 3.20

In the context involving two atomic statements, a single atomic statement can be written as a *product of max terms* as follows:

$$\begin{aligned} P &\Leftrightarrow P \vee (Q \wedge \sim Q) \\ &\Leftrightarrow (P \vee Q) \wedge (P \vee \sim Q). \end{aligned}$$

■

PRACTICE 3.14

Compute an equivalent statement in the form of *sum of min terms* to the atomic statement P involving atomic statements P and Q.

Example 3.21

In the context involving three atomic statements, a sum of min terms equivalent to P is computed as follows:

$$
\begin{aligned}
P &\Leftrightarrow P \wedge (Q \vee \sim Q) \\
&\Leftrightarrow (P \wedge Q) \vee (P \wedge \sim Q) \\
&\Leftrightarrow ((P \wedge Q) \vee (P \wedge \sim Q)) \wedge (R \vee \sim R) \\
&\Leftrightarrow (P \wedge Q \wedge R) \vee (P \wedge Q \wedge \sim R) \vee (P \wedge \sim Q \wedge R) \vee \\
&\quad (P \wedge \sim Q \wedge \sim R).
\end{aligned}
$$

■

PRACTICE 3.15

Write P as a product of max terms involving three atomic statements P, Q and R.

Principal Disjunctive Normal Forms: A Principal Disjunctive Normal Form is the sum of min terms.

The phrase *Principal Disjunctive Normal Form* is abbreviated as PDNF. The procedure for obtaining an equivalent PDNF is illustrated in the next example.

Example 3.22

Obtain the PDNF of $(P \vee (R \to \sim (Q \vee P))) \wedge R$ without using the truth table.

$$
\begin{aligned}
(P \vee (R \to \sim (Q \vee P))) \wedge R &\Leftrightarrow (P \vee (\sim R \vee \sim (Q \vee P))) \wedge R \\
&\Leftrightarrow (P \vee (\sim R \vee (\sim Q \wedge \sim P))) \wedge R \\
&\Leftrightarrow (P \wedge R) \vee ((\sim R \vee (\sim Q \wedge \sim P)) \wedge R) \\
&\Leftrightarrow (P \wedge R) \vee (\sim R \wedge R) \vee (\sim Q \wedge \sim P \wedge R) \\
&\Leftrightarrow (P \wedge R) \vee (\sim Q \wedge \sim P \wedge R).
\end{aligned}
$$

So far, we have obtained the DNF. Now, to get the PDNF, we have to rewrite $(P \wedge R)$ using min terms. So, we get

$$(P \vee (R \to \sim (Q \vee P))) \wedge R \Leftrightarrow (P \wedge R \wedge Q) \vee (P \wedge R \wedge \sim Q) \vee (\sim Q \wedge \sim P \wedge R).$$

In the next example, we show how to obtain the PDNF of a given statement formula using the truth table. ■

Example 3.23

Obtain the PDNF of $(p \wedge q) \to \sim q$.

SOLUTION

Let us first construct the truth table for the given statement:

p	q	$p \wedge q$	$\sim q$	$p \wedge q \rightarrow q$
T	T	T	F	F
T	F	F	T	T
F	T	F	F	T
F	F	F	T	T

Now, examine the rows in which the given statement is true. Find the min terms that are true for the corresponding rows. The sum of these min terms is the required PDNF.

p	q	$\sim p$	$\sim q$	$p \wedge q$	$p \wedge \sim q$	$\sim p \wedge q$	$\sim p \wedge \sim q$
T	T	F	F	T	F	F	F
T	F	F	T	F	T	F	F
F	T	T	F	F	F	T	F
F	F	T	T	F	F	F	T

So, the PDNF is given by

$$(p \wedge \sim q) \vee (\sim p \wedge q) \vee \sim p \wedge \sim q. \qquad \blacksquare$$

Principal Conjunctive Normal Forms: A principal conjunctive normal form is a product of max terms. It is abbreviated as PCNF.

PRACTICE 3.16

Below is the truth table for max terms. What is the procedure to get the PCNF of a statement formula?

p	q	$\sim p$	$\sim q$	$p \vee q$	$p \vee \sim q$	$\sim p \vee q$	$\sim p \vee \sim q$
T	T	F	F	T	T	T	F
T	F	F	T	T	T	F	T
F	T	T	F	T	F	T	T
F	F	T	T	F	T	T	T

PRACTICE 3.17

Find the PCNF of $P \rightarrow (\sim Q \vee R)$. Its truth table is given below.

P	Q	R	$\sim Q$	$\sim Q \vee R$	$P \rightarrow (\sim Q \vee R)$
T	T	F	F	F	F
T	F	T	T	T	T
T	F	F	T	T	T
F	T	T	F	T	T
F	T	F	F	F	T
F	F	T	T	T	T
F	F	F	T	T	T

Example 3.24

Let us denote the statement formula $(p \wedge q) \rightarrow \sim q$ of Example 3.24 by X. What is the PDNF of $\sim X$? As $\sim X$ is true whenever X is false and vice-versa, one can easily see that the PDNF of $\sim X$ is nothing but *the sum of the remaining min terms.* Hence, the PDNF of $\sim ((p \wedge q) \rightarrow \sim q)$ is $p \wedge q$. ■

Proposition 3.1

The PDNF of $\sim X$ is the sum of all the min terms, which are not present in the PDNF of X. The PCNF of $\sim X$ is the product of all the max terms that are not present in the PCNF of X.

In the next example, we find the PCNF of a statement formula and use it to get its PDNF.

Example 3.25

Find the PDNF and PCNF of $((Q \vee \sim R) \rightarrow P) \wedge (P \longrightarrow (Q \vee R))$.

P	Q	R	$P \vee Q \vee R$	$P \vee Q \vee \sim R$	$P \vee \sim Q \vee R$	$P \vee \sim Q \vee \sim R$
T	T	T	T	T	T	T
T	T	F	T	T	T	T
T	F	T	T	T	T	T
T	F	F	T	T	T	T
F	T	T	T	T	T	F
F	T	F	T	T	F	T
F	F	T	T	F	T	T
F	F	F	F	T	T	T

P	Q	R	$\sim P \vee Q \vee R$	$\sim P \vee Q \vee \sim R$	$\sim P \vee \sim Q \vee R$	$\sim P \vee \sim Q \vee \sim R$
T	T	T	T	T	T	F
T	T	T	T	T	F	T
T	T	T	T	F	T	T
T	F	F	F	T	T	T
F	T	T	T	T	T	T
F	T	F	T	T	T	T
F	F	T	T	T	T	T
F	F	F	T	T	T	T

Now we prepare the truth table for the given statement:

P	Q	R	$\sim R$	$Q \vee \sim R$	$Q \vee R$
T	T	T	F	T	T
T	T	F	T	T	T
T	F	T	F	F	T
T	F	F	T	T	F
F	T	T	F	T	T
F	T	F	T	T	T
F	F	T	F	F	T
F	F	F	T	T	F

Find the rows where the given statement has the truth value F and examine the max terms that are true for the corresponding rows. We obtain

$$((Q \vee R) \rightarrow P) \wedge (P \rightarrow (Q \vee R)) \Leftrightarrow (\sim P \vee Q \vee R) \wedge (P \vee \sim Q \vee \sim R) \wedge (P \vee \sim Q \vee R) \wedge (P \vee Q \vee R).$$

$(Q \vee \sim R) \rightarrow P$	$P \rightarrow (Q \vee R)$	$((Q \vee \sim R) \rightarrow P) \wedge (P \rightarrow (Q \vee R))$
T	T	T
T	T	T
T	T	T
T	F	F
F	T	F
F	T	F
T	T	T
F	T	F

Now let X be the given formula $((Q \vee R) \rightarrow P) \wedge (P \rightarrow (Q \vee R))$. Its negation is the product of the remaining max terms. Hence

$$\sim X \Leftrightarrow (P \vee Q \vee \sim R) \wedge (P \vee \sim Q \vee R) \wedge (\sim P \vee Q \vee \sim R) \wedge (\sim P \vee \sim Q \vee \sim R).$$

Now,

$$
\begin{aligned}
X &\Leftrightarrow \sim(\sim X) \\
&\Leftrightarrow \sim\big((P \vee Q \vee \sim R) \wedge (P \vee \sim Q \vee R) \\
&\quad \wedge (\sim P \vee Q \vee \sim R) \wedge (\sim P \vee \sim Q \vee \sim R)\big) \\
&\Leftrightarrow (\sim P \wedge \sim Q \wedge R) \vee (\sim P \wedge Q \wedge \sim R) \\
&\quad \vee (P \wedge \sim Q \wedge R) \vee (P \wedge Q \wedge R)
\end{aligned}
$$

is the PDNF of X. ■

Example 3.26

Find the PDNF and the PCNF of $P \to (\sim Q \vee R)$.

Its truth table is constructed as follows:

P	Q	R	$\sim Q$	$\sim Q \vee R$	$P \to (\sim Q \vee R)$
T	T	T	F	T	T
T	T	F	F	F	F
T	F	T	T	T	T
T	F	F	T	T	T
F	T	T	F	T	T
F	T	F	F	F	T
F	F	T	T	T	T
F	F	F	T	T	T

P	Q	$\sim P$	$\sim Q$	$P \vee Q$	$P \vee \sim Q$	$\sim p \vee q$	$\sim p \vee \sim q$
T	T	F	F	T	T	T	F
T	F	F	T	T	T	F	T
F	T	T	F	T	F	T	T
F	F	T	T	F	T	T	T

■

3.5 PROOF TECHNIQUES

In this section we consider various methods of proving logical arguments. Inference Theory is a systematic method to prove or disprove logical inferences or equivalences without using truth tables. Automatic theorem proving is a mechanical way of proving arguments.

3.5.1 Inference Theory for Statement Calculus

Example 3.27

Examine the validity of the following argument:

Kannan either travels by car or by train. Whenever he travels by train, he reads *Readers Digest*. Kannan reads *Readers Digest* or he eats a 'samosa'. Kannan does not eat samosa and hence he travels by train.

We want to know if the above argument is logically valid. We show that it is not valid as follows:

SOLUTION I: (by Verbal Argument)

From the statements *Everyday Kannan reads Readers Digest or he eats a Samosa* and *Kannan did not eat Samosa*, we can conclude that Kannan must have read *Readers Digest* today. But from this we cannot conclude that he must have travelled by train, because we only know that he reads *Readers Digest* when he travels by train. It does not logically follow that when he reads *Readers Digest*, he travels by train. So, this argument is not valid.

This type of verbal argument may lead to errors sometimes. Formal logic is the best way is to handle these kind of problems. To use formal logic, we need to write the above argument symbolically:

P : Kannan travels by car.
Q : Kannan travels by train.
R : Kannan reads *Readers Digest*.
S : Kannan eats a 'samosa'.

Now, we have to check the validity of the implication:

$$(P \vee Q) \wedge (Q \rightarrow R) \wedge (R \vee S) \wedge \sim S \Rightarrow Q. \tag{3.7}$$

Drawing a truth table and finding set a of truth values for the four statements such that $(P \vee Q) \wedge (Q \rightarrow R) \wedge (R \vee S) \wedge \sim S$ is true but Q is false would show that the above implication is invalid. It is clear that there are $2^4 = 16$ possible truth values for these four statements involved. Instead of drawing the big truth table, we guess the set of truth values.

SOLUTION II:

Let P be true. Q be false, R be true and S be false. Now, $P \vee Q$ is true. $Q \rightarrow R$ is true, $R \vee S$ is true and $\sim S$ is true. Hence, the RHS of equation (3.6) is true while the the LHS is false. Thus, the inference is not valid.

Next, we consider an example of a valid argument and prove its validity without using a truth table. ■

Example 3.28

Show that the hypotheses $(P \rightarrow Q) \rightarrow R$, $P \wedge A$ and Q imply the conclusion R. Q is a given hypothesis. We know that $Q \rightarrow \sim P \vee Q$ and hence we have shown that $\sim P \vee Q$ is true. Now, $\sim P \vee Q \Leftrightarrow (P \rightarrow Q)$. Thus, we have proved

$$P \to Q. \tag{3.8}$$

It is given that

$$(P \to Q) \to R. \tag{3.9}$$

Now, from equations (3.8) and (3.9), R follows.

Let us note that in Example 3.28, the following implications/equivalences were used.

(i) $A \Rightarrow A \vee B$.

(ii) $(A \to B) \Leftrightarrow \sim A \vee B$.

(iii) $A \wedge (A \to B) \Rightarrow B$.

A typical logical inference involves some hypotheses or assumptions and a conclusion. If we denote the hypotheses by $H_1, H_2, \ldots, H_n$ and the conclusion by C, then the logical inference has the following form:

$$H_1 \wedge H_2 \wedge \ldots \wedge H_n \to C.$$

A derivation or a proof of this inference consists of a sequence of statement formulas

$$\begin{array}{c} A_1 \\ A_2 \\ \vdots \\ A_n \\ \vdots \\ C \end{array}$$

Note that the last line is the conclusion. Each A_i could either be a hypothesis or a statement formula, which logically follows from one or more A_j's *above* A_i.

Now, we give a complete list of basic implications and equivalences that would be used in the process of deducing a given line from the lines above:

I_1	$P \wedge Q \Rightarrow P$	Simplification
I_2	$P \wedge Q \Rightarrow Q$	Simplification
I_3	$P \Rightarrow P \vee Q$	Addition
I_4	$Q \Rightarrow P \vee Q$	Addition
I_5	$\sim P \Rightarrow P \to Q$	
I_6	$Q \Rightarrow P \to Q$	
I_7	$\sim (P \to Q) \Rightarrow P$	
I_8	$\sim (P \to Q) \Rightarrow \sim Q$	
I_9	$P, Q \Rightarrow P \wedge Q$	
I_{10}	$\sim P, P \vee Q \Rightarrow Q$	Disjunctive Syllogism
I_{11}	$P, P \to Q \Rightarrow Q$	Modus Ponens
I_{12}	$\sim Q, P \to Q \Rightarrow \sim P$	Modus Tollens
I_{13}	$P \to Q, Q \to R \Rightarrow P \to Q$	Hypothetical Syllogism

I_{14}	$P \vee Q, P \to R, Q \to R \Rightarrow R$	Dilemma
E_1	$\sim\sim P \Leftrightarrow P$	Double Negation
E_2	$P \wedge Q \Leftrightarrow Q \wedge P$	Commutative Law
E_3	$P \vee Q \Leftrightarrow Q \vee P$	Commutative Law
E_4	$P \wedge (Q \wedge R) \Leftrightarrow (P \wedge Q) \wedge R$	Associative Law
E_5	$P \wedge (Q \wedge R) \Leftrightarrow (P \wedge Q) \wedge R$	Associative Law
E_6	$P \vee (Q \wedge R) \Leftrightarrow (P \vee Q) \wedge (P \vee R)$	Distributive Law
E_7	$(P \wedge (Q \vee R) \Leftrightarrow (P \wedge Q) \vee (P \wedge R)$	Distributive Law
E_8	$\sim P(\wedge Q) \Leftrightarrow \sim P \vee \sim Q$	De Morgan's Law
E_9	$\sim P(\vee Q) \Leftrightarrow \sim P \wedge \sim Q$	
E_{10}	$P \vee P \Leftrightarrow P$	
E_{11}	$P \wedge P \Leftrightarrow P$	
E_{12}	$R \vee (P \wedge \sim P) \Leftrightarrow R$	
E_{13}	$R \wedge (P \vee \sim P) \Leftrightarrow R$	
E_{14}	$R \vee (P \vee \sim P) \Leftrightarrow \mathbf{T}$	
E_{15}	$R \wedge (P \wedge \sim P) \Leftrightarrow \mathbf{F}$	
E_{16}	$P \to Q \Leftrightarrow \sim P \vee Q$	
E_{17}	$\sim (P \to Q) \Leftrightarrow P \wedge \sim Q$	
E_{18}	$P \to Q \Leftrightarrow \sim Q \to \sim P$	
E_{19}	$P \to (Q \to R) \Leftrightarrow (P \wedge Q) \to R$	
E_{20}	$\sim (P \leftrightarrow Q) \Leftrightarrow P \leftrightarrow \sim Q$	
E_{21}	$P \leftrightarrow Q \Leftrightarrow (P \to Q) \wedge (Q \to P$	
E_{22}	$P \leftrightarrow Q \Leftrightarrow (P \wedge Q) \vee (\sim P \wedge \sim Q)$	

We illustrate the systematic method of inference by an example. ■

Example 3.29

Show the validity of the following arguments: $B \wedge C$, $(B \leftrightarrow C) \to H \vee G \Rightarrow G \vee H$.

SOLUTION

$\{1\}$	(1)	$B \wedge C$	Rule **P**
$\{1\}$	(2)	B	Rule **T**, (1), I_1
$\{1\}$	(3)	$C \to B$	Rule **T**, (2), I_6
$\{1\}$	(4)	C	Rule **T**, (1), I_2
$\{1\}$	(5)	$B \to C$	Rule **T**, (4), I_6
$\{1\}$	(6)	$B \leftrightarrow C$	Rule **T**, (3), (5), E_{21}
$\{7\}$	(7)	$(B \leftrightarrow C) \to H \vee G$	Rule **P**
$\{1,7\}$	(8)	$H \vee G$	Rule **T**, (6), (7), I_{11}
$\{1,7\}$	(9)	$G \vee H$	Rule **T**, (8) E_3

In inference theory, we learn to write a systematic proof of a statement assuming that certain statements are true. A statement, which is assumed to be true, is called a hypothesis. A proof is a vertical coloumn of statements in which every statement is either a hypothesis, or a statement that has been logically proved using hypotheses and applying inference rules given earlier. The right-most coloumn explains how the current statement has been arrived at. If it is a hypothesis, then the right most entry declares that 'Rule P' has been applied. Rule P is just the rule that a given hypothesis can be used in the proof.

When the current statement is a new statement and not a part of the hypotheses, the right-most entry indicates the previous statements used in arriving at the current statement. It also indicates the inference rules applied. Further, it says that Rule T has been used. Rule T is just a rule that the statement which follows from a hypothesis or a previously proved statement can be used in a proof.

Each line is assigned a serial number in the second coloumn. This number is used in the right-most coloumn. Further, the first coloumn indicates only the hypothesis which is used in arriving at the current statement. ■

Example 3.30

$\sim J \to (M \vee N)$, $(H \vee G) \to \sim J$, $H \vee G \Rightarrow M \vee N$.

SOLUTION

$\{1\}$	(1)	$H \vee G$	Rule **P**
$\{2\}$	(2)	$(H \vee G) \to \sim J$	Rule **P**
$\{1,2\}$	(3)	$\sim J$	Rule **T**, (1), (2), I_{11}
$\{4\}$	(4)	$\sim J \to (M \vee N)$	Rule **P**
$\{1,2,4\}$	(5)	$M \vee N$	Rule **T**, (3), (4),I_{11}

■

Example 3.31

Show that Q is a valid conclusion from the premises $\sim S$, $P \vee (Q \wedge R)$, $P \leftrightarrow S$.

SOLUTION

$\{1\}$	(1)	$\sim S$	Rule **P**
$\{2\}$	(2)	$P \leftrightarrow S$	Rule **P**
$\{2\}$	(3)	$P \to S \wedge S \to P$	Rule **T** ,(2), E_{21}
$\{2\}$	(4)	$P \to S$	Rule **T**, (3),I_1
$\{2\}$	(5)	$\sim S \to \sim P$	Rule **T**,(4), E_{18}
$\{1,2\}$	(6)	$\sim P$	Rule **T**, (1),(5) I_{11}
$\{7\}$	(7)	$P \vee (Q \wedge R) \to B$	Rule **P**
$\{1,2,7\}$	(8)	$Q \wedge R$	Rule **T** (6),(7), I_{10}
$\{1,2,7\}$	(9)	Q	Rule **T** (8) I_1

■

PRACTICE 3.18

Show that $S \to P$, $S \wedge R$, $P \wedge \sim Q \wedge R$ imply $Q \to P \vee (Q \wedge \wedge - \sim S)$.

PRACTICE 3.19

Examine the validity of the following argument after writing the sentences in symbolic form: Jayanthi watches Moon TV and she drinks tea. When Jayanthi drinks tea, she is happy. When Jayanthi watches Moon TV, she is angry. Jayanthi is not happy and hence she is angry.

Proof by Contradiction: Sometimes, it is convenient to prove an inference by contradiction. In this procedure the conclusion is assumed to be false. It is shown that this assumption leads to a contradiction. This method is also known as *the method of indirect proof or reductio ad absurdum*. Formally, this method can be described as follows: To prove,

$$H_1 \wedge H_2 \wedge \ldots \wedge H_n \Rightarrow C,$$

we show that

$$\sim C \Rightarrow H_1 \wedge H_2 \wedge \ldots \wedge H_n$$

is false. ■

Example 3.32

Show by contradiction that Q is a valid conclusion from the hypotheses $P \vee Q$, $P \to R$, $S \wedge \sim R$.

SOLUTION

We need to show that the hypotheses $\sim Q$, $P \vee Q$, $P \to R$, $S \wedge \sim R$ lead to a contradiction.

$\{1\}$	(1) $S \wedge \sim R$	Rule P	
$\{1\}$	(2) $\sim R$	Rule T (1), I_2	
$\{3\}$	(3) $P \to R$	Rule P	
$\{3\}$	(4) $\sim R \to \sim P$	Rule T (3), E_{18}	
$\{1, 3\}$	(5)	$\sim P$	Rule T (2),(4), I_{11}
$\{6\}$	(6)	$\sim Q$	Rule P
$\{7\}$	(7)	$P \vee Q$	Rule P
$\{1, 3, 7\}$	(8)	Q	Rule T,(5) (7) I_{1C}
$\{1, 3, 6, 7\}$	(9)	$Q \wedge \sim Q$	Rule T, (6),(7) I_9

■

3.6 AUTOMATIC THEOREM PROVING

We have seen how to prove logical inferences and equivalences using a list of basic inferences and equivalences. The proof involves some thinking to decide on which hypothesis to use, and at what stage. Only after some trial and error, we arrive at the final sequence of deductions. Is there a straightforward method to prove these inferences? Automatic theorem proving is such a method.

Definition 3.6: A string formula is defined as below:

(a) An empty string is a string of formulas.
(b) Any statement formula is a string of formulas.
(c) If α and β are strings of formula, then α, β and β, α are strings of formula.

Definition 3.7: $\alpha \xrightarrow{s} \beta$ is called a sequent. α is known as an antecedent and β is known as a consequent. Let $\alpha = A_1, A_2, ..., A_m$ and $\beta = B_1, B_2, ..., B_n$. The sequent $\alpha \xrightarrow{s} \beta$ is true if and only if

$$A_1 \wedge A_2 \wedge \cdots \wedge A_m \longrightarrow B_1 \vee B_2 \vee \cdots \vee B_n$$

is true.

It is clear that the inference

$$P \wedge A_1 \wedge \cdots \wedge A_m \longrightarrow P \vee B_1 \vee B_2 \vee \cdots \vee B_m$$

is valid. In other words, the sequent $\alpha \xrightarrow{s} \beta$ is valid if there is a statement formula common to the strings α and β.

Automatic theorem proving involves simplification of the given inference into a sequent $A_1, A_2, ..., A_m \xrightarrow{s} B_1, B_2, ..., B_n$. After simplification, if there is a common statement on both the sides, the inference is valid. This simplification is just the process of getting rid of the connectives. We have rules that help us do this job. We give the proof of two of these rules for illustration and other rules are only stated.

The following rule tells us how to get rid of the connective $\sim$ from the antecedent part of a sequent:

$$\text{If } X, \alpha \overset{s}{\Rightarrow} \beta, \gamma \text{ then } \alpha \overset{s}{\Rightarrow} \sim X, \beta, \gamma. \tag{3.10}$$

This rule tells us that if we want to prove $\alpha \overset{s}{\Rightarrow} \sim X, \beta, \gamma$, then we may prove $X, \alpha \overset{s}{\Rightarrow} \beta, \gamma$. The latter sequent does not involve $\sim$. Other rules tell us how to get rid of connectives in a similar fashion. Equation (3.10) is called 'Rule $\Rightarrow \sim$'. We shall now prove it.

Proof of equation (3.10): This rule is simply

$$X \wedge A_1 \wedge A_2 \wedge \cdots \wedge A_m \rightarrow B_1 \vee \cdots \vee B_n \Rightarrow A_1 \wedge A_2 \wedge \cdots \wedge A_m \rightarrow \sim X \vee B_1 \vee B_2 \vee \cdots \vee B_n.$$

Let

$$X \wedge A_1 \wedge A_2 \wedge \cdots \wedge A_m \rightarrow B_1 \vee \cdots \vee B_n \tag{3.11}$$

hold. We need to prove

$$A_1 \wedge A_2 \wedge \cdots \wedge A_m \rightarrow \sim X \vee B_1 \vee B_2 \vee \cdots \vee B_n. \tag{3.12}$$

Assume that $A_1 \wedge A_2 \wedge \cdots \wedge A_m$ is true. If X is true, then $X \wedge A_1 \wedge A_2 \wedge \cdots \wedge A_m$ is true and hence by equation (3.11), $B_1 \vee \cdots \vee B_n$ is true. This in turn implies that $\sim X \vee B_1 \vee \cdots \vee B_n$ is true.

If X is false, $X \wedge A_1 \wedge \cdots \wedge A_m$ is false, and hence the implication equation (3.12) is true.

We next prove the rule

$$\text{If } X, \alpha, \beta \overset{s}{\Rightarrow} \gamma \text{ and } Y, \alpha, \beta \overset{s}{\Rightarrow} \gamma \text{ then } (X \vee Y), \alpha, \beta \overset{s}{\Rightarrow} \gamma. \tag{3.13}$$

This rule is called 'Rule $\vee \Rightarrow$'. ■

Proof of equation (3.13): We need to prove

$$(X \wedge A_1 \wedge A_2 \wedge \ldots A_m \rightarrow B_1 \vee \cdots \vee B_n)$$
$$\wedge(Y \wedge A_1 \wedge A_2 \wedge \ldots A_m \rightarrow B_1 \vee \cdots \vee B_n)$$
$$\Rightarrow (X \vee Y) \wedge (A_1 \wedge A_2 \wedge \ldots A_m) \rightarrow B_1 \vee \cdots \vee B_n.$$

The above follows from

$$((X \wedge P) \rightarrow Q) \wedge ((Y \wedge P) \rightarrow Q) \Rightarrow ((X \vee Y) \wedge P) \rightarrow Q.$$

The following rules tell us how to eliminate connectives from the consequent:

Rule $\Rightarrow\sim$ If $\alpha, X \overset{s}{\Rightarrow} \beta, \gamma$ then $\alpha \overset{s}{\Rightarrow} \beta, \sim X, \gamma$.
Rule $\Rightarrow \vee$ $\alpha \overset{s}{\Rightarrow} \beta, X, Y, \gamma$ then $\alpha \overset{s}{\Rightarrow}, X \vee Y, \beta\gamma$.
Rule $\Rightarrow \wedge$ $\alpha \overset{s}{\Rightarrow} \beta, X\gamma$ and $\alpha \overset{s}{\Rightarrow} \beta, X\gamma$ then $\alpha \overset{s}{\Rightarrow}, X \wedge Y, \beta\gamma$.
Rule $\Rightarrow\rightarrow$ $\alpha, X \overset{s}{\Rightarrow} \beta, X \rightarrow YY, \gamma$ then $\alpha \overset{s}{\Rightarrow} \beta, X \rightarrow Y, \gamma$.
Rule $\Rightarrow\leftrightarrow$ $\alpha, X \overset{s}{\Rightarrow} \beta, X \rightarrow YY, \beta\ \gamma$ and $\alpha, Y \overset{s}{\Rightarrow} X, \beta\gamma$ then $\alpha \overset{s}{\Rightarrow} \beta, X \leftrightarrow Y, \gamma$.

The following rules tell us how to eliminate connectives from the antecedent:

Rule $\sim\Rightarrow$ If $\alpha, \beta \overset{s}{\Rightarrow} X, \gamma$ then $\alpha, \sim X, \beta \overset{s}{\Rightarrow} \gamma$.
Rule $\vee \Rightarrow$ $X, \alpha, \beta \overset{s}{\Rightarrow} \gamma$ and $Y, \alpha, \beta \overset{s}{\Rightarrow} \gamma$ then $\alpha, X \vee Y, \beta \overset{s}{\Rightarrow} \gamma$.
Rule $\wedge \Rightarrow$ $\alpha, X, Y, \beta \overset{s}{\Rightarrow} \beta, X\gamma$ then $\alpha, X \wedge Y, \beta \overset{s}{\Rightarrow} \gamma$.
Rule $\rightarrow\Rightarrow$ $\alpha, Y, \beta \overset{s}{\Rightarrow} X\gamma$ then $\alpha, X \rightarrow Y, \beta \overset{s}{\Rightarrow} \gamma$.
Rule$\leftrightarrow\Rightarrow$ X,Y $\alpha, \beta \overset{s}{\Rightarrow} \beta, X \rightarrow Y\gamma$ and $\alpha, \beta \overset{s}{\Rightarrow} X, Y, \gamma$ then $\alpha X \leftrightarrow Y, \beta \overset{s}{\Rightarrow}, \gamma$. ■

Example 3.33

Show

$$P \Rightarrow (\sim P \rightarrow Q)$$

by automatic theorem proving.

SOLUTION

We need to prove $\overset{s}{\Rightarrow} P \Rightarrow (\sim P \rightarrow Q)$.

	$\overset{s}{\Rightarrow} P \Rightarrow (\sim P \rightarrow Q)$	
if	$P \overset{s}{\Rightarrow} (\sim P \rightarrow Q)$	Rule $\Rightarrow\rightarrow$
if	$P, \sim P \overset{s}{\Rightarrow} Q$	Rule $\Rightarrow\rightarrow$
if	$P \overset{s}{\Rightarrow} P, Q$	Rule $\sim\Rightarrow$

■

PRACTICE 3.20

Prove by automatic theorem proving: $R \Rightarrow (P \vee \sim P \vee Q)$.

PRACTICE 3.21

Prove by automatic theorem proving: $p \wedge (q \vee r) = (p \wedge q) \vee (p \wedge r)$.

3.7 PREDICATE CALCULUS

3.7.1 Introduction

Consider the sentence:

$$\text{Any engineer can repair a faulty plug point.} \tag{3.14}$$

This sentence tells us that if someone is an engineer, then that person can repair a faulty plug point. This sentence encompasses millions of statements like

(i) Sultana is an engineer → Sultana can repair a faulty plug point.
(ii) Kannan is an engineer → Kannan can repair a faulty plug point.

Let us consider the following statements:
P_1: Sultana is an engineer.
Q_1: Sultana can repair a faulty a plug point.
P_2: Kannan is an engineer.
Q_1: Kannan can repair a faulty a plug point.

Statement (3.14) includes the truth of $P_1 \rightarrow Q_1$, $P_2 \rightarrow Q_2$, etc.

If we want to write statement (3.14) in a symbolic manner, it is clear that we will have to introduce an *unknown*.

Consider the sentences:

$P(x)$: x is an engineer.

$Q(x)$: x can repair a faulty plug point.

Finally, we write statement (3.14) as

$$P(x) \rightarrow Q(x).$$

If we omit x from $P(x)$, we get the following portion of a sentence: 'is an engineer.' *is an engineer* and *can repair a faulty plug point* are what are known as *predicates*.

3.7.2 Statement Functions

We know what we mean by a statement formula. For example,

(Mani is a Physicist or whenever Ekalvya gets good marks Dhronacharya scolds him.) is a statement formula.

If

P	:	Mani is a physicist
Q	:	Ekalvya gets good marks
R	:	Dhronacharya scolds Ekalvya

then the above is nothing but

$$P \vee (Q \rightarrow R).$$

Now let us consider the following sentence:

x **is a Physicist or whenever** y **gets good marks** z **scolds him.**

Now, this is certainly not a statement. Unless we know who these x, y or z are, we cannot assign a truth value. But when substituting for x, y and z some known individuals we get a statement.

Let us consider some more examples. 3 **is a factor of** 312 is a statement. But

x **is a factor of** y

is not a statement. But when we take $x = 5$ and $y = 56$, we get the following statement:

$$5 \text{ is a factor of } 56. \tag{3.15}$$

This is false; nevertheless, (3.15) is a statement.

When we substitute $x = 100$ and $y = 10{,}000$, we get the following true statement:

100 **is a factor of** 10,000.

Roughly speaking, a statement function is a sentence involving *variables* or *unknowns*. Such a sentence might involve one, two or more variables. When variables are assigned specific values, these sentences become statements. The variables must assume values from a specific set D. Thus, these sentences define a function whose domain is D and the co-domain is a set of statements.

A statement function of one variable is denoted by $H(x)$. A statement function of two variables is denoted by $H(x, y)$ and so on. $H(x_1, x_2, ..., x_n)$ is a statement function of n variables. The formulation of and analysis of the truth values of statement functions is called predicate calculus.

Example 3.34

Consider:

x is the chief minister of Bihar.

Obviously, it is meaningless to assign an integer value for x. Hence, the domain for the above statement function must be a subset of human beings. ■

3.7.3 Quantifiers, Free and Bound Variables

Consider the following two statements:

(a) All the students of Krishna College of Engineering are wearing green shirts.
(b) There is at least one student of Krishna College of Engineering who is wearing a green shirt.

Next, consider the statement function:

$P(x)$: x is wearing a green shirt.

Let S be the set of all students of Krishna College of Engineering. The above statements can now be rewritten as

(a) For all $x \in S$, $P(x)$.
(b) There is $x \in S$ such that $P(x)$.

If the domain for a particular context is clear, we omit to mention it explicitly.

We use $\forall x$ to denote for all x and $\exists x$ to denote there exists x. These symbols $\forall$ and $\exists$ are known as *quantifiers*.

Example 3.35

Consider the following statement function x is a prime $\rightarrow$ x is odd.

The addition of a quantifier makes it a statement. For example, the following is a statement, provided the domain is specified:

$$\forall x,\ x \text{ is a prime} \rightarrow x \text{ is odd.}$$

If we take the domain as the set of all natural numbers, N, then the above statement is false. The reason is that $2 \in N$, which is a prime and which is not odd.

But if the domain is taken as $\{10, 11, 12, \ldots\}$, then the above statement is true. ■

Example 3.36

Consider $\exists x$, x is a woman $\wedge$ x had been a chief minister.

The above statement is true if the domain is taken as the set of all Indians. We can take x as Ms Rabri Devi, Ms Jayalalitha, Smt Vasundhara Raje Scindia or Ms Sheila Dixit.

Is it true if the domain is taken as a set of all Keralites? ■

Example 3.37

Consider

$$\forall x,\ x \text{ has been a prime minister.} \tag{3.16}$$

Let $A = \{$Nehru, Indira Gandhi, V. P. Singh, Deve Gowda$\}$ and $B = \{$Nehru, Indira Gandhi, V. P. Singh, Deve Gowda, Arjun Singh$\}$. If the domain is taken as A, then statement (3.16) is true but statement (3.16) is false if B is taken as the domain.

To be very specific,

$$\forall x \in A,\ x \text{ has been a prime minister} \tag{3.17}$$

is a true statement and

$$\forall x \in B,\ x \text{ has been a prime minister} \tag{3.18}$$

is a false statement. The reason being that Mr Arjun Singh has not been a prime minister.

Now, since statement (3.18) is false, its negation

$$\sim (\forall x \in B,\ x \text{ has been a prime minister}) \tag{3.19}$$

must be true. It is tempting to conclude that statement (3.19) is equivalent to

$$\forall x \in B,\ x \text{ has not been a prime minister.} \tag{3.20}$$

But obviously statement (3.20) is false. Because, for $x = V.P.Singh \in B$, x *had not been a prime minister* is false. Statement (3.18) is false not because no one in B had been a prime minister but because *there is someone* in B who has not been a prime minister.

In short, we conclude that the negation of

$$\forall x, P(x)$$

is

$$\exists x, \sim (P(x)).$$ ■

PRACTICE 3.22

What is the negation of $\exists x, P(x)$?

Example 3.38

Consider the statements

$$\exists x\ \forall y\ P(x, y) \tag{3.21}$$

and

$$\forall y\ \exists x\ P(x, y). \tag{3.22}$$

$P(x, y)$ is a statement function of two variables. Hence, its domain must be a cartesian product $(A \times B)$.

How are statements (3.21) and (3.22) related to each other?

Let us examine these statements for various domains:

Let $D_1 = \{1, 2, 3, 4, 5\} \times \{1, 2, 3, 4, 5\}$ and $D_2 = N \times N$.

We take $P(x, y)$ as $x \geq y$. Since $5 \geq y$ for all $y \in \{1, 2, 3, 4, 5\}$, statement (3.21) is true with domain as D_1.

If we take the domain as D_2, then to make statement (3.21) true, we must find $x\ inN$ such that $x \geq y$ for all $y \in N$. We cannot find such an x. Hence, statement (3.21) is false with domain D_2.

Next, we consider statement (3.22). To make statement (3.22) true, for each y we must find x with $x \geq y$. It is clear that statement (3.22) is true for both the domains because of the following:

With D_1, $x = 5$ can be chosen for each y.

With D_2, for each y, we can take $x = y$.

It is easy to observe that irrespective of the domain, if statement (3.21) is true then statement (3.22) is true. In other words,

$$\exists x\, \forall y\, P(x, y) \Rightarrow \forall y\, \exists x\, P(x, y).$$

Its converse is false. ■

Definition 3.8: The part of a statement function to which a quantifier refers to is called *the scope of the quantifier.*

Example 3.39

Consider

$$\forall z(\exists y(P(x, y, z)) \vee \forall x(Q(x, y, z))).$$

Here the scope of $\forall x$ is $Q(x, y, z)$ and the scope of $\exists y$ is $P(x, y, z)$ and the scope of $\forall z$ is $\exists y(P(x, y, z)) \vee \forall x(Q(x, y, z))$. ■

3.8 INFERENCE THEORY FOR PREDICATE CALCULUS

In the previous section, we noted that the inference

$$\exists x\, \forall y\, P(x, y) \Rightarrow \forall y\, \exists x\, P(x, y)$$

is valid and that equivalence

$$\sim (\forall x P(x)) \Leftrightarrow \exists x(\sim (P(x))$$

is true.

In this section, we shall see how such inferences and equivalences can be proved in a systematic procedure. The methods of proof are similar to what we have done in the inference theory for statement calculus. But we must have extra rules involving the quantifiers. These rules are stated after we discuss some examples of descriptive proofs.

Let A and B be predicate formulas. Let $H(x_1, x_2, ..., x_n)$ be a statement function of n variables. Now, if one of the variables, say x_2, is replaced by a constant like a, we obtain another statement function $H(x_1, a, ..., x_n)$. Of course, this new statement function is a statement function of $n - 1$ variables. Now, $H(a_1, a_2, ..., a_n)$ is a simple statement.

Q, $P(x, y)$, $P(a, x)$, $R(x, y, c)$, $H(x_1, a, ..., x_n)$ are all examples of atomic formula of predicate calculus.

PRACTICE 3.23

Define the idea of a well-formed formula for predicate calculus.

An inference or an equivalence of predicate calculus is valid provided it is valid *for all possible domains and for all possible interpretations assigned to the statement functions involved.* But to disprove an inference or an equivalence, it is enough to exhibit one domain over which it is invalid.

Example 3.40

Prove: $\exists x P(x) \to \forall x Q(x) \Rightarrow \forall x(P(x) \to Q(x))$.

SOLUTION

Case (i): $P(x)$ is false for all x. Then,

$$\forall x(P(x) \to Q(x)).$$

Case (ii): There is a such that $P(a)$ is true. Then, from the hypothesis, $Q(x)$ is true for all x. So, $\forall x(P(x) \to Q(x))$. ■

Example 3.41

Proof by Contradiction

Show that:

$$\forall x(P(x) \vee Q(x)) \Rightarrow \forall x P(x) \vee \exists x Q(x).$$

SOLUTION

Suppose that the conclusion is false. Then, we have

$$\begin{aligned} &\sim (\forall x P(x) \vee (\exists x Q(x))) \\ &\Leftrightarrow \sim (\forall x P(x)) \wedge \sim (\exists x Q(x)) \\ &\Leftrightarrow \exists x(\sim P(x)) \wedge (\forall x \sim Q(x)). \end{aligned}$$

In particular, for some a, $\sim P(a)$ holds. Since $Q(x)$ is false for all x, $\sim Q(a)$ is also true. Thus, $\sim P(a) \wedge \sim Q(a)$ is true, which is equivalent to $\sim (P(a) \vee Q(a))$. Now, this is a contradiction to the hypothesis and hence the proof is complete. ■

Example 3.42

Show that:

$$\exists x(P(x) \vee Q(x)) \Leftrightarrow \exists P(x) \vee \exists Q(x).$$

Proof.

Assume that $\exists(P(x) \vee Q(x))$. There is one element, say a, for which $P(a) \vee Q(a)$ holds. This implies that $P(a)$ is true or $Q(a)$ is true. If $P(a)$ is true, then $\exists x P(x)$ is true and if $Q(a)$ is true, then $\exists x Q(x)$ is also true. So, $\exists P(x) \vee \exists Q(x)$.

Next, we assume that $\exists P(x) \vee \exists Q(x)$ is true. This implies that either $\exists x P(x)$ is true or $\exists x Q(x)$ is true. Then, for some a, $P(a)$ is true or for some b, $Q(b)$ is true. Hence, $P(a) \vee Q(a)$ is true or $P(b) \vee Q(b)$ is true. Thus, there exists x such that $P(x) \vee Q(x)$ is true. ■

PRACTICE 3.24

Show that

$$\forall x(P(x) \wedge Q(x)) \Leftrightarrow \forall x P(x) \wedge \forall x Q(x).$$

Example 3.43

Show that:

$$\exists x(P(x) \wedge Q(x)) \Rightarrow \exists x P(x) \wedge \exists x Q(x).$$

Further, show that:

$$\exists x P(x) \wedge \exists x Q(x) \Rightarrow \exists x(P(x) \wedge Q(x))$$

is not valid.

SOLUTION

From $\exists x(P(x) \wedge Q(x))$, we deduce that for some a, $P(a) \wedge Q(a)$ is true. Hence, $P(a)$ is true and $Q(a)$ is true. That is, $\exists x P(x)$ is true and $\exists x Q(x)$ is true. Hence, $\exists P(x) \wedge \exists Q(x)$.

Next, we disprove the converse as follows:

$P(x)$ is the statement that x is even. $Q(x)$ is the statement that x is odd. Let N be the domain. It is clear that $\exists P(x) \wedge \exists Q(x)$ is true. Under this interpretation, $\exists x(P(x) \wedge Q(x))$ means that there is a natural number, which is both odd and even. This is clearly not valid. Hence, the converse implication is not true. ■

PRACTICE 3.25

Show that:

$$\forall x P(x) \vee \forall x Q(x) \Rightarrow \forall x(P(x) \vee Q(x))$$

is valid and the converse is not valid.

Rules of Inference in Predicate Calculus: If every one is wearing a green shirt in a class, it is obvious that an arbitrary person in the class is wearing a green shirt. That is, from the hypothesis $\forall x P(x)$ we can derive $P(y)$. Here y could be a constant or a variable. This rule is known as *Universal Instantiation*. We denote it by **UI**.

Example 3.44

Consider:

$$\forall x \; \exists y \; x < y$$

which is valid for the domain $N \times N$. Now an application **UI** gives us

$$\exists y \; y < y$$

which is false. The above is an example of a wrong use of **UI**. We must use it under the following restriction:

If y is a variable, it must not fall within the scope of a quantifier for y. ■

Example 3.45

Let $\exists x P(x)$ be true. Let us denote the particular value of x, for which $P(x)$ by a. Then $P(a)$ is valid. Thus from the hypothesis $\exists P(x)$, $P(a)$ is a valid conclusion. This rule is known as Existential Instantization.

The following example illustrates a wrong use of this rule:

Let us imagine that in a class there is someone wearing a green shirt. Further, let us also imagine that someone in the class who is wearing a cap. There is no logical reason to believe that the same person is wearing both the green shirt and the cap.

So, if we conclude $P(a)$ from $\exists P(x)$, then we must not conclude $Q(a)$ from $\exists Q(x)$. We have to use a different symbol, say b and can conclude $Q(b)$.

Next, we tabulate the inference rules and restrictions for their use:

Hypothesis	Conclusion	Name of the rule	Restrictions on use
$\forall x P(x)$	$P(y)$	Universal Instantiation (**UI**)	If y is a variable it must not fall within the scope of a quantifier for y
$\exists x P(x)$	$P(a)$ where a is a constant	Existential Instantiation (**EI**)	Must be the first time this rule is used to introduce a
$P(x)$	$\forall x P(x)$	Universal Generalization (**UG**)	$P(x)$ has not been deduced from any hypothesis in which x is free. P(x) should not have been deduced by **EI** from any wff in which x is free
$P(x)$ or $P(a)$	$\exists x P(x)$	Existential Generalization (**EG**)	To go from $P(a)$ to $\exists P(x)$, x must not appear in $P(a)$

■

PRACTICE 3.26

Illustrate why the restrictions for **EG** and **UG** are necessary.

Example 3.46

Prove: $\forall x(P(x) \rightarrow Q(x)) \Rightarrow \forall x P(x) \rightarrow \forall x Q(x)$. Let $P(x)$ be true for all x.

Proof.
From the hypothesis,

$\{1\}$	(1)	$\forall x P(x)$	Temporary Hypothesis
$\{1\}$	(2)	$P(a)$	1, **US**
$\{3\}$	(3)	$\forall x(P(x) \to Q(x))$	**P**
$\{3\}$	(4)	$P(a) \to Q(a)$	3, **US**
$\{1, 3\}$	(5)	Q(a)	2, 4, Modus Ponens
$\{1, 3\}$	(6)	$\forall x Q(x)$	5, **UG**
$\{1, 3\}$	(7)	$\forall x P(x) \to \forall x Q(x)$	1, 6, **CP**

■

Example 3.47

Show that:
$\forall x P(x) \wedge \exists Q(x) \Rightarrow \exists(P(x) \wedge Q(x))$.

Proof.

$\{1\}$	(1)	$\forall x P(x) \wedge \exists Q(x)$	**P**
$\{1\}$	(2)	$\forall x P(x)$	1, Simplification
$\{1\}$	(3)	$P(a)$	2, **US**
$\{1\}$	(4)	$\exists q(x)$	1, Simplification
$\{1\}$	(5)	$Q(a)$	4, bf ES
$\{1\}$	(6)	$P(a) \wedge Q(a)$	3, 5
$\{1\}$	(7)	$\exists x(P(x) \wedge Q(x))$	EG

■

Example 3.48

Show that:

$$\forall x(P(x) \to Q(x)) \wedge \exists(P(x) \wedge R(x)) \to \exists x(Q(x) \wedge R(x).$$

(1)	$\forall x(P(x) \to Q(x))$	**P**, Simplification.
(2)	$P(a) \to Q(a)$	1 **US**
(3)	$\exists(P(x) \wedge R(x))$	**P**, Simplification.
(4)	$P(a) \wedge R(a)$	3, **ES**
(5)	$P(a)$	(4) Simplification
(6)	$Q(a)$	2, 5, Modus Ponens
(7)	$R(a)$	(4), Simplification
(8)	$Q(a) \wedge R(a)$	6, 7
(9)	$\exists x(Q(x) \wedge R(x))$	8, **EG**

■

PRACTICE 3.27

Show that the conclusion $\forall x(F(x) \rightarrow \sim S(x))$ follows from the hypotheses $\exists x(F(x) \wedge S(x)) \rightarrow \forall y(M(y) \rightarrow W(y))$ and $\exists y(M(y) \wedge \sim W(y))$.

Example 3.49

Show that:

$$\exists x P(x) \rightarrow \forall x((P(x) \vee Q(x)) \rightarrow R(x)), \exists x P(x), \exists Q(x)$$

$$\Rightarrow \exists x \exists y(R(x) \wedge R(y)).$$

$\{1\}$	(1)	$\exists x P(x)$	**P**
$\{1\}$	(2)	$P(a)$	1, **ES**
$\{3\}$	(3)	$\exists x P(x) \rightarrow \forall x((P(x) \vee Q(x)) \rightarrow R(x))$	**P**
$\{1,3\}$	(4)	$\forall x((P(x) \vee Q(x)) \rightarrow R(x))$	1, 3 Modus Ponens
$\{1,3\}$	(5)	$P(a) \vee Q(a) \rightarrow R(a)$	4, US
$\{1\}$	(6)	$P(a) \vee Q(a)$	2, Add
$\{1,3\}$	(7)	$R(a)$	5, 6, Modus Ponens
$\{8\}$	(8)	$\exists x Q(x)$	**P**
$\{8\}$	(9)	$Q(b)$	8, **ES**
$\{8\}$	(10)	$P(b) \vee Q(b)$	9, Add
$\{1,3\}$	(11)	$P(b) \vee Q(b) \rightarrow R(b)$	4, **US**
$\{1,3,8\}$	(12)	$R(b)$	10, 11 Modus Ponens
$\{1,3,8\}$	(13)	$R(b) \wedge R(a)$	7, 12
$\{1,3,8\}$	(14)	$\exists y(R(b) \wedge R(y))$	13, **EG**
$\{1,3,8\}$	(14)	$\exists x \exists y(R(x) \wedge R(y))$	14, **EG** ■

Example 3.50

Examine the validity of the argument:

It is not true that everyone knows C++. If someone does not know C++, that person likes music. A person who likes music is happy, and so there is someone who is happy.

SOLUTION

We shall write these statements symbolically.
$P(x)$: x knows C++.
$Q(x)$: x likes music.
$R(x)$: x is happy.

We have to examine the validity of

$$\sim(\forall P(x)), \forall x(\sim P(x) \rightarrow Q(x)), \forall x(Q(x) \rightarrow R(x)) \Rightarrow \exists x R(x).$$

We prove it as follows:

$\{1\}$	(1)	$\sim(\forall P(x))$	**P**
$\{1\}$	(2)	$\exists \sim P(x)$	1
$\{1\}$	(3)	$\sim P(a)$	2, **ES**
$\{4\}$	(4)	$\forall(\sim P(x) \rightarrow Q(x))$	**P**
$\{4\}$	(5)	$\sim P(a) \rightarrow Q(a)$	4, **US**
$\{1, 4\}$	(6)	$Q(a)$	3, 5, Modus Ponens
$\{7\}$	(7)	$\forall x(Q(x) \rightarrow R(x))$	**P**
$\{7\}$	(8)	$Q(a) \rightarrow R(a)$	7, **US**
$\{1, 4, 7\}$	(9)	$R(a)$	6, 8, Modus Ponens
$\{1, 4, 7\}$	(10)	$\exists R(x)$	9, **EG**

■

PRACTICE 3.28

Show that the hypotheses $\forall x(P(x) \rightarrow (Q(y) \wedge R(x)))$, $\exists x P(x)$ imply that $Q(y) \wedge \exists x(P(x) \wedge R(x))$.

Example 3.51

Show by an interpretation that

$$\exists A(x) \wedge \exists B(x) \Rightarrow \exists(A(x) \wedge B(x))$$

is invalid.

SOLUTION

Consider the domain $S = \{6, 7, 8, 9\}$.

Let $P(x)$ be the statement '*x is a prime*' and $Q(x)$ be the statement '*x is even*'. Since $P(7)$ is true and $Q(6)$ is true, we have the following true statements:

$$\exists A(x) \wedge \exists B(x).$$

But obviously, there is no prime that is also even in S. Hence

$$\exists(A(x) \wedge B(x)).$$

is false. ■

SUMMARY

- Statements are either true or false.
- Complex statements can be obtained by combining atomic statements with the help of logical connectives.
- Tautology is a statement that is true, irrespective of the truth values of the atomic statements constituting it.

SUMMARY (Continued)

- Contradiction is a statement that is false irrespective of the truth values of the atomic statements constituting it.
- Disjunctive normal form is a sum of elementary products.
- Conjunctive normal form is a product of elementary sums.
- A principal disjunctive normal form is a sum of min terms.
- A principal conjunctive normal form is a product of max terms.
- Logical implications can be proved or disproved by the methods of truth table, inference theory or automatic theorem proving.
- A statement function is an assertion involving variables. Its truth value depends on the domain of its application.
- The phrases *for all* and *there exists* are called quantifiers, which are denoted by $\forall$ and $\exists$, respectively.
- The validity of logical implications of predicate calculus can be proved by the method of inference theory for predicate calculus.

PROBLEMS

1. Identify statements among the following sentences:
   (i) I am an honest person.
   (ii) Mr Bill Gates weighs 70 kg.
   (iii) z is a complex number.
   (iv) $\sqrt{2}$ is rational.
   (v) $\sqrt{25}$ is rational.
   (vi) Everest is 7000 m high and Greenland is a tropical country.
   (vii) $(x+y+z)^2 = x^2+y^2+z^2+2xy+2yz+2zx$ for all $x, y, z \in R$.
   (viii) The boiling point of water is 100°C.
   (ix) He is a good person.
   (x) $x+3=5$.
   (xi) $x+3=5$ for all $x \in N$.
   (xii) The freezing point of water is 2°C.
   (xiii) Every prime integer is odd.
2. Among the following statements, identify the true and false statements. Further, whenever converse of a given statement can be written down, identify whether the converse is true:
   (i) A square is a rectangle.
   (ii) A parallelogram is a rectangle.
   (iii) For an integer x, if the unit digit of x is even, then x is even.
   (iv) All Californians are Americans.
   (v) $((x \text{ is an integer}) \wedge (x > 2) \wedge (x \text{ is a prime})) \rightarrow x$ is odd.
   (vi) For an integer x, if the unit digit of x is divisible by 3, then x is divisible by 3.
   (vii) For an integer x, if the unit digit of x is 5, then x is divisible by 5.
   (viii) For an integer x, if the unit digit is 0, then x is divisible by 10.
   (ix) The tiger is a wild animal.
   (x) The Earth moves around the Sun.

3. Write the following in symbolic form:

   (i) Whenever the US market crashes, so does the Indian market.
   (ii) Mr Arul is gentle, but firm.
   (iii) Ms Vani is kind, but not weak.
   (iv) When it rains and when there is no thunder, Selvi sings.
   (v) Ram uses his bow and arrow only during an emergency.
   (vi) It is not true that John is dark and handsome.
   (vii) If it rains, crops grow well and when crops grow well farmers are happy.

4. Find a statement formula X, which involves the atomic statements P and Q whose truth table is given as below:

P	Q	X
T	T	F
T	F	T
F	T	T
F	F	F

5. Draw the truth tables for

   (i) $P \rightarrow (Q \rightarrow R)$.
   (ii) $P \vee \sim (P \wedge Q)$.
   (iii) $P \vee (Q \rightarrow R)$.
   (iv) $(P \vee Q) \rightarrow R$.
   (v) $(\sim (X \vee Y)) \wedge Z$.
   (vi) $(P \rightarrow Q) \vee (\sim Q \wedge P)$.
   (vii) $P \wedge (Q \vee \sim P)$.
   (viii) $\sim (P \wedge Q \wedge R)$.
   (ix) $(P \vee \sim Q) \rightarrow R$.
   (x) $\sim ((P \vee Q) \leftrightarrow R)$.

6. A set of connectives is said to be functionally complete if any statement formula can be written in a logically equivalent manner using only the connectives in it. Show that $\{\vee, \sim\}$ is functionally complete. Is $\{\vee, \wedge\}$ functionally complete?

7. Examine which of the following are tautologies:

   (i) $((\sim P \wedge \sim Q) \wedge (\sim P \vee \sim Q)) \vee P \vee Q$.
   (ii) $(P \vee (Q \wedge R)) \rightarrow ((P \vee Q) \wedge R)$.
   (iii) $(P \wedge \sim R) \vee (P \rightarrow (Q \wedge R)) \vee (P \wedge \sim Q)$.
   (iv) $(P \rightarrow Q) \rightarrow ((P \vee Q) \leftrightarrow Q)$.
   (v) $P \vee (Q \wedge R) \vee \sim Q$.
   (vi) $(P \rightarrow (P \wedge Q)) \rightarrow (P \rightarrow Q)$.

8. Show that $(p \rightarrow (p \wedge q)) \vee (p \wedge (\sim p \vee \sim q))$ is a tautology without using the truth table.

9. Show that $((P \to Q) \vee R) \vee (P \wedge \wedge \sim Q \wedge \sim R)$ is a tautology with and without using the truth table.
10. Examine which of the following are contradictions:
    (i) $P \wedge \sim Q \wedge (P \to (Q \vee R)) \wedge \sim R$.
    (ii) $((P \vee Q) \to R)) \wedge (P \wedge Q \wedge \sim R)$.
    (iii) $(P \leftrightarrow Q) \wedge ((\sim P \wedge Q) \vee (P \wedge \sim Q))$.
    (iv) $(P \to Q) \wedge (P \to \sim Q)$.
    (v) $((P \wedge Q) \to (P \vee Q)) \wedge P \wedge \sim Q$.
11. Prove the set theoretic De Morgan laws using De Morgan Laws of Statment Calculus.
12. Examine the validity of the following inferences:
    (i) $P \to (Q \vee R) \Leftrightarrow (P \to Q) \vee (P \to R)$.
    (ii) $(P \vee Q) \leftrightarrow (P \wedge Q) \Leftrightarrow (P \leftrightarrow Q)$.
13. Show that $((p \vee \sim r) \to (q \wedge r)) \Rightarrow \sim p \vee q \vee r$.
14. Check the validity of the following argument:
    (i) On Sundays Mr Murthy does not read his newspaper or reads *Readers Digest*. This Sunday he read his newspaper. So, he did not read *Readers Digest*.
    (ii) Mary cooks after every rain. John sings whenever Mary cooks. Now, John is singing. Hence, it must have rained.
15. Find the DNF of the following
    (a) $(p \to (q \wedge \sim r)$.
    (b) $p \wedge (q \vee (\sim p \to r))$.
    (c) $(p \vee q) \wedge r \wedge (\sim q \to p)$.
    (d) $(p \wedge \sim r) \to (\sim q \vee \sim p)$.
16. Without using the truth table find the PCNF of $(P \vee \sim Q) \wedge (R \vee Q)$.
17. Let $X = (\sim P \wedge Q \wedge R) \vee (P \wedge Q \sim R) \vee (\sim P \wedge \sim Q \wedge \sim R)$. What is the PDNF of $\sim X$ and PCNF of X?
18. Find the disjunctive normal form of $(P \wedge Q) \to (\sim P \vee R)$.
    Using inference theory prove that R is a valid conclusion following from the hypotheses $(P \wedge Q) \to R$, $\sim P \to S$, $Q \vee S$, $\sim S$.
19. Prove or disprove the validity of the following argument after rewriting the statements in the symbolic form:
    (i) Whenever one of the students is wearing a pink dress, the teacher teaches well. If the teacher teaches well, then every student in the class is happy. Ramu, who is a student of this class is not happy. Therefore, there is no one in this class who is wearing a pink dress.
    (ii) For every human being there is some human being who loves the former. Therefore, there are two human beings who love each other.
    (iii) Only when streets are wet, accidents happen. Only when it rains streets get wet. Accidents do not happen. Therefore, it does not rain.

20. Use automatic theorem proving to show:

    (i) $(P \to P \wedge Q) \Rightarrow P \to Q$.

FURTHER READING

Gerstring, J. L. (2004), *Mathematical Structures for Computer Science,* W. H. Freeman and Company, New York.

Tremblay, J. P. and Manohar, R. (1997), *Discrete Mathematical Structures with Applications to Computer Science,* Tata McGraw-Hill, New Delhi.

CHAPTER

4

Counting Principles

Learning Objectives

On completing this chapter, you should be able to:

- state the fundamental counting principle
- distinguish between permutations and combinations
- identify and solve problems involving the use of fundamental counting principle, permutations and combinations
- understand the idea of a circular permutation and state the formula for the number of all circular permutations of n objects
- identify and solve problems based on permutation of identical objects
- state and prove the multi-nomial theorem
- solve problems based on distribution of identical objects, which include problems on compositions of a positive integer

4.1 A FUNDAMENTAL COUNTING PRINCIPLE

Let us think of a simple password consisting of just two characters. The first character should be an English letter and the second is a digit. How many such passwords are possible? We have 26 choices for the character and for each of these choices, 10 choices are available for the second character. Hence, the total number of such passwords is $26 \times 10 = 260$. Next, let us imagine that your grandmother has brought 15 *mangoes* and 12 *oranges*. You are supposed to choose one *mango* and one *orange*. What are the number of choices available to you? As you can choose any one of the 15 *mangoes* and any one of the 12 *oranges*, you have $15 \times 12 = 180$ choices to make.

Let us assume that there are m flights between Chennai and Mumbai and n flights between Mumbai and Kolkata. A person from Chennai who wants to visit Mumbai before reaching Kolkata can do so in mn different ways.

All these three examples suggest that the total number of ways of completing a task can be obtained by multiplying the number ways of completing each stage of the total task. We give a more clear formulation **of this fundamental counting principle** as follows:

Consider an activity consisting of k independent subtasks. Assume that the first task can be performed in m_1 ways, the second task can be performed in m_2 ways and so on upto the kth task, which can be completed in m_k ways. Then, this activity can be completed in $m_1 m_2 \ldots m_k$ different number of ways.

Example 4.1

There are 15 true or false types of questions in an examination. In how many ways can a student answer the question if she or he can choose not to answer some of the questions?

SOLUTION

The first question can be attempted in three different ways. That is, by answering 'true' or by answering 'false' or by not answering. The second question can also be attempted in the same way and so on. Thus each question can be attempted in three different ways. With reference to the counting principle, we have $k = 15$ and $m_1 = m_2 = \ldots m_{15} = 3$. Hence, the number of ways of doing this exam is $3 \times 3 \times \ldots .15$ times. That is 3^{15}. ■

Example 4.2

Find the number of all possible six-digit numbers that can be formed by using digits 2, 6, 8 and 9.

SOLUTION

The unit place can be filled by any of the four given digits, the ten's place can be filled by any of the given four digits and so on upto to the hundred thousand's place. Hence, the required answer is $4 \times 4 \times 4 \times 4 \times 4 \times 4 = 4^6$. ■

Example 4.3

How many symmetric 3×3 matrices can be formed using the entries $0, 1, 2, 3, 4, 5, 6, 7, 8, 9$?

SOLUTION

As the matrix is symmetric, we have to choose only those entries that have been indicated as $\circ$ below.

$$\begin{pmatrix} \circ & \bullet & \bullet \\ \circ & \circ & \bullet \\ \circ & \circ & \circ \end{pmatrix}.$$

So, there are six places, each of which can be filled in nine ways. So, the required answer is 9^6. ■

PRACTICE 4.1

What is the number of all possible 3×3 matrices, whose entries come from $0,1,\ldots,9$?

Example 4.4

How many three-digit numbers can be formed using the digits 0, 1, 2 and 3?

SOLUTION

The hundredth place can be filled only with 1, 2 or 3 and the remaining places can be filled with any of the given four digits. Hence, the required answer is $3\times4\times4=48$. ■

Example 4.5

Find the three-digit numbers divisible by 5, which can be formed by using the digits 1, 2, 3, 4 and 5?

SOLUTION

The unit place can only be 5. Now, the tenth place digit can be any of the five given digits. The hundredth place digit can also be any of the five given digits. So, the required number is $1 \times 5 \times 5 = 25$. ■

PRACTICE 4.2

How many four-letter words can be formed by using all the 26 letters of the English alphabet? How many of them begin with vowels? How many of them do not end with vowels?

Example 4.6

How many seven-letter words can be formed using the letters A, B, C, D and E if

(i) the word must contain exactly one D?

(ii) the word must contain at least one D?

SOLUTION

(i) We can allot D to any of the seven places and the remaining six places can be filled by the remaining four letters in 4^6 ways. Hence, the required number of words is 7×4^6.

(ii) The total number of words of length seven, which can be formed by using the given five letters is 5^7. The number of words of length seven without D is the number of words of length seven, which can be formed with the remaining four letters, which is 4^7. Hence, the number of words with at least one D is 5^7-4^7. ■

PRACTICE 4.3

Consider an alternative method to solve (ii) above: as the word must contain at least one D, we can fill one of the seven places by D and the remaining six places with any of the five letters. The first step in this method can be done in seven ways (choose one place out of seven for D) and the second step can be done in 5^6 ways. Hence, the required answer is 7×5^6. But, there is a flaw in this method. Find the flaw in this argument.

Example 4.7

Show that the number of ways of placing n distinct objects into m distinct boxes is $= m^n$.

SOLUTION

The first object can go into any one of the m boxes. Clearly , there are m ways of doing this. Similarly, the second object can also be placed in any of the m boxes and so on. Hence, the required answer is $m \times m \times \ldots. n$ times $= m^n$.

Let us now compare Example 4.2 and Example 4.7. In Example 4.2, if we think of the places as boxes and the numbers as objects, a wrong application of Example 4.7 gives us the answer 6^4. To resolve this confusion, we prove the following mathematical principle. ■

Proposition 4.1

Let A and B be finite sets with $|A| = n$ and $|B| = m$. The number of all functions from A into B is m^n.

Proof

Let $A = \{x_1, x_2, \ldots, x_n\}$. $f(x_1)$ can be defined in m possible ways, $f(x_2)$ can be defined in m possible ways and so on. Thus, the number of all possible functions = $m \times m \times \ldots n$ times = m^n. ■

Example 4.8

Solve Example 4.2 and Example 4.7 using Proposition 4.1.

SOLUTION

Let us recall that if $f : A \longrightarrow B$ is a function, for each x, $f(x)$ is a unique element of B.

First, we take Example 4.7. Consider one specific distribution of the n objects into m distinct boxes. This specific distribution defines a function f from the set of n objects into the set of boxes as follows: $f(x)$ is the box into which object x has been placed. Thus, the required number is nothing but the number functions from an n set into an m set. Hence, it is m^n.

In Example 4.2, note that a digit can occupy more than one place. Thus, we cannot define $f(x)$ as the place taken by x. So we take A as the set of all places and $B = \{2, 6, 8, 9\}$. For a specific six-digit number, we define a function $f : A \longrightarrow B$ as follows: For a place x, $f(x)$ is the number from the set B, which goes into x. Thus, the required answer is the number of functions from a six-element set into a four-element set, which is 4^6. ■

PRACTICE 4.4

Show that the number of ways of distributing n different *laddus* to m children is $= m^n$.

Example 4.9

In a certain implementation of the programming language Pascal, an identifier consists of a single letter or a letter followed by up to seven symbols, which may be letters or digits.We assume that the computer does not distinguish between capital and lowercase letters; there are 26 letters and 10 digits. Certain key words, however, are reserved for commands; consequently these key words may not be used as identifiers. If this implementation has 36 reserved words, how many distinct identifiers are possible in this version of Pascal?

SOLUTION

Consider an identifier consisting of a letter followed by k symbols. There are 26 possibilities for the first position and for each of the remaining k positions, we have 36 possibilities. Hence, there are $26(36)^k$ such identifiers. So, the required answer is $\sum_{k=0}^{7} 26(36)^k - 36$. ■

PRACTICE 4.5

In a particular coding system, a message is a string of nine symbols made up of the 26 English letters and the digits 0, 1,...,9. How many messages are possible? If the digit is permitted only in the first and/or last symbol of the message, how many messages are possible?

Example 4.10

How many license plates can be made using either two letters followed by four digits or two digits, followed by four letters?

SOLUTION

$(26)^2(10)^4 + (10)^2(26)^4 = 67{,}60{,}000 + 45{,}697{,}600 = 5{,}24{,}57{,}600$. ■

PRACTICE 4.6

There are seven multiple choice questions in an examination. How many sequences of answers are possible if the first three questions have four choices and the next four questions have five each?

In all these examples, we counted the number of objects satisfying specific predetermined conditions, or counted the number of ways of completing a given task. Further, when we wanted to count the number of objects, we did so by finding the number of ways of creating such an object. We identified the subtasks and the number of ways of completion of each subtask. The fundamental counting principle tells us that the product of the number of ways of completing the subtasks is the required answer.

4.1.1 Arrangements

Imagine a set of r boxes, each of which can hold only one object. We also have a collection of r objects to be placed in these boxes. Thus, each object will get a box to be placed in and every box will be filled. But there may be special boxes into which only certain special objects can be placed. In this section, we count the number of ways of placing r objects into r boxes with or without restrictions regarding special boxes or objects.

In how many ways can three objects be arranged in some order? Let us call these objects as a, b, c. We can arrange them as abc, acb, bac, bca, cab, cba. These are all the possible ways and hence three objects can be arranged in six different ways.

In how many different ways can 10 things be arranged? We can easily say that it is a difficult job to list all the arrangements and count them. So, we take a different approach.

Proposition 4.2

r objects can be arranged in $r!$ ways.

Proof

We have to find the number of all possible arrangements. Imagine an array of r boxes. **Consider the filling of each of the r boxes as a subtask.** We can place any of the r objects in the first box. **Thus, the number of ways of completing this first task is** r. Now we are left with $r - 1$ objects. So, the second box can be filled in $r - 1$ ways and so on. Figure 4.1 illustrates this.

r	r−1	r−2	.....	2	1

Figure 4.1

Thus, by using the counting principle of the previous section, the number of different arrangements of r things is $r(r - 1)(r - 2)....1 = r!$ ■

In the following examples, you will find different situations where the idea of arranging objects plays a role.

Example 4.11

How many seven-digit numbers can be formed using 2, 3, 4, 5, 6, 7 and 9 without using any of the digits more than once?

SOLUTION

The answer is just the number of ways of arranging seven objects and hence is 7! ■

Example 4.12

In how many ways can Maran, Gujral and Yadav be allotted the portfolios of Home, Defense and Information and Broadcasting? (A minister is allowed to have only one portfolio.)

SOLUTION

Each portfolio can be thought of as a box, for which we have a minister. Hence, the required answer is $3! = 6$. ■

PRACTICE 4.7

How many words (not necessarily meaningful) can be formed by rearranging the letters of *VIKRAM*?

In the following, we consider problems involving special boxes into which only certain special objects can be placed. Consider a row of r boxes. Assume that only m_1 of the objects can be placed in the first box and after which only m_2 of the remaining objects can be filled in the second box and so on (Fig. 4.2).

m_1	m_2	.	.	.	m_r

Figure 4.2

Thus, the required number of arrangements is $m_1 \times m_2 \times \cdots \times m_r$.

Example 4.13

Imagine a function that is attended by five ministers including the chief minister. Now, all of them have to be seated on the dias. The ornamental middle seat must go to the chief minister and the remaining four ministers can be seated in any of the remaining four seats. Find the the total number of possible seating arrangements.

SOLUTION

The first seat can be filled by any minister other than the chief minister and hence there are four different ways of doing this. The next seat can be filled in three ways. The middle seat can be filled by only one way. The fourth seat can be filled in two different ways and the last seat in only one way. This is represented by Figure 4.3.

4	3	1	2	1

Figure 4.3

So, the required number is $4 \times 3 \times 1 \times 2 \times 1 = 4! = 24$. ■

Example 4.14

How many five-digit numbers, which are divisible by 10, can be formed using 0, 1, 2, 3, 4 without repeating any digit?

SOLUTION

We can get the answer with the help of Figure 4.4.

1	2	3	4	1

Figure 4.4

(Here we are filling the boxes from the left end.) ■

PRACTICE 4.8

Find the total number of all six-digit even numbers in which every digit occurs at most once.

Example 4.15

A family consisting of a father mother and a child engage a taxi and travel. Other than the driver's seat there are three seats in the car one in the front and two at the back. If the child is not allowed to sit in the front seat, in how many ways can the family seat themselves (Fig. 4.5)?

SOLUTION

2 (Father or Mother)	1 (Driver)
2 (Child or the remaining parent)	1 (The last person)

Figure 4.5

The required number is $2 \times 1 \times 2 \times 1 = 4$. ■

PRACTICE 4.9

(1) How many three-digit numbers can be formed by using the digits 0, 5, 9 and 8?
(2) What is the total number of all the five-digit numbers? How many of them are divisible by 10? How many five-digit numbers can be formed by using all the nine digits without repetition?
(3) There are eight courses to be taught and eight faculty members in a department. Except the course on computer architecture, any faculty member can handle any of the courses. If computer architecture can be handled only by Prof. Subramaniyam or by Prof. Valliyammai, in how many ways can the courses be allotted?
(4) How many odd numbers less than 1,000 can be formed using digits 0, 1, 8, 9?

PRACTICE 4.10

Explain why the answers to the following problems are *not the same:*

(1) In how many different ways can Ramasamy, Kuppusamy and Muthusamy be seated so that the last seat has to be occupied only by Kuppusamy?
(2) How many three-digit even numbers can be formed using digits 2, 3 and 5?

4.2 PERMUTATIONS AND COMBINATIONS

The following example helps in understanding the essential difference between Permutations and Combinations.

Example 4.16

A mother and a son visit a toy shop. The mother tells her son that he can choose any three toys out of a set consisting of *a tiger, an elephant, a jeep, a helicopter and a soldier*. Let us list the number of choices available to the son.

$$\begin{array}{lll} \{T,E,J\} & \{T,E,H\} & \{T,E,S\} \\ \{T,J,H\} & \{T,J,S\} & \{T,H,S\} \\ \{J,H,S\} & & \end{array}$$

We find that he has 10 different choices to make.

But now, consider a package consisting of three boxes. After deciding to buy T, E and J, *these toys can be packed in* 3! ways in the package. For each choice of three toys, there are 3! ways of packing them. Thus, the boy can *choose and pack* three toys out of the five toys in $10 \times 3!$ ways.

Similarly, we can easily see that the number of ways of choosing 3 executive members of an association consisting of 5 members is 10.

After choosing three, consider the problem of allotting the posts of President, Vice-President and Secretary. In how many ways can this be done? Now, let us say that Murugan, Peter and Begum are going to be selected. Now, these three people can hold the posts in the following ways:

President	Vice-President	Secretary
M	P	B
M	B	P
P	M	B
P	B	M
B	P	M
B	M	P

Thus, for every group of three selected persons, there are $3! = 6$ different ways of them holding the posts. In all, there are $10 \times 6 = 60$ different possibilities of doing this. ■

Definition 4.1: Let there be n objects. The number of ways of choosing r objects out of n objects is called the combination of r objects out of n objects. This is denoted by ${}^nC_r$. The number of ways of *choosing and arranging* r objects out of the n objects is called the permutation of r objects out of n and is denoted by ${}^nP_r$.

Proposition 4.3

${}^nP_r = n(n-1)(n-2)...(n-r+1) = \frac{n!}{(n-r)!}$.

Proof

Consider an array of r boxes. In the first box, we can place any of the n objects, in the second we can place any one of the remaining $n-1$ objects and so on. In other words, we can place

$n-0$ objects in the first box
$n-1$ objects in the second box
$n-2$ objects in the third box
....
$n-r$ objects in the $(r+1)$th box and finally $n-(r-1)$ objects in the rth box. Hence, the number of ways this can be achieved is $n(n-1)(n-2)...(n-r+1)$.

Now,

$$n(n-1)(n-2)....(n-r+1)$$
$$= \frac{n(n-1)(n-2)...(n-r+1)(n-r)(n-r-1)....1}{(n-r)(n-r-1)....1} = \frac{n!}{(n-r)!}. \quad ■$$

Example 4.17

(i) How many five-digit numbers can be formed by using digits 1, 2, 3, 4, 5, 6, 7, 8 without repetition?

(ii) A father distributes exactly one orange to each of his five children. If he has eight oranges, in how many ways can he do this?

SOLUTION

For both the problems, the answer is ${}^8P_5$. ■

PRACTICE 4.11

How many numbers are there between 1,000 and 10,000 in which all the digits are distinct?

Example 4.18

How many five-letter words can be formed using the letters from the list $A, B, ...I$, if repetitions are not allowed? How many of them do not contain A and I? How many of them contain A and I?

SOLUTION

The number of five-letter words that can be formed using the given letters without repetition is nothing but the number of permutations of five things out of nine.

Hence it is $^9P_5$. The number of words among them, not containing A and I, is nothing but the number of permutations of five things out of seven. Hence it is $^7P_5$.

Next, we have to find the number of five-letter words formed using the given letters without repetition, which must contain A and I. First, we shall allot two places out of five to A and I. This can be done in $^5P_2$ ways. Next, we shall fill the remaining three places out of the remaining seven letters, which can be done in $^7P_3$. Hence, the required answer is $^5P_2 \times ^7P_3$. ■

In Example 4.18, as we are interested in *words*, the order in which the letters appear is important. Hence, the idea of permutation is used in the solution. But in the next example, letters are just selected without any consideration for the order. Hence, its solution uses the idea of combination.

Example 4.19

In how many ways can five letters be chosen from the list $A, B, \ldots, I$? In how many ways can five letters be chosen, if the choice must contain both A and I?

SOLUTION

There are nine letters and hence the number of ways of choosing five out of them is $^9C_5$. If our choice must include both A and I, then we need to choose only three out of seven. Hence, this can be done in $^7C_3$. ■

PRACTICE 4.12

How many different numbers below 1000 can be formed from digits 3, 5, 8 and 9 if the digits are not allowed to be repeated?

Proposition 4.4

$$^nC_r = \frac{n!}{r!(n-r)!}.$$

Proof

Let us choose r objects and then we can arrange them in $r!$ ways. So, the number of ways of choosing and arranging r objects out of n is $= {^nC_r} \times r!$ Thus $^nP_r = {^nC_r} \times r!$ So,

$$^nC_r = \frac{^nP_r}{r!} = \frac{n!}{r!(n-r)!}.$$ ■

Example 4.20

A wife wants to present three shirts to her husband. At the shop, the husband finds seven shirts of his liking. What is the number of choices available to the wife?

SOLUTION

The wife has to select three shirts out of seven. This can be done in $^7C_3$ ways.

$$\begin{aligned} {}^7C_3 &= \frac{7!}{(3!)(4!)} \\ &= \frac{1\times2\times3\times4\times5\times6\times7}{1\times2\times3\times1\times2\times3\times4} \\ &= \frac{5\times6\times7}{1\times2\times3} \\ &= 35. \end{aligned}$$

■

Example 4.21

If a team has to have 11 players, how many different teams can be formed out of a group of 17 players?

SOLUTION

This problem involves choosing 11 things out of 17 things, and hence the number of different teams that can be formed is ${}^{17}C_{11}$. ■

Example 4.22

A book shop has 10 copies of the Thirupukazh, seven copies of the Bhagvath Geetha, eight copies of the Bible and nine copies of the Koran. A lady wants to buy three copies of the Thirupukazh, five copies of the Bhagvath Geetha, seven copies of the Bible and seven copies of the Koran. In how many ways can she make a purchase?

SOLUTION

The lady can choose three copies of the Thirupukazh in ${}^{10}C_3$ ways, five copies of the Bhagvath Geetha in in ${}^7C_5$ ways, seven copies of the Bible in ${}^8C_7$ ways and seven copies of the Koran in ${}^9C_7$ ways.
Hence, the number of ways she can make a purchase is

$${}^{10}C_3 \times {}^7C_5 \times {}^8C_7 \times {}^9C_7 = \frac{10\times9\times8}{1\times2\times3} \times \frac{7\times6}{1\times2} \times \frac{8}{1} \times \frac{9\times8}{1\times2}.$$

■

Example 4.23

Mr Chezhiyan has five Tamil books, six Telugu books, three Kannada books and four Malayalam books.

(i) In how many ways can these books can be arranged?
(ii) What is the number of possible arrangements if books of the same language should be together?
(iii) What is the number of possible arrangements if books of the same language should be together and the first bunch consists only of Kannada books?

SOLUTION

(i) There are 18 books that can be arranged in 18! ways.

(ii) The four languages can be arranged in 4! ways. Within the Tamil bunch, Tamil books can be arranged in 5! ways; within the Telugu bunch, Telugu books can be arranged in 6! ways; within the Kannada bunch, Kannada books can be arranged in 3! ways and within the Malayalam bunch, Malayalam books can be arranged in 4! ways. So, the required number is $4! \times 5! \times 6! \times 3! \times 4!$

(iii) Now, the Kannada bunch should be the first. So, the arrangement of langauges is possible in 3! ways. Proceeding as in (ii), the required answer is $3! \times 5! \times 6! \times 3! \times 4!$ ■

Example 4.24

Eight scientists and five politicians take part in a conference. In how many ways can they be seated in a single row if

(i) No two politicians must sit together?

(ii) No two scientists must sit together?

SOLUTION

For (i), first we will arrange the scientists in a row with a gap between them as shown below:

	S		S		S		S		S		S		S		S	

Now, we have nine places where the five politicians can be seated. The eight scientists can be seated in 8! ways and the nine places can be allotted to the five politicians in $^9P_5$ ways. So the required answer is $8! \times {}^9P_5$.

For (ii), there is no such possible arrangement. ■

Example 4.25

Three female ministers and three male ministers have to be chosen from a group of MLAs consisting of 10 female MLAs and nine male MLAs. In how many ways can this be done?

SOLUTION

Three female ministers can be chosen in $^{10}C_3$ ways and three male ministers can be chosen in $^9C_3$ ways. Thus, the required number is $^{10}C_3 \times {}^9C_3$. ■

Example 4.26

How many matrices of order 2×3 can be formed, in which the digits from 0 to 9 occur not more than once?

SOLUTION

There are six places to be filled and we have 10 objects to choose from. This can be done in $^{10}P_6$ ways. ■

Example 4.27

A book has to be presented to each of five students. There are only two copies of Gandhi's 'My Experiments with Truth'. Both the copies of Gandhi's book must be presented. The remaining three presents have to be chosen from six other books of which any number of copies are available. If the same book can be presented to more than one student, in how many ways can the five students be presented with books?

SOLUTION

First, out of five students, two students who would be presented with 'My Experiment with Truth' can be selected in ${}^5C_2$ ways. For each of the the remaining students, the six books can be distributed in ${}^6C_3$ ways. So, the total number of ways in which books can be presented is ${}^6C_3 \times {}^5C_2$. ■

Example 4.28

In how many ways can letters in the English alphabet be arranged so that there are exactly seven letters between letters a and b?

SOLUTION

Now, a required permutation can be thought of as the permutation of ax_1x_2 $x_3x_4x_5x_6x_7b$ and the remaining $26 - 9 = 17$ letters. Now, the seven letters x_1, $x_2, \ldots, x_7$ can be chosen in ${}^{26}C_7$ ways. These seven letters can be arranged in 7! ways. The remaining 17 letters and $ax_1x_2x_3x_4x_5x_6x_7b$ can be arranged in 18! ways. So, the required answer is $18! \times {}^{26}C_7 \times 7! = 18! \times {}^{26}P_7$. ■

Example 4.29

How many four-digit numbers can be formed using the seven digits 0, 1,...,6 if repetitions are not allowed?

SOLUTION

Method I: The first place can be filled with six digits (since it cannot be filled with zero), the second place with six, the third place with five and the fourth place with four digits. Hence, the required number of digits is $6 \times 6 \times 5 \times 4 = 720$.
Method II: Consider the number of permutations of four digits out of seven, which is ${}^7P_4 = 7 \times 6 \times 5 \times 4 \times 3 = 840$. Among them, 0 will be the first digit in ${}^6P_3 = 6 \times 5 \times 4 = 120$. Hence, the required number $= 840 - 120 = 720$. ■

Example 4.30

How many ways are there to distribute five cards to each of four players from a standard deck of 52 cards?

SOLUTION

The first player gets 5 out of 52, which can be done in ${}^{52}C_5$ ways. The second player gets 5 out of 47 cards and so on. Thus, the number of ways of distributing 5 to each 4 out of 52 is

$${}^{52}C_5 \, {}^{47}C_5 \, {}^{42}C_5 \, {}^{37}C_5.$$ ■

Example 4.31

In how many ways can we distribute 10 toys to Amar, 15 toys to Akbar and seven toys to Anthoni out of a collection of 100?

SOLUTION

$$^{100}C_{10}\ ^{90}C_{15}\ ^{75}C_7.$$

Remark 4.1

We shall see that the answer for the above is nothing but the coefficient of $x^{10}y^{15}z^7$ in the expansion of $(x+y+z)^{100}$.

4.2.1 Circular Permutation

Imagine five people sitting around a table. Consider the following four different seating arrangements (Fig. 4.6).

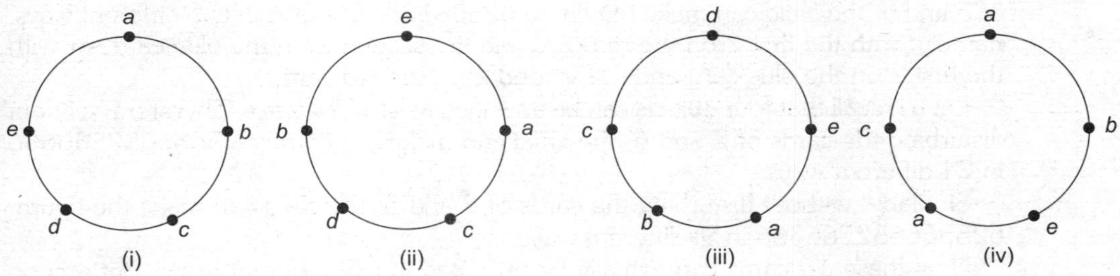

Figure 4.6 Circular permutation

We note that in arrangements (*i*) and (*iii*), the left and the right neighbours of *c* are *d* and *b*, respectively. In fact, (*iii*) is simply a rotation of (*i*). Also note that (*ii*) is a rotation of (*iv*). Further, we cannot obtain (*ii*) by rotating (*i*).

This kind of arrangement in mathematical terms is called circular permutation. A circular permutation is the number of ways you can arrange n distinct items in a circle. A circular arrangement and another circular arrangement that can be obtained by the rotation of the former are treated as identical. Thus, in the above figure, (*i*) and (*iii*) represent the same circular permutation.

The following proposition gives the number of circular permutations of n objects.

Proposition 4.5

The number of circular permutations of n objects is $(n-1)!$

Proof

Imagine n chairs placed round a circular table. We shall number the chairs as 1, 2,...,n. Let a_1 be seated in the first chair, a_2 be seated in the second chair, etc. upto a_n being seated in the nth chair. Now, let a_1 move to the second chair, a_2 moves to the third chair,...,a_{n-1} move to the nth and finally a_n move to the first chair. We get a circular arrangement that represents the same circular permutation. We can easily see that after repeating the above process n times, we get back the original circular arrangements. Therefore, a single circular permutation is represented by n circular arrangements. As there are n objects and n places to be filled, there are $n!$ different circular arrangements. Thus, the number of circular permutations is $n!/n = (n-1)!$

4.2.2 Permutation of Identical Objects

Let us consider the following different arrangements of the three girls Jaya, Maya and Sheila: JMS, JSM, MJS, etc. Now, if Maya and Jaya are identical twins and they are identically dressed, then we cannot differentiate between the arrangements like JMS, MJS, etc. This is an example of permutation of identical objects. If a collection of objects contains identical objects, we would like to count only those permutations that can be distinguished from one another.

Illustrative Example: Imagine a child playing with cards with numerals on them. Let us say she picks up two cards of 2, four cards of 5 and nine cards of 6. Let us say that the two cards of 2 are of colours red and blue. Without disturbing the cards of 5 and 6, the child can make the figure 625566652566666 in two different ways, namely, with the first 2 on the red card and the second 2 on the blue card, **or** with the first 2 on the blue card and the second 2 on the red card.

Let us recall that four objects can be arranged in $4! = 24$ ways. Therefore, without disturbing the cards of 2 and 6, the child can make the figure 625566652566666 in 24 different ways.

Similarly, without disturbing the cards of 2 and 5, the child can make the figure 625566652566666 in 9! different ways.

Now these 15 cards can actually be arranged in 15! different ways. But among them, many of them would represent the same figures. **From the above discussion, it is clear that each figure will be counted** $2! \times 4! \times 9!$ times. Hence, the number of distinct figures that can be formed is

$$\frac{15!}{2!\,4!\,9!}.$$

Proposition 4.6

In a collection there are m_1 identical articles of type 1, m_2 articles of type 2 and so on upto m_k articles of type k. The number of permutations of this collection is

$$\frac{(m_1 + m_2 + \ldots + m_k)!}{m_1!\, m_2! \ldots m_k!}.$$

Example 4.32

How many different words can be formed by rearranging the letters of the word LALITHAKAMALAM?

SOLUTION

Let us list how many times each letter in this word is repeated: L-3, A-5, I-1, T-1, H-1, K-1, M-2. Thus, the required number is

$$\frac{14!}{3!\ 5!\ 1!\ 1!\ 1!\ 1!\ 2!}.$$ ■

PRACTICE 4.13

Find the number of arrangements of the letters in CHAMMANNARCH. Also, find how many of these arrangements have no adjacent A's.

Example 4.33

A hotel has six single rooms, six double rooms, and four rooms, in each of which three persons can stay. In how many ways can 30 persons be accommodated in this hotel?

SOLUTION

We shall denote the single rooms by S_1, $S_2, \ldots, S_6$, the double rooms by D_1, $D_2, \ldots, D_6$ and the rooms that would accommodate three persons by T_1 T_2, T_3 and T_4.

Now, we shall allot a unique room to each person. Let us say that the first person stays in S_6, the second person stays in T_2, the third person stays in D_4, etc. This is listed as $S_6, T_2, D_4, \ldots$. Now, the double room will be allotted to two individuals and hence each of the D's will occur twice in the list; similarly each of the T's will occur thrice in the list. Hence, this is a problem of permutation of identical things. The required number is $\frac{30!}{1!\,1!\,1!\,1!\,1!\,1!\,2!\,2!\,2!\,2!\,2!\,2!\,3!\,3!\,3!\,3!}$. ■

4.3 APPLICATIONS OF PERMUTATIONS AND COMBINATIONS

In this section, we shall learn about the multinomial theorem, compositions of a positive integer, number of ways of distributing identical objects in distinct boxes, etc.

We are familiar with the following expansions from elementary algebra:

$$\begin{aligned}(x+y)^n &= \binom{n}{0}x^n + \binom{n}{1}x^{n-1}y + \ldots + \binom{n}{n}y^n \\ (x+y+z)^2 &= x^2+y^2+z^2+2xyb+2yzc+2zx.\end{aligned}$$

On the lines of the above examples, one would like to know the expansion for

$$(x_1 + x_2 + \ldots x_m)^n.$$

The answer is provided by the multinomial theorem given below.

Illustrative Example: Consider the expansion $(x+y+z)^{10}$. The terms in this expansion look like $x^i y^j z^k$ with $i+j+k=10$. Take $i=5$, $j=2$ and $z=3$. The term $x^5y^2z^3$ occurs, for example, as *xyxxxyzxzyy*, *xxxxxxyzyzz* and *yxyxzxzxzx*, etc. Thus, the number of times the term $x^5y^2z^3$ occurs is the number of permutations of 10 objects of 3 kinds, each of them appearing 5, 2 and 3 times. Hence it is $\frac{10!}{5!2!3!}$. This is the coefficient of $x^5y^2z^3$ in the expansion of $(x+y+z)^{10}$.

Generalizing, we see that the the coefficient of $x_1^{i_1}x_2^{i_2}...x_m{}^{i_m}$ in the expansion of $(x_1+x_2+...x_m)^n$ is $\frac{m!}{i_1!\, i_2!...i_m!}$. So, we have the following theorem known as Multinomial Theorem.

Theorem 4.1

$$(x_1+x_2+...x_m)^n = \sum_{i_1+i_2+...i_m=n} \frac{m!}{i_1!\, i_2!...i_m!} x_1^{i_1}x_2^{i_2}...x_m^{i_m}.$$

PRACTICE 4.14

Prove the special case of binomial theorem:

$$(x+y)^n = \sum_{r=0}^{n} \binom{n}{i} x^r y^{n-r}.$$

Example 4.34

If $|A| = n$, then $|P(A)| = 2^n$.

SOLUTION

The number of subsets of A having r elements is nothing but the number of ways of choosing r objects out of n objects and hence it is ${}^nC_r$. Hence, the number of all subsets of A is $\sum_{r=0}^{n} \binom{n}{r}$. By taking $x=1$ and $y=1$ in the binomial theorem, we get $(1+1)^n = \sum_{r=0}^{n} \binom{n}{r}$. Hence, the number of all subsets of A or the the number of elements of $P(A) = 2^n$. ■

Remark 4.2

Using the notation $P(A) = 2^A$, we get $|2^A| = 2^{|A|}$ if A is a finite set.

Let A and B be finite sets with $|A| = n$ and $|B| = m$. We have already learnt that the number of all functions from A into B is given by m^n. In the next proposition, we count the number of *one-to-one* functions from A into B.

Proposition 4.7

Let A and B be finite sets with $|A| = n$ and $|B| = m$ and $n \leq m$. The number of all one-to-one functions from A into B is

$$m(m-1)(m-2)....(m-n+1) = {}^{m}P_{n}.$$

Proof
Let $A = \{x_1, x_2, ..., x_n\}$. $f(x_1)$ can be defined in m possible ways. Since for a one-to-one function, $f(x_1) \neq f(x_2)$, $f(x_2)$ can be defined in $m-1$ possible ways and so on. Thus, the number of all possible functions $= m \times (m-1) \times \cdots \times (m-n+1) = {}^{m}P_{n}$. ■

PRACTICE 4.15

The number of ways of placing n objects into m boxes such that *there is only one object in each box* = the number of ways of distributing n *different* laddus to m children such that every child gets exactly one laddu is $= {}^{m}P_{n}$.
= The number of one-to-one functions from a set containing n elements into a set containing m elements. The reader is urged to explain this fact.

The proofs of the following propositions are easy.

Proposition 4.8

The number of binary sequences with p ones and q zeros is

$$\frac{(p+q)!}{p!q!}.$$

Proof
Apply Proposition 4.6. ■

Proposition 4.9

The number of binary sequences with p ones and q zeros such that no two zeros are together is

$${}^{p+1}C_{q+1}.$$

Proof
First, arrange the ones. The zeros can be placed in the $p-1$ gaps or at the beginning or at the end. Thus, the total number of places is $p+1$. Thus, the required number is

$${}^{p+1}C_{q+1}.$$ ■

PRACTICE 4.16

Find the number of binary sequences containing p ones and q zeros, starting and ending with 1 such that no zeros are together.

Proposition 4.10

The number of ways of distributing n *identical* laddus to two children is $n + 1$.

Proof
The different possibilities are

$$\begin{array}{cc} 0 & n \\ 1 & n-1 \\ .. & .. \\ n & 0. \end{array}$$

■

Remark 4.3

Note that here we are **not** distributing the laddus in a **fair** manner. We can be completely partial to just one of the two children and give all the laddus to him/her. Next, we consider the problem of distributing identical laddus to many children, again not necessarily in a fair manner.

Proposition 4.11

The number of ways of distributing n identical objects in k distinct boxes is

$$^{n+k-1}C_{k-1}.$$

This proposition is analogous to the statement that the number of ways of distributing n identical laddus to k children is

$$^{n+k-1}C_{k-1}.$$

Before presenting the proof, we give some thought to the problem as follows: Imagine seven laddus and three children, say, Kannan, Radha and Fatima. We list some possible ways of distribution:

K	R	F
2	2	3
0	5	2
4	2	1
0	0	7

Now, with each of these arrangements we associate a sequence of zeros and ones as follows:

$$\begin{array}{cccc} 2+2+3 & 0+5+2 & 4+2+1 & 0+0+7 \\ 110110111 & 011111011 & 111101101 & 001111111 \end{array}$$

The number of ones to the left of the first zero is the number of laddus given to Kannan, the number of ones between the first and second zero is the number of laddus allotted to Radha and the number of ones present after the last zero is the number of laddus given to Fatima. Note that in the second distribution Kannan gets no laddus and hence there are no ones to the left of the first zero. Now, we can give the proof.

Proof
Arrange n ones and place $k-1$ zeros among them. The number of ones placed before the first zero is the number of laddus given to the first child, the number of ones placed before the second zero is the number of laddus given to the second child etc. Thus, the required number is the number of binary sequences of n ones and $k-1$ zeros, which is

$$^{n+k-1}C_{k-1}.$$ ■

PRACTICE 4.17

In how many ways can 10 identical Appu toys be distributed to Ramya, Divya, Vageesh and Swanika?

Example 4.35

Krishna's grandmother brings an unlimited supply of jangiries, laddus and appams. She orders Krishna to eat exactly 10 sweets. He is free to choose any number of any of the three kinds but the total must be exactly 10. For example, he can take 10 appams; or three jangiries, four laddus and three appams; or eight laddus and two appams, etc. Find the number of all possible choices for Krishna.

SOLUTION

Let us pretend that each type of sweet represents a box. If Krishna chooses three jangiries, four laddus and three appams, then this is similar to placing three objects in the jangiri box, four in the laddu box and three in the appam box. Thus, this problem is mathematically identical to that of placing 10 identical objects in three boxes. Hence, the required number of choices is $^{10+3-1}C_{3-1}$. ■

In the following proposition, the above example is generalized.

Proposition 4.12

The number of ways of choosing k things out of n kinds of unlimited things is

$$^{n+k-1}C_{n-1}.$$

Proof I
As in the previous case, this problem is analogous to that of placing k identical objects in n distinct boxes. Hence its solution is

$$^{n+k-1}C_{n-1}.$$ ■

Proof II Arrange k ones in a row and place $n-1$ zeros among them. The number of ones placed before the first zero is the number of objects of the first type, the number of ones placed before the second zero is the number of second type of objects selected, etc. Thus, the required number is the number of binary sequences of length $= n-1+k$ with k ones and $n-1$ zeros. Thus, the required number is

$$^{n-1+k}C_{n-1} = {}^{n-1+k}C_k.$$ ■

In Proposition 4.11, we considered the problem of distributing n identical laddus to k children. In that situation, some children may not get any laddus. In the next proposition, we consider the problem of distributing n laddus to k children such that each child gets at least one laddu.

Proposition 4.13

The number of ways of distributing n identical laddus to k children such that every child gets at least one laddu is

$$^{n-1}C_{k-1}.$$

Proof

Arrange n ones and place $k-1$ zeros in the $n-1$ gaps with *not more than a zero in each gap.* The number of ones placed before the first zero is the number of laddus given to the first child; the number of ones placed before the second zero is the number of laddus given to the second child, etc. It is clear that every child gets at least one laddu.

Thus, the required number is

$$^{n-1}C_{k-1}. \qquad \blacksquare$$

PRACTICE 4.18

Find the number of ways of choosing k things out of n kind of unlimited things such that at least one object of each kind is chosen.

Compositions of Positive Integers: We can write 10 as a sum of positive integers in many ways. For example,

$$\begin{aligned} 10 &= 1+9 \\ 10 &= 2+3+5 \\ 10 &= 1+1+1+2+3+2. \end{aligned}$$

We are interested in finding the total number of ways of expressing 10 as a sum of positive integers. We shall treat the expression 1 + 1 + 1 + 2 + 3+ 2 as different from 1 + 2 + 3 + 1 + 1 + 2.

Now, consider a composition of n into k positive integers. That is, we are interested in a solution of $x_1 + x_2 + \cdots + x_k = n$ with $x_1, x_2, \ldots, x_k \in N$. It is clear that such a composition corresponds to a distribution of n identical objects to k persons such that each one gets at least one object. Hence, the number of such compositions is $^{n-1}C_{k-1}$.

So, the total number of compositions is

$$\sum_{k=1}^{n} \binom{n-1}{k-1} = \sum_{k=0}^{n-1} \binom{n-1}{k} = 2^{n-1}.$$

PRACTICE 4.19

There are six children, seven men and 10 women. In how many ways can they be seated such that a child is always between a man and a woman and no man is adjacent to a woman?

Example 4.36

In Hotel Vinayaka Muruga, one can order a dosa with none, any, some or all of coconut chutney, pudhina chutney, onion chutney, chilli sauce or red chilli powder. How many possible orders of a dosa are possible?

SOLUTION

As there is no restriction in placing the order, the required number is nothing but the number of all subsets of a five-element set. Hence the answer is 2^5. ■

Example 4.37

In how many ways can nine persons be made to sit around a table, if two of them do not want to sit next to each other? (Here two seating arrangements are regarded as the same if one can be obtained from the other by a rotation.)

SOLUTION

Seven persons can be arranged around a table in 6! ways. Now, there are seven places where the remaining two persons can sit. This can be done in ${}^7P_2$ ways. So, the required number of arrangements is $6! \times 7 \times 6$.

Now, we can do this problem in a different way. The number of arrangements in which the two are together is $7! \times 2$. The total number of arrangements for nine persons is 8! So, the required number of arrangements is $8! - 2(7!)$. ■

Example 4.38

Find the number of staircase paths from $(2, 5)$ and $(8, 9)$.

SOLUTION

A typical path is shown in Figure 4.7.

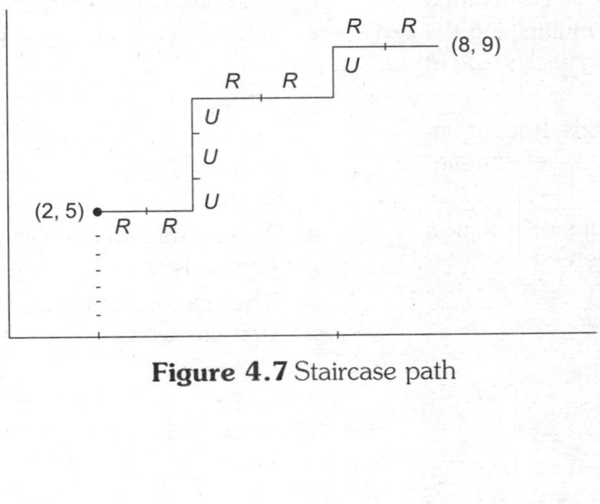

Figure 4.7 Staircase path

We give the precise meaning of such a path. From a given junction (x, y), we can either go to $(x+1, y)$ (moving one unit right horizontally) or to $(x, y+1)$ moving one unit up vertically.

Now to reach (8, 9) from (2, 5) we have to move $8 - 2 = 6$ units horizontally and $9 - 5 = 4$ units vertically. Thus, with every staircase path, we can associate a sequence of R's and U's consisting of six R's and four U's. For example, for the staircase path shown in the figure, we have the sequence $RRUUURRURR$. So, the required number is the number of permutations of six R's and four U's, which is given by $\frac{10!}{6!\,4!}$. ■

PRACTICE 4.20

Let p, q, m, n be integers such that $p \leq m$ and $q \leq n$. Then, show that the number of staircase paths in R^2 joining points (p, q) and (m, n) is given by

$$\frac{(n-q+m-p)!}{(n-q)!(m-p)!}.$$

PRACTICE 4.21

A linguist has seven Telugu books and nine Gujarati books. In how many ways can the linguist arrange these books on her book case if the languages should alternate?

PRACTICE 4.22

How many words can be formed from the eight letters a, c, f, g, i, t, w, x? How many start with i? How many start with t and end with c?

SUMMARY

- The total number of ways of completing a task can be obtained by multiplying the number ways of completing each stage of the total task.
- The number of all functions from a set containing n elements into a set containing m elements is m^n.
- The number of permutations of r objects taken from n objects is given by

$$^nP_r = \frac{n!}{(n-r)!}.$$

- The number of combinations of r objects taken from n objects is given by

$$^nC_r = \frac{n!}{(n-r)!r!}.$$

- The number of circular permutations of n objects is $(n-1)!$
- The number of permutations of a collection in which there are m_1 identical

SUMMARY (Continued)

articles of type 1, m_2 articles of type 2 and so on upto m_k articles of type k is

$$\frac{(m_1+m_2+\cdots+m_k)!}{m_1!\, m_2! \cdots m_k!}.$$

- The number of ways of distributing n identical objects in k distinct boxes is

$$^{n+k-1}C_{k-1}.$$

- The number of ways of distributing n identical objects in k distinct boxes such that each box gets at least one object is

$$^{n-1}C_{k-1}.$$

- The total number of compositions of a positive integer n is 2^{n-1}.

PROBLEMS

1. A coin and a six-faced die are tossed simultaneously. How many different possible outcomes are there?
2. Kannan knows only Tamil, Malayalam and Marathi. Roy knows only Bengali and Punjabi. Vijay knows all these languages. In how many different ways can Vijay get a message from Kannan and pass it on to Roy?
3. Five roads lead to the top of a mountain. In how many ways can a tourist go up and down the mountain? What is the answer if the tourist must not retrace his steps?
4. In a class there are 53 girls and 47 boys. In how many ways, can two representatives, one of each sex, be selected?
5. Consider the suburban stations Beach, Fort, Park, Nungambakkam and Kodambakkam in Chennai. If there are 1500 passengers boarding at Beach, assuming that everyone of these passengers would get down at or before Kodambakkam, in how many possible ways can they get down?
6. Find the number of seven-letter words that can be formed by using the letters A, B, C, X, Y and Z.
7. At a dinner party, an individual must choose a roti item, a rice item, a subji item and a dessert item. The roti item must be one of butter naan, rumaali roti, or chapaathi. The rice item must be one of lemon rice, vegetable pulav and Chinese fried rice. The subji item must be one of vendaikaay poriyal (fried bhendi), alu mutter, malai kofta, baingan bartha (brinjal curry) or paneer palak. The dessert item must be chosen from ice-cream and fruit salad. How many possible dinners are there?
8. A group of six people includes Selvi and Lakshman. Find the number of ways of seating them provided there should be at least one person between Selvi and Lakshman.
9. In a buffet you must choose a total of seven dishes out of 10 South Indian dishes and eight North Indian dishes by choosing not less than two dishes from the two set of dishes. Find the number of ways of doing this.
10. A rich woman owns three engineering colleges and four medical colleges. She has six daughters and a son. Before she dies, she wants to give one college to each of her offspring. If the son wants to get only a medical college, in how many ways can the rich women distribute her colleges to her children?
11. From an assembly consisting of 132 MLAs, in how many ways can ministers be alloted for 15 portfolios?
12. How many five-letter words can be formed by using the letters of the word RIDAGNES without repetition?
13. How many six-digit numbers can be formed by using the digits $1, 2, 3, 4, 5, 6, 7, 8$ without repetition?
14. A class consists of 42 students. In how many ways can a sports committee having the posts of President, Secretary, Joint Secretary and Treasurer be formed? If there is one class representative and if she or he has to be the president of the sports committee, in how many ways can this be achieved? If the class representative need not be the president but only has to be one of the committee members, in how many ways can such committees be formed?

15. Five children and four adults go for camping. There are eight sleeping bags inside the tent, and one sleeping bag is kept outside the tent. In how many ways can the campers occupy the sleeping bags if only an adult is allowed to use the sleeping bag outside?
16. How many even nine-digit numbers can be formed using the digits from 1 to 9 without repetition of the digits?
17. How many one-to-one functions can be formed using a set containing three elements into a set containing seven elements?
18. A committee consisting of five people has to be formed out of a class consisting of 25 students. In how many ways can this be achieved?
19. A class consists of 16 girls and 14 boys.

    (i) In how many ways can a committee of seven students be formed?
    (ii) In how many ways can a committee of four girls and three boys be formed?
    (iii) In how many ways can a committee of seven students having at least two girls be formed?.

20. How many numbers between 1 and 10^4 contain exactly one 8 and one 9?
21. Fifteen girls and six boys have to be seated in a row in such a way that no two boys sit next to each other. How many seating arrangements are possible?
22. How many six-digit integers are divisible by 4?
    Hint: For a number to be divisible by 4 the last two digits must be one of the following pairs: (00,04,08,...,96).
23. In how many ways can the letters in SANGAILAKKIYAM be arranged? In these arrangements how many have all the A's together?
24. Andaal and Meera are friends and they keep talking to each other. A group of 10 students including these two meet. The group decides that there must be five persons between Andaal and Meera. How many seating arrangements are possible?
25. From a choice of 12 colours, a flag consisting of three bands of different colours has to be created? How many such flags can be formed?
26. In all there are 735 teachers employed by five engineering colleges, including the principals. The managements of the five colleges would like to form an academic council consisting of 50 teachers from this group. How many such councils can be formed? If the council must necessarily include all the principals, how many councils can be constituted?
27. Twenty-eight persons are to be partitioned into five teams such that the first team has seven members, the second and the third team each has three members, the fourth team has 10 members and the fifth team has three members. How many partitions are possible?
28. How many automobile licence plates are possible if the inscription of each contains two different letters followed by three different digits?
29. The licence plates for vehicles registered in Delhi consists of three letters of the English alphabet followed by 1, 2, 3 or 4 digits. The letter on the extreme left has to be D. Zero is not allowed immediately after the letters. Find the number of all possible licence plates.
30. Rani has a 10-day holiday. She has 15 friends. She would like to meet exactly one of her friends during each of these 10 days. In how many ways can she arrange the meetings? If she does not want to meet the same friend twice, how many arrangements are possible?
31. A committee of 12 is to be selected from 10 men and 10 women. In how many ways can the selection be carried out if

    (a) there are no restrictions?
    (b) there must be six men and six women?

(c) there must be an even number of women?
(d) there must be more women than men?
(e) there must be at least eight men?

32. In computers, each character is represented by a sequence of eight zeros and ones called a bit pattern. How many bit patterns are there? How many of them will have five ones and how many of them will have an even number of zeros?
33. A gym coach must select 11 seniors to play on a football team. If he can make his selection in 12,376 ways, how many seniors are eligible to play?
34. In how many ways can we place 10 identical green cubes, six identical red cubes and four identical yellow cubes in eight boxes?

FURTHER READING

Grimaldi, R. P. (1999), *Discrete and Combinatorial Mathematics, An Applied Introduction*, Pearson Education, Singapore.

Krishnamurthy, V. (1985), *Combinatorics, Theory and Applications*, Affiliated East West Press Private Ltd, New Delhi.

Vilenkin, N.Ya. (1971), *Combinatorics*, Acadamic Press, New York and London.

CHAPTER

5

Mathematical Induction, Principle of Inclusion and Exclusion, and Pigeon-hole Principle

Learning Objectives

On completing this chapter, you should be able to:

- state the principle of mathematical induction
- prove formulas that are valid for all $n \in \mathbb{N}$ by using the principle of mathematical induction
- state the principle of inclusion and exclusion
- solve counting problems using the principle of inclusion and exclusion
- state and prove the result on the number of functions from a finite set onto another finite set
- state the pigeon-hole principle
- solve simple counting problems using the pigeon-hole principle

Mathematical induction is a technique, which is used in proving statements that are valid for all $n \in N$. We are familiar with the formula

$$1 + 2 + 3 + \ldots n = \frac{n(n+1)}{2}. \tag{5.1}$$

We can easily verify this formula for small values of n by actual computation. Perhaps a very patient person might verify it for $n = 1, 2, \ldots, 1{,}000$. Does it mean that the formula is valid for all $n \in N$? There is no logical reason to conclude so. But we can easily show that

$$1 + 2 + 3 + \ldots k + (k+1) = \frac{(k+1)(k+2)}{2}$$

by assuming that

$$1 + 2 + 3 + \ldots k = \frac{k(k+1)}{2}$$

is valid.

Now, if we denote the assertion (5.1) by $P(n)$, we know that $P(1)$ is true and that

$$P(k) \Rightarrow P(k+1).$$

The principle of mathematical induction allows us to conclude that $P(n)$ is true for all $n \in N$.

The principle of inclusion and exclusion is a counting technique that is proved by mathematical induction. In Chapter 4 we calculated the number of one-to-one functions between finite sets. In this chapter, we shall calculate the number of onto functions between finite sets using the principle of inclusion and exclusion. Pigeonhole principle is yet another counting technique, which asserts that when $|B| < |A|$, there cannot be a one-to-one function $f : A \longrightarrow B$. The proof of this principle also uses mathematical induction.

Induction is the process of generalization starting from observations in some special cases. Historically, mathematical induction has been used in proving statements that have been *guessed* by looking at some special cases. But in a strict sense, mathematical induction is actually a process of deduction.

5.1 MATHEMATICAL INDUCTION

Before we state the principle of mathematical induction, we shall discuss an example useful in computer science. We know that there are many ways to solve a mathematical problem using computers. It is not enough if a computer program just solves the problem correctly but it should solve it efficiently. For example, a programmer would like to minimize the time needed to run the program. To estimate this time duration, it is necessary to find the number of arithmetic operations used in the computations. We now consider the problem of finding the total number of arithmetic operations involved in computing a determinant of order n. Let $u(n)$ be the number of arithmetic operations involved in computing a determinant of order n. Consider a 2×2 determinant

$$\begin{vmatrix} a & b \\ c & d \end{vmatrix} = ad - bc.$$

As we perform two multiplications and one subtraction, $u(2) = 3$. Let us now consider a 3×3 determinant.

$$\begin{vmatrix} a & b & c \\ p & q & r \\ l & m & n \end{vmatrix} = a\begin{vmatrix} q & r \\ m & n \end{vmatrix} - b\begin{vmatrix} p & r \\ l & n \end{vmatrix} + c\begin{vmatrix} p & q \\ l & m \end{vmatrix}.$$

In the above computations, we need to calculate three 2×2 determinants. This involves $9 = 3 \times u(2)$ arithmetic operations. Further, we perform three multiplications and a subtraction and an addition. Hence $u(3) = 9 + 3 + 2 = 14$. While calculating a determinant of order $n + 1$, we need to calculate $n + 1$ determinants of order n. Further, each of this determinant is multiplied by an element in the first row. This step gives us $n+1$ multiplications. Further we need to perform n additions/substractions. Thus, we get the relation

$$u(n+1) = (n+1)u(n) + 2n + 1. \tag{5.2}$$

Equation (5.2) does not give $u(n)$ explicitly. We now give the explicit expression for $u(n)$.

$$u(n) = n!\left[\sum_{k=0}^{n-1} \frac{2k+1}{(k+1)!} - 1\right]. \tag{5.3}$$

We shall prove equation (5.3) in Example 5.1.

Now we state the principle of mathematical induction in the form of a theorem.

Theorem 5.1

Let $P(n)$ be a statement for $n \in N$. If $P(1)$ is true and $P(k) \Rightarrow P(k+1)$, then $P(n)$ is true for all $n \in N$.

Example 5.1

Prove that equation (5.1) holds for all $n \in N$.

SOLUTION

Let us denote equation (5.1) by $P(n)$. The statement $P(1)$ simplifies to

$$1 = \frac{1 \times 2}{2}.$$

which is obvious. Now, let us assume that $P(k)$ is true. Under this assumption, we shall prove that $P(k+1)$ is true. $P(k)$ is

$$1 + 2 + 3 + ... k = \frac{k(k+1)}{2}. \tag{5.4}$$

Adding $k + 1$ to both sides, we get .

$$\begin{aligned} 1 + 2 + 3 + ... k + k + 1 &= \frac{k(k+1)}{2} + k + 1 \\ &= \frac{k(k+1) + 2(k+1)}{2} \\ &= \frac{(k+1)(k+2)}{2}. \end{aligned}$$

Hence, we have

$$1 + 2 + 3 + ... k + k + 1 = \frac{(k+1)(k+2)}{2}.$$

That is, $P(k+1)$ is true. By the principle of mathematical induction $P(n)$ holds for all $n \in N$. ■

Example 5.2

Prove by induction that

$$3 + 11 + \ldots + (8n - 5) = 4n^2 - n.$$

SOLUTION

We have

$$\begin{aligned} P(1) : 3 &= 4 - 1 \\ 3 &= 3. \end{aligned}$$

Hence, $P(1)$ is true. Assume that $P(k)$ is true. That is

$$3 + 11 + \ldots (8k - 5) = 4k^2 - k. \tag{5.5}$$

To prove:

$$3 + 11 + \ldots (8k - 5) + [8(k+1) - 5] = 4(k+1)^2 - (k+1). \tag{5.6}$$

Adding $[8(k+1) - 5]$ to both sides of equation (5.5), we get

$$\begin{aligned} 3 + 11 + \ldots (8k - 5) + [8(k+1) - 5] &= 4k^2 - k + 8k + 8 - 5 \\ &= 4k^2 + 7k + 3. \end{aligned}$$

Now, $4(k+1)^2 - (k+1) = 4k^2 + 8k + 4 - k - 1 = 4k^2 + 7k + 3$. So, we get

$$\begin{aligned} 3 + 11 + \ldots (8k - 5) + [8(k+1) - 5] &= 4k^2 + 7k + 3 \\ &= 4(k+1)^2 - (k+1). \end{aligned}$$

So, equation (5.6) is true. By induction, $P(n)$ is true for all $n \in N$. ■

PRACTICE 5.1

Show by induction that

$$2 + 9 + \cdots + (7n - 5) = n(7n - 3).$$

Example 5.3

Prove by mathematical induction that

$$\sum_{k=1}^{n} k^2 - 3k = \frac{n(n+1)(n-4)}{3} \tag{5.7}$$

for all natural numbers n.

SOLUTION

Let the assertion (5.7) be denoted by $P(n)$. Let us assume that $P(n)$ is true. By adding $(n+1)^2 - 3(n+1)$ to both sides of $P(n)$, we obtain

$$\begin{aligned}
\sum_{k=1}^{n}(k^2 - 3k) + (n+1)^2 - 3(n+1) &= \frac{n(n+1)(n-4)}{3} + (n+1)^2 - 3(n+1) \\
&= \frac{n(n+1)(n-4)}{3} + (n+1)(n+1-3) \\
&= \frac{n(n+1)(n-4)}{3} + (n+1)(n-2) \\
&= \frac{n(n+1)(n-4) + 3(n+1)(n2)}{3} \\
&= \frac{(n+1)[n(n-4) + 3(n-2)]}{3} \\
&= \frac{(n+1)[n^2 - 4n + 3n - 6]}{3} \\
&= \frac{(n+1)(n^2 - n - 6)}{3} \\
&= \frac{(n+1)(n+2)(n-3)}{3} \\
&= \frac{(n+1)(n+1+1)(n+1-4)}{3}.
\end{aligned}$$

Thus,

$$\sum_{k=1}^{n+1}(k^2 - 3k) = \frac{(n+1)(n+1+1)(n+1-4)}{3}$$

which is nothing but $P(n+1)$. Hence the proof follows from induction. ■

PRACTICE 5.2

Show that

$$1 \times 2 + 2 \times 3 + 3 \times 4 + \cdots + n(n+1) = \frac{2n(n+1)(n+2)}{6}$$

for all $n \in N$ using induction.

Example 5.4

Show by induction that

$$\frac{1}{1.2} + \frac{1}{2.3} + \cdots + \frac{1}{n(n+1)} = 1 - \frac{1}{n+1}.$$

SOLUTION

Here, we only give the algebraic steps. The actual proof has to be completed by the reader. Consider

$$\begin{aligned} \frac{1}{1.2} + \frac{1}{2.3} + \cdots + \frac{1}{k(k+1)} + \frac{1}{(k+1)(k+2)} &= 1 - \frac{1}{k+1} + \frac{1}{(k+1)(k+2)} \\ &= 1 + \frac{-(k+2)+1}{(k+1)(k+2)} \\ &= 1 + \frac{-(k+1)}{(k+1)(k+2)} \\ &= 1 - \frac{1}{k+2}. \end{aligned}$$

■

Example 5.5

Show that the number $u(n)$ of arithmetic operations involved in the computation of an $n \times n$ determinant is given by equation (5.3).

SOLUTION

A 1×1 determinant is just a single number and hence there are no arithmetic operations involved in its computation. In other words $u(1) = 0$ and equation (5.3) gives the same result for $n = 1$. Now, let us assume that $u(n)$ is given by equation (5.3) and we shall prove that

$$u(n+1) = (n+1)!\left[\sum_{k=0}^{n} \frac{2k+1}{(k+1)!} - 1\right]. \tag{5.8}$$

From the previous discussion on this problem, we know that $u(n+1) = (n+1)u(n) + 2n + 1$. Refer equation (5.2). Hence, we have

$$
\begin{aligned}
u(n+1) &= (n+1)n!\left[\sum_{k=0}^{n-1}\frac{2k+1}{(k+1)!}-1\right]+2n+1 \\
&= (n+1)!\left[\sum_{k=0}^{n-1}\frac{2k+1}{(k+1)!}+\frac{2n+1}{n+1)!}-1\right] \\
&= (n+1)!\left[\sum_{k=0}^{n}\frac{2k+1}{(k+1)!}-1\right].
\end{aligned}
$$

■

Example 5.6

Show that $8^n - 3^n$ is divisible by 5.

SOLUTION

Let $P(n)$ be the statement: $8^n - 3^n$ is divisible by 5. Now $P(1)$ is the statement: $8^1 - 3^1$ is divisible by 5, which is true.

Let us assume that $P(k)$ is true. That is, we are assuming that $8^k - 3^k$ is divisible by 5. We get the following equivalent statement for $P(k)$:

There is $m \in N$ such that $8^k - 3^k = 5m$. (5.9)

Next, consider

$$
\begin{aligned}
8^{k+1} - 3^{k+1} &= 8(8^k) - 3(3^k) = 8(5m + 3^k) - 3(3^k) \text{ from equation (5.6)} \\
&= 5(8m) + 5(3^k) \\
&= 5(8m + 3^k).
\end{aligned}
$$

■

PRACTICE 5.3

Show that for any positive integer n, $2^{3n} - 7n - 1$ is divisible by 49.
Hint: Let $2^{3n} - 7n - 1 = 49q$. Then, $2^{3n} = 49q + 7n + 1$.

$$
\begin{aligned}
2^{3(n+1)} - 7(n+1) - 1 &= 8(2^{3n}) - 7n - 7 - 1 \\
&= 8(49q + 7n + 1) - 7n - 8 \\
&= 49(8q) + 49n.
\end{aligned}
$$

Example 5.7

Show by induction that $N(N^2 + 5)$ is divisible by 6 for all $N \in N$.

SOLUTION

Define $P(N)$: $N(N^2+5)$ is divisible by 6.

$P(1)$: $1(1+5)$ is divisible by 6.

That is, $P(1)$ is the statement that 6 is divisible by 6. Thus, $P(1)$ is true.

Let us assume that $P(k)$ is true. That is,

$$k(k^2+1) \text{ is divisible by } 6.$$

To prove that $P(k+1)$ is true, we have to show:

$$(k+1)[(k+1)^2+5)] \text{ is divisible by } 6.$$

Now,

$$\begin{aligned}
(k+1)[(k+1)^2+5] &= (k+1)(k^2+2k+1+5) \\
&= k(k^2+2k+1+5)+(k^2+2k+1+5) \\
&= k(k^2+5)+k(2k+1)+(k^2+2k)+6 \\
&= k(k^2+5)+k(2k+1)+k(k+2)+6 \\
&= k(k^2+5)+k(2k+1+k+2)+6 \\
&= k(k^2+5)+3k(k+1)+6.
\end{aligned}$$

Now, either k or $(k+1)$ is even. Hence $3k(k+1)$ is divisible by 6. By inductive hypothesis, $k(k^2+5)$ is divisible by 6. Now, it is clear that $(k+1)[(k+1)^2+5]$ is divisible by 6 as it is the sum of numbers divisible by 6. So, by induction the result follows. ■

Example 5.8

Prove by induction that for every $n \geq 4$, $n! > 2^n$.

SOLUTION

$$\begin{aligned}
P(n) &: n! > 2^n \\
P(4) &: 4! > 2^4 \\
& \quad 24 > 16.
\end{aligned}$$

So, $P(4)$ is true. Now, let us assume that $P(k)$ is true for $k \geq 4$. That is

$$k! > 2^k. \tag{5.10}$$

To prove:

$$(k+1)! > 2^{k+1}. \tag{5.11}$$

Consider $(k+1)! = k!(k+1)$. Because of equation (5.10), we have

$$(k+1)! = k!(k+1) > 2^k(k+1).$$

Now, as $k \geq 4$, $k+1 > 2$, So,

$$(k+1)! > 2^k(k+1) > 2^k 2 = 2^{k+1}.$$

Hence equation (5.11) is proved and the result follows from induction. ■

The following theorem is intuitively obvious. Nevertheless, we need to prove it. It is an important theorem in the theory of finite sets. (See Section 2.5.)

Theorem 5.2

Let $n, m \in N$ be such that $m < n$. Then, there is no bijection $f : \{1, 2, ..., m\} \longrightarrow \{1, 2, ..., n\}$.

Proof

It is clear that $n \geq 2$. For $n = 2$, the only possible value of $m = 1$. It is obvious that there is no bijection between the sets $\{1\}$ and $\{1, 2\}$. Now we assume that for $k, m \in N$ with $m < k$, there is no bijection between $\{1, 2, ..., m\}$ and $\{1, 2, ..., k\}$. We shall prove that for $m < k+1$, there is no bijection between $\{1, 2, ..., m\}$ and $\{1, 2, ..., k+1\}$. By induction, the theorem would follow.

Let us suppose that there is a bijection $f : \{1, 2, ..., m\} \longrightarrow \{1, 2, ..., k+1\}$. We shall derive a contradiction. First let $m = 1$. Since $1 < k+1$, $k+1 \geq 2$.

If $f(1) = 1$, then $2 \in \{1, 2, ..., k+1\}$ does not have a pre-image. If $1 < f(1)$, then $1 \in \{1, 2, ..., k+1\}$ does not have a pre-image. The proof for $m = 1$ is complete.

Let $m \geq 2$. We shall define $g : \{1, 2, ..., m-1\} \longrightarrow \{1, 2, ..., k\}$, which is a bijection. This will be a contradiction to the inductive hypothesis and the proof will be complete.

If $f(m) = k+1$, then g can be defined as $g(x) = f(x)$ for $x \in \{1, 2, ..., m-1\}$. It is easy to show that g is a bijection.

If $f(m) = j \neq k+1$, then $f^{-1}(k+1) \in \{1, 2, ..., m-1\}$. Let $i = f^{-1}(k+1)$. Define $g : \{1, 2, ..., m-1\} \longrightarrow \{1, 2, ..., m-1\}$ as

$$\begin{aligned} g(x) &= f(x), \quad x \neq i \\ g(i) &= j. \end{aligned}$$

It is easy to check that g is a bijection and the proof is complete. ■

PRACTICE 5.4

Let A be a finite set and let B be a non-empty proper subset of B. Then prove that B is finite and $0 < |B| < |A|$.
Hint: Use induction on the cardinality of $|A|$. Further, use the following lemma, which can be easily proved.

Lemma 5.1

Let A be a set, $n \in N$ and $a \in A$. There is a bijection $f : \{1, 2, 3, ..., n+1\}$ if and only if there is a bijection from $A - \{a\} \longrightarrow \{1, 2, ..., n\}$.

Example 5.9

Show by induction that *every* set having n elements contains 2^n subsets.

SOLUTION

Consider a set containing one element $\{a\}$. The subsets are clearly $\{a\}$ and the empty set. Thus, the number of subsets of a one element subset is $2 = 2^1$.

Now, let us assume that the number of subsets of a set having k elements is 2^k. Consider a set $A = \{a_1, a_2, ..., a_{k+1}\}$, which contains $k+1$ elements. Now, the subsets of A can be divided into two disjoint groups, namely, subsets containing a_{k+1} and those that do not contain a_{k+1}. Now, the subsets not containing a_{k+1} are precisely the subsets of $B = \{a_1, a_2, ..., a_k\}$. They are 2^k in number. Any subset of A containing a_{k+1} is obtained by the union of a subset of B and $\{a_{k+1}\}$. Hence, the number of subsets of A containing a_{k+1} is also 2^k in number.

So, the total number of subsets of A is $2^k + 2^k = 2(2^k) = 2^{k+1}$.

We have proved that the number of subsets of a set containing $k+1$ elements is 2^{k+1}. The result follows from induction. ■

Example 5.10

If $A = \begin{pmatrix} 3 & -4 \\ 1 & -1 \end{pmatrix}$, show that for all $n \in N$,

$$A^n = \begin{pmatrix} 1+2n & -4n \\ n & 1-2n \end{pmatrix}.$$

SOLUTION

Let $P(n)$ be the statement

$$A^n = \begin{pmatrix} 1+2n & -4n \\ n & 1-2n \end{pmatrix}.$$

$P(1)$ reduces to

$$\begin{pmatrix} 3 & -4 \\ 1 & -1 \end{pmatrix} = \begin{pmatrix} 3 & -4 \\ 1 & -1 \end{pmatrix}$$

which is obviously true. Let $P(k)$ be true. That is

$$A^k = \begin{pmatrix} 1+2k & -4k \\ k & 1-2k \end{pmatrix}. \tag{5.12}$$

Now,

$$\begin{aligned} A^{k+1} &= A^k A \\ &= \begin{pmatrix} 1+2k & -4k \\ k & 1-2k \end{pmatrix}\begin{pmatrix} 3 & -4 \\ 1 & -1 \end{pmatrix} \quad \text{(using the inductive hypothesis (5.12))} \end{aligned}$$

$$= \begin{pmatrix} (1+2k)3-(4k)(1) & (1+2k)(-4)+4k \\ 3k+1-2k & 1-2(k+1) \end{pmatrix}$$

$$= \begin{pmatrix} 3+2k & -4-4k \\ 1+k & -1-2k \end{pmatrix}$$

$$= \begin{pmatrix} 1+2(k+1) & -4(k+1) \\ k+1 & 1-2(k+1) \end{pmatrix}.$$

We get

$$A^{k+1} = \begin{pmatrix} 1+2(k+1) & -4(k+1) \\ k+1 & 1-2(k+1) \end{pmatrix}$$

which implies that $P(k+1)$ is true. Hence, by induction the required theorem is proved. ■

In Chapter 4, we proved the multinomial theorem, which is the generalization of the binomial theorem. Here, we give an inductive proof of the binomial theorem.

Example 5.11

Prove the binomial theorem for positive integral index by mathematical induction.

SOLUTION

We need to prove

$$(x+y)^n = {}^nC_0x^n + {}^nC_1x^{n-1}y + \cdots + {}^nC_rx^{n-r}y^r + \cdots + {}^nC_ny^n.$$

Denote the given statement as $P(n)$. ${}^1C_0 = {}^1C_1 = 1$ and so the RHS of $P(1)$ becomes $(x+y)^1 = x+y$, which is true. Assume that $P(k)$ is true. That is,

$$(x+y)^k = {}^kC_0x^k + {}^kC_1x^{k-1}y + \cdots + {}^kC_rx^{k-r}y^r + \cdots + {}^kC_ky^k.$$

Next, using the above, we get

$$\begin{aligned}(x+y)^{k+1} &= (x+y)^k(x+y) \\ &= ({}^kC_0x^k + {}^kC_1x^{k-1}y + \cdots + {}^kC_rx^{k-r}y^r + \cdots + {}^kC_ky^k)(x+y).\end{aligned}$$

The term containing $x^{k+1-r}y^r$ in the expansion of $(x+y)^{k+1}$ is the sum of the two terms containing $(x^{k-r}y^r)x$ and $(x^{k+1-r}y^{r-1})y$.

Hence, the coefficient of $x^{k+1-r}y^r$ is ${}^kC_r + {}^kC_{r-1}$, but ${}^kC_r + {}^kC_{r-1} = {}^{k+1}C_r$. Thus,

$$(x+y)^{k+1} = {}^{k+1}C_0x^{k+1} + {}^{k+1}C_1x^ky + \cdots + {}^{k+1}C_rx^{k+1-r}y^r + \cdots - {}^{k+1}C_{k-1}y^{k+1}.$$

Here we have used ${}^kC_0 = {}^{k+1}C_0$ and ${}^kC_k = {}^{k+1}C_{k+1}$. As we have proved $P(k+1)$ is true, the theorem is proved. ■

Example 5.12

Show by induction that

$$1^2 - 2^2 + 3^2 + \cdots + (-1)^{n+1}n^2 = \frac{(-1)^{n+1}n(n+1)}{2}.$$

SOLUTION

We only show the algebraic steps and the reader has to supply the complete proof.

$$\begin{aligned} &1^2 - 2^2 + 3^2 + \cdots + (-1)^{k+1}k^2 + (-1)^{k+2}(k+1)^2 \\ &= \frac{(-1)^{k+1}k(k+1)}{2} + (-1)^{k+2}(k+1)^2 \\ &= \frac{(-1)^{k+1}(k+1)}{2}[k - 2(k+1)] \\ &= \frac{-1^{k+1}(k+1)}{2}(-k-2) \\ &= \frac{-1^{k+2}(k+1)(k+2)}{2}. \end{aligned}$$

■

5.2 PRINCIPLE OF INCLUSION AND EXCLUSION

Imagine a group of music lovers. Let us also presume that everyone in this group likes either Carnatic music or Hindusthani music. Let us say that there are 15 persons who like Carnatic and 21 persons who like Hindusthani. Can we now assert that the total number of persons in this group is 36? Obviously no. For, there may be some persons who like both. Unless we know the number of people who like both kinds of music, we cannot find the total number of people in this group. If seven persons like both kinds of music, then the total number of persons in this group is $21 + 15 - 7$ (Fig. 5.1).

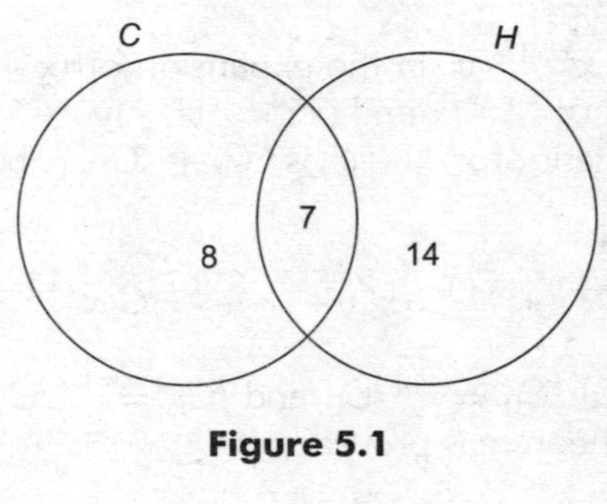

Figure 5.1

As the seven persons are *included* twice when 21 and 15 are added, we *exclude* it to find the precise total. Keeping this in mind, we proceed to formally state the principle of inclusion and exclusion.

Principle of inclusion and exclusion for two sets: Let A and B be finite sets. Then,

$$|A \cup B| = |A| + |B| - |A \cap B|.$$

Example 5.13

Everyone in a group of 19 students have passed in at least one of the subjects, Discrete Mathematics and Mathematics III. Nine students have passed in Discrete Mathematics and three have passed in both. How many students have passed in Mathematics III?

SOLUTION

Let A be the set of students who have passed in Discrete Mathematics and B be the set of students who have passed in Mathematics III. Given that $|A \cup B| = 19$, $|A| = 9$ and $|A \cap B| = 3$.

By the principle of inclusion and exclusion,

$$\begin{aligned} |A \cup B| &= |A| + |B| - |A \cap B| \\ 19 &= 9 + |B| - 3 \\ |B| &= 13. \end{aligned}$$

■

Example 5.14

Out of 200 students, 50 take the course Discrete Mathematics (DM), 140 take the course Economics and 24 take both the courses. How many of them do not take either of these courses?

SOLUTION

Let A be the set of students taking DM and B be the set of students taking Economics. Now,

$$\begin{aligned} |A \cup B| &= |A| + |B| - |A \cap B| \\ &= 50 + 140 - 24 = 166. \end{aligned}$$

We want to find the number of elements in $(A \cup B)^c$, which is equal to 200 − 166 = 34. ■

PRACTICE 5.5

If A, B, C are finite sets, prove that

$$|A \cup B \cup C| = |A| + |B| + |C| - |A \cap B| - |A \cap C| - |B \cap C| + |A \cap B \cap C|.$$

Hint: $(A \cup B \cup C) = (A \cup B) \cup C$.

PRACTICE 5.6

Amongst a group of 50 persons, 22 persons like Ilayaraja's music, 20 persons like R.D. Burman's music and 17 persons like Bupendra Hasarika's music. Ten of them like both Ilayaraja and Burman, four of them like Burman and Bupendra and five of them like Ilayaraja and Bupendra. Three of them like the music of all these three music directors. How many in the group do not like the music of any of these three music directors? How many like only R.D. Burman's music?

Example 5.15

Let us consider the following problem, which can be solved by using ideas from Chapter 4.

A group of five persons includes Mr J and Ms K. Find the number of ways in which this group can be seated so that Mr J and Ms K do not sit next to each other.

We can do this in two different ways. First arrange the remaining three persons. J and K can now be seated at the ends or in the two vacant places between the three already seated . This can be done in $3! \times 4P_2 = 72$ ways.

Now, using a different approach, let us calculate the number of ways in which J and K are together. This can be done in $4! \times 2$ ways. The required answer is $5! - (4! \times 2!) = 72$. ■

Next, imagine a group that includes *two pairs* of rivals. In the next example, we use the principle of inclusion and exclusion to find out the number of ways in which rivals do not sit next to each other.

Example 5.16

Consider a group of five, which includes the persons J, K, I and G. Here J and K should not sit next to each other and I and G should not sit next to each other. As in the previous case, the number of ways in which J and K can sits together is 48 and the number of ways in which I and G can sit together is 48. But the total number of ways in which at least one pair sits together is not 96. The number of ways in which J and K sit together and I and G sit together is $2! \times 2! \times 3! = 24$.

So, by the principle of inclusion and exclusion, the number of ways in which at least one pair sit together is $48+48-24 = 72$. Hence, the number ways in which the enemies do not sit next to each other is $5!-72 = 48$. ■

Example 5.17

In how many ways can five letters be chosen out of the 26 English letters, if the choice must contain either A or I?

SOLUTION

Let P be the set of selections of five letters that includes A and let Q be the set of selections of five letters, which include I. $|P| = {}^{25}C_4$ and $|Q| = {}^{25}C_4$. Now, the required number of selections is $|P \cup Q|$. To calculate this, we should find $|P \cap Q|$. But, $P \cap Q$ is the set of selections of five letters, which include both A and I.

Hence, $|P \cap Q| = {}^{24}C_3$. Thus, the required answer is

$$\begin{aligned} |P \cup Q| &= |P| + |Q| - |P \cap Q| \\ &= {}^{25}C_4 + {}^{25}C_4 - {}^{24}C_3. \end{aligned}$$

■

Example 5.18

Find how many numbers are there between 1 and 65, which are divisible by any one of 2, 3 and 5.

SOLUTION

The number of positive integers between 1 and N that are divisible by n is given by $\left\lceil \frac{N}{n} \right\rceil$. Here $\lceil k \rceil$ is the largest integer less than or equal to k. Now, let A_1 be the set of numbers between 1 and 65, which are divisible by 2; let A_2 be the set of numbers between 1 and 65, which are divisible by 3 and let A_3 be the set of numbers between 1 and 65, which are divisible by 5.

Now, $|A_1| = \frac{65}{2} = 32$, $|A_2| = \frac{65}{3} = 21$, $|A_3| = \frac{65}{5} = 13$. Now $A_1 \cap A_2$ is the set of numbers between 1 and 65, which are divisible by both 2 and 3. Hence, it is the set of numbers between 1 and 65, which are divisible by 6. Hence, $|A_1 \cap A_2| = \frac{65}{6} = 10$. Similarly, $|A_2 \cap A_3| = \frac{65}{15} = 4$. $|A_1 \cap A_3| = \frac{65}{10} = 6$ and finally $|A_1 \cap A_2 \cap A_3| = \frac{65}{30} = 2$. The required answer is $|A_1 \cup A_2 \cup A_3|$.

$$\begin{aligned} |A_1 \cup A_2 \cup A_3| = &|A_1| + |A_2| + |A_3| - |A_1 \cap A_2| \\ &- |A_2 \cap A_3| - |A_1 \cap A_3| + |A_1 \cap A_2 \cap A_3| \\ &= 32 + 21 + 13 - 10 - 4 - 6 + 2 = 48. \end{aligned}$$

■

PRACTICE 5.7

State the principle of inclusion and exclusion for four sets. Using it, find the number of integers between 1 and 250 that are divisible by any one of 2, 3, 5 and 7.

PRACTICE 5.8

State and prove the principle of inclusion and exclusion for n sets using induction

PRACTICE 5.9

Find the number of permutations of $a, b, c, d...z$ in which the words Tamil, pun and horse do not occur.

5.2.1 Number of Onto Functions from an n-Set into an m-Set

Suppose Rani has 10 days holidays and that she has 15 friends. Suppose she does not want to meet more than one friend on any day. It is obvious that she cannot meet all her friends during her holidays. She can arrange her meetings in 10^{15} ways. Now, the maximum number of friends she can meet is 10 and this would be achieved if she does not meet the same friend twice. She can carry this out in ${}^{15}P_{10}$ ways.

Now, let us consider the situation that she has 15 days holidays and 10 friends. Again, if she wants to meet only one friend per day, she can do so in 15^{10} ways. But unlike the previous situation, it is possible that she can meet all her friends at least once. Now, we would like to find the number of possible ways of doing this.

If A is the set of holidays and B is the set of friends, we can define a function $f : A \longrightarrow B$ as $f(x)$ is the friend to be met on day x. If this function is onto, then Rani meets all her 10 friends. Thus, the problem is equivalent to finding the number of onto functions from H into S. We solve this problem in the next theorem.

Theorem 5.3

Let $|A| = n$ and $|B| = m$ and $n \geq m$. The number of onto functions $f : A \longrightarrow B$ is given by

$$m^n - [n(m-1)^n - {}^nC_2(m-2)^n + {}^nC_3(m-3)^n \ldots + (-1)^m m].$$

Proof

Let us denote the elements of B by $y_1, y_2, \ldots, y_m$.

Let S_i be the set of all functions from A into B such that their ranges do not contain y_i. Thus, the number of elements of S_i is the number of all functions from $A \longrightarrow B - \{y_i\}$; hence, $|S_i| = (m-1)^n$. Now, $S_i \cap S_j$ is the set of functions whose ranges do not contain either of y_i and y_j. Hence $|S_i \cap S_j| = (m-2)^n$.

Similarly, we have $|S_i \cap S_j \cap S_k| = (m-3)^n$.

Now, the set of all functions whose range does not contain at least one of y_i is the union of S_i's. By the principle of inclusion and exclusion, we have

$$
\begin{aligned}
|S_1 \cup S_2 \cup \ldots S_m| &= |S_1| + |S_2| \ldots + |S_m| - \left(\sum_{i<j} |S_i \cap S_j| \right) \\
&\quad + \sum_{i<j<k} |S_i \cap S_j \cap S_k| \ldots + (-1)^{m+1} |S_1 \cap S_2 \cap \ldots S_m| \\
&= m(m-1)^n - {}^m C_2(m-2)^n + {}^m C_3(m-3)^n \ldots \\
&\quad + [(-1)^m]\,{}^mC_{m-1}(m-(m-1))^n + [(-1)^{m+1}]\,{}^mC_m(m-m)^n.
\end{aligned}
$$

Thus, the number of onto functions is

$$m^n - [m(m-1)^n - {}^mC_2(m-2)^n + {}^mC_3(m-3)^n \ldots + (-1)^m m]. \quad \blacksquare$$

PRACTICE 5.10

Mani has a 12-day holiday. He and his friends know football, cricket and badminton and they want to play exactly one game each day such that each of the three games is played on at least one day. How many such game schedules be formed?

5.3 PIGEON-HOLE PRINCIPLE

Consider a group of seven persons. Can we say that at least two of them would be born on the same day of the week? No. It is possible that each one is born on a different day of the week. But now let us imagine a group of eight persons. Clearly, it is possible that two of them could have been born on the same day of the week. Even if seven persons were to be born on seven different days, the eighth person should share the day of her/his birth with one of the previous persons. This is an example of the consequence of the *pigeon-hole principle*. This principle is also known as the Shoebox principle or Dirichlet's Drawer principle.

Pigeon-hole principle: If there are m boxes, and n objects are distributed in them, where $m < n$, at least one box should contain at least two objects.

A mathematically precise statement is as follows:

Let A and B be finite sets with $|A| = n$ and $|B| = m$. Further, let $m < n$. If $f : A \longrightarrow B$ is a function, then it cannot be one-to-one. That is, there must exist *distinct* elements $a, b \in A$ with $f(a) = f(b)$.

PRACTICE 5.11

Prove the above statement by induction on n.
Consider m boxes and n objects that are distributed in them with $m < n$. Let A be the set of objects and B be the set of boxes. Define an appropriate function $f : A \longrightarrow B$, which tells us that at least one box must contain at least two objects.

Example 5.19

How many cards from a standard pack of 52 must be picked to get two cards of the same denomination?

SOLUTION

Ace, 2, 3,..., Jack, King and Queen are the denominations that are 13 in number. Hence, by the pigeon-hole principle, if we select 14 cards, then two of them will be of the same denomination. ■

PRACTICE 5.12

How many cards from a standard pack of 52 must be picked to get two cards of the same suit? (Club, Spade, Heart and Diamond are the suits.)

PRACTICE 5.13

Suppose we pick up nine cards. Can we say that at least three of them would be of the same suit?

Example 5.20

Show that among Kannan, Kamala, Arulmozhi, Bharath, Peter, Mary, Ahmad, Begum, John, Ramasamy, Vignesh, Swanika and Yazhini, at least two of them must have been born on the same month.

There are 13 persons and 12 months. Hence, by the pigeon-hole principle at least two of them must have been on the same month. (Not necessarily in the same year.) ■

PRACTICE 5.14

Among how many people can at least two have the same date of birth?

The above examples might give the impression that not much can be achieved using the pigeon-hole principle, which is simple and obvious. Below, we shall give examples of interesting and difficult problems that can be solved by a clever use of this simple principle.

Example 5.21

Show that if any 11 numbers are chosen from the set $\{1, 2, 3, ..., 20\}$, then at least one of them will be a multiple of the other.

Proof

Using prime factorization, any positive integer n can be written as $2^k m$ where $k \leq 0$ and m is odd. We call m as the odd part of n. Now, for numbers between 1 and 20, the only possible odd parts are 1, 3, 5, 7, 9, 11, 13, 15, 17, 19. These are 10 in number. So, by the pigeon-hole principle, when 11 numbers are chosen, two of them must have the same odd part. If these two numbers are n_1 and n_2, we can write them as $n_1 = 2^{k_1} m$ and $n_2 = 2^{k_2} m$. If $k_1 > k_2$, then n_1 is a multiple of n_2 and if $k_2 > k_1$, then n_2 is a multiple of n_1. ■

PRACTICE 5.15

Show that among four integers at least two of them will give the same remainder when divided by three.

Example 5.22

Show that if any five numbers from 1 to 8 are chosen, then two of them will add up to 9.

SOLUTION

Consider the pairs $(1, 8)$, $(2, 7)$, $(3, 6)$, $(4, 5)$, which are *four* in number. If we choose *five* numbers from 1 to 8, by the pigeon-hole principle two of them must belong to one of the above pairs. Hence, their sum must be nine. ■

PRACTICE 5.16

How many numbers must be selected from the set $\{2, 4, 6, 8, 10, 12, 14, 16, 18, 20\}$ to guarantee that at least one pair adds up to 22?

PRACTICE 5.17

Consider the equilateral triangular region shown in Figure 5.2. The side of the triangle has three units of length. Show that if any seven points are chosen in this region, then two of them must be no farther apart than 1 unit.

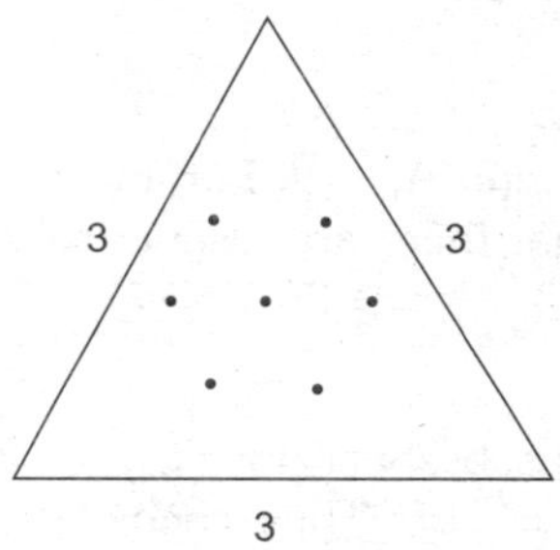

Figure 5.2

Example 5.23

Let S be a subset of N containing at least three elements. Show that there exists at least two distinct numbers in S such that $x + y$ is even.

SOLUTION

Let A be the set of odd natural numbers and B be the set of even natural numbers. S contains at least three elements and hence by the pigeon-hole principle, either A or B must contain at least two elements of S. Now, as these two elements are either both odd or both even, their sum is even. ■

PRACTICE 5.18

Let n be a positive integer. Show that in any set of n consecutive integers there is exactly one integer divisible by n.

PRACTICE 5.19

Let d be a positive integer. Show that among any group of $d+1$ integers there are at least two with the same remainder when they are divided by d.

Example 5.24

Show that for any positive integer n, there is a number composed of only digits 0 and 1, which is a multiple of n.

SOLUTION

Consider the sets $S = \{1, 11, 111, 1111, ...\}$,
$S_k = \{x \in S :$ the remainder when x is divided by n is $k\}$, $k = 0, 1, 2...n-1$.

Every element of S must belong to one S_k. Since S is an infinite set, by the pigeon-hole principle, one of S_k must contain two elements of S. Let $p, q \in S_k \cap S$, and without loss of generality assume that $p > q$. Then, $p - q$ is a multiple of n as p and q leave the same remainder when divided by k. Further, it is clear that $p - q$ is composed only of digits 0 and 1. ■

Example 5.25

Let $A \subseteq \{1, 2, 3, ..., 25\}$ where $|A| = 9$. For any subset B of A let S_B be the sum of elements of B. Prove that there are distinct subsets C and D of A such that $|C| = |D| = 5$ and $S_C = S_D$.

SOLUTION

The number of five-element subsets of A is ${}^9C_5 = 126$. Now, for any $B \subset A$, $1 \leq S_B \leq 25+24+23+22+21 = 105$. Thus, each of the 126 sets have to be assigned a total, which comes from a collection of 105 numbers. Thus, at least two of them must have the same total. ■

SUMMARY

- Let $P(n)$ be a statement function with $n \in N$. The principle of mathematical induction asserts that

$$[P(1) \wedge (P(n) \rightarrow P(n+1))] \rightarrow [\forall n P(n)].$$

- For finite sets A, B and C, we have

$$n(A \cup B \cup C) = n(A) + n(B) + n(C) - n(A \cap B) - n(A \cap C) - n(B \cap C) + n(A \cap B \cap C).$$

- If A contains n elements and B contains m elements and $m < n$, then for any function $f : A \longrightarrow B$, there must be two elements x and y of A such that $f(x) = f(y)$.

PROBLEMS

1. Prove the following by induction:

   (i) $(\cos\theta + i\sin\theta)^n = \cos n\theta + i\sin n\theta$ for all $n \in \mathbb{N}$.

   (ii) $1^3 + 2^3 + 3^3 + \cdots + n^3 = \left(\frac{n(n+1)}{2}\right)^2$.

   (iii) $1^2 + 3^2 + 5^2 + \cdots (2n-1)^2 = \frac{n(4n^2-1)}{3}$.

   (iv) $\sum_{i=1}^{n} \frac{i}{i^4 + i^2 + 1} = \frac{n(n+1)}{2(n^2+n+1)}$.

   (v) $\sum_{k=1}^{n} \frac{2}{4k^2 + 8k + 3} = \frac{2n+1}{4n+6}$.

   (vi) $\sum_{k=1}^{n} 2k^2 + 5k = \frac{n(n+1)(4n+17)}{6}$.

   (vii) $\begin{pmatrix} \cos\theta & -\sin\theta \\ \sin\theta & \cos\theta \end{pmatrix}^n = \begin{pmatrix} \cos n\theta & -\sin n\theta \\ \sin n\theta & \cos n\theta \end{pmatrix}$.

   (viii) $\begin{pmatrix} 1-a & a \\ b & 1-b \end{pmatrix}^n = \begin{pmatrix} \frac{b + a(1-a-b)^n}{a+b} & \frac{a - a(1-a-b)^n}{a+b} \\ \frac{b - a(1-a-b)^n}{a+b} & \frac{a + b(1-a-b)^n}{a+b} \end{pmatrix}$.

2. Let m be a fixed natural number. Further, let $P(m)$ be true and $P(n) \rightarrow P(n+1)$. Defining $Q(n) = P(n-m+1)$ for all $n \geq m$, and using the principle of mathematical induction for $Q(n)$, show that $P(n)$ is true for all $n \geq m$.
3. Show by induction that $n! > 3^n$ for all $n \geq 7$.
4. Prove by induction that the number of multiplications involved in the computation of a determinant of order n is given by $n!$
5. Find the number of permutations of $a, b, c, d \ldots z$ in which the words *john* and *ali* do not occur.
6. Among the set of bytes, how many either begin with 1 or end with 1?
7. In a class of 37 students, 17 can speak Spanish and 21 can speak German. Seven can speak both. How many of them can speak neither of these two languages? How many of them can speak only one of these two languages?
8. In a group of 37 people, 14 are Americans, 20 have visited Mahabalipuram and 25 like pani puri. Among the people who have visited Mahabalipuram 11 like pani puri and 9 are Americans. Four persons are Americans who have visited Mahabalipuram and they like pani puri. In this group, how many Americans like pani puri?
9. In a group of people, each one is either a woman or a professor or a singer. Seventeen of them are professors, and 13 are singers. Nine professors are singers, five woman are professors and six singers are ladies. How many males are there in this group?
10. Let $A = \{x_1, x_2, \ldots, x_n\}$ and $B = \{y_1, y_2, \ldots, y_m\}$. Find the number of functions $f : A \longrightarrow B$ such that $f(A)$ contains either y_1 or y_2.
11. Let O be the origin in the xy plane and P, Q, R, S, T, U and V be, respectively, the points (8, 7), (1, 4), (2, 4), (2, 5), (3, 5), (4, 6) and (5, 6). Find the number of

    (i) shortest OP paths.

    (ii) shortest OP paths that pass through QR.

(iii) shortest OP paths that pass through at least one of the segments QR, ST, UV.
(iv) shortest OP paths that do not pass through any one of the segments QR, ST, UV.

12. Let A be a set containing n elements. Show that the number of onto functions $f : A \longrightarrow A$ is $n!$
13. There is a group of 15 children and 7 teachers. Every child should have precisely one of the teachers as a guardian and every teacher should be a guardian of at least one child. Find the number of all possible ways of assigning guardians to the children.
14. Find the number of ways of distributing eight objects in three boxes in such a way that no box is empty.
15. Show that if any 14 integers are selected from the set $S = \{1, 2, 3, ..., 25\}$, there are at least two whose sum is 26.
16. Given eight Pascal books, 17 Fortran books, six Java books, 12 C books and 20 C++ books, how many of these books must we select to ensure that we have 10 books dealing with the same computer language?
17. From the integers 1, 2, 3,..., 200, 101 of them are chosen randomly. Show that, among the chosen numbers, there exists at least two such numbers such that one number divides the other.
18. Show that if there are n persons (with $n \geq 2$) who can arbitrarily shake hands with one another, there is always a pair of persons who shake the same number of hands.

FURTHER READING

Grimaldi, R. P. (1999), *Discrete and Combinatorial Mathematics, An Applied Introduction*, Pearson Education, Singapore.

Joshi, K. D. (1989), *Foundations of Discrete Mathematics*, New Age International (P) Limited, New Delhi.

CHAPTER

6

Recurrence Relations

Learning Objectives

On completing this chapter, you should be able to:

- formulate recurrence relations to solve problems involving an unknown sequence
- solve certain first order and second order recurrence relations
- define the generating function of a sequence
- solve second order recurrence relations by finding the corresponding generating functions

6.1 INTRODUCTION

Can you guess the next number in the following sequence?

$$4, 11, 18, \ldots$$

A fairly intelligent fifth standard girl can answer this question. The child perceives that the successive terms differ by 7, and so she adds 7 to the last term given and gives the answer as 25. Mathematically speaking, the child realizes that the difference between the $(n+1)$th term and nth term is 7. If we denote the nth term by x_n, then the above sequence has the property

$$x_{n+1} - x_n = 7. \tag{6.1}$$

Now, let us consider the sequence

$$4, 13, 28, 49, \ldots$$

Based on our experience with the earlier example, it is tempting to look at the difference between successive terms. But if we do so, we find that the difference is not a constant. In fact we get the following sequence:

$$9, 15, 21, \ldots$$

Let us denote this sequence by d_n. Now, we find that the *sequence of differences* satisfy

$$d_{n+1} - d_n = 6.$$

In terms of the given sequence, which we denote by y_n, $d_n = y_{n+1} - y_n$ and so the original sequence satisfies

$$y_{n+2} - y_{n+1} - (y_{n+1} - y_n) = 6$$

which on simplification gives

$$y_{n+2} - 2y_{n+1} + y_n = 6. \tag{6.2}$$

Equations (6.1) and (6.2) are examples of recurrence relations. A recurrence relation is also known as a difference equation. Equation (6.1) is an example of a first order recurrence relation. Equation (6.2) is an example of a second order recurrence relation.

Let us take equation (6.1). We note that if we want to find the value of x_{10}, then it is enough to know x_9. But in the case of equation (6.2), if we want to find the value of x_{10} we need the values of both x_9 and x_8. This observation tells us why equation (6.1) is known as a first order recurrence relation and equation (6.2) is known as a second order recurrence relation.

A general first order recurrence relation is of the form

$$f(x_n, x_{n+1}) = 0.$$

But in this chapter we are interested only in *linear* recurrence relations. A general linear first order recurrence relation is given by

$$p_n x_{n+1} + q_n x_n = r_n. \tag{6.3}$$

Here p_n, q_n and r_n are *known* sequences. Under the assumption that $p_n \neq 0$ for all n, we divide equation (6.3) by p_n and let $a_n = \frac{q_n}{p_n}$ and $f_n = \frac{r_n}{p_n}$. Thus, equation (6.3) now becomes

$$x_{n+1} + a_n x_n = f_n. \tag{6.4}$$

In equation (6.1), a_n and f_n are the constant sequences $a_n = -1$ and $f_n = 0$. In equation (6.2), $q_n = -1$ and $f_n = 7$. We shall solve equation (6.4) in the next section.

A general second order linear recurrence relation is given by

$$a_n x_{n+1} + b_n x_n + c_n x_{n-1} = f_n. \tag{6.5}$$

Now, it is easy to see that the sequence x_n defined as $x_n = 4 + 7(n-1) = 7n - 3$ satisfies equation (6.1). The sequence y_n defined as $y_n = 1 + 3n^2$ satisfies equation (6.2). These formulae give explicit expressions to find the nth term of a sequence.

Given that an unknown sequence satisfies the recurrence relation (6.5), can we find an explicit formula for it?

$x_{n+1} + n^2 x_n = 0$ is an example of a linear recurrence relation with variable coefficients.

$x_{n+1} = x_n^2$ is an example of a nonlinear recurrence relation.

In Section 6.4, we shall learn procedures for solving equation (6.5).

6.2 MOTIVATION

Why do we have to study recurrence equations? So far, our motivating examples have been nothing more than puzzles and one could have easily guessed the explicit formulas for these puzzles. Is it useful to study recurrence relations and their solutions? In the sequel, we present some examples to answer this question.

In Chapter 5, we discussed the necessity to know the number of arithmetic operations in computations. This helps us in writing *efficient computer programs*. As an example, we studied the number of arithmetic operations involved in the computation of an $n \times n$ determinant. We proved that if $u(n)$ is the number of arithmetic operations involved in the computation of an $n \times n$ determinant, then

$$u(n+1) = (n+1)u(n) + 2n + 1. \tag{6.6}$$

We also proved using mathematical induction that

$$u(n) = n!\left[\sum_{k=0}^{n-1} \frac{2k+1}{(k+1)!} - 1\right]. \tag{6.7}$$

Let us observe the fact that induction could not have helped us in *discovering* equation (6.7). After writing down equation (6.7), induction can prove it. But how does one *obtain* equation (6.7) starting from equation (6.6)? We note that equation (6.6) is a recurrence relation for $u(n)$ and in the next section we shall see how one can obtain equation (6.7) from equation (6.6).

We know that differential equations are very important in many applied fields and for many differential equations, it is impossible to obtain their exact solutions. Hence, we have to calculate the solutions *numerically*. Most of the time the derivative of the unknown function $\frac{dy}{dx}$ is replaced by *finite differences* like $\frac{y_{n+1}-y_n}{h}$. Here y_n is the approximate value of the function y at some discrete point $x_o + nh$. Here h is called the step-size. As a result of these manipulations, *differential equations get transformed into recurrence relations or difference equations*.

Example 6.1

Consider the initial value problem:

$$\begin{aligned} y'(x) &= -3y(x) + x \\ y(0) &= 1. \end{aligned} \tag{6.8}$$

Now, we are interested in finding the approximate values of $y(x)$ for $x = 0, 0.1, 0.2, \ldots$. In other words, we would like to approximate $y(x_n)$ where $x_n = (0.1)n$. Let y_n be the approximation of $y(x_n)$. This approximation is calculated by assuming that the derivative $y'(x_n)$ is approximately equal to $\frac{y_{n+1}-y_n}{0.1}$. Thus, the differential equation (6.8) gives rise to the *difference equation*

$$y_{n+1} - y_n = 0.1[-3y_n + (0.1)n].$$ ■

Example 6.2

In the numerical analysis of higher order differential equations, we are required to calculate determinants of the following type:

$$\begin{vmatrix} 2 & \lambda & 0 & 0 & 0 & \cdot & \cdot & \\ 1-\lambda & 2 & \lambda & 0 & 0 & \cdot & \cdot & \\ 0 & 1-\lambda & 2 & \lambda & 0 & \cdot & \cdot & \\ \cdot & \cdot & \cdot & \cdot & \cdot & \cdot & \cdot & \cdot \\ \cdot & \cdot & \cdot & \cdot & \cdot & \cdot & \cdot & \\ \cdot & \cdot & \cdot & \cdot & 1-\lambda & 2 & \lambda & \\ \cdot & \cdot & \cdot & \cdot & \cdot & 1-\lambda & 2 & \end{vmatrix}$$

SOLUTION

Let us denote the above $n \times n$ determinant by D_n. We can easily calculate D_1, D_2, etc. Sometimes we can guess the value of D_n and prove our guess by induction. But guessing is not always possible. The interesting thing is that *we can find a recurrence relation satisfied by D_n*.

Now, we find the recurrence relation satisfied by D_n.

Let

$$D_n = \begin{vmatrix} 2 & \lambda & 0 & 0 & 0 & \cdot & \cdot & \\ 1-\lambda & 2 & \lambda & 0 & 0 & \cdot & \cdot & \\ 0 & 1-\lambda & 2 & \lambda & 0 & \cdot & \cdot & \\ \cdot & \cdot & \cdot & \cdot & \cdot & \cdot & \cdot & \cdot \\ \cdot & \cdot & \cdot & \cdot & \cdot & \cdot & \cdot & \\ \cdot & \cdot & \cdot & \cdot & 1-\lambda & 2 & \lambda & \\ \cdot & \cdot & \cdot & \cdot & 0 & 1-\lambda & 2 & \end{vmatrix}_n$$

By expanding the above determinant by the first row, we get

$$D_n = 2\begin{vmatrix} 2 & \lambda & 0 & 0 & 0 & \cdot & \cdot & \cdot \\ 1-\lambda & 2 & \lambda & 0 & 0 & \cdot & \cdot & \cdot \\ \cdot & \cdot & \cdot & \cdot & \cdot & \cdot & \cdot & \cdot \\ \cdot & \cdot & \cdot & \cdot & \cdot & \cdot & \cdot & \cdot \\ \cdot & \cdot & \cdot & \cdot & \cdot & \cdot & \cdot & \cdot \\ \cdot & \cdot & \cdot & \cdot & \cdot & 1-\lambda & 2 & \lambda \\ \cdot & \cdot & \cdot & \cdot & \cdot & 0 & 1-\lambda & 2 \end{vmatrix}_{n-1} - \lambda\begin{vmatrix} 1-\lambda & \lambda & 0 & 0 & 0 & \cdot & \cdot & \cdot \\ 0 & 2 & \lambda & 0 & 0 & \cdot & \cdot & \cdot \\ \cdot & \cdot & \cdot & \cdot & \cdot & \cdot & \cdot & \cdot \\ \cdot & \cdot & \cdot & \cdot & \cdot & \cdot & \cdot & \cdot \\ \cdot & \cdot & \cdot & \cdot & \cdot & \cdot & \cdot & \cdot \\ \cdot & \cdot & \cdot & \cdot & \cdot & 1-\lambda & 2 & \lambda \\ \cdot & \cdot & \cdot & \cdot & \cdot & 0 & 1-\lambda & 2 \end{vmatrix}_{n-1}$$

$$= 2a_{n-1} - \lambda(1-\lambda)\begin{vmatrix} 2 & \lambda & 0 & 0 & 0 & \cdot & \cdot & \cdot \\ 1-\lambda & 2 & \lambda & 0 & 0 & \cdot & \cdot & \cdot \\ \cdot & \cdot & \cdot & \cdot & \cdot & \cdot & \cdot & \cdot \\ \cdot & \cdot & \cdot & \cdot & \cdot & \cdot & \cdot & \cdot \\ \cdot & \cdot & \cdot & \cdot & \cdot & \cdot & \cdot & \cdot \\ \cdot & \cdot & \cdot & \cdot & \cdot & 1-\lambda & 2 & \lambda \\ \cdot & \cdot & \cdot & \cdot & \cdot & 0 & 1-\lambda & 2 \end{vmatrix}_{n-2}$$

$$= 2D_{n-1} - \lambda(1-\lambda)D_{n-2}.$$

So, we get the recurrence relation,

$$D_n - 2D_{n-1} + (\lambda)(1-\lambda)D_{n-2} = 0. \tag{6.9}$$

Thus, we can find the value of D_n by solving the recurrence relation (6.9). ■

PRACTICE 6.1

A particle is moving in the horizontal direction. The distance it travels in each second is equal to two times the distance it traveled in the previous second. Let a_r denote the position of the particle in the rth second. Find the recurrence relation satisfied by a_r.

6.3 FIRST ORDER RECURRENCE RELATION

As we have already seen, a general first order linear recurrence relation is given by

$$x_{n+1} + a_n x_n = f_n. \tag{6.10}$$

Before getting into the solution of a general first order recurrence relation, let us consider a simple recurrence relation whose solution can be easily obtained. This solution would help us to solve equation (6.10). Consider

$$x_{n+1} - x_n = f_n. \tag{6.11}$$

Now, for $n = 0$, we obtain

$$x_1 - x_0 = f_0.$$

Hence $x_1 = f_0 + x_0$. Now, taking $n = 2$ and using the value of x_1, we get $x_2 = f_1 + f_0 + x_0$. It is easy to guess that

$$x_n = f_n + f_{n-1} + \ldots + f_0 + x_0.$$

It is customary to use the symbol $\sum$ to denote such summations. For any sequence $\alpha_0, \alpha_1, \alpha_2, \ldots$, we have the following notation:

$$\sum_{i=k}^{n} \alpha_i = \alpha_k + \alpha_{k+1} + \ldots + \alpha_n.$$

Using this notation, we have the following proposition:

Proposition 6.1

The solution of equation (6.11) is given by

$$x_n = x_0 + \sum_{i=0}^{n} f_n. \tag{6.12}$$

PRACTICE 6.2

Prove Proposition (6.1) by induction.

When f_n is zero, equation (6.10) is known as a homogenous equation. Now, for such an equation

$$x_{n+1} + a_n x_n = 0 \tag{6.13}$$

we have the following proposition.

Proposition 6.2

The solution of equation (6.13) is given by

$$x_n = (-1)^n x_0 \prod_{i=0}^{n-1} a_i. \tag{6.14}$$

Sketch of the Proof: Taking $n = 1$, we get $x_1 + a_0 x_0 = 0$. Hence $x_1 = -a_0 x_0$. Proceeding in this manner,

$$\begin{aligned} x_2 + a_1 x_1 &= 0 \\ x_2 &= -a_1 x_1 \\ &= -a_1(-a_0 x_0) \\ &= a_1 a_0 x_0. \\ x_3 + a_2 x_2 &= 0 \\ x_3 &= -a_2 x_2 \\ &= -a_2 a_1 a_0 x_0. \end{aligned}$$

Let us now introduce a notation for a continued product like $a_0 a_1 a_2$.

For any sequence $\alpha_k, \alpha_{k+1}, \alpha_{k+2}, \ldots,$ the product $\alpha_k \alpha_{k+1} \ldots \alpha_n$ is denoted by $\prod_{i=k}^{n} \alpha_i$.

Now, from the array of calculations shown above, we guess the solution to equation (6.13) as

$$x_n = (-1)^n x_0 \prod_{i=0}^{n-1} a_i.$$

PRACTICE 6.3

Prove Proposition (6.2) by induction.

Let us now solve equation (6.10). To avoid confusion, let us denote the solution of the homogenous equation (6.13) by u_n. That is, we have

$$u_n = (-1)^n u_0 \prod_{i=0}^{n-1} a_i.$$

Further, to make the calculations simple, let us assume that $u_0 = 1$. Under this assumption and notation, we have the following theorem.

Theorem 6.1

All the solutions of equation (6.10) are given by

$$x_n = \left(c + \sum_{i=0}^{n} \frac{f_n}{(-1)^n \prod_{i=0}^{n} a_i} \right) (-1)^n \prod_{i=0}^{n-1} a_i. \tag{6.15}$$

where u_n is the solution of equation (6.13) with $u_0 = 1$ and c is an arbitrary constant.

Proof

The proof is similar to the variation of the parameter used in the solution of differential equations. We assume that the solution of equation (6.10) is given by $x_n = v_n u_n$ and find v_n. First, let us note that

$$1 + \frac{a_n u_n}{u_{n+1}} = 0. \tag{6.16}$$

This follows by dividing equation $u_{n+1} + a_n u_n = 0$ by u_{n+1}.

If $v_n u_n$ is a solution to equation (6.10), then we must have

$$v_{n+1} u_{n+1} + a_n v_n u_n = f_n. \tag{6.17}$$

Next, our aim is to rewrite equation (6.17) in the form of equation (6.11) whose solution is given by Proposition 6.1. Dividing equation (6.17) by u_{n+1} and adding and subtracting v_n, we get

$$v_{n-1} - v_n + v_n + \left(\frac{a_n u_n}{u_{n+1}}\right) v_n = \frac{f_n}{u_{n+1}}$$
$$v_{n+1} - v_n = -v_n \left(1 + \frac{a_n u_n}{u_{n+1}}\right) + \frac{f_n}{u_{n+1}}$$
$$v_{n+1} - v_n = \frac{f_n}{u_{n+1}}. \tag{6.18}$$

Equation (6.18) follows from equation (6.16).

Using Proposition 6.1, we obtain

$$v_n = v_0 + \sum_{i=0}^{n} \frac{f_n}{u_{n+1}}.$$

Thus,

$$\begin{aligned} x_n &= v_n u_n \\ &= \left(v_0 + \sum_{i=0}^{n} \frac{f_n}{u_{n+1}}\right) (-1)^n \prod_{i=0}^{n-1} a_i \\ &= \left(v_0 + \sum_{i=0}^{n} \frac{f_n}{(-1)^n \prod_{i=0}^{n} a_i}\right) (-1)^n \prod_{i=0}^{n-1} a_i. \end{aligned}$$

Theorem 6.1 follows by taking $c = v_0$ whose value can be arbitrarily assigned. ■

6.4 HOMOGENEOUS RECURRENCE RELATION OF SECOND ORDER

Consider recurrence relation equation (6.3). The theory of the solutions to equation (6.3) is complicated, and hence we consider the simpler case where the sequences a_n, b_n and c_n reduce to constant sequences. That is, we consider

$$ax_{n+1} + bx_n + cx_{n-1} = f_n. \tag{6.19}$$

If $f_n = 0$, it is called a homogeneous recurrence relation. In this section we consider

$$ax_{n+1} + bx_n + cx_{n-1} = 0. \tag{6.20}$$

Theorem 6.2

If a solution to equation (6.20) is given by m^n, then m is a root of the quadratic equation $am^2 + bm + c = 0$.

Proof

We assume that a solution to equation (6.20) is given by $x_n = m^n$, where $m \neq 0$ and find the value of m. Substituting $x_n = m^n$ in equation (6.20), we get

$$\begin{aligned} u(n+1) &= (n+1)u(n)+2n+1 \\ am^{n+1}+bm^{n}+cm^{n-1} &= 0 \\ m^{n-1}(am^2+bm+c) &= 0 \\ (am^2+bm+c) &= 0 \quad (m \neq 0). \end{aligned}$$

Thus, if m^n is a solution, then m satisfies the equation

$$am^2+bm+c=0. \tag{6.21}$$

Now, we can easily reverse the above arguments and prove the following theorem. ■

Theorem 6.3

The sequence $x_n = m^n$ is a solution to equation (6.20) if and only if m is a solution of the auxiliary equation (6.21). Equation (6.21) is also referred to as the characteristic equation.

PRACTICE 6.4

Prove Theorem 6.3.

Example 6.3

Consider the recurrence relation:

$$x_{n+1}+2x_n-15x_{n-1}=0. \tag{6.22}$$

Its auxiliary equation is given by

$$m^2+2m-15=0.$$

That is, we get $(m-3)(m+5)=0$. So, we get the solutions $m=3$ and $m=-5$. Using the above propositions, we know that the only solutions of equation (6.22), which are of the type $x_n = m^n$, are given by 3^n and $(-5)^n$.

Now, the following questions arise:

(i) Are these the only solutions?

(ii) If there are other solutions, what are they and how many solutions are possible?

The answer to the first question is a big **no**. For, we can easily check that $10(3^n) - 123[(-5)^n]$ is a solution. In fact, we have the following theorem. ■

Theorem 6.4

If y_n and z_n are solutions of equation (6.20), the sequence $x_n = c_1 y_n + c_2 z_n$ is also a solution of equation (6.20).

Proof

By hypothesis, we get

$$ay_{n+1} + by_n + cy_{n-1} = 0az_{n+1} + bz_n + cz_{n-1} = 0.$$

Now, multiplying the first equation by c_1, the second by c_2 and adding we get

$$c_1(ay_{n+1} + by_n + cy_{n-1}) + c_2(az_{n+1} + bz_n + cz_{n-1} = 0$$

$$a(c_1y_{n+1} + c_2z_{n+1}) + b(c_1y_n + c_2z_n) + c(c_1y_{n-1} + c_2z_{n-2}) = 0$$

$$ax_{n+1} + bx_n + cx_{n-1} = 0.$$ ■

Example 6.4

Consider the recurrence relation

$$4x_{n+1} - 12x_n + 9x_{n-1} = 0.$$

Its auxiliary equation is given by

$$4m^2 - 12m + 9 = 0.$$

That is, $(2m - 3)^2 = 0$. Thus, $m = \frac{3}{2}$ is a double root of the auxiliary equation. Thus, Proposition 6.2 and Theorem 6.4 give us the solution $c\left(\frac{3}{2}\right)^n$. But we can check that $n\left(\frac{3}{2}\right)^n$ is also a solution. This solution cannot be obtained with the help of the previous results.

Now, we state a theorem (without proof), which completely solves the problem of finding all the solutions of equation (6.20). ■

Theorem 6.5

If the roots α and β are distinct then the general solution to equation (6.20) is given by

$$x_n = c_1\alpha^n + c_2\beta^n.$$

If $x = \alpha$ is a double root, then the general solution is given by

$$x_n = (c_1 + nc_2)\alpha^n.$$

Example 6.5

Solve $4x_{n+1} + 8x_n + 3x_{n-1} = 0$.

SOLUTION

The auxiliary equation is $4m^2 + 8m + 3 = 0$.

Now,

$$4m^2 + 8m + 3 = 0$$

$$4m^2 + 2m + 6m + 3 = 0$$

$$2m(2m + 1) + 3(2m + 1) = 0$$

$$(2m + 3)(2m + 1) = 0$$

$$m = -\frac{3}{2} \text{ or } m = -\frac{1}{2}.$$

Thus, the general solution is given by

$$x_n = c_1\left(\frac{-3}{2}\right)^n + c_2\left(\frac{-1}{2}\right)^n.$$

■

Example 6.6

Solve $x_{n+1} + 14x_n + 49x_{n-1} = 0$.

SOLUTION

The auxiliary equation is $m^2 - 14m + 49 = 0$.

Now,

$$m^2 - 14m + 49 = 0$$
$$(m-7)^2 = 0$$
$$m = 7.$$

So, $m = 7$ is a double root. Thus, the general solution is given by

$$x_n = (c_1 + c_2 n)7^n.$$

■

Example 6.7

Find the solution to

$$x_{n+1} - 8x_n + 15x_{n-1} = 0$$

which satisfies the conditions $x_0 = 5$ and $x_1 = 21$.

SOLUTION

As in the previous examples, we find the general solution as

$$c_1(3)^n + c_2(5^n).$$

The conditions $x_0 = 5$ and $x_1 = 21$ imply that c_1 and c_2 satisfy the system

$$\begin{aligned} c_1 + c_2 &= 5 \\ 3c_1 + 5c_2 &= 21. \end{aligned} \tag{6.23}$$

Solving equation (6.23), we obtain $c_1 = 2$ and $c_2 = 3$. Hence, the required solution is

$$x_n = 2(3^n) + 3(5^n).$$

■

PRACTICE 6.5

Find the solution to

$$x_{n+1} + 14x_n - 240x_{n-1} = 0$$

with $x_0 = 0$ and $x_1 = 1$.

PRACTICE 6.6

Find the determinant D_n of Example 6.2 of the previous section.

SOLUTION

We have already seen that D_n satisfies the recurrence relation (6.5). Hence, the auxiliary equation is

$$m^2 - 2m + (\lambda)(1-\lambda) = 0.$$

Using the formula for solving a quadratic equation, we get the roots of the auxiliary equation as $\frac{2\pm\sqrt{4-4(\lambda)(1-\lambda)}}{2}$. Hence, the general solution of the recurrence relation is

$$c_1[1+\sqrt{1-\lambda+\lambda^2}]^n + c_2[1-\sqrt{1-\lambda+\lambda^2}]^n.$$

In this problem, we are not interested in the general solution. To find c_1 and c_2, we directly calculate D_1 and D_2. We see that $D_1 = 2$ and $D_2 = 4 - \lambda(1-\lambda)$.

So, we get the equations

$$c_1[1+\sqrt{1-\lambda+\lambda^2}] + c_2[1-\sqrt{1-\lambda+\lambda^2}] = 1$$
$$c_1[1+\sqrt{1-\lambda+\lambda^2}]^2 + c_2[1-\sqrt{1-\lambda+\lambda^2}]^2 = 4 - \lambda(1-\lambda).$$

From this, c_1 and c_2 can be calculated.

6.5 NON-HOMOGENEOUS RECURRENCE RELATION

A recurrence relation of the type

$$ax_{n+1} + bx_n + cx_{n-1} = f_n \tag{6.24}$$

is called a non-homogeneous recurrence relation.
We prove a simple proposition, which relates the solutions of equations (6.20) and (6.24).

Proposition 6.3

If x_n and y_n are solutions of equation (6.24), then $z_n = x_n - y_n$ is a solution of equation (6.20). Hence, it follows that

General solution to equation (6.24) = General solution to equation (6.19) + A particular solution to equation (6.24).

Proof
We have

$$ax_{n+1} + bx_n + cx_{n-1} = f_n$$
$$ay_{n+1} + by_n + cy_{n-1} = f_n.$$

Subtracting the second from the first equation above, we get

$$a(x_{n+1} - y_{n+1}) + b(x_n - y_n) + c(x_{n-1} - y_{n-1} = 0$$
$$az_{n+1} + bz_n + cz_{n-1} = 0.$$

Since x_n and y_n are arbitrary, the result follows.

Unlike the first order recurrence relation (6.10) whose solution can be completely determined irrespective of the nature of f_n, the second order recurrence relation (6.24) cannot be easily solved for an arbitrary f_n. Table 6.1 lists the special cases of f_n for which we can find a particular solution, and the procedure to obtain it. ■

Methods to find particular solution: We can solve equation (6.10) for polynomial functions

$f_n = a_0 + a_1 n + ...a_m n^m$ and exponential functions $f_n = cp^n$.

If f_n is a polynomial, we can find a particular solution g_n, which is also a polynomial, and if f_n is an exponential function we can find a particular solution g_n, which is a product of an exponential function and a polynomial function. We summarize the procedures in Table 6.1.

Table 6.1

Nature of roots	Form of $f(n)$	Particular solution
$\alpha \neq \beta$ $\alpha \neq 1$ and $\beta \neq 1$	$f(n) = a_0 + a_1 n + ...a_m n^m$	$g(n) = b_0 + b_1 n + \cdots b_m n^m$
$\alpha \neq \beta$ $\alpha = 1$	$f(n) = a_0 + a_1 n + \cdots + a_m n^m$	$g(n) = b_0 + b_1 n + \cdots + b_{m+1} n^{m+1}$
$\alpha = \beta = 1$	$f(n) = a_0 + a_1 n + \cdots + a_m n^m$	$g(n) = b_0 + b_1 n + \cdots + b_{m+2} n^{m+2}$
$\alpha \neq \beta$ $\alpha \neq p$ and $\beta \neq p$	$f(n) = cp^n$	$g(n) = dp^n$
$\alpha \neq \beta$ $\alpha = p$	$f(n) = cp^n$	$g(n) = dnp^n$
$\alpha = p = \beta$	$f(n) = cp^n$	$g(n) = dn^2 p^n$

After choosing the correct form of the particular solution $g(n)$, we have to find the unknown constants in $g(n)$ by substituting in the given equation. We illustrate the procedure using the following examples:

Example 6.8

Solve $x_{n+1} + 10x_n + 22x_{n-1} = 4$.

SOLUTION

This is a non-homogeneous equation. The auxiliary equation is $m^2 + 10m + 22 = 0$. The roots are given by

$$m = \frac{-10 \pm \sqrt{100 - 4 \times 22}}{2} = -5 \pm \sqrt{3}.$$

Thus, the general solution z_n for the homogeneous recurrence equation is given by

$$z_n = c_1 \left(-5 + \sqrt{3}\right)^n + c_2 \left(-5 - \sqrt{3}\right)^n.$$

Next, we have to find a particular solution to the non-homogeneous equation. Since 4 is a constant, it is a zero-degree polynomial. Since 1 is not a characteristic root, the trial solution is also a constant. Let $y_n = d$. Since y_n is a constant sequence, $y_{n+1} = y_n = y_{n-1} = d$. Further, as y_n is a solution to the given recurrence relation, we must have

$$d + 10d + 22d = 4.$$

Thus, $d = \frac{4}{33}$. By Proposition 6.3, if x_n is the general solution of the given recurrence relation, then $x_n = z_n + y_n$. Hence the general solution is given by

$$c_1\left(-5+\sqrt{3}\right)^n + c_2\left(-5-\sqrt{3}\right)^n + \frac{4}{33}.$$ ■

Example 6.9

Solve $3x_{n+1} - 7x_n + 4x_{n-1} = 3 + 2n$.

SOLUTION

The auxiliary equation is $3m^2 - 7m + 4 = 0$. Now,

$$\begin{aligned} 3m^2 - 7m + 4 &= 0 \\ 3m^2 - 3m - 4m + 4 &= 0 \\ 3m(m-1) - 4(m-1) &= 0 \\ (3m-4)(m-1) &= 0. \end{aligned}$$

$m = 1$ and $m = \frac{4}{3}$ are the roots. The general solution to the homogeneous part is

$$c_1 1^n + c_2\left(\frac{4}{3}\right)^2 = c_1 + c_2\left(\frac{4}{3}\right)^2.$$

The given function on the RHS is a polynomial of degree 1; 1 is a root of the auxiliary equation. From row 2 of Table 6.1, the trial solution is a polynomial of degree 2.

Let $y_n = p + qn + rn^2$ be a particular solution. We have $y_{n+1} = p + q(n+1) + r(n+1)^2$ and $y_{n-1} = p + q(n-1) + r(n-1)^2$.

Substituting in the given equation:

$$\begin{aligned} &3\left(p + qn + q + rn^2 + 2rn + r\right) - 7(p + qn + rn^2) \\ &+ 4(p + qn - q + rn^2 - 2rn + r) = 3 + 2n \\ &n^2(3r - 7r + 4r) + n(3q + 6r - 7q + 4q - 8r) \\ &+ 3p + 3q + 3r - 7p + 4p - 4q + 4r = 3 + 2n. \end{aligned}$$

Equating the coefficients, we get the system of equations

$$0 = 0$$

$$-2r = 2$$
$$-q + 7r = 3.$$

We get the solution p is arbitrary, $r = -1$ and $q = -10$.

Hence, $y_n = p - 10n - n^2$ and the general solution of the given non-recurrence equation is

$$c_1 + c_2\left(\frac{3}{4}\right)^n + p - 10n - n^2 = (c_1 + p) + c_2\left(\frac{3}{4}\right)^n - 10n + n^2$$

which can be rewritten as

$$x_n = A + B\left(\frac{3}{4}\right)^n + -10n - n^2.$$ ■

Example 6.10

Solve $x_{n+1} - 2x_n + x_{n-1} = 1 + 6n$.

SOLUTION

The auxiliary equation is $m^2 - 2m + 1 = 0$. That is

$$(m-1)^2 = 0.$$

Hence, the equation has double root $m = 1$. The general solution to the homogenous equation is given by $c_1 1^n + c_2 n 1^n = c_1 + nc_2$.

$1 + 6n$ is a polynomial of degree 1 and $m = 1$ is a double root of the auxiliary equation. Hence, the particular trial solution has to be of the form $y_n = p + qn + rn^2 + sn^3$. But we can see that $p + qn$ is already a part of the above solution. Thus, it is enough to assume a solution of the form $y_n = rn^2 + sn^3$.

$$\begin{aligned} y_{n+1} &= r(n+1)^2 + s(n+1)^3 \\ &= rn^2 + 2rn + r + sn^3 + 3sn^2 + 3sn + s \\ &= sn^3 + n^2(r+3s) + n(2r+3s) + r + s \\ y_{n-1} &= r(n-1)^2 + s(n-1)^3 \\ &= rn^2 - 2rn + r + sn^3 - 3sn^2 + 3sn - s \\ &= sn^3 + n^2(r-3s) + n(2r+3s) + r - s. \end{aligned}$$

Substituting in the given equation,

$$\begin{aligned} &sn^3 + n^2(r+3s) + n(2r+3s) + r + s - 2rn^2 - 2sn^3 \\ &+ sn^3 + n^2(r-3s) + n(2r+3s) + r - s = 1 + 6n. \end{aligned}$$

Equating the coefficients of n and the constant term,

$$2r + 3s + 2r + 3s = 6$$

$$r + s + r - s = 1.$$

We get $r = \frac{1}{2}$ and $s = \frac{2}{3}$. ■

Example 6.11

Solve $x_{n+1} - 12x_n + 35x_{n-1} = 6^n$.

SOLUTION

The reader can easily verify that the general solution to the homogeneous part is

$$c_1 5^n + c_2 7^n.$$

As one of the roots is equal to 6, the trial solution has to be of the form $y_n = a6^n$. $y_{n+1} = a6^{n+1} = 6a6^n$
$y_{n-1} = a6^{n-1} = a6^{n-1} = \frac{a}{6}6^n$.
Substituting in the given equation

$$6a6^n - 12a6^n + \frac{35a}{6}6^n = 6^n$$

$$a6^n \frac{36 - 72 + 210}{6} = 6^n$$

$$a = \frac{144}{6} = 24.$$

Thus, the general solution to the given equation is

$$c_1 5^n + c_2 7^n + 24(6)^n.$$ ■

Example 6.12

Solve $x_{n+1} + 2x_n - 15x_{n-1} = 4(3)^n$.

SOLUTION

The auxiliary equation is $m^2 + 2m - 15 = 0$. The roots are -5 and 3. The general solution for the recurrence part is

$$A(-5)^n + B3^n.$$

As one of the roots of the auxiliary equation is 3 and the RHS is $4(3)^n$, the particular trial solution has to be taken as $y_n = an3^n$.
$y_{n+1} = a(n+1)3^{n+1} = 3a(n+1)3^n = 3an3^n + 3a3^n$
$y_{n-1} = a(n-1)3^{n-1} = \frac{an}{3}3^n + \frac{a}{3}3^n$.

Substituting in the given equation,

$$3an3^n + 3a3^n + 2an3^n - 15\left(\frac{an}{3}3^n + \frac{a}{3}3^n\right) = 4(3^n)$$

$$3^n\left(3a - \frac{15a}{3}\right) + n3^n\left(3a + 2a - \frac{15a}{3}\right) = 4(3^n).$$

Equating the coefficients of 3^n and $n3^n$, we get $-2a = 4$ and $0 = 0$. Thus, $a = -2$ and so a particular solution is $y_n = -2n3^n$. Hence, the general solution is

$$x_n = A(-5)^n + B3^n - 2n3^n.$$

■

Example 6.13

Solve $x_{n+1} - 6x_n + 9x_{n-1} = 3^n$.

SOLUTION

The auxiliary equation is $m^2 - 6m + 9 = 0$ and hence $m = 3$ is a double root. The general solution to the homogeneous part is $(A + nB)3^n$. Further, the RHS is 3^n. Hence the particular solution $y_n = an^2 3^n$. Now,

$$\begin{aligned} y_{n+1} &= a(n+1)^2 3^{n+1} \\ &= 3a(n^2 + 2n + 1)3^n \\ y_{n-1} &= a(n-1)^2 3^{n-1} \\ &= \frac{a}{3}(n^2 - 2n + 1)3^n. \end{aligned}$$

Substituting in the given equation,

$$3a(n^2 + 2n + 1)3^n - 6an^2 3^n + 9\left(\frac{a}{3}(n^2 - 2n + 1)3^n\right) = 3^n$$

$$3^n\left(3a + \frac{9a}{3}\right) + n3^n\,(6a - 6a) + n^2 3^n\,(3a - 6a + 3a) = 3^n.$$

Hence, equating the coefficients of 3^n, $n3^n$ and $n^2 3^n$, we get $6a = 1$, $0 = 0$ and $0 = 0$. Thus, $a = \frac{1}{6}$. Hence, a particular solution is $\frac{1}{6}n^2 3^n$ and so the general solution is given by

$$x_n = (A + nB)3^n + \frac{1}{6}n^2 3^n.$$

■

Next, we solve an interesting puzzle using recurrence relations. How to get a total of Rs 5 using only coins of Re 1 and Rs 2? Let us tabulate the possibilities. Hence, there are three different ways of building Rs 5. The following example gives us the number of ways of building Rs n using only Re 1 and Rs 2 coins.

Table 6.2

No. of Re 1 coins	No. of Rs 2 coins
5	0
3	1
1	2

Example 6.14

Let a_n be the number of ways of getting Rs n with 1 and 2 coins. We get Rs $n+1$ by adding a Re 1 coin to a collection of coins totalling Rs n, or by adding a Rs 2 coin to a collection of coins totalling Rs $n-2$. Hence

$$a_{n+1} = a_n + a_{n-1}.$$

Its auxiliary equation is

$$m^2 - m - 1 = 0$$

whose roots are

$$\frac{-1 \pm \sqrt{5}}{2}.$$

Hence

$$a_n = c_1\left(\frac{-1+\sqrt{5}}{2}\right)^n + c_2\left(\frac{-1-\sqrt{5}}{2}\right)^n.$$

The number of ways of getting a total of Re 1 is 1 and the number of ways of getting a total of Rs 2 is 2 (two Re 1 coins or one Rs 2 coin). So we can easily see that $a_1 = 1$ and $a_2 = 2$. ■

PRACTICE 6.7

Solve for a_n in Example 14.

PRACTICE 6.8

We now consider the famous Fibonacci sequence. Assume that a pair of rabbits has a pair of children every year. These children take two years to grow to be able to give rise to a new pair. For simplicity, we assume that rabbits never die. Now let a_n denote the number of *pairs* at the end of the nth year. Show that

$$a_n = a_{n-1} + a_{n-2}.$$

a_n are called Fibonacci numbers.

Example 6.15

Let us now consider the Fibonacci recursive algorithm to compute the Fibonacci numbers. In the previous example, if we start with a pair of baby rabbits in the first year, we have $a_1 = a_2 = 1$. Now, from equation (6.26), we get $a_3 = 1 + 1 = 2$, $a_4 = 1 + 2 = 3$, etc. Now, the following algorithm gives the steps in the computation of a_n.

(i) Check if $n = 1$.
(ii) If $n = 1$, $a_1 = 1$.
(iii) If $n \neq 1$, then check if $n = 2$.
(iv) If $n = 2$, $a_2 = 1$.
(v) If $n \neq 2$, then $a_n = a_{n-1} + a_{n-2}$.

This is a recursive algorithm in the sense that we compute a_n in terms of a_k for $k < n$. Our aim is to find the number of steps involved in the calculation of a_n. It is interesting that the solution of this problem involves another recurrence relation. Let N_n be the number of steps involved in the calculation of a_n. For $n > 2$, we note that checking whether $n \leq 2$ gives us two steps. The calculation of a_{n-1} gives us N_{n-1} steps and the calculation of a_{n-2} gives us N_{n-2} steps. The addition of a_{n-1} and a_{n-2} gives us yet another step. So, N_n satisfies

$$N_n = 3 + N_{n-1} + N_{n-2}$$
$$N_1 = 1, \; N_2 = 1.$$

∎

PRACTICE 6.9

Solve for N_n in Example 6.15.

6.6 GENERATING FUNCTIONS

The mathematical pre-requisites needed for this section are binomial expansions and the method of partial fractions to simplify rational functions.

Consider the sequence $a_n = \frac{1}{3^{n+1}}$ and the power series

$$\frac{1}{3} + \frac{x}{9} + \frac{x^2}{27} + \frac{x^3}{81} + \ldots.$$

What is the relationship between these two?

The nth term of the given sequence is the *coefficient of* x^n in the given power series. In general, given a sequence a_n we can construct a power series such that the coefficient of x^n is a_n.

Definition 6.1 The generating function (G.F.) of a sequence is the *formal* power series defined as

$$\sum_{n=0}^{\infty} a_n x^n = a_0 + a_1 x + a_2 x^2 + \ldots$$

Remark 6.1

We use the word *formal* because we do not worry about the convergence of the infinite series. If the above power series converges and the sum of the power series is a function $g(x)$, then $g(x)$ is also called the generating function of the sequence. Again, we do not worry much about the domain of the definition of $g(x)$.

Example 6.16

Find the generating functions of the following sequences:
(i) $a_n = 1$.
(ii) $b_n = n + 1$.
(iii) $c_n = n$.
(iv) $d_n = n^2$.
(v) $e_n = 5^n$.

SOLUTION

The generating functions for the first four sequences, respectively, are as follows

$$\sum_{n=0}^{\infty} x^n = 1 + x + x^2 + x^3 + ...$$

$$\sum_{n=0}^{\infty} (n+1)x^n = 1 + 2x + 3x^2 + 4x^3....$$

$$\sum_{n=0}^{\infty} nx^n = x + 2x^2 + 3x^3 + 4x^4....$$

$$\sum_{n=0}^{\infty} n^2x^n = x + 4x^2 + 9x^3 + 27x^4....$$

Using the sum of an infinite geometric series, we find that

$$1 + x + x^2 ... = \frac{1}{1-x}.$$

Using the binomial expansion, we find that

$$1 + 2x + 3x^2 + ... = \frac{1}{(1-x)^2}.$$

Next,

$$\begin{aligned} x + 2x^2 + 3x^3 + 4x^4 .... &= x(1 + 2x + 3x^2 + ...) \\ &= \frac{x}{(1-x)^2}. \end{aligned}$$

Hence, the generating functions for the first three sequences are given by

$$g_1(x) = \frac{1}{1-x}$$
$$g_2(x) = \frac{1}{(1-x)^2}$$
$$g_3(x) = \frac{x}{(1-x)^2}.$$

To find the summation $x + 4x^2 + 9x^3 + 16x^4 \ldots$, we use differentiation and some algebraic manipulations.

$$\sum_{n=0}^{\infty} n^2 x^n = \sum_{n=0}^{\infty} [n(n-1) + n]x^n$$
$$[0.5ex] = \sum_{n=0}^{\infty} n(n-1)x^n + \frac{x}{(1-x)^2}$$
$$= \sum_{n=2}^{\infty} n(n-1)x^n + \frac{x}{(1-x)^2}$$
$$= x^2 \sum_{n=2}^{\infty} n(n-1)x^{n-2} + \frac{x}{(1-x)^2}$$
$$= x^2 \sum_{n=2}^{\infty} \frac{d^2}{dx^2} x^n + \frac{x}{(1-x)^2}$$
$$= x^2 \frac{d^2}{dx^2} \sum_{n=2}^{\infty} x^n + \frac{x}{(1-x)^2}$$
$$= x^2 \frac{x^2}{(1-x)} + \frac{x}{(1-x)^2}$$
$$= \frac{x^4}{(1-x)} + \frac{x}{(1-x)^2}.$$

Now the generating function of e_n is

$$\sum_{n=0}^{\infty} 5^n x^n = \sum_{n=0}^{\infty} (5x)^n = \frac{1}{1-5x}$$ ■

PRACTICE 6.10

Find the generating functions of

(i) $a_n = \frac{1}{n!}$.
(ii) $a_n = (2n + 1)$.
(iii) $a_n = n^3$.
(iv) $a_n = n^2 - 3n + 6$.
(v) $a_n = 4^{n+1}$.
(vi) $a_n = n + 2^n$.

If we know the generating function $g(x)$ of a sequence a_n and an expansion for $g(x)$ in powers of x, then we know the sequence. In Example 6.17, we illustrate how a recurrence relation can be used to compute the generating function of an unknown sequence. In certain cases, the unknown sequence can be computed using the generating function.

The following simple result of Proposition 6.4 will be used very often.

Proposition 6.4

Let k be a fixed positive integer. Then,

$$\sum_{n=0}^{\infty} A_n = \sum_{n=k}^{\infty} A_{n-k}.$$

Proof

Consider the summation on the LHS. Its expanded version is $A_0 + A_1 + A_2 + \ldots$. Next, let us take the summation on the RHS. The first term is $A_{k-k} = A_0$, the second term is $A_{k+1-k} = A_1$, etc. Hence this summation is also equal to $A_0 + A_1 + A_2 \ldots$. The proof is complete. ■

Example 6.17

Solve

$$y_{n+1} - 4y_n + 3y_{n-1} = 4 \tag{6.25}$$

$$y_0 = 1 \quad y_1 = 0 \tag{6.26}$$

by finding the generating function of y_n.

SOLUTION

Multiplying both sides of equation (6.28) by x^{n+1}, we obtain

$$y_{n+1}x^{n+1} - 4y_nx^{n+1} + 3y_{n-1}x^{n+1} = 4x^{n+1}.$$

Summing both the sides of the above equation as n ranges from 1 to ∞, we get

$$\sum_{n=1}^{\infty}(y_{n+1}x^{n+1} - 4y_n x^{n+1} + 3y_{n-1}x^{n+1}) = \sum_{n=1}^{\infty} 4x^{n+1}.$$

Let $g(x) = \sum_{n=0}^{\infty} y_n x^n$ be the generating function of the sequence y_n. So, we have

$$\sum_{n=1}^{\infty}(y_{n+1}x^{n+1} - 4y_n x^{n+1} + 3y_{n-1}x^{n+1}) = \sum_{n=1}^{\infty} 4x^{n+1}$$

$$\Rightarrow g(x) - y_1 x - y_0 - 4x\sum_{n=1}^{\infty} y_n x^n + 3x^2\sum_{n=1}^{\infty} y_{n-1}x^{n-1} = 4x\sum_{n=1}^{\infty} x^n$$

$$\Rightarrow g(x) - y_1 x - y_0 - 4x[g(x) - y_0] + 3x^2\sum_{n=0}^{\infty} y_n x^n = \frac{4x^2}{1-x}$$

$$\Rightarrow g(x) - y_1 x - y_0 - 4x[g(x) - y_0] + 3x^2 g(x) = \frac{4x^2}{(1-x)}$$

$$\Rightarrow \left[1 - 4x + 3x^2\right] g(x) - 1 - 4x = \frac{4x^2}{1-x}$$

$$\Rightarrow g(x) = \frac{8x^2 - 5x + 1}{(3x^2 - 4x + 1)(1-x)}$$

$$\Rightarrow g(x) = \frac{8x^2 - 5x + 1}{1 - 3x)(1-x)^2}.$$

Now, to find an equivalent expansion for $g(x)$, we have to use *the method of partial fractions.* This method gives us

$$g(x) = \frac{5}{2}\frac{1}{1-x} + \frac{(-1)}{2}\frac{1}{(1-x)^2} + \frac{1}{2}\frac{1}{(1-3x)}.$$

Now,

$$\frac{1}{1-x} = \sum_{n=0}^{\infty} x^n$$

$$\frac{1}{(1-x)^2} = \sum_{n=0}^{\infty}(n+1)x^n$$

$$\frac{1}{1-3x} = \sum_{n=0}^{\infty} 3^n x^n.$$

Hence,

$$g(x) = \sum_{n=0}^{\infty}\left(\frac{5}{2} - \frac{1}{2}(n+1) + \frac{1}{2}3^n\right)x^n.$$

Thus,

$$y_n = \frac{5}{2} - \frac{1}{2}(n+1) + \frac{1}{2}3^n. \qquad \blacksquare$$

PRACTICE 6.11

Solve $a_{n+1} + 4a_{n-1} = 2^n$ by the method of generating functions.

SUMMARY

- $x_{n+1} + a_n x_n = f_n$ is the general first order recurrence relation whose general solution is given by

$$x_n = \left(c + \sum_{i=0}^{n} \frac{f_n}{(-1)^n \prod_{i=0}^{n} a_i} \right) (-1)^n \prod_{i=0}^{n-1} a_i.$$

- $ax_{n+1} + bx_n + cx_{n-1} = 0$ is the general first order homogenous recurrence relation whose general solution is given by

$$\begin{aligned} x_n &= c_1\alpha^n + c_2\beta^n, \quad \text{if } \alpha \neq \beta \\ x_n &= (c_1 + c_2 n)\alpha^n, \quad \text{if } \alpha = \beta. \end{aligned}$$

- The general solution of the non-homogenous recurrence relation $ax_{n+1} + bx_n + cx_{n-1} = f_n$ can be found when f_n is an exponential or a polynomial function of n.
- $g(x) = \sum_{n=0}^{\infty} a_n x^n$ is called the generating function of x_n.

PROBLEMS

1. Let P be the principal amount borrowed at the interest rate of $R\%$ per annum. Let S_n and C_n, respectively, be the amounts payable at the end of n years under simple and compound interest schemes. Show that

$$\begin{aligned} S_n &= S_{n-1} + \frac{RP}{100} \\ C_n &= C_{n-1} + \frac{RC_{n-1}}{100}. \end{aligned}$$

Derive the formulae for S_n and C_n using the methods of Section 3.

2. Solve $x_n - 4x_{n+1} = 1$.

3. Find the recurrence relation satisfied by the sequence 3, 4, 8, 17, 33....
4. Solve $x_n - nx_{n+1} = 0$ with $x_3 = 1$.
5. Find the following determinants using recurrence relations:

$$\begin{vmatrix} b & b & 0 & 0 & 0 & . & . & . \\ b & b & b & 0 & 0 & . & . & . \\ 0 & b & b & b & 0 & . & . & . \\ . & . & . & . & . & . & . & . \\ . & . & . & . & . & . & . & . \\ . & . & . & . & . & . & . & . \\ . & . & . & . & 0 & b & b & b \\ . & . & . & . & 0 & 0 & b & b \end{vmatrix}$$

$$\begin{vmatrix} 2 & 1 & 0 & 0 & 0 & . & . & . \\ 1 & 2 & 1 & 0 & 0 & . & . & . \\ 0 & 1 & 2 & 1 & 0 & . & . & . \\ . & . & . & . & . & . & . & . \\ . & . & . & . & 1 & 2 & 1 & 0 \\ . & . & . & . & . & 1 & 2 & 1 \\ . & . & . & . & . & 0 & 1 & 2 \end{vmatrix}$$

6. Solve the following recurrence relations:

   (i) $x_{n+1} - 11x_n + 28x_{n-1} = 0$.

   (ii) $x_{n+1} - 2x_n - 3x_{n-1} = 2^n$.

   (iii) $x_{n+1} + 18x_n + 32x_{n-1} = 3n + 3$.

   (iv) $x_{n+1} + 2x_n - 80x_{n-1} = n^2 + 2n + 1$.

7. Solve the recurrence relation $a_{n+1} - 5a_n + 6a_n = 10$ with $a_0 = 5$ and $a_1 = 10$.
8. Find the generating function of the following sequences:

   (i) $n^2 - n$.

   (ii) $2^n - 6^n$.

9. If $g(x)$ and $h(x)$ are the generating functions of a_n and b_n, respectively, show that the generating function of $Aa_n + Bb_n$ is $Ag(x) + Bh(x)$.
10. Solve the following recurrence relation using the generating function method:

   (i) $y_{n+1} + 8y_n + 12y_{n-1} = 2n$, $y_0 = 0$, $y_1 = 0$.

   (ii) $y_{n+1} - 15y_n + 150y_{n-1} = 2^n$, $y_0 = 1$, $y_1 = 2$.

FURTHER READING

Kelley, W. G. and Peterson, A. C. (2001), *Difference Equations, An Introduction with Applications*, Academic Press, San Diego.

Milne, L. M. (1933), *The Calculus of Finite Differences*, Macmillan and Co., London.

CHAPTER

7

Number Theory

Learning Objectives

On completing this chapter, you should be able to:

- state the definitions of a factor and a multiple of an integer
- state and prove the division algorithm
- define the ideas of the greatest common divisor (gcd) and the least common multiple (lcm)
- define a pair of relatively prime numbers
- use the Euclidean algorithm to find the gcd
- find the gcd and lcm of pairs of numbers by prime factorization
- define a prime number
- define congruence classes and congruence relations
- solve linear congruence equations for small numbers
- state and prove Fermat's theorem and its generalization due to Euler

7.1 INTRODUCTION AND MOTIVATION

There was a time when Number Theory was considered a very pure subject in the sense that it did not have any applications to real life. G. H. Hardy, the famous number theorist, who was a mentor to the equally famous Indian mathematician Srinivasa Ramanujan held the extreme view that Applied Mathematics was ugly. Number Theory was beautiful because it did not have any applications. If he were to be alive today, he would be a very sad person because Number Theory has turned out to be highly useful in the modern world. Network security and cryptography are two fields in which number theory plays a great role. We shall now present a simple example of how an elementary idea may be useful in a computer application.

Example 7.1

We know that computers use strings of 0's and 1's for calculations and other purposes. For example, if we want to store the number 256, we may use the binary system and rewrite it as 100000000, which requires 9 bits. Now, $256 = 2^8$. Under the binary system 2 is 10 and 8 is 100. If we store these two numbers, we have to use only $2 + 3 = 5$ bits. Of course, to get back 256 we will have to do some calculation, but we have reduced the amount of storage required. Here we have used the elementary notion of factorization of a number into its prime factors.

In Section 7.5, we shall see how number theory is useful in cryptography. ■

7.1.1 Divisibility

Number theory is essentially the study of integers. The most important fact about integers is that they can be factorized into prime numbers. Prime numbers are integers p, which cannot be factorized except in the trivial manner $p = 1 \times p$. Let us now recall some elementary facts regarding integers. Unless specified otherwise, the numbers considered in this chapter are all integers.

Definition 7.1: If there exists k such that $m = kn$, then we say that

(i) m is divisible by n.
(ii) m is a multiple of n.
(iii) n divides m.
(iv) n is a factor of m and.
(v) n is a divisor of m.

The fact that n divides m is denoted by $n|m$.

Example 7.2

We make the following simple observations:

(i) $10|2345670$.
(ii) 8796 is a multiple of 3.
(iii) 105 is a factor of 8925.
(iv) 457 is not a multiple of 2.
(v) 1 is a factor of any integer.
(vi) Every integer is a divisor of itself. ■

PRACTICE 7.1

Answer the following questions:

(i) Is 203 a factor of 267964?
(ii) Is 123 a factor of 4428?
(iii) Is 8796 a multiple of 13?
(iv) Is 1350 a multiple of 27?

The following proposition is elementary, but we prove it rigourously.

Proposition 7.1

We have

(a) $1|n$ for all n.
(b) If $a|b$ and $a|c$ then $a|(b+c)$.

Proof

(a) To show that 1 divides n, we have to find k such that $n = k \times 1$. It is obvious that we have to have $k = n$.
(b) Since $a|b$, there is k_1 such that $b = k_1 a$. Similarly, there is k_2 such that $c = k_2 a$. Now, $b + c = k_1 a + k_2 a = (k_1 + k_2)a$. This shows that $a|(b+c)$. ■

PRACTICE 7.2

If $a|b$ and $b|c$, show that $a|c$.

Example 7.3

If $a|b$ and $b|a$, then $a = \pm b$.

SOLUTION

Here, let us note that 0 is a factor of only 0 and not any other integer. Thus, if $a = 0$, then $b = 0$. Let us now assume that $a \neq 0$ and $b \neq 0$. There exists k_1 and k_2 such that $a = k_1 b$ and $b = k_2 a$. Multiplying, we obtain $ab = k_1 k_2 ab$. Since a and b are non-zero integers, we get $k_1 k_2 = 1$. Hence, we must have either $k_1 = k_2 = 1$ or $k_1 = k_2 = -1$ and the proof follows. ■

PRACTICE 7.3

Show that if $a|b$ and $a|c$, then $a|mb + nc$.

Example 7.4

If $a|b$, then $|a| \leq |b|$.

SOLUTION

We have $b = ka$ for some $k \in Z$. Noting that $|k| \geq 1$, we get $|b| = |k||a| \geq |a|$. ■

Example 7.5

Let us divide 6478 by 72. The quotient is 72 and the remainder is 70. This fact can be written as

$$6478 = 72 \times 89 + 70.$$

In general, when m is divided by $n > 0$, if the quotient is q and the remainder is r we can write

$$m = qn + r.$$

Let us note that $0 \leq r < n$ and $q \geq 0$.

What are the quotient and remainder if 23 is divided by 786? Since we can write $23 = 0 \times 786 + 23$, we take the quotient as 0 and the remainder as 23. In general, when m is divided by n and $0 < m < n$, the quotient is 0 and the remainder is m.

What are the quotient and the remainder if -6478 is divided by 72? We may write

$$-6478 = 72 \times -89 - 70 \tag{7.1}$$

and take the quotient as -89 and the reminder as -70. But for theoretical convenience we prefer to work with non-negative remainders. We now rewrite equation (7.1) as follows:

$$\begin{aligned} -6478 &= 72 \times -89 - 70 \\ &= 72 \times (-89 - 1) + 72 - 70 \\ &= 72 \times -90 + 2. \end{aligned}$$

Now, we declare -90 as the quotient and 2 as the remainder.

Let $m = 1256874952097563655555678$ and $n = 4627631298$. It is intuitively clear there must exist integers q and r such that $1256874952097563655555678 = 4627631298q + r$ where $0 \leq r < 4627631298$. A patient high school student can find them. But what about bigger and bigger numbers? One can easily imagine huge integers whose division cannot be performed on the latest supercomputer. So, though intuitively obvious, mathematicians would like to give a formal proof of Theorem 7.1, which asserts that for any pair m and $n > 0$, q and r exist. The proof of Theorem 7.1 is based on the following proposition known as the **Well Ordering Principle.** ■

Proposition 7.2

If $T \subseteq N$ is a non-empty subset, there exists $r \in T$ such that $r \leq x$ for all $x \in T$.

Theorem 7.1

The Division Algorithm: Given any integers m and $n > 0$, there exist unique integers q and r such that $m = qn + r$ and $0 \leq r < n$. If n is not a divisor of m, then $0 < r < n$.

Proof

If n is a divisor of m, then there exists q such that $m = nq$, and we take $r = 0$.

Next, we prove that for a negative integer M, if $P > |M|$, then $M + P > 0$. Since M is negative $-M = |M| < P$, we have $-M - P < 0$, which proves $M + P > 0$.

Let $S = \{m - kn : k \in Z\}$. We assume that m is not a multiple of n. Hence $0 \notin S$. S can also be written as

$$\{...m - 3n, m - 2n, m - n, m, m + n, m + 2n, m + 3n, ...\}.$$

If m is positive, since $m \in S$, $S \cap N \neq \phi$. Let us consider the case where m is negative. It is clear that for some $k \in N$, $kn > |m|$. Hence, $m + kn > 0$ and $m + kn = m - (-k)n \in S$. Thus, $S \cap N \neq \phi$.

Now applying Proposition 7.2, there is $r \in S \cap N$ such that $r \leq m - kn$ for all k with $m - kn \geq 0$. Since $r \in S$, $r = m - qn$ for some q. Now, we have to show that $0 < r < n$. $0 < r$ is obvious. If $r \geq n$, then

$$m - (q+1)n = r - n \geq 0$$

and $m - (q+1)n < r$, which is a contradiction to the definition of r. ■

Example 7.6

Show that if n is odd, $n^2 - 1$ is divisible by 8.

SOLUTION

n is odd implies that the remainder we get when we divide n by 2 is 1. Thus, $n = 2q + 1$ for some q.

Now, $n^2 - 1 = (2q+1)^2 - 1 = 4q^2 + 4q + 1 - 1 = 4q(q+1)$. The numbers q and $q+1$ are consecutive and so one of them has to be even. If q is even, then $q = 2m$ for some m. So, $n^2 - 1 = 8m(q+1)$, which is divisible by 8. The proof is similar for the case that $q + 1$ is even. ■

PRACTICE 7.4

Show that $n^3 - n$ is divisible by 6.

Example 7.7

How many integers between 200 and 500 are divisible by 23?

SOLUTION

It is clear that in the list

$$mk, (m+1)k, \ldots, nk,$$

there are $n - m + 1$ numbers present.

Now, let us find the smallest multiple of 23 greater than 200. Dividing 200 by 23, we find that $200 = 23 \times 8 + 16$. Thus, to get the first multiple of 23 bigger than 200, we should add the difference $(23 - 16) = 7$ to 200. Further, $207 = 23 \times 9$. Let us now find the biggest multiple of 23 smaller than 500. Now $500 = 21 \times 23 + 17$. Thus, the biggest multiple of 23 smaller than 500 is $500 - 17 = 483 = 23 \times 21$. Thus the multiples of 23 between 200 and 500 are listed as

$$9 \times 23, 10 \times 23, \ldots, 21 \times 23.$$

So, their count is just $21 - 9 + 1 = 13$. ■

7.2 LCM AND HCF

In this section, we present the computations of LCM and HCF of a given set of numbers.

Example 7.8

In a school, the mathematics club meets after every 25 days and the literary club meets after every 30 days. The first meeting of both the clubs took place on the same day. How many days later, shall both the clubs meet again on the same day?

SOLUTION

Let us prepare a table of meetings for both the clubs as shown in Table 7.1.

Table 7.1

	1st	2nd	3rd	4th	5th	6th	7th	8th
Maths club	0	25	50	75	100	125	150	175
Literary club	0	30	60	90	120	150	180	210

It is clear that after 150 days both the clubs will meet again on the same day as for the first time. The first rows are multiples of 25 and the second rows are multiples of 30. The 'first' that is, the smallest common multiple is 150. This example motivates the following definition. ■

Definition 7.2: Let m and n be positive integers. Their least common multiple (LCM) is defined as the smallest positive integer, which is a common multiple of both m and n. It is denoted by $[m, n]$.

Example 7.9

Let us recall the addition of fractions $\frac{a}{b}$ and $\frac{c}{d}$. To add two fractions, first we rewrite them as fractions having the same denominator. This denominator has to be a common multiple of the given denominators. The simplest common multiple is bd. So, we add these fractions as follows

$$\frac{a}{b}+\frac{c}{d}=\frac{ad}{bd}+\frac{bc}{bd}=\frac{ad+bc}{bd}.$$

But in many cases, choosing the least common multiple of b and d as the common denominator simplifies the calculations. ■

PRACTICE 7.5

Find the LCM of 14 and 35 by listing their multiples.

PRACTICE 7.6

Find the length of the smallest square that can be formed by a combination of rectangular pieces of length l and breadth b.

PRACTICE 7.7

Consider a rectangle of length 236 units and breadth 78 units. It is clear that we can fill this rectangle by squares of length 2 units. We cannot do so with squares of length 3 units. Though 26 squares can be accommodated along the breadth, an integral number of squares cannot fit along the length. What are the possible lengths of squares that would do the job? It is clear that the length of the square must be a divisor of both 236 and 78. Can you find the largest such square?

The solution to the above problem is the greatest common divisor (gcd) of 236 and 78. We now formally define the notion of gcd.

Definition 7.3: Let m and n be integers with at least one of them being non-zero The largest of the common divisors of m and n is called greatest common divisor of m and n. It is denoted by (m, n).

Example 7.10

Find the gcd of 56 and 70.

SOLUTION

Let us list the divisors of the given numbers. The divisors of 56 are $1, 2, 4, 7, 8, 14$, 28 and 56. The divisors of 70 are $1, 2, 5, 7, 10, 14, 35$ and 70. Their common divisors are $1, 2, 7$ and 14. Hence the $(56, 70) = 14$.

In Example 7.10, we find that all the common divisors are divisors of the gcd. We shall prove this observation after looking at gcd from a different point of view. ■

Theorem 7.2

(m, n) is the least positive integer of the set

$$S = \{mx + ny : x, y \in \mathbf{Z}\}.$$

Proof

Since $|m| = \pm m + 0 \times n$, $S \cap N \neq \phi$. Thus, by Proposition 7.2, there is $g \in S$ such that $g \leq k$ for any $k \in S \cap N$. This is the element that is called as the least positive element of S. Let $g = ma + nb$.

Now, we shall show that $g = (m, n)$. First, we have to show that $g|m$ and $g|n$.

If g does not divide a, then by division algorithm (Theorem 7.1), we obtain q and r such that $m = g \times q + r$ where $0 < r < g$. Now, $r = m - gq = m - (ma + nb)q$ $= m(1 - aq) - n(-bq \in S)$. Thus, g cannot be the least positive element of S, which is a contradiction. Thus, we must have $g|m$. Similarly, $g|n$.

Next, we have to show that g is the greatest common factor. If d is any other common factor, then $d|m$ and $d|n$ and, we have $d|ma + nb = g$ (refer Practice 7.3). Hence, by Example 7.4, $d \leq |d| \leq |g| = g$. So, g is the gcd of m and n.

The last part of the proof of Theorem 7.2 proves the following proposition. ■

Proposition 7.3

(m, n) is the common factor of m and n such that if d is any common factor of m and n, then $d|g$.

PRACTICE 7.8

Let $k > 0$. Using the fact that the least positive integer of kS is ka where a is the least positive integer of S, show that $(km, kn) = k(m, n)$.

7.2.1 The Euclidean Algorithm

Next, we shall learn about the Eulidean Algorithm to find the gcd of two numbers.

Suppose that we are interested in finding a common divisor of 1782 and 2146. Finding the divisors of both the numbers is a tedious process. Let us make the following observation. By division algorithm we write 2146 as $2146 = 1782 \times q + r$. If d is a common divisor of the given numbers, it is clear that d is also a divisor of $r = 2146 - 1782 \times q$. Further, if d is a common divisor of 1782 and r, then it is a divisor of 2146. Thus, d is a common divisor of the given numbers if and only if it is a common divisor of the smaller number and the remainder we get when the smaller number divides the other number. The advantage is that now we can work with smaller numbers. This process can be continued further till we get a zero remainder. The following proposition is the building block of the Euclidean Algorithm.

Proposition 7.4

$(a, b) = (a, b + ax)$ for any x.

Proof

Let $(a, b) = m$ and $(a, b + ax) = n$ us $m|a$,$m|b$ and hence $m|b + ax$. So, $m|n$. By writing $b = (b + ax) - xa$, we can show that $n|m$. By Example 7.3, $m = \pm n$. Being gcd's of pairs of integers, both m and n are positive. Hence $m = n$. ■

The following example motivates the proof of the Euclidean Algorithm.

Example 7.11

Find the gcd of 2146 and 1782.

SOLUTION

Let $m = nq + r$ where q and r are the quotient and the remainder, respectively, when m is divided by n. Then $r = m - nq$ and hence $(n, m) = (n, m - nq) = (n, r)$. By Proposition 7.4,

$$2146 = 1782 \times 1 + \times 364$$
$$\rightarrow (1782, 2146) = (1782, 364))$$
$$1782 = 364 \times 4 + 326$$
$$\rightarrow (1782, 364) = (364, 326)$$
$$364 = 326 \times 1 + 38$$
$$\rightarrow (364, 326) = (326, 38)$$
$$326 = 38 \times 8 + 22$$
$$\rightarrow (326, 38) = (38, 22)$$

$$38 = 22 \times 1 + 16$$
$$\rightarrow (38, 22) = (22, 16)$$
$$22 = 16 \times 1 + 6$$
$$\rightarrow (22, 16) = (16, 6)$$
$$16 = 6 \times 2 + 4$$
$$\rightarrow (16, 6) = (6, 4)$$
$$6 = 4 \times 1 + 2$$
$$\rightarrow (6, 4) = (4, 2)$$
$$4 = 2 \times 2.$$

Hence the required gcd is $(4, 2) = 2$ ■

Theorem 7.3

The following steps give the gcd of a and b.

$$\begin{aligned} a &= bq_1 + r_1 \\ b &= r_1 q_2 + r_2 \\ r_1 &= r_2 q_3 + r_3. \end{aligned}$$

The general step is $r_i = r_{i+1} q_i + r_{i+2}$ and the algorithm stops when for some n,

$$r_n = r_{n+1} q_n$$

and the gcd is given by $(r_n, r_{n+1}) = r_{n+1}$.

7.2.2 Relatively Prime Number s

We know that any fraction $\frac{m}{n}$ may be reduced to its simplest form $\frac{a}{b}$ where a and b do not have any common divisor other than 1. For example,

$$\frac{336}{390} = \frac{168}{195} = \frac{56}{65}.$$

The divisors of 56 are $1, 2, 4, 7, 8, 14, 28, 56$ and the divisors of 65 are $1, 5, 13, 65$. Their gcd is 1. Such numbers are called relatively prime.

Definition 7.4: m and n are said to be relatively prime if $(m, n) = 1$.

PRACTICE 7.9

Using Practice 7.8 show that $\frac{m}{(m,n)}$ and $\frac{n}{(m,n)}$ are relatively prime.

We have learnt many divisibility properties of integers in high school. For example, $0, 2, 4, 6$ and 8 are the single digit even numbers and for any other integer to be divisible by 2, it is enough if the unit digit is one of these.

An integer is divisible by 3 if the sum of its digits is divisible by 3.The reader shall be able to prove this after going through Section 7.3.

An integer is divisible by 5 if the unit digit is either 0 or 5. This can be easily proved as follows: If n is the unit digit of N, then $N - n$ is divisible by 10 and hence $N - n$ is divisible by 5. If $n = 5$ or 0, then n is divisible by 5. Thus, $(N - n) + n$ is divisible by 5. Here we have used Proposition 7.1.

Now, we consider the divisibility by 15.

Example 7.12

Show that $N > 0$ is divisible by 15 if the unit digit is 0 or 5 and if the sum of its digits is divisible by 3.

SOLUTION

We have $3|N$ and $5|N$. Thus, for some m and n, $N = 3m = 5n$. We get $5|3m$. Since $(5, 3) = 1$, $(5m, 3m) = m = (5m, 5n)$ (refer Practice 7.2). Since 5 is a common divisor of $5m$ and $5n$, $5|m$. Thus, $m = 5k$ for some k. We finally obtain, $N = 3m = 15k$ and hence N is divisible by 15

Example 7.12 illustrates the importance of relatively prime numbers. The next theorem is another important result for relatively primes. This theorem says that if an integer m divides a product of two integers and m has no common divisor with one of the factors of the product, then m has to divide the other factor. ■

Theorem 7.4

If $m|xy$ and $(m, x) = 1$, then $m|y$.

Proof

Since $(m, x) = 1$, $(my, xy) = (m, x)y = y$. Further, m is a common divisor of my and xy. Hence $m|y$. If $y < 0$, we can show that $(my, xy) = |y|$ and $m|y$, which again proves the result. ■

7.3 PRIMES

For every integer m, $1, -1, m, -m$ are all its divisors. There are some integers for which these are the only divisors. For example, the only divisors of 17 are -1, 1, -17 and 17 Such numbers are called *primes*.

Definition 7.5: A positive number p is said to be a prime, if it has exactly two positive divisors – p and 1. A negative number p is said to be a prime if $-p$ is a prime.

Example 7.13

1 is not a prime because the only positive divisor of 1 is 1. 2 is a prime as it has the two positive divisors 1 and 2. 3 is a prime. 6 is not a prime since its positive

divisors are 1, 2, 3 and 6, which are four in number. No even number *other than* 2 is a prime. Since 2 is a divisor of e, e has at least three positive divisors including 1, 2 and e.

The first few positive prime numbers are 2, 3, 5, 7, 11, 13, 17, 19, 23, 29, etc. ■

Example 7.14

If p is a prime, show that for any m with $p \nmid m$, $(m, p) = 1$. In other words, any integer m that is not a multiple of p is relatively prime to it.

SOLUTION

Assume that $p \nmid m$ and $(m, p) = g > 1$. Then $g|p$ and hence $g = \pm p$. But since $g|m$, $p|m$, which is a contradiction. Hence $(m, p) = 1$. ■

Definition 7.6: A positive number n is not a prime if there is a divisor k with $1 < k < n$. Such a number is called a *composite* number. A negative integer n is composite if $|n|$ is composite.

Note that 1 is neither a prime nor a composite number.

Euclid's Theorem: The number of primes is infinite. That it, there is no end to the sequence of primes, 2, 3, 5, 7, 11, 13,....

Proof

Let there be only finite number of primes and their total number be N. Thus, we can list these finitely many primes as $p_1, p_2, \ldots, p_N$ (refer Section 8, Chapter 2). Now, consider $Q = p_1 p_2 \ldots p_N + 1$. As per hypothesis, Q cannot be a prime. By prime factorization theorem Q must have a prime factor, which is less than Q. Hence, there is $1 \leq i \leq N$ such that p_i is a factor of Q. But p_i divides $p_1 p_2 \ldots p_N$. Thus, p_i must divide the difference $1 = Q - p_1 p_2 \ldots p_N$, which is a contradiction. Hence, the number of primes cannot be finite. ■

Factorization and Prime Factorization: Consider the number 75. It can be written as a product of positive integers in many ways. For example, $75 = 1 \times 75$, $75 = 3 \times 25$, $75 = 3 \times 5 \times 5$. Expressing a number n as a product of other numbers is called *factorization*. If all the numbers involved in the product are primes or their powers, then it is called a prime factorization.

Example 7.15

We have

(a) $78 = 2 \times 3 \times 13$.

(b) $1024 = 2 \times 2 \times 2 \times 2 \times 2 \times 2 \times 2 \times 2 \times 2 \times 2$.

(c) $18375 = 3 \times 5 \times 5 \times 5 \times 7 \times 7$.

How do we know that any integer can be factorized into prime powers? The following theorem proves this fact. ■

Theorem 7.5

Every non-zero integer N can be factorized as a product of prime numbers.

Proof

Since p is a prime if and only if $-p$ is a prime, without loss of generality, let us assume that $N > 0$. If N is not a prime, then there is $0 < n < N$ such that $N = nm$. Next, we take n. If n is a prime, we consider m. If a factor is not a prime, then there are numbers k_1 and k_2 such that it can be written as k_2k_2. This procedure is bound to stop since every time we get a prime number or numbers x with $1 < x$ and less than the non-prime factor. We repeat this procedure till all the factors become prime. These primes need not be distinct.

Next, we consider the uniqueness of the factorization. Is it possible that a number can be factorized in two different ways? For example, consider

$$140 = 2 \times 5 \times 7 \times 2 = 2^2 \times 7 \times 5 = 5 \times 7 \times 2 \times 2.$$

It is clear that all these factorizations differ only in the manner of groupings and arrangements of factors and are essentially same. We shall prove the uniqueness of prime factorization in Theorem 7.7. The following results and examples help us to prove and understand Theorem 7.7.

Let $k|ab$. Can we say that k must divide one of a or b? The answer is no. Consider the fact $12|40 \times 30$. But neither $12|40$ nor $12|30$ is true. If a number is a prime, this will not occur. ■

Proposition 7.5

If p is a prime and $p|ab$, then either $p|a$ or $p|b$.

Proof

Suppose $p \nmid a$, then by Example 7.13, $(a, p) = 1$ and by Theorem 7.3, $p|b$. Similarly, we can show that if $p \nmid b$, then we must have $p|a$. ■

PRACTICE 7.10

Using Proposition 7.5 and mathematical induction prove that if a prime number p divides the product $m_1m_2...m_n$, then $p|m_i$ for some i.

Example 7.16

Find the number of all divisors of 648.

SOLUTION

$648 = 2^3 3^4$. The factors of 648 are of the form $2^m 3^n$ where $0 \le m \le 3$ and $0 \le n \le 4$. The possible values of m are four in number and the possible values of n are 5 in number. Hence, by the fundamental counting principle, the total number of divisors is $4 \times 5 = 20$. ■

In Example 7.15, we have used the following result.

Theorem 7.6

If $k|p_1^{\alpha_1} p_2^{\alpha_2} ... p_n^{\alpha_n}$, then $k = p_1^{\beta_1} p_2^{\beta_2} ... p_n^{\beta_n}$ where each $\beta_i \le \alpha_i$.

Proof

Let $N = p_1^{\alpha_1} p_2^{\alpha_2} ... p_n^{\alpha_n}$ and q be a prime factor of k. Since $q|k$ and $k|N$, we have $q|N$. But $q|p_i$ for some i. Hence $q = p_i$ for some i (refer Practice 7.10). Hence $k = p_1^{\beta_1} p_2^{\beta_2} ... p_n^{\beta_n}$. We now show that for all i, $\beta_i \leq \alpha_i$. Suppose not. (Proof by contradiction). Then for some i, $\beta_i - \alpha_i = m > 0$. We have

$$p_1^{\beta_1} p_2^{\beta_2} ... p_n^{\beta_n} | p_1^{\alpha} p_2^{\alpha_2} ... p_n^{\alpha_n}.$$

From this we obtain

$$\left(\Pi_{j \neq i} p_j^{\beta_j}\right) p_i^m | \left(\Pi_{j \neq i} p_j^{\alpha_j}\right).$$

From this we obtain $p_i | p_j$ with $p_i \neq p_j$ a contradiction to the distinctness of p_i. The proof is now complete. ■

The next theorem proves the **Uniquess of prime factorization.**

Theorem 7.7

Let

$$\Pi_{i=1}^{m} p_i^{\alpha_i} = \Pi_{j=1}^{n} q_j^{\beta_j} \tag{7.2}$$

where p_i's and q_j's are distinct. Then we have $m = n$, each p_i is equal to some q_j and vice versa. Further, the power of a prime factor on both the sides of the equation is the same.

Proof

From an argument in the proof of Theorem 7.6, for each $1 \leq i \leq m$, $p_i = q_j$ for some $1 \leq j \leq n$ and for each $1 \leq j \leq n$, $q_j = p_i$ for some $1 \leq i \leq m$. This implies that $m = n$ and the set $\{p_1, p_2, ..., p_n\} = \{q_1, q_2, ..., q_n\}$ are identical.

Now by renaming and rearranging the LHS of equation (7.2), we get

$$\Pi_{i=1}^{m} p_i^{\alpha_i} = \Pi_{j=1}^{m} p_i^{\gamma_i}.$$

Using an argument in the proof of Theorem 7.6, we obtain $\alpha_i \leq \gamma_i$ and $\gamma_i \leq \alpha_i$. Thus, the result is proved. ■

Example 7.17

Find the gcd of 5500 and 217800 by the method of prime factorization.

SOLUTION

We factorize the given numbers into primes.

$$5500 = 2^2 \times 5^3 \times 11$$

and

$$217800 = 2^4 \times 3^2 \times 5^2 \times 11^2.$$

It is clear that

$$2^{i_1} \times 3^{i_2} \times 5^{i_3} \times 11^{i_4}$$

is a divisor of $2, 17, 800$ if $0 \le i_1 \le 4$, $0 \le i_2 \le 2$, $0 \le i_3 \le 2$, $0 \le i_4 \le 2$. Next, we write 5500 as $2^2 \times 3^0 \times 5^3 \times 11$. So,

$$2^{j_1} \times 3^{j_2} \times 5^{j_3} \times 11^{j_4}$$

is a divisor of 5500 if $0 \le j_1 \le 2$, $0 \le j_2 \le 0$, $0 \le j_3 \le 3$, $0 \le j_4 \le 1$. Further, any divisor of the given numbers is of the type given above.

Hence

$$2^a \times 3^b \times 5^c \times 11^d$$

is a common factor of 5500 and $2, 17, 800$ if and only if $0 \le a \le 2$ and $a \le 4$, $b \le 0$ and $b \le 2$, $c \le 3$ and $c \le 2$, $d \le 1$ and $d \le 2$. ■

The greatest integer less than or equal to both p and q is the minimum of p and q. Hence the gcd of the given numbers is $2^2 \times 3^0 \times 5^2 \times 11^1$.

PRACTICE 7.11

Find the lcm of 246 and 426 by the method of prime factorization.

PRACTICE 7.12

Let

$$M = p_1^{i_1} p_2^{i_2} \ldots p_n^{i_n}$$

and

$$N = p_1^{j_1} p_2^{j_2} \ldots p_n^{j_n}.$$

Here $p_1, p_2, \ldots, p_n$ are all primes. Show that

$$(M, N) = p_1^{k_1} p_2^{k_2} \ldots p_n^{k_n}$$

where

$$k_1 = \min(i_1, j_1)$$
$$k_2 = \min(i_2, j_2)$$

etc.

Example 7.18

Find the gcd of 4340 and 2821 by Euclid's Algorithm. Also express it as $4340a + 2821b$

SOLUTION

$$4340 = 2821 \times 1 + 1519$$
$$2821 = 1519 \times 1 + 1302$$

$$1519 = 1302 \times 1 + 217$$
$$1302 = 217 \times 6 + 0.$$

Thus, 217 is the gcd of the given pair of numbers. Now, starting from the last but one equation above, we obtain

$$\begin{aligned} 217 &= 1519 - 1302 \\ &= 1519 - (2821 - 1519) \\ &= 2 \times 1519 - 2821 \\ &= 2 \times (4340 - 2831) - 2831 \\ &= 2 \times 4340 - 3 \times 2831. \end{aligned}$$ ■

7.4 CONGRUENCE

We know that integers can be classified into odd and even numbers. An odd number leaves the remainder 1 when divided by 2, and an even number leaves the remainder 0 when divided by 2. Similarly, we can classify the natural numbers into three parts based on their divisibility properties with respect to three. Consider the subsets of **Z**:

$$\begin{aligned} A_0 &= \{\ldots, -9, -6, -3, 0, 3, 6, 9, \ldots\} \\ A_1 &= \{\ldots, -8, -5, -2, 1, 4, 7, 10, \ldots\} \\ A_2 &= \{\ldots, -7, -4, -1, 2, 5, 8, 11, \ldots\}. \end{aligned}$$

A_0 is the set of multiples of 3, A_1 is the set of all natural numbers leaving the remainder 1 when divided by 3 and A_2 is the set of all natural numbers leaving the remainder 2 when divided by 3. For example, $3672 \in A_0$, $6790 \in A_1$ and $9602 \in A_2$.

Hence, two integer numbers belong to the same set if they leave the same remainder when divided by 3. We can also say that two numbers belong to the same set if their difference is a multiple of 3.

Let us recall the idea of a relation introduced in Section 2 of Chapter 2. Define R on **Z** as $a\,R\,b$ if $a - b$ is a multiple of 3. We have already learnt that this relation is an equivalence relation and we recognize the sets A_0, A_1 and A_2 as the equivalence classes [0],[1] and [2], respectively.

Example 7.19

Let us recall that a number is divisible by 3 if the sum of its digits is divisible by 3. This can be stated as follows: If $n \in A_0$, then the sum of the digits of n is also in A_0. Now, consider 2347. It leaves a remainder of 1 when divided by 3. Hence $2347 \in A_1$. The sum of its digits is 7 and 7 is also in A_1. In fact, we shall see that if $n \in A_i$, then $s(n) \in A_i$ where the sum of the digits of n is denoted by $s(n)$.

Next, let $a \in A_1$ and $b \in A_2$. What about $a + b$? By taking some examples, we can see that $a + b \in A_0$. The reason for this is clear. When a is divided by

3, it leaves a remainder 1 and when b is divided by 3 it leaves a remainder of 2 and hence $a+b$ would be a multiple of 3. The following observations can also be made:

(a) If $a \in A_0$ and $b \in A_1$, then $a+b \in A_1$.
(b) If $a \in A_0$ and $b \in A_2$, then $a+b \in A_2$.
(c) If $a \in A_1$ and $b \in A_2$, then $a+b \in A_0$.
(d) If $a \in A_0$ and $b \in A_0$, then $a+b \in A_0$.
(e) If $a \in A_1$ and $b \in A_1$, then $a+b \in A_2$.
(f) If $a \in A_2$ and $b \in A_2$, then $a+b \in A_1$.

Based on divisibility by 8, we can classify the integers into eight parts. Let $[i]$ be the set of all integers that leave a remainder of i when divided by 8 for $i = 0, 1, 2, 3, 4, 5, 6, 7$.

Let $a \in [5]$ and $b \in [6]$. When a is divided by 8 it leaves a remainder of 5 and when b is divided by 8 it leaves a remainder of 6. A little bit of thinking tells us that $a+b$ would leave a remainder of $11 - 8 = 3$ when divided by 8. We summarize the properties related to the sums of elements of these sets in Table 7.2.

Table 7.2

	[0]	[1]	[2]	[3]	[4]	[5]	[6]	[7]
[1]	B_1	[2]	[3]	[4]	[5]	[6]	[7]	[0]
[2]	[2]	[3]	[4]	[5]	[6]	[7]	[0]	[1]
[3]	B_3	[4]	[5]	[6]	[7]	[0]	[1]	[2]
[4]	[4]	[5]	[6]	[7]	[0]	[1]	[2]	[3]
[5]	[5]	[6]	[7]	[0]	[1]	[2]	[3]	[4]
[6]	B_6	[7]	[0]	[1]	[2]	[3]	[4]	[5]
[7]	[7]	[0]	[1]	[2]	[3]	[4]	[5]	[6]

We find that [3] is found in the row corresponding to [5] and the coloumn corresponding to [6]. This means that if $a \in B_5$ and $b \in [6]$, then $a+b \in [3]$.

Let us now imagine a clock as shown in Figure 7.1. Suppose the 'hour' now is 5 as per this clock. After 6 'hours', it would be 3 O'clock as per this clock. This is the reason why the calculations shown in Table 7.2 are known as clock arithmetic.

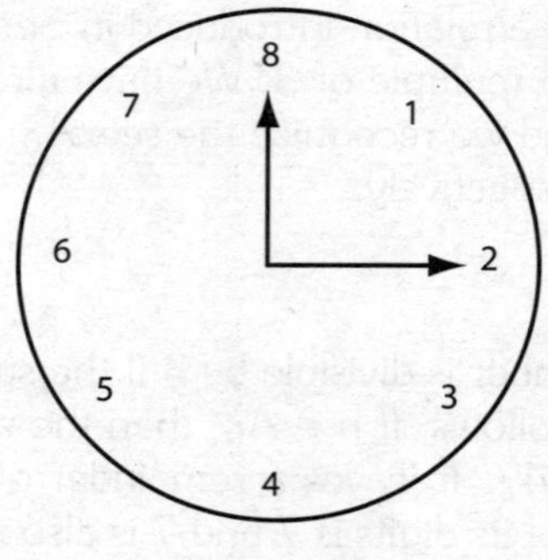

Figure 7.1 Clock arithmetic

Note that a number m that leaves a remainder of 5 when divided by 8 is of the form $l + 5$, where l is a multiple of 8. Now, l being a multiple of 8 is of the form $8k$. Hence, the elements of [5] are numbers of the form $8k + 5$. In fact, for $k = 0, 1, 2, ..., 7$

$$[i] = \{8k + i : k = 0, \pm 1, \pm 2, ..., \}.$$ ■

In the following example, we show the usefulness of this notation.

Example 7.20

Show that:
(a) If $a \in [3]$ and $b \in [4]$, then $a + b \in [7]$.
(b) If $a \in [7]$ and $b \in [6]$, then $a+b \in [5]$. ■

Proof

(a) $a = 8k_1 + 3$ for some $k_1 \geq 0$ and $b = 8k_2 + 4$ for some $k_2 \geq 1$. Now, $a+b = 8k_1+3+8k_2+4 = 8(k_1+k_2)+7$, and hence it is clear that $a+b \in B_7$.
(b) $a = 8k_1 + 7$ for some $k_1 \geq 0$ and $b = 8k_2 + 6$ for some $k_2 \geq 1$. Now,

$$\begin{aligned} a + b &= 8k_1 + 7 + 8k_2 + 4 \\ &= 8(k_1 + k_2) + 11 \\ &= 8(k_1 + k_2) + 8 + 3 \\ &= 8(k_1 + k_2 + 1) + 3 \end{aligned}$$

and hence it is clear that $a + b \in B_3$.

The idea of **congruence** defined below is nothing but a proper mathematical way of studying the various types of clock arithmetic. ■

Definition 7.7 If $m \neq 0$, then we say that a is congruent to b modulo m and write

$$a \cong b \pmod{m}$$

if $a - b$ is a multiple of m.

Example 7.21

The reader can easily check the following:

$$9 \cong 7 \bmod 2 \qquad 8 \cong 114 \bmod 53 \qquad 50 \cong 64 \bmod 7$$
$$34 \ncong 37 \bmod 5 \quad 67 \ncong 17 \bmod 10 \quad 9 \ncong 10 \pmod{7}.$$ ■

Proposition 7.6

For integers a, b, c, d, m, n, we have

(i) $a \cong b \pmod{m} \leftrightarrow b \cong a \pmod{m} \leftrightarrow a - b \cong 0 \pmod{m}$.
(ii) $a \cong b \pmod{m} \wedge c \cong d \pmod{m} \rightarrow ax + by \cong cx + dy \pmod{m}$.

(iii) $a \cong b \pmod{m} \wedge c \cong d \pmod{m} \rightarrow ac \cong bd \pmod{m}$.
(iv) $a \cong b \pmod{m} \wedge d|m \rightarrow a \cong b$ ((mod d) m).

Proof
We prove (i), (iii), (iv) and the proof of (ii) is left as a practice question for the readers. The proof of (i) is as follows:

$$\begin{aligned} & a \cong b \mod m \\ & \leftrightarrow \exists\, k \in Z \text{ such that } a - b = km \\ & \leftrightarrow b - a = -km \\ & \leftrightarrow b \cong a \mod m. \end{aligned}$$

Further,

$$\begin{aligned} & a \cong b \mod m \\ & \leftrightarrow \exists k \in Z \text{ such that } a - b = km \\ & \leftrightarrow (a - b) - 0 = km \\ & \leftrightarrow a - b \cong 0 \pmod{m}. \end{aligned}$$

The proof of (iii) is as follows:
If $a - b = k_1 m$ and $c - d = k_2 m$, we get $a = b + k_1 m$ and $c = d + k_2 m$.
Now $ac = (b + k_1 m)(d + k_2 m) = bd + k_1 md + k_2 mb + k_1 k_2 m^2$
$= bd + (k_1 d + k_2 b + k_1 k_2 m)m$. This implies that $ac - bd$ is a multiple of m.
Now (iii) follows from the definition.
The proof of (iv) is as follows: Since $d|m$, there is $l \in Z$ such that $m = ld$. Next,

$$\begin{aligned} & a \cong b \pmod{m} \\ & \rightarrow \exists k \in Z \text{ such that } a - b = km \\ & \rightarrow a - b = kld \\ & \rightarrow a \cong b \pmod{d}. \end{aligned}$$

■

PRACTICE 7.13

Prove (ii) of Proposition 7.6.

In high school, we learn that a number is divisible by 9 if and only if the sum of the digits is divisible by 9. The sum of the digits of 20305 is $2+0+3+0+5 = 10$. 10 is not divisible by 9 and hence 20305 is also not divisible by 9. Proposition 7.7 proves this observation using the principles of Proposition 7.6.

We are familiar with expansions of the kind

$$20305 = 2 \times 10^4 + 0 \times 10^3 + 3 \times 10^2 + 0 \times 10 + 5.$$

In general, every positive integer k can be written as

$$k = a_0 \times 10^n + a_1 \times 10^{n-1} + ... + a_n$$

where $a_0, a_1, ..., a_n \in \{0, 1, 2, ..., 9\}$. This idea is used in the proof of Proposition 7.7

Proposition 7.7

Prove that an integer is divisible by 9 if and only if the sum of its digits is divisible by 9.

Proof

Let $k = a_0 \times 10^n + a_1 \times 10^{n-1} + ... + a_n$. The sum of the digits of k is just $a_0 + a_1 + ... + a_n$. We show that $m \cong a_0 + a_1 + ... a_n \pmod 9$. As $10 - 1 = 9$, $10^2 - 1 = 99$, etc., we note that $10^m - 1$ is divisible by 9. k can be written as

$$\begin{aligned} k &= a_0(10^n - 1 + 1) + a_1(10^{n-1} - 1 + 1) + ... + a_{n-1}(10 - 1 + 1) + a_n \\ &= a_0(10^n - 1) + a_1(10^{n-1} - 1) + ... + a_{n-1}(10 - 1) + a_0 + a_1 + ... + a_{n-1} + a_n \\ &\cong a_0 + a_1 + ... + a_{n-1} + a_n \pmod 9 \end{aligned}$$

since $a_0(10^n - 1) + a_1(10^{n-1} - 1) + ... + 9a_{n-1}$ is divisible by 9.

PRACTICE 7.14

Prove that an integer is divisible by 3 if and only if the sum of its digits is divisible by 3.

As a special case of (iii) of Proposition 7.6, (taking $c = d = x$), we have the following result

$$a \cong b \pmod m \rightarrow ax \cong bx \pmod m.$$

This tells us that 'we can multiply a congruence by an integer on both the sides.' The next natural question is to ask if one can divide a congruence by an integer on both the sides? The question can now be posed as follows: Is it true that

$$a \cong b \pmod m \rightarrow \frac{a}{x} \cong \frac{b}{x} \pmod m? \quad (7.3)$$

Obviously, we must take $x \neq 0$ and further $\frac{a}{x}$ and $\frac{b}{x}$ have to be integers. But even if these conditions are satisfied, (7.3) need not be true. Consider the following example:

Example 7.22

$$84 \cong 252 \pmod{12}$$

is true. Division by 6 gives us

$$14 \cong 42 \pmod{12}$$

which is clearly false.

But division by 7 gives us

$$12 \cong 36 \pmod{12}$$

which is true.

As the reader might have guessed, division by 7 gives us a correct congruence because $(7, 12) = 1$

Let us examine this problem closely before stating the relevant result:

To avoid the possibility of fractions, we start with $ax \cong bx \pmod m$ and ask if $a \cong b \pmod m$. Given that $ax - bx$ is a multiple of m, is it true that $a - b$ is a multiple of m?

Now if $ax - bx = km$, then $a - b = \frac{ax-bx}{x} = \frac{km}{x}$. We now want $\frac{km}{x}$ to be a multiple of m. Since $\frac{km}{x} = a - b$, $x|km$. So, if $(x, m) = 1$, then by Theorem 7.4, $x|k$. Thus $\frac{km}{x} = \frac{k}{x}m$ is a multiple of m. ■

This discussion has actually proved the following theorem:

Theorem 7.8

If $(x, m) = 1$ and $ax \cong bx \pmod m$ then $a \cong b \pmod m$.

Example 7.23

Consider the following questions:

(i) If the time now is 8 O' clock, what will be the time after 5 hours?

(ii) If the time now is 2 O' clock, before how many hours was it 7 O'clock?

Let us consider the solution of (i). We add 8 and 5 and get the answer as 13. Since there is no hour like 13 and since $13 = 12 + 1$, the answer is 1 O'clock. It is obvious that the solution to (i) involves modulo 12 arithmetic. That is, (i) can be formulated as finding $x \leq 12$ such that

$$x \cong 8 + 5 \pmod{12}.$$

Similarly, the solution of (ii) involves finding $x \leq 12$ such that

$$7 + x \cong 2 \pmod{12}.$$

So, the answer for (ii) is $x = 8$.

Example 7.23 motivates us to consider the solution of the congruence equation

$$ax + b \cong 0 \pmod m.$$ ■

Theorem 7.9

The solution of

$$x + b \cong 0 \pmod m$$

is given by an x such that $x \in [m-b]$ where $[y]$ is the equivalence class of y under the equivalence relation R defined as $x\,R\,y$ if and only if $x \cong y \pmod{m}$.

Proof

If $x \in [m-b]$, then $x \cong (m-b) \pmod{m}$. But then we have $x+b \cong m \pmod{m}$ and finally $x+b \cong 0 \pmod{m}$. ■

Theorem 7.10

The congruence $ax \cong 1 \pmod{m}$ has a solution if and only if $(a, m) = 1$.

Proof

If $(a, m) = g > 1$ and x is a solution of the required congruence, then there is k such that $ax - 1 = km$ and hence $1 = ax - mk$. But from Theorem 7.2, we know that g is the least positive number, which can be expressed as $at + ms$ with t and s as integers. Since $g > 1 = ax - mk$, this is a contradiction. Thus, if $(a, m) \neq 1$, there is no solution.

Let us now assume that $(a, m) = 1$. We can find x and y such that $ax + my = 1$ and hence $ax - 1 = my$. In other words, $ax \cong 1 \pmod{m}$ and this solves the congruence. ■

Example 7.24

Examine which of the following congruences have solutions and find the solution if it exists:

(i) $6x \cong 1 \mod 4$.
(ii) $15x \cong 1 \mod 8$.
(iii) $8x \cong 1 \mod 13$.

SOLUTION

Since $(6, 4) = 2$, (i) has no solution.
For (ii), we have $(15, 8) = 1$. Thus, there is a solution. We find it by evaluating $15 \times x \mod 8$ by giving various values to x. We have

$$\begin{aligned} 15 \times 1 &= 15 \cong 7 \mod 8 \\ 15 \times 2 &= 30 \cong 6 \mod 8 \\ 15 \times 3 &= 45 \cong 5 \mod 8 \\ 15 \times 4 &= 60 \cong 4 \mod 8 \\ 15 \times 5 &= 75 \cong 3 \mod 8 \\ 15 \times 6 &= 90 \cong 2 \mod 8 \\ 15 \times 7 &= 105 \cong 1 \mod 8. \end{aligned}$$

Hence, the required solution is $x \cong 7 \mod 8$.

We can simplify the calculation if we notice that $15 \cong 7 \mod 8$ and hence the required solution is also a solution of $7x \cong 1 \mod 8$.

Further simplification takes place with the observation that $7 \cong -1 \mod 8$.

Next, let us solve (iii), which has a solution since $(8, 13) = 1$. We start with $8 \times 1 \cong 8 \mod 13$. We keep on adding 8.

$$\begin{aligned} 8+8 &= 16 \cong 3 \mod 13 \\ 8+8+8 &\cong 3+8 \cong -2 \mod 13 \\ 8+8+8+8 &\cong -2+8 \cong 6 \mod 13 \\ 8+8+8+8+8 &\cong 6+8 \cong 14 \cong 1 \mod 13. \end{aligned}$$

Hence, the required solution is $x \cong 5 \mod 13$. ■

PRACTICE 7.15

Show that the congruence

$$ax \cong b \mod m$$

has a solution if and only if $(a, m)|b$.

We have noted the importance of the condition $(a, m) = 1$ in the solution of the linear congruences such as $ax \cong 1 \pmod m$. We have also observed that it is enough to consider this equation for $0 < a < m$. If $a > m$, then we choose $0 < \alpha < m$ such that $\alpha \cong a \bmod (m)$ $(\alpha > m)$. The solutions of $ax = b \pmod m$ and $\alpha x = b \pmod m$ are the same. In this context, the importance of the following definition is obvious.

Definition 7.8: The number of a's with $0 < a < m$ and $(a, m) = 1$ is denoted by $\varphi(a)$ and φ is known as Euler's phi function.

Example 7.25

Find $\varphi(a)$ for $a = 1, 2, \ldots, 10$.

SOLUTION

It is clear that $\varphi(1) = 1$, $\varphi(2) = 1$ and $\varphi(3) = 2$. Among 1, 2, 3, 1 and 3 are relatively prime to 4. Hence $\varphi(4) = 2$. We have $\varphi(5) = 4$. Among $1, 2, 3, 4, 5$, the numbers 1 and 5 are relatively prime to 6. So, $\varphi(6) = 2$, $\varphi(7) = 6$. Among $1, 2, 3, 4, 5, 6, 7$, the numbers $1, 3, 5, 7$ are relatively prime to 8. So, $\varphi(8) = 4$. Among $1, 2, 3, 4, 5, 6, 7, 8$, the numbers $1, 2, 4, 5, 7, 8$ are relatively prime to 9. Hence $\varphi(9) = 5$. Among $1, 2, 3, 4, 5, 6, 7, 8, 9$ the numbers $1, 3, 7, 9$ are relatively prime to 10. Hence $\varphi(10) = 4$. ■

PRACTICE 7.16

Find $\varphi(a)$ for $a = 11, 12, 13, 14, 15$.

PRACTICE 7.17

Show that for a prime number p, $\varphi(p) = p - 1$.

Example 7.26

Let $N = pq$ where p and q are distinct primes. Find $\varphi(N)$.

SOLUTION

One can show that among the numbers $1, 2, ..., N - 1$, the only integers not relatively prime to N are p, $2p,...,(q - 1)p$ and q, $2q,...,(p - 1)q$. These are $q - 1 + p - 1 = p + q - 2$ in number. Hence, $\varphi(N) = N - 1 - (p + q - 2) = N - p - q + 1 = pq - p - q - 1 = (p-1)(q-1)$. ■

7.5 FERMAT'S THEOREM

In the introduction, we have mentioned about the applications of number theory to cryptography. In this section, we state and prove Fermat's theorem and indicate its application to cryptography.

Let us recall that the relation congruence modulo m is an equivalence relation and the equivalence classes are given by

$$[x] = \{y \in \mathbf{Z} : x \cong y \pmod{}\}.$$

There are m distinct equivalence classes. An arbitrary $n \in \mathbf{Z}$ belongs to precisely one equivalence class.

Let us take $m = 3$. If we take the three numbers 0, 1 and 2, then we have

$$\mathbf{Z} = [0] \cup [1] \cup [2].$$

But if we take 0, 2 and 5, we have

$$\mathbf{Z} \neq [0] \cup [2] \cup [5].$$

The reason is that $[2] = [5]$ and $1 \notin [0] \cup [2] \cup [5]$. Since $2 \cong 5 \pmod 3$, after having chosen 2, we should not have chosen 5.

Now if we choose 0, 2 and 4, then

$$\mathbf{Z} = [0] \cup [2] \cup [4].$$

We say that $\{0, 1, 2\}$ is a complete residue system modulo 3, whereas $\{0, 2, 5\}$ is not a complete residue system modulo 3. Of course, $\{0, 2, 4\}$ is a complete residue system modulo 3.

Let us now define a complete residue system.

Definition 7.9: If $a \cong b \pmod m$, then b is called a residue of x modulo m. A set of numbers $\{x_1, x_2, ..., x_m\}$ is called a complete residue system modulo m if for every integer n there is only one x_i such that $n \cong x_i \pmod m$.

It is clear that a set of m numbers $x_1, x_2, \ldots, x_m$ form a complete residue system modulo m if $x_i \not\cong x_j \pmod{m}$ for $i \neq j$. This follows from the fact that $[x_i] \neq [x_j]$ and there are precisely m distinct equivalence classes.

Example 7.27

7, 12, 20, 27, 71, 82 is a complete set of residues for 6. Because, we note that

$$\begin{aligned} 7 &\cong 1 \pmod 6 \\ 12 &\cong 0 \pmod 6 \\ 20 &\cong 2 \pmod 6 \\ 27 &\cong 3 \pmod 6 \\ 71 &\cong 5 \pmod 6 \\ 82 &\cong 4 \pmod 6. \end{aligned}$$ ■

Example 7.28

The set of numbers 7, 12, 20, 27, 32, 71 is not a complete set of residues because $20 \cong 32 \pmod 6$. ■

Example 7.29

The simplest complete set of residues modulo m is $\{0, 1, 2, \ldots, m-1\}$.

SOLUTION

Let $x_i = i - 1$, $i = 1, 2, \ldots, m$. Let $x_i \cong x_j$. Then $x_i - x_j = i - j$ is a multiple of m. There exists k such that $i - j = km$. Taking $i \neq j$, we obtain $k \neq 0$. Since $|k| > 1$ for any integer $k \neq 0$, $|km| > m$. But this implies that $|i - j| > m$, which is impossible. ■

PRACTICE 7.18

Show that for any integer x, $\{x+1, x+2, \ldots, x+m\}$ is a complete set of residues modulo m.

Example 7.30

$\{1, 2, 3, 4\}$ is a complete residue class modulo 4. Now consider the set $\{2, 4, 6, 8\}$. The new set is obtained by multiplying each element of the given set by 2. The set is not a complete residue class since $4 \cong 8 \pmod 4$. But the set $\{3, 6, 9, 12\}$ obtained by multiplication by 3 is a complete residue class modulo 4. ■

The above example gives rise to the following proposition.

Proposition 7.8

If S is a complete residue system modulo m, then $T = \{ax : x \in S\}$ is a complete residue system modulo m if $(a, m) = 1$.

Proof

Let $y_1, y_2 \in T$ be distinct elements. We have to show that $y_1 \not\cong y_2$. We have $y_1 = ax_1$ and $y_2 = ax_2$ for some $x_1, x_2 \in S$ with $x_1 \not\cong x_2$. S is a complete residue system and hence $x_1 \not\cong x_2$. If $y_1 \cong y_2$, then $ax_1 \cong ax_2$. Since $(a, m) = 1$, we get $x_1 \cong x_2$ by Theorem 7.8, which is a contradiction. Hence $y_1 \not\cong y_2$ and the result is proved.

In the last section, we examined the solutions of linear congruences. Naturally, we would like to study congruences like

$$x^n \cong b.$$

Fermat's theorem and its generalization by Euler is related to the congruence

$$x^n \cong 1. \blacksquare$$

Example 7.31

Let us compute the powers of integers under the modulo 5 multiplication and observe what happens. We define $x^0 \cong 1$.

$$\begin{array}{lllll} 1^0 \cong 1 & 1^1 \cong 1 & 1^2 \cong 1 & 1^3 \cong 1 & 1^4 \cong 1 \\ 2^0 \cong 1 & 2^1 \cong 2 & 2^2 \cong 4 & 2^3 \cong 3 & 2^4 \cong 1 \\ 3^0 \cong 1 & 3^1 \cong 3 & 3^2 \cong 4 & 3^3 \cong 2 & 3^4 \cong 1 \\ 4^0 \cong 1 & 4^1 \cong 4 & 4^2 \cong 1 & 4^3 \cong 4 & 4^4 \cong 1. \end{array}$$

Now an arbitrary a, which is not a multiple of 5, is congruent to one of $1, 2, 3$ and 4. If x is any one of $1, 2, 3, 4$, then we have $x^4 \cong 1$ and hence $a^4 \cong x^4 \cong 1$. $\blacksquare$

Fermat's theorem is a generalization of Example 7.31.

Fermat's Theorem: If a is not a multiple of p, then

$$a^{p-1} \cong 1 \pmod p.$$

As Fermat's theorem can be obtained as a special case of Theorem 7.11, we omit the proof of Fermat's theorem.

Example 7.32

Consider the powers of 2 under modulo 6 multiplication. We obtain $2^0 \cong 2$, $2^1 \cong 2$, $2^2 \cong 4$, $2^3 \cong 2$, $2^4 \cong 4$, $2^5 \cong 2$. Higher powers of 2 would reproduce $4, 2, 4, 2, \ldots$. Hence no power of 2 will be congruent to 1 modulo 6. But let us consider the powers of 5. $5^0 \cong 1$, $5^1 \cong 5$, $5^2 \cong 1$. $\blacksquare$

The following theorem is the generalization of Fermat's theorem by Euler. For a prime number p, $\varphi(p) = p - 1$ and any a with $0 < a < p$ is such that $(a, p) = 1$. These observations make Fermat's theorem a special case of Theorem 7.11.

Theorem 7.11

$$x^{\varphi(m)} \cong 1 \pmod{m}.$$

The proof of Theorem 7.11 uses the idea of a reduced residue system whose definition is given below.

Definition 7.10: A set of numbers $\{x_1, x_2, ..., x_{\varphi(m)}\}$ is called a reduced residue system modulo m if $(x_i, m) = 1$ and for every integer n with $(n, m) = 1$ there is only one x_i such that $n \cong x_i \pmod{m}$.

PRACTICE 7.19

If S is a reduced residue system modulo m, then $T = \{ax : x \in S\}$ is a reduced residue system modulo m if $(a, m) = 1$.

Proof of Theorem 7.11: Let $S = \{r_1, r_2, ..., r_{\varphi(m)}\}$ be a reduced residue class modulo m. Then by Practice 7.19, $\{ar_1, ar_2, ..., ar_{\varphi(m)}\}$ is also a reduced residue class modulo m. Hence for each $1 \leq i \leq \varphi(m)$, there is a unique $1 \leq j \leq \varphi(m)$ such that $r_i \cong ar_j \pmod{m}$. We denote this j by $\alpha(i)$. We now have

$$\begin{aligned} r_1 &\cong ar_{\alpha(1)} \\ r_2 &\cong ar_{\alpha(2)} \\ &\vdots \\ r_{\varphi(m))} &\cong ar_{\alpha(\varphi(m))}. \end{aligned}$$

Now, by Proposition 7.6, we get

$$\begin{aligned} \Pi_{i=1}^{\varphi(m)} r_i &\cong \Pi_{i=1}^{\varphi(m)} ar_{\alpha(i)} \\ &\cong \Pi_{i=1}^{\varphi(m)} ar_i \\ &\cong a^{\varphi(m)} \Pi_{i=1}^{\varphi(m)} ar_i. \end{aligned}$$

This gives us

$$r_1 r_2 ... \cong r_1 r_2 ... a^{\varphi(m)} \pmod{m}. \tag{7.4}$$

Since each $(r_i, m) = 1$, we can cancel each of r_i in equation (7.4) successively using Theorem 7.5 to obtain the required result.

7.6 APPLICATION OF NUMBER THEORY IN CRYPTOGRAPHY

Example 7.33

Consider $S = \{1, 2, ..., 16\}$, which is a complete residue system modulo 17. We define a function $C: \{1, 2, ..., 16\} \longrightarrow \{1, 2, ..., 12\}$ as $C(x) \cong x^5 \pmod{17}$. Here x^k denotes the number in S obtained by multiplying x with itself k times under modulo 17 multiplication. For illustration, let us compute the 'powers' of 3:

$$3^1 \cong 3 \pmod{17}$$
$$3^2 \cong 9 \pmod{17}$$
$$3^3 \cong 10 \pmod{17}$$
$$3^4 \cong 13 \pmod{17}$$
$$3^5 \cong 5 \pmod{17}.$$

Now we shall treat the set S as a set of code words. You may imagine that each number represents an income tax evader and the income tax officials discuss about these people using these code words. Next, let us imagine that a spy has come to know about these codes by some means. The income tax officials will have to change these codes. They have a secret understanding as given below:

When they want to talk about x, they talk about $C(x)$.

For example, instead of 3, they will use $3^5 \pmod{17}$. Instead of 7 they will use $7^5 \cong \pmod{17}$. Let us now consider the following problem. If they mention y, whom are they really talking about? It is clear that we have to find $x \in S$ such that $x^5 \cong y \pmod{17}$. Now, let us observe that $5 \times 13 \cong 1 \pmod{16}$. Hence

$$y^{13} \cong (x^5)^{13} \cong x^{65} \cong [x^{16}]^4 x \cong x$$

by Fermat's theorem. Thus, to decode y, we will have to raise it to the power 13

Generalizing from this example, let us take $\{1, 2, ..., p-1\}$ as a set of code words. We choose e relatively prime to $p - 1$ and find d such that $e \times d \cong 1 \bmod p-1$. Now, the secret code y for x is taken as x^e. For decoding y, we raise it to the power d. This gives back x. **Here the interesting fact is that for a large prime p, finding d is a time-consuming process. Even if a spy knows about the coding procedure and the number e, he will have to spend such an enormous time to find d that he would not attempt to do so.** This is an elementary example of cryptography in which the theory of numbers is used. ■

SUMMARY

- a is a multiple of b if there is k such that $a = bk$. Also, b is a divisor of a.
- Given integers $m, n > 0$, there exist integers q and r such that $m = nq + r$ with $0 \leq r < n$.
- The gcd of m and n is the greatest number that divides both m and n.
- The lcm of m and n is the smallest number that is divisible by both m and n.
- Two numbers m, n are said to be relatively prime if their gcd is 1.
- If m divides ab and m and a are relatively prime, then m divides b.
- An integer p is said to be a prime if it has precisely four divisors: $\pm p, \pm 1$.
- Every integer can be uniquely factorized into powers of primes.
- $a \cong b \pmod{m}$ if $a - b$ is divisible by m.
- For a given integer $m > 0$, the set of integers can be divided into m classes such that all the numbers in the same class leave the same remainder when divided by m.
- $\varphi(m)$ is the count of all positive numbers less than m and relatively prime to m.
- The congruence $ax \cong b \pmod{m}$ can be solved if and only if b divides the gcd of a and m.
- For a prime number p, $\varphi(p) = p - 1$.
- $a^{\varphi(m)} \cong 1 \pmod{m}$ if a is relatively prime to m.

PROBLEMS

1. Find all the divisors of 1374.
2. Find the lcm of 15 and 24 by listing their multiples.
3. Factorize 4578 into primes.
4. Examine which of the following are primes: 117, 323, 713, 12354675439868.
5. Find the gcd and lcm of the following pairs:

   (i) 45, 99.
   (ii) 345, 247.
   (iii) 1000, 100,000.
   (iv) 678, 3987.

6. Show that an integer is divisible by 4 if and only if the number formed by its last two digits is divisible by 4.
7. Devise a divisibility test for 6 and prove your result.
8. Find the multiples of 18 between 1000 and 10,000.
9. Let x be a real number. $[x]$ denotes the largest integer less than or equal to x. Show that the number of multiples of m between 1 and M is $\left[\frac{M}{m}\right]$.
10. Show that $n^5 - n$ is divisible by 2, 5 and 6.
11. Examine which of the following pairs are relatively prime.

    (i) 45, 89.
    (ii) 453, 178.
    (iii) 2345, 9843.

12. Show that n and $n + 1$ are always relatively prime.
13. Examine which of the following numbers are prime: 117, 239, 1428, 1427.
14. Let

$$M = p_1^{i_1} p_2^{i_2} ... p_n^{i_n}$$

and

$$N = p_1^{j_1} p_2^{j_2} ... p_n^{j_n}.$$

Here $p_1, p_2, ..., p_n$ are all primes. Show that

$$[M, N] = p_1^{k_1} p_2^{k_2} ... p_n^{k_n}$$

where, $k_1 = \max(i_1, j_1)$, $k_2 = \max(i_2, j_2)$, etc.
15. Using the previous problem show that $(M, N)[M, N] = MN$.
16. What are the gcd and lcm of m^i and m^j?
17. If $a|m$, $b|m$ and $(a, b) = 1$, then show that $ab|m$.
18. Find $\varphi(200)$.
19. If $N = pq$ where p and q distinct primes, show that $\varphi(N) = (p - 1)(q - 1)$.
20. Solve the following congruences:

(i) $45x \cong 27 \pmod{18}$.
(ii) $21x \cong 28 \pmod{14}$.

21. Show that the congruence $12x \cong 13 \pmod 6$ has no solution.

FURTHER READING

Niven I. and Zuckerman, H.S. (1985), *An Introduction to the Theory of Numbers*, Wiley Eastern Limited, New Delhi.

Apostol T. M. (1995), *Introduction to Analytic Number Theory*, Narosa Publishing House, New Delhi.

Neal Kolblitz (1994), *A course in Number Theory and Cryptography*, Springer Verlog, New York.

Hsiung, C. Y. (1992), *Elementary Theory of Numbers*, Allied Publishers Ltd, Bombay.

CHAPTER

8

Groups, Rings and Fields

Learning Objectives

On completing this chapter, you should be able to:

- recall the definitions of the algebraic structures *semigroup, monoid and group*
- give examples of semigroups, monoids and groups
- prove elementary properties of groups
- recall the definition of a subgroup
- identify subgroups of standard groups
- compute the cosets of subgroups in simple examples
- state and prove the Lagrange Theorem
- define a normal subgroup
- give examples of normal subgroups
- define the notion of quotient group and give examples
- state the definitions of homomorphism and isomorphism between groups
- identify isomorphic groups
- show that every group G is isomorphic to and subgroup of the group of one-to-one and onto functions on G.
- recall the definition of a ring
- give examples of rings
- define subrings and ideals
- define the notion of a quotient ring and give examples
- define a ring homomorphism and prove simple properties
- recall the definition of a field

8.1 INTRODUCTION

Example 8.1

The expression $\frac{1}{\sqrt{(x-3)(7-y)(9+z)}}$ has real value when one of the following things happen:

(i) $x > 3, y < 7, z > -9.$
(ii) $x < 3, y > 7, z > -9.$
(iii) $x < 3, y < 7, z < -9.$
(iv) $x > 3, y > 7, z < -9.$

The following is a part of a C program

```
If ((x > 3) && (y < 7) && (z > −9)) || ((x < 3) && (y > 7) &&
                (z > −9)) ||
   ((x < 3) && (y < 7) && (z < −9)) || ((x > 3) && (y > 7) &&
                (z > −9))
          E = 1/(sqrt((x-3)(7-y)(9+z));
                else
          print("E is imaginary");
```

$(x > 3)$ is an example of a Boolean Expression. A Boolean Expression takes the two possible values *True* or *False*.

In Chapter 3 we have learnt about logical connectives $\wedge$ and $\vee$. && is just the C language version of $\wedge$ and || is the C language version of $\vee$.

Now, if $x = 9$, $y = 9$ and $z = 10$, let us find out the output of the above C program. When $x = 9$, the value of the Boolean expression $(x > 3)$ is *True*. When $y = 9$, the value of the Boolean Expression $(y < 7)$ is *False* and when $z = 10$ the value of the Boolean Expression $(z > -9)$ is *True*. Denoting a true statement by T and a false statement by F, let us find out the value of the long Boolean expression given in the program. This amounts to finding the truth value of

$$(T \wedge F \wedge T) \vee (F \wedge T \wedge T) \vee (F \wedge F \wedge F) \vee (T \wedge T \wedge F).$$

If P is true and Q is false, we know that $P \vee Q$ is true and $P \wedge Q$ is false. Using the eight rules summarized as

$\vee$	T	F
T	T	T
F	T	F

$\wedge$	T	F
T	T	F
F	F	F

we carry out the following simplifications to find the truth value:

$$\begin{aligned}(T \wedge F \wedge T) \vee (F \wedge T \wedge T) \vee (F \wedge F \wedge F) \vee (T \wedge T \wedge F)\\ = F \vee F \vee F \vee F\\ = F.\end{aligned}$$

Hence, for the given data, the computer output will be the printed statement *E is imaginary*.

Addition and multiplication are rules that associate a new number with a given pair of numbers. Similarly, $\vee$ and $\wedge$ can be thought of as rules that associate a symbol with a given pair of symbols from the set $\{T, F\}$.

The simplification of a long expression based on rules is analogous to the simplification of long arithmetic expressions. A set S for which there are rules to associate an element with a given pair of elements is known as an algebraic structure. This chapter is a study of many algebraic structures. ■

Example 8.2

We know that computer processes and calculations use strings of 0's and 1's. Each of 0 and 1 is known as a bit, and a byte is a string of eight bits. For example, 01001101 is a byte.

We can use *addition modulo 2* to design a rule, which gives a byte for a given pair of bytes.

If $x + y \cong z \pmod{2}$, we write $x \oplus_2 y = z$.

The table for addition modulo 2 is as follows:

$\oplus_2$	0	1
0	0	1
1	1	0

Let us note that this is not the binary addition. In binary addition $1 + 1 = 10$ and not 0.

Taking a pair of bytes, we produce another byte by adding the first bit of one byte with the first bit of the other byte, the second bit with the second bit, etc. Here instead of the usual addition we use modulo 2 addition. For example,

$$01000101 \oplus_2 00010011 = 01010110.$$

What is the use of this kind of addition? If a byte differs from another byte in its kth position, then the value at the kth position of the new byte is 1. If the given bytes agree at the kth position, then at the kth position of the new byte we shall have 0. So, the number of 1's in the new byte is the number of positions where the given two bytes differ from each other. This fact is useful in coding theory. The set of all bytes with the above-mentioned addition is an example of an algebraic structure. ■

8.2 MOTIVATION

The solution to the equation

$$x + a = b$$

is

$$x = b + (-a).$$

The solution to the equation

$$x \times a = b$$

is

$$x = b \times \frac{1}{a}.$$

Now, consider the following simple problem:

How many hours after 7 O'clock would it be 3 O'clock?

The answer is 8. Now, this problem can be posed as

$$x \oplus_{12} 7 = 3.$$

Here $\oplus_{12}$ is the clock addition as defined by Table 8.1.

TABLE 8.1
CLOCK ADDITION TABLE

$\oplus_{12}$	1	2	3	4	5	6	7	8	9	10	11	12
1	2	3	4	5	6	7	8	9	10	11	12	1
2	3	4	5	6	7	8	9	10	11	12	1	2
3	4	5	6	7	8	9	10	11	12	1	2	3
4	5	6	7	8	9	10	11	12	1	2	3	4
5	6	7	8	9	10	11	12	1	2	3	4	5
6	7	8	9	10	11	12	1	2	3	4	5	6
7	8	9	10	11	12	1	2	3	4	5	6	7
8	9	10	11	12	1	2	3	4	5	6	7	8
9	10	11	12	1	2	3	4	5	6	7	8	9
10	11	12	1	2	3	4	5	6	7	8	9	10
11	12	1	2	3	4	5	6	7	8	9	10	11
12	1	2	3	4	5	6	7	8	9	10	11	12

We note that 12 is like 'zero' for clock addition. That is, $a \oplus_{12} a = a$ for any a. Now let us imagine that the time is a O'clock. **Let a^{-1} be the number of hours we have to wait until it becomes** 12 **O'clock.** For example, if $a = 5$,then $a^{-1} = 7$ and if $a = 11$, $a^{-1} = 1$ and so on. What is the use of all this?

Let us try to solve the equation

$$x \oplus_{12} a = b.$$

Now if $b > a$, $x = b - a$. If $b < a$, how do we solve this? We want to remove a from the right-hand side of the above equation. If we 'add' a^{-1} to both the sides, we get

$$x \oplus_{12} a \oplus_{12} a^{-1} = b \oplus_{12} a^{-1}$$

$$x \oplus_{12} 12 = b \oplus_{12} a^{-1}$$

$$x = b + \oplus_{12} a^{-1}.$$

Perhaps some of the readers are wondering why we are complicating a simple problem by introducing things like a^{-1}. The aim here is not necessarily to solve a simple problem, but to understand the *similarity* between the solutions to three different types of problems.

Consider the operations $+$, $\times$ and $\oplus_{12}$. For each one of these operations there is a special element that does not change any element when operated upon itself. To be precise, we have

$$a + 0 = a$$
$$a \times 1 = a$$
$$a \oplus_{12} 12 = a.$$

Formally speaking,

0 is the identity element for $+$
1 is the identity element for $\times$
12 is the identity element for $\oplus_{12}$.

With respect to these operations, we also have elements that 'destroy' elements.

$$a + (-a) = 0 = \text{the identity of } +$$
$$a \times \frac{1}{a} = 1 = \text{the identity for } \times$$
$$a \oplus_{12} a^{-1} = 12 = \text{the identity for } \oplus_{12}.$$

Now, if $*$ is any new operation and we want to solve

$$x * a = b$$

then, perhaps its solution is

$$x = b * \text{ destroyer of } a \text{ with respect to} *$$

Consider the following matrix equation:

$$\begin{pmatrix} a & b \\ c & d \end{pmatrix}\begin{pmatrix} 8 & 3 \\ 5 & 2 \end{pmatrix} = \begin{pmatrix} 1 & 2 \\ -2 & 4 \end{pmatrix}.$$

Let $X = \begin{pmatrix} a & b \\ c & d \end{pmatrix}$ be the 'unknown' matrix, $A = \begin{pmatrix} 8 & 3 \\ 5 & 2 \end{pmatrix}$ and $B = \begin{pmatrix} 1 & 2 \\ -2 & 4 \end{pmatrix}$. Thus, our equation becomes $XA = B$.

Now,

$$\begin{pmatrix} 1 & 0 \\ 0 & 1 \end{pmatrix}$$

is the identity for matrix multiplication. And so, a destroyer for $\begin{pmatrix} 8 & 3 \\ 5 & 2 \end{pmatrix}$ would be a matrix such that

$$\begin{pmatrix} 8 & 3 \\ 5 & 2 \end{pmatrix}\begin{pmatrix} x & y \\ u & v \end{pmatrix} = \begin{pmatrix} 1 & 0 \\ 0 & 1 \end{pmatrix}.$$

We can easily verify that

$$\begin{pmatrix} 8 & 3 \\ 5 & 2 \end{pmatrix}\begin{pmatrix} 2 & -3 \\ -5 & 8 \end{pmatrix} = \begin{pmatrix} 1 & 0 \\ 0 & 1 \end{pmatrix}$$

That is, $A^{-1} = \begin{pmatrix} 2 & -3 \\ -5 & 8 \end{pmatrix}$. So, the solution is possibly $X = BA^{-1}$ and we can easily verify that

$$X = \begin{pmatrix} 1 & 2 \\ -2 & 4 \end{pmatrix}\begin{pmatrix} 2 & -3 \\ -5 & 8 \end{pmatrix}$$

is our required solution.

Now, consider the equation

$$\begin{pmatrix} a & b \\ c & d \end{pmatrix}\begin{pmatrix} 8 & 4 \\ 4 & 2 \end{pmatrix} = \begin{pmatrix} 1 & 2 \\ -2 & 4 \end{pmatrix}.$$

This equation does not have any solution. The reason is that there is no destroyer for the matrix $\begin{pmatrix} 8 & 4 \\ 4 & 2 \end{pmatrix}$.

In the subsequent sections, we shall explain the concepts of an identity, inverse, etc. in a proper manner.

8.3 BINARY OPERATIONS

Addition and multiplication are examples of binary operations. Loosely speaking, binary operations are operations that require two numbers as inputs to produce the required answer. For example, the clock addition $\oplus$ is a binary operation on the set $\{1, 2, 3, 4, 5, 6, 7, 8, 9, 10, 11, 12\}$, which is given by Table 8.1. Now, consider the set N of natural numbers and $+$. If we add any two natural numbers we get another natural number. Let us consider a general set G. Given $a, b \in G$, a binary operation gives a third element $c \in G$. Mathematically speaking, we have the following definition.

Definition 8.1: Binary operation $*$ on a set G is a mapping $* : G \times G \longrightarrow G$. The image of (a, b) under $*$ is denoted by $a * b$.

Example 8.3

Let us note the following facts:

(a) Addition is a binary operation on each of the sets $\mathbb{N}$, **Z**, **Q**, $\mathbb{R}$.

(b) Subtraction is a binary operation on Z, **Q** and $\mathbb{R}$ but not on $\mathbb{N}$. *Note that the difference between two natural numbers need not be a natural number.*

(c) If $* : \mathbb{N} \times \mathbb{N} \to \mathbb{N}$ is defined as $a * b = a^b$, then $*$ is a binary operation on $\mathbb{N}$. Further, $3*2 = 9$, $2*5 = 32$ and $5*2 = 25$, etc. ■

PRACTICE 8.1

Answer the following questions:

(1) Is multiplication a binary operation on $\mathbb{N}$?
(2) Is division a binary operation on $\mathbf{Q}$?
(3) Is division a binary operation on non-zero rationals?
(4) Is division a binary operation on $\mathbf{Z}$?

Example 8.4

In number theory, we have learnt the definition of $x + y \cong z \pmod{m}$. We also know that every integer n is congruent to k modulo m where $0 \le k \le m - 1$. These facts give us a nice algebraic structure described as follows:

Let $m \in \mathbb{N}$. Consider the equivalence relation R on $\mathbf{Z}$ defined as $a\ R\ b$ if $a - b$ is a multiple of m. $[a]$ is defined as the equivalence class of a. Now, let $Z_m = \{[0], [1], [2], [3], \ldots, [m-1]\}$. We define a binary operation $\oplus_m$ on Z_m as $[x] \oplus_m [y] = [x + y]$. This is called addition (mod m). Whenever there is no cause for confusion, we use the symbol $+$ instead of $\oplus_m$.

We define the binary operation multiplication (mod m) as $[x][y] = [xy]$.

The following tables describe addition mod (5) and multiplication mod (6):

+ (mod 5)	[0]	[1]	[2]	[3]	[4]
[0]	[0]	[1]	[2]	[3]	[4]
[1]	[1]	[2]	[3]	[4]	[0]
[2]	[2]	[3]	[4]	[0]	[1]
[3]	[3]	[4]	[0]	[1]	[2]
[4]	[4]	[0]	[1]	[2]	[3]

× (mod 6)	[0]	[1]	[2]	[3]	[4]	[5]
[0]	[0]	[0]	[0]	[0]	[0]	[0]
[1]	[0]	[1]	[2]	[3]	[4]	[5]
[2]	[0]	[2]	[4]	[2]	[3]	[4]
[3]	[0]	[3]	[0]	[3]	[0]	[3]
[4]	[0]	[4]	[2]	[0]	[4]	[2]

■

Example 8.5

The following table is the the table of usual addition for the set $\{-5, 0, 5\}$:

+ (mod 5)	−5	0	5
−5	−10	−5	10
0	−5	0	5
5	0	5	10

The following table is the table of multiplication (mod 6) for the set $\{[0], [2], [4]\}$:

$\times$ (mod 6)	[0]	[2]	[4]
[0]	[0]	[0]	[4]
[2]	[2]	[4]	[2]
[4]	[4]	[2]	[2]

Notice that the second table *does not* contain any new element other than [0], [2] and [4]. So, we say that the set $\{[0], [2], [4]\}$ is closed with respect to multiplication (mod 6).

Notice that the first table contains elements that are not members of $\{-5, 0, 5\}$. Thus, $\{-5, 0, 5\}$ is *not closed* with respect to the binary operation of addition.

The next definition tells us what we mean by a set $S \subseteq G$ to be closed with respect to a binary operation on G. ■

Definition 8.2: Let $*$ be a binary operation on a set G. We say that a subset $S \subset G$ is said to be closed with respect to $*$ if whenever $a, b \in S$, $a * b \in S$.

Proposition 8.1

Let $*$ be a binary operation on a set G and $S \subset G$ be closed with respect to $*$. Then, $*$ is a binary operation on S.

Proof

Let $a, b \in S$. Since $a, b \in G$, $a * b$ is defined and $a * b \in G$. Since S is closed with respect to $*$, $a * b \in S$. The restriction of $*$ to $S \times S$ gives a function from $S \times S \to S$. Thus $*$ is a binary operation on S. ■

Example 8.6

Consider the binary operation of subtraction on $\mathbb{R}$. If $x, y \in \mathbf{Z}$, then $x - y \in \mathbf{Z}$ and hence $\mathbf{Z}$ is closed with respect to subtraction. So, subtraction gives a binary operation on $\mathbf{Z}$.

But taking $\mathbb{N}$, we find that $4 \in \mathbb{N}$ and $7 \in \mathbb{N}$ but $4 - 7 \notin \mathbb{N}$ and hence $\mathbb{N}$ is not closed with respect to subtraction. ■

Example 8.7

Let G be the set of all 2×2 matrices. Consider the binary operation of matrix addition on G. Let

$$X = \left\{ \begin{pmatrix} a & b \\ c & a \end{pmatrix} : a, b, c \in R \right\}$$

and

$$Y = \left\{ \begin{pmatrix} 1 & a \\ b & c \end{pmatrix} : a, b, c \in R \right\}.$$

We show that A is closed with respect to $+$ but B is not closed with respect to $+$. Let $A = \begin{pmatrix} a & b \\ c & a \end{pmatrix}$ and $B = \begin{pmatrix} x & y \\ z & x \end{pmatrix}$ be elements of X. Now,

$$\begin{pmatrix} a & b \\ c & a \end{pmatrix} + \begin{pmatrix} x & y \\ z & x \end{pmatrix} = \begin{pmatrix} a+x & b+y \\ c+z & a+x \end{pmatrix}.$$

Thus, we find that the entries of the sum $A + B$ are also real numbers and the diagonal elements of $A + B$ are equal. Thus, $A + B \in X$. So, X is closed with respect to matrix addition.

Next, we show that Y is not closed with respect to $+$. Let $A = \begin{pmatrix} 1 & 0 \\ 3 & 5 \end{pmatrix}$ and $B = \begin{pmatrix} 1 & 9 \\ -5 & 8 \end{pmatrix}$. Clearly, $A, B \in Y$. $A + B = \begin{pmatrix} 2 & 9 \\ -2 & 13 \end{pmatrix} \notin Y$ since the element in the first row first column is not 1. Thus, Y is not closed. ■

PRACTICE 8.2

Fill in the blanks in the required places and answer the questions:

(1) The set of even natural numbers $= \{2, 4, 6, ...\} = \{2n : n \in N\}$ is closed with respect to ———————— and ——————————————.

(2) The set $\{2, 4, 8, ...\} = \{2^n : n = 1, 2, 3...\}$ is closed with respect to ——————————————————.

(3) The binary operation $\odot$ is defined on R as $a \odot b = a + b + ab$. Which of the following sets are closed with respect to $\odot$?

(i) $\mathbb{N}$.
(ii) **Z**.
(iii) Q.

(4) Let G be the set of all 2×2 matrices. Consider the binary operation of matrix multiplication on G. Let

$$X = \left\{ \begin{pmatrix} a & 0 \\ 0 & b \end{pmatrix} : a, b, c \in R \right\}$$

and

$$Y = \left\{ \begin{pmatrix} a & b \\ a & c \end{pmatrix} : a, b, c \in R \right\}.$$

Show that X is closed, but not Y.

8.4 ASSOCIATIVITY

The addition and multiplication of numbers are *associative operations*. That is

$$(a + b) + c = a + (b + c)$$

and

$$(ab)c = a(bc).$$

Subtraction is a binary operation on Z but it is *not associative*. Because,

$$5 - (9 - 9) = 5$$

but

$$(5 - 9) - 9 = -13.$$

Division is a binary operation on R^* but it is not associative. $\frac{5}{\left(\frac{9}{3}\right)} = \frac{5}{3}$ but $\frac{\left(\frac{5}{9}\right)}{3} = \frac{5}{27}$.

Definition 8.3: Let $\odot$ be a binary operation on G. We say that $\odot$ is associative if

$$a \odot (b \odot c) = (a \odot) \odot c.$$

Addition and multiplication of matrices are associative. Addition and multiplication of complex numbers are also associative.

As mentioned in the introduction, an algebraic structure is a set along with one or more binary operations defined on the set. A typical algebraic structure is denoted as $(S, *)$ or $(S, *, \odot)$, etc. The simplest algebraic structure we shall learn about is a semigroup.

Definition 8.4: Let G be a set and $\odot$ be a binary operation on G. We say that $(G, \odot)$ is a semigroup if $\odot$ is associative.

For $(G, \odot)$ to be a semigroup, the only requirement is the associativity of $\odot$.

Example 8.8

$(\mathbb{N}, +)$, $(\mathbb{N}, .)$, $(Z, +)$, $(\mathbb{R}, +)$ are all semigroups. $(Z, -)$ and $(\mathbb{R}, -)$ are not semigroups.

SOLUTION

$(\mathbb{N}, +)$ is a semigroup because $+$ is associative. Similarly, $(\mathbb{N}, .)$, $(Z, +)$, $(\mathbb{R}, +)$ are all semigroups.

We have already seen that for $a, b, c \in Z$, $((a-b)-c) \neq (a-(b-c))$ in general and hence $-$ is not associative. Thus, $(Z, -)$ is not a semigroup. The same is true of $(\mathbb{R}, -)$. ■

PRACTICE 8.3

Does $a \odot b = a^b$ define a binary operation on N? Is $(N, \odot)$ a semigroup?

Example 8.9

Define $\odot$ on $\mathbb{R}$ as $a \odot b = a + b - ab$. Show that $(\mathbb{R}, \odot)$ is a semigroup.

SOLUTION

If a and b are real numbers $a + b - ab$ is also a real number. So, $\odot$ is a binary operation on R. Now, we have to show

$$a \odot (b \odot c) = (a \odot b) \odot b.$$

Now,

$$\begin{aligned} a \odot (b \odot c) &= a \odot (b + c - bc) \\ &= a + (b + c - bc) - a(b + c - bc) \\ &= a + b + c - bc - ab - ac + abc \end{aligned}$$

and

$$\begin{aligned} (a \odot b) \odot c &= (a + b - ab) \odot c \\ &= (a + b - ab) + c - (a + b - ab)c \\ &= a + b + c - ab - ac - bc + abc. \end{aligned}$$

Hence, $a \odot (b \odot c) = (a \odot b) \odot c$. Thus, $\odot$ is associative and $(R, \odot)$ is a semigroup. ∎

8.5 IDENTITY ELEMENTS

Let us recall that

$$a + 0 = 0 + a = a$$
$$a \times 1 = 1 \times a = a$$
$$a \oplus_{12} 12 = 12 \oplus_{12} a = a$$
$$\begin{pmatrix} a & b \\ c & d \end{pmatrix} + \begin{pmatrix} 0 & 0 \\ 0 & 0 \end{pmatrix} = \begin{pmatrix} a & b \\ c & d \end{pmatrix} + \begin{pmatrix} 0 & 0 \\ 0 & 0 \end{pmatrix} = \begin{pmatrix} a & b \\ c & d \end{pmatrix}$$

and

$$\begin{pmatrix} a & b \\ c & d \end{pmatrix} \begin{pmatrix} 1 & 0 \\ 0 & 1 \end{pmatrix} = \begin{pmatrix} 1 & 0 \\ 0 & 1 \end{pmatrix} \begin{pmatrix} a & b \\ c & d \end{pmatrix} = \begin{pmatrix} a & b \\ c & d \end{pmatrix}.$$

We find that the addition of 0 to a real number gives back the same real number. Further, multiplication by 1 of a real number gives back the same number.

Informally speaking, an identity element is an element that does not change another element when the binary operation is applied. We define an identity element below.

Definition 8.5: Let G be a set and $\oplus$ be a binary operation. An element e is said to be an identity element if $e \oplus a = a \oplus e = a$.

Example 8.10

In this example we look at some **obvious identity elements:** 0 is the identity for the operation of addition of numbers. 1 is the identity for the operation of multiplication of numbers.

$\begin{pmatrix} 1 & 0 \\ 0 & 1 \end{pmatrix}$ is the identity for the multiplication of 2×2 matrices. $\begin{pmatrix} 0 & 0 \\ 0 & 0 \end{pmatrix}$ is the identity for the addition of 2×2 matrices.

For the set $Z_n = \{[0], [1], [2], ..., [n-1]\}$ under addition modulo n, $[0]$ is the identity element.

For the set $Z_n = \{[0], [1], [2], ..., [n-1]\}$ under multiplication modulo n, $[1]$ is the identity element. ■

Example 8.11

Let G be the set of all even integers and consider the binary operation of multiplication. It is clear that G *does not have any identity.* ■

Example 8.12

We know that $[1]$ is an identity for the operation of multiplication modulo 6. Now, under the same operation is there an identity for the set $G = \{[0], [2], [4]\}$? If you think that the answer is no, because $[1] \notin G$, you would be surprised. Let us prepare the table of operation for G

$\otimes_6$	[0]	[2]	[4]
[0]	[0]	[0]	[0]
[2]	[0]	[4]	[2]
[4]	[0]	[2]	[4]

From the table we find that

$[0] \times [4] = [0]$.

$[2] \times [4] = [2]$.

$[4] \times [4] = [4]$.

Thus, $[4]$ is an identity for $G = \{[0], [2], [4]\}$.

Note that $[4] \times [3] = [0]$ and hence $[4]$ is *not* an identity for the set $\{[0], [1], [2], [3], [4], [5]\}$. ■

Example 8.13

Find the identity for the binary operation $*$ defined on $\mathbb{R}$ as $a * b = a + b + 3$.

SOLUTION

If e is the identity, then $a * e = a$. So, $a + e + 3 = a$. Thus, $e + 3 = 0$. So, the identity for this binary operation is -3. ■

PRACTICE 8.4

Let

$$G = \left\{ \begin{pmatrix} a & a \\ a & a \end{pmatrix} : a \in \mathbb{R} \right\}.$$

Find the identity element of G for the matrix multiplication.

PRACTICE 8.5

Recall Example 8.1. Find the identity elements of $(G, \vee)$ and $(G, \wedge)$ where $G = \{T, F\}$.

Definition 8.6: A semigroup $(G, \odot)$ is said to be a monoid if there is an identity element $e \in G$ for $\odot$.

Example 8.14

(i) $G = \{[0], [2], [4]\}$ under multiplication (mod 6) is a monoid.
(ii) $(\mathbb{N}, +)$ is a semigroup but it is not a monoid. There is no identity.
(iii) $(Z, -)$ is not a monoid since subtraction is not associative and hence $(Z, -)$ is not even a semigroup.
(iv) $(\mathbb{N}, .)$ is a monoid. 1 is the identity. ■

8.6 INVERSE ELEMENTS

Definition 8.7: Let $(G, \odot)$ be a monoid and let e be the identity. An element b is called an inverse of a if $a \odot b = b \odot a = e$. It is denoted by a^{-1}. An element a that has an inverse is called invertible.

Example 8.15

(i) $(\mathbb{R}, \times)$ is a monoid. 1 is the identity. Every $x \neq 0$ has the inverse element $1/x$. The element 0 does not have an inverse.
(ii) In the monoid $(\mathbf{Z}, \times)$, no element has an inverse.
(iii) Let $\mathbb{R}^* = \{x \in \mathbb{R} : x \neq 0\}$. Then, $(\mathbb{R}^*, \times)$ is a monoid. 1 is the identity. Further, every $x \in \mathbb{R}^*$ has the inverse element $1/x$. ■

Example 8.16

Let us construct the table of multiplication for the set of all fifth roots of unity = $\{1, \omega, \omega^2, \omega^3, \omega^4\}$.

$\times$	1	ω	ω^2	ω^3	ω^4
1	1	ω	ω^2	ω^3	ω^4
ω	ω	ω^2	ω^3	ω^4	1
ω^2	ω^2	ω^3	ω^4	1	ω
ω^3	ω^3	ω^4	1	ω	ω^2
ω^4	ω^4	1	ω	ω^2	ω^3

Clearly, 1 is the identity. From the table, we can observe that $\omega \times \omega^4 = 1$. So, the inverse of ω is ω^4. Similarly, the inverse of ω^2 is ω^3, etc. ■

Example 8.17

Let $G = \{1, \omega, \omega^2, ..., \omega^{n-l}\}$ be set of all *nth* roots of unity. The inverse of ω^l is ω^{n-l}. ■

Example 8.18

Consider the table for the binary operation of addition (mod 6) on Z_6:

$\oplus_6$	[0]	[1]	[2]	[3]	[4]	[5]
[0]	[0]	[1]	[2]	[3]	[4]	[5]
[1]	[1]	[2]	[3]	[4]	[5]	[0]
[2]	[2]	[3]	[4]	[5]	[0]	[1]
[3]	[3]	[4]	[5]	[0]	[1]	[2]
[4]	[4]	[5]	[0]	[1]	[2]	[3]
[5]	[5]	[0]	[1]	[2]	[3]	[4]

Thus, for addition (mod 6), the inverse of [0] is [0], the inverse of [1] is [5], the inverse of [2] is [4], the inverse of [3] is [3], the inverse of [4] is [2] and the inverse of [5] is [1]. ■

Example 8.19

Let $Z_m = \{[0], [1], ..., [m]\}$. In $(G, \oplus_m)$ the inverse of $[k]$ is $[m-k]$. ■

Example 8.20

Let $Z_8 = \{[0], [1], ..., [7]\}$. Let us prepare the multiplication table for $\otimes_8$.

$\otimes_8$	[0]	[1]	[2]	[3]	[4]	[5]	[6]	[7]
[0]	[0]	[0]	[0]	[0]	[0]	[0]	[0]]	[0]
[1]	[0]	[1]	[2]	[3]	[4]	[5]	[6]	[7]
[2]	[0]	[2]	[4]	[6]	[0]	[2]	[4]	[6]
[3]	[0]	[3]	[6]	[1]	[4]	[7]	[2]	[5]
[4]	[0]	[4]	[0]	[4]	[0]	[4]	[0]	[4]
[5]	[0]	[5]	[2]	[7]	[4]	[1]	[6]	[3]
[6]	[0]	[6]	[4]	[2]	[0]	[6]	[4]	[2]
[7]	[0]	[7]	[6]	[5]	[4]	[3]	[2]	[1]

Obviously, [1] is the identity element. Now, to find the inverse of $[x]$ we examine the row (or coloumn) corresponding to $[x]$. If [1] is in the row, then $[x]$ has an inverse. The $[y]$ such that $[x][y] = [1]$ is the inverse. Using this procedure, we find that

- [0] has no inverse.
- [1] has the inverse [1].
- [2] has no inverse.
- [3] has the inverse [3].

- [4] has no inverse.
- [5] has the inverse [5].
- [6] has no inverse.
- [7] has the inverse [7]. ■

PRACTICE 8.6

Let $Z_9 = \{[0], [1], ..., [7], [8]\}$. Find all the invertible elements of $(G, \otimes_9)$ and their inverses.

PRACTICE 8.7

Using Theorem 7.9 of Chapter 7, show that the invertible elements of $(\mathbf{Z}_m, \otimes_m)$ are $[a]$ such that $(a, m) = 1$.

8.7 GROUPS

Informally speaking, a group is an algebraic structure with an identity for the algebraic operation and with an inverse for each of its elements. In other words, it is a monoid where every element is invertible. Let us now give a formal definition of a group.

Definition 8.8: Let G be a set and $\odot$ be a binary operation on G. We say that $(G, \odot)$ is a *group* if the following holds:

(i) $\odot$ is associative.
That is, $a \odot (b \odot c) = (a \odot b) \odot c$ for all $a, b, c \in G$.
(ii) There exists an element $e \in G$ with the property that $a \odot e = e \odot a = a.e$ is called the identity of G.
(iii) For every element $a^{-1} \in G$, there is an element such that $a \odot a^{-1} = a^{-1} \odot a = \epsilon$.

We say that G is an abelian group if $a \odot b = b \odot a$ for all $a, b \in G$.

Example 8.21

Z is a group under $+$. It is an abelian group. ■

Example 8.22

$(\mathbb{R}, \times)$ is not a group. 0 does not have an inverse under multiplication. ■

Example 8.23

$(\mathbb{R}^*, \times)$ is an abelian group. ■

Example 8.24

The set of all 2×2 matrices under multiplication is not a group. If the determinant of A is zero, the matrix does not have an inverse. But the set of all 2×2 matrices whose determinant is non-zero is a group under matrix multiplication. For

$$A = \begin{pmatrix} a & b \\ c & d \end{pmatrix}$$

with $ad - bc \neq 0$, A^{-1} is given by

$$A^{-1} = \frac{1}{ad - bc} \begin{pmatrix} d & -b \\ -c & a \end{pmatrix}.$$

But this group is not abelian. ■

Example 8.25

Suppose $M_{m \times n}$ is the set of all $m \times n$ matrices. Defining addition of matrices as the binary operation, show that $M_{m \times n}$ is a group.

SOLUTION

We shall verify the conditions of the definition for a group one by one:

(i) Let A, B, C be $m \times n$ matrices. We have $(A+B)+C = A+(B+C)$. Associativity holds.

(ii) Let

$$O = \begin{pmatrix} 0 & 0 & \dots & 0 \\ 0 & 0 & \dots & 0 \\ \dots & \dots & \dots & \dots \\ 0 & 0 & \dots & 0 \end{pmatrix}.$$

Then, $A + O = O + A = A$ and hence O is the identity element.

(iii) Let $-A$ be the matrix whose entries are negatives of the corresponding entries of A. Then, $A + -A = -A + A = 0$. Thus, every element has an inverse. So, $M_{m \times n}$ is a group.

Further, since $A + B = B + A$, this is also an abelian group. ■

Example 8.26

Let

$$G = \left\{ \begin{pmatrix} x & y \\ z & w \end{pmatrix} : x, y, z, w \text{ are even integers} \right\}$$

i.e. G is the set of all 2×2 matrices whose entries are even integers. Defining the addition of matrices as a binary operation, show that G is a group.

SOLUTION

Addition is a binary operation on the set of all 2×2 matrices with real entries. Now, we have to show that addition is a binary operation on G. If A and B are matrices whose entries are even, then $A + B$ is also a matrix with even entries.

Thus, $A, B \in G \to A + B \in G$. So, G is closed. Hence by Proposition 8.1, $+$ is a binary operation on G. Next, we verify the conditions of Definition 8.8.

(i) Let $A, B, C \in G$. We have $(A + B) + C = A + (B + C)$. Associativity holds.

(ii) Let

$$O = \begin{pmatrix} 0 & 0 \\ 0 & 0 \end{pmatrix}.$$

Since 0 is an even integer, $O \in G$. Further, $A + O = O + A$. Thus, O is the identity element.

(iii) If $A = \begin{pmatrix} x & y \\ z & w \end{pmatrix} \in G$, then x, y, z, w are even integers. But then, $-x, -y, -z, -w$ are also even. Hence $-A = \begin{pmatrix} -x & -y \\ -z & -w \end{pmatrix} \in G$. Further as $A + -A = -A + A = O$, every element has an inverse.

Clearly, $A + B = B + A$.

Hence G is an abelian group. ■

Example 8.27

Let

$$G = \left\{ \begin{pmatrix} x & y \\ z & w \end{pmatrix} : x, y, z, w \text{ are odd integers} \right\}$$

i.e G is the set of all 2×2 matrices whose entries are odd integers. Examine whether G is a group under matrix addition.

SOLUTION

We show that G is not closed with respect to addition. Let $A = \begin{pmatrix} 1 & 3 \\ -41 & 17 \end{pmatrix}$ and $B\begin{pmatrix} 1 & 7 \\ -1 & 7 \end{pmatrix}$. Then, $A + B = \begin{pmatrix} 2 & 10 \\ -42 & 24 \end{pmatrix}$. Clearly, the entries of $A + B$ are even and hence $A+B \notin G$. Thus, $(G, +)$ is not a group. ■

Example 8.28

Show that the set $GL_n(\mathbb{R})$ of all non-singular $n \times n$ matrices is a group under matrix multiplication.

$$GL_n(\mathbb{R}) = \{A : \text{ is a } n \times n \text{ matrix } |A| \neq 0\}.$$

SOLUTION

Let $A, B \in GL_n(\mathbb{R})$. AB is a matrix of order $n \times n$. Now $|AB| = |A||B|$. Since $|A| \neq 0$ and $|B| \neq 0$, $|AB| \neq 0$. Thus, $GL_n(\mathbb{R})$ is closed under matrix multiplication.

(i) $(AB)C = A(BC)$ for any matrices $A, B, C \in GL_n(\mathbb{R})$. Therefore, associativity is true.

(ii) Let I be the unit matrix whose diagonal entries are 1 and the remaining entries are zero. We have $AI = IA = A$. Thus, I is the identity element.

(iii) Every non-singular matrix has an inverse. Thus, inverse exists.

So, $GL_n(R)$ is a group.
Since $AB \neq BA$, in general, $GL_n(\mathbb{R})$ is not an abelian group. ■

Example 8.29

Let T be any set. Consider the set $F(T)$ of all functions from A into T. Define a binary operation on $F(T)$ as follows:

$$(f \circ g)(x) = f(g(x)).$$

Now, $(F(T), \circ)$ is a monoid, which is **not** a group, in general.

SOLUTION

The composition of two functions on T is another function on T. Hence $\circ$ is a binary operation on $F(T)$.

(i) Associativity:

$$[f \circ (g \circ h)](x) = f[(g \circ h)(x)] = f[g(h(x))].$$

Now,

$$[(f \circ g) \circ h](x) = [f \circ g](h(x)] = f[g(h(x))].$$

Thus,

$$(f \circ g) \circ h = f \circ (g \circ h).$$

(ii) Consider the identity function $I : T \longrightarrow T$ defined as $I(x) = x$. Clearly, $I \circ f = f \circ I$ and hence I is the identity of $F(T)$. Thus $(F(T), \circ)$ is a monoid.
For functions that are not one-to-one and onto, f^{-1} does not exist. (See Section 5 of Chapter 2.) Hence $F(T)$ *is not a group.* ■

PRACTICE 8.8

Let $S(T)$ be a set of all one-to-one and onto functions on T. Show that $S(T)$ is a group, which is not abelian.

By compiling the already proven facts regarding Z_n, the reader can complete Practice 8.9. Z_n is an important example of a group.

PRACTICE 8.9

Show that $Z_n = \{[0], [1], ..., [n-1]\}$ is a group under addition (mod n).

In the following, we describe another important example of group theory, namely, the *permutation* group on a set of n elements.

Example 8.30

Let $T = \{1, 2, ..., n\}$. If a_1, a_2, ..., a_n are distinct elements of T then

$$\sigma = \begin{pmatrix} 1 & 2 & \dots & n \\ a_1 & a_2 & \dots & a_n \end{pmatrix}$$

is known as a permutation of T. The second row is just a rearrangement of the symbols of T and hence the name permutation. We define the binary operation $*$ on the set of all permutations S_n as follows:

$$\begin{pmatrix} 1 & \dots & k & \dots & n \\ a_1 & \dots & a_k & \dots & a_n \end{pmatrix} * \begin{pmatrix} 1 & \dots & a_k & \dots & n \\ b_1 & \dots & c & \dots & a_n \end{pmatrix}$$

$$= \begin{pmatrix} 1 & \dots & k & \dots & n \\ & \dots & c & \dots & \end{pmatrix}.$$

That is, if a_k is below k in the first permutation σ_1 and c is below a_k in the second permutation σ_2, then we place c below k in the permutation $\sigma_1 * \sigma_2$.

It is easy to see that $\sigma_1 * (\sigma_2 * \sigma_3) = (\sigma_1 * \sigma_2) * \sigma_3$. Clearly,

$$\begin{pmatrix} 1 & 2 & \dots & n \\ 1 & 2 & \dots & n \end{pmatrix}$$

is the identity. Now for

$$\sigma = \begin{pmatrix} 1 & 2 & \dots & n \\ a_1 & a_2 & \dots & a_n \end{pmatrix}$$

its inverse is computed as follows.

Exchanging the rows in σ, we get

$$\begin{pmatrix} a_1 & a_2 & \dots & a_n \\ 1 & 2 & \dots & n \end{pmatrix}$$

which is not a permutation. We rewrite the first row in the usual increasing order without changing the corresponding elements below. For $n = 4$, let us illustrate σ^{-1}.

Let

$$\sigma = \begin{pmatrix} 1 & 2 & 3 & 4 \\ 4 & 3 & 1 & 2 \end{pmatrix}.$$

By interchanging the rows we get

$$\begin{pmatrix} 4 & 3 & 1 & 2 \\ 1 & 2 & 3 & 4 \end{pmatrix}.$$

Now, σ^{-1} is given by

$$\begin{pmatrix} 1 & 2 & 3 & 4 \\ 3 & 4 & 2 & 1 \end{pmatrix}.$$

Thus, $(S_n, *)$ is a group and is known as the *permutation* group on n elements.

As indicated earlier, the set of all one-to-one and onto functions $f : T \longrightarrow T$ is a group under the binary operation of composition of functions. The groups $(S_n, *)$ and $(S(T), \circ)$ are closely related when $T = \{1, 2, ..., n\}$. The relationship is explained below:

If $f \in S(T)$, then we associate the permutation given by

$$\sigma_f = \begin{pmatrix} 1 & 2 & \dots & n \\ f(1) & f(2) & \dots & f(n) \end{pmatrix}$$

with f.

Then, we have

$$\sigma_f * \sigma_g = \sigma_{g \circ f}.$$

■

PRACTICE 8.10

Show that the set of 2×2 matrices with integer entries is an abelian group with respect to matrix addition.

PRACTICE 8.11

Is the set of 2×2 non-singular matrices with integer entries a group with respect to matrix multiplication?

PRACTICE 8.12

Show that the set of 2×2 non-singular matrices with rational entries is a group with respect to matrix multiplication.

Example 8.31

The set $Z_9 = \{[0], [1], ..., [8]\}$ is a monoid under multiplication (mod 9). Its identity is $[1]$. This is not a group since $[0][x] = [0] \neq [1]$ and hence $[0]$ does not have an inverse. ■

PRACTICE 8.13

Is the set $\{[1], [2], ..., [8]\}$ a group under multiplication (mod 9)?

Example 8.32

Show that the set of all positive integers less than 9 and relatively prime to 9 forms a group under multiplication (mod 9).

SOLUTION

Positive integers relatively prime to 9 are $1, 2, 4, 5, 7, 8$. Let $G = \{1, 2, 4, 5, 7, 8\}$. Let us form the table of multiplication.

$\odot_9$	1	2	4	5	7	8
1	1	2	4	5	7	8
2	2	4	8	1	5	7
4	4	8	7	2	1	5
5	5	1	2	7	8	4
7	7	5	1	8	4	2
8	8	7	5	4	2	1

From the table, we find that the given set is closed under multiplication mod 9. Multiplication mod 9 is associative. 1 is the identity. Further

- The inverse of 1 is 1.
- The inverse of 2 is 5.
- The inverse of 4 is 7.
- The inverse of 5 is 2.
- The inverse of 7 is 4.
- The inverse of 8 is 8.

Thus, the given set is a group. ■

PRACTICE 8.14

(1) Consider a set of all $n \times n$ matrices with integer entries. Is it closed with respect to addition? Is it closed with respect to matrix multiplication?

(2) Is the operation $a \odot b = a(b + 1)$ defined on the set of all real numbers associative?

(3) Show that the set of all matrices of the type

$$\begin{pmatrix} x & x \\ x & x \end{pmatrix}$$

forms a group under multiplication. Is this group abelian?

Example 8.33

Let $(G_1, \odot)$ and $(G_2, \oplus)$ be groups. Show that $G = G_1 \times G_2$ is a group under the binary operation defined as $(a, x) * (b, y) = (a \odot b, x \oplus y)$.

SOLUTION

If $(a, x), (b, y) \in G$, then we have $a, b \in G_1$ and $x, y \in G_2$. Hence $a \odot b \in G_1$ and $x \oplus y \in G_2$. Thus $(a, x) * (b, y) = (a \odot b, x \oplus y) \in G_1 \times G_2 = G$. Hence $*$ is a well-defined binary operation.

$$\begin{aligned}((a,x)*(b,y))*(c,z) &= (a\odot b, x\oplus y)*(c,z)\\ &= ((a\odot b)\odot c, (x\oplus y)\oplus c)\\ &= (a\odot(b\odot c), x\oplus(y\oplus z))\\ &= (a,x)*((b,y)*(c,z)).\end{aligned}$$

Hence * is associative.

Let e_1 be the identity in G_1 and e_2 be the identity in G_2. We can easily check that $(e_1, e_2) \in G$ is the identity for $*$. Further, we can show that the element (a^{-1}, x^{-1}) is the inverse of (a, b) under $*$. Hence G is a group. ■

PRACTICE 8.15

Show that $(R^2, +)$ is an abelian group where $+$ is defined as $(x, y) + (a, b) = (x + a, y + b)$.

Practice 8.16 is useful in coding theory.

PRACTICE 8.16

Let $B = (\mathbf{Z_2}, \oplus_2)$ where $\oplus_2$ denotes the addition (mod 2). Define $B^n = B \times B \times \ldots n$ times and define the binary operation on B^n as $(a_1, a_2, \ldots, a_n) * (b_1, b_2, \ldots, b_n) = (a_1 \oplus_2 b_1, a_2 \oplus_2 b_2, \ldots, a_n \oplus_2 b_n)$. For example, $(1, 1, 0, 1), (0, 1, 0, 0) \in B^4$ and

$$(1, 1, 0, 1) * (0, 1, 0, 0) = (1, 0, 0, 1).$$

Show that $(B^n, *)$ is an abelian group.

8.7.1 Geometric Transformations

Let P be the point (x, y) in the xy plane and O be the origin. Let us rotate OP by an angle θ in the anti-clockwise direction.

The co-ordinates of Q are given by $(x\cos\theta - y\sin\theta, x\sin\theta + y\cos\theta)$

In some calculations, it is convenient to think of a point in the xy plane as a coloumn vector $\begin{pmatrix} x \\ y \end{pmatrix}$.

Since

$$\begin{pmatrix} \cos\theta & -\sin\theta \\ \sin\theta & \cos\theta \end{pmatrix}\begin{pmatrix} x \\ y \end{pmatrix} = \begin{pmatrix} x\cos\theta - y\sin\theta \\ x\sin\theta + y\cos\theta \end{pmatrix}$$

the matrix

$$R_\theta = \begin{pmatrix} \cos\theta & -\sin\theta \\ \sin\theta & \cos\theta \end{pmatrix}$$

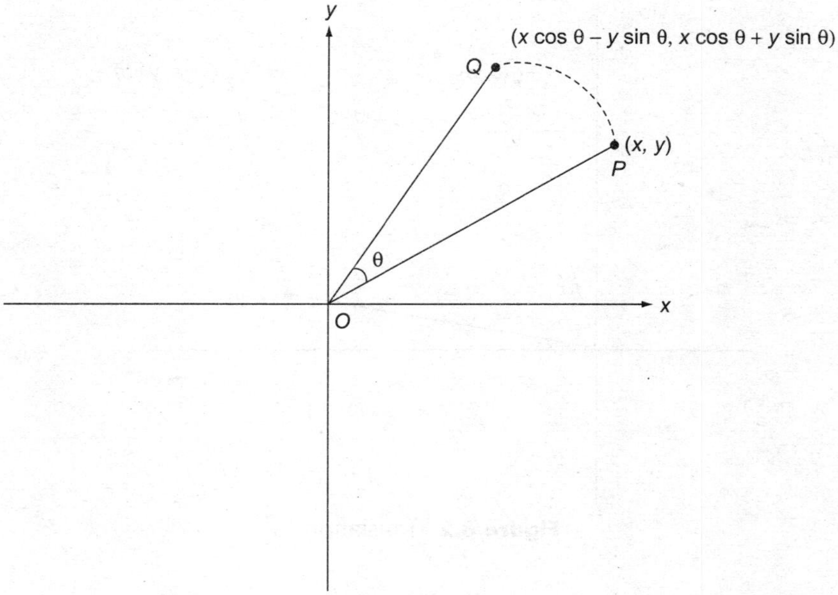

Figure 8.1 Rotation by an angle θ

represents the geometrical transformation of rotation by an angle of θ. Now, a rotation by an angle of θ followed by a rotation by an angle of ϕ in the same plane is another rotation by an angle of $\theta + \phi$. Thus, the set of all rotation matrices is closed under matrix multiplication.

Rotation by an angle of 0 does not change anything. Hence, this is the identity element.

Next, if a rotation by an angle of θ in the anti-clockwise direction is followed by a rotation by an angle θ but in the clockwise direction, nothing changes. Hence, the inverse of

$$\begin{pmatrix} \cos\theta & -\sin\theta \\ \sin\theta & \cos\theta \end{pmatrix}$$

is

$$\begin{pmatrix} \cos(-\theta) & -\sin(-\theta) \\ \sin(-\theta) & \cos(-\theta) \end{pmatrix} = \begin{pmatrix} \cos\theta & \sin\theta \\ -\sin\theta & \cos\theta \end{pmatrix}.$$

Thus, we have geometrically proved that the set of all rotations forms a group. Problem 8 at the end of this chapter demands an algebraic proof of the same.

At this stage, let us define a transformation T as a one-to-one and onto map from $\mathbb{R}^2$ into $\mathbb{R}^2$. We describe some simple transformations and explain them with figures.

Let $(a, b) \in \mathbb{R}^2$ be a given point. The transformation $T_1 : \mathbb{R}^2 \longrightarrow \mathbb{R}^2$ defined as $T_1(x, y) = (x + a, y + b)$. T_1 is a translation. It moves all the points in the same direction by the same distance.

Define $T_2 : \mathbb{R}^2 \longrightarrow \mathbb{R}^2$ as $T_2(x, y) = (x, -y)$. T_2 represents the reflection on the x axis.

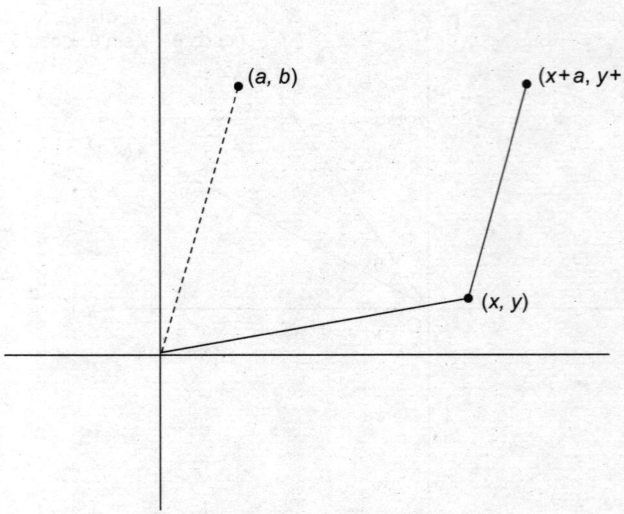

Figure 8.2 Translation

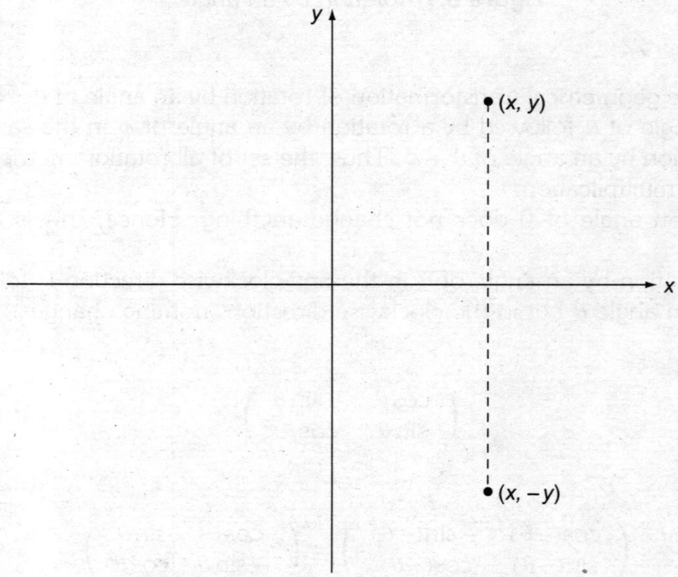

Figure 8.3 Reflection on the x axis

Define $T_3 : \mathbb{R}^2 \longrightarrow \mathbb{R}^2$ as $T_2(x, y) = (-x, y)$. T_2 represents the reflection on the y axis.

Consider a line passing through the origin making an angle ϕ with the x-axis. The transformation of reflection on this line is given by

$$T(x, y) = (x \cos 2\phi + y \sin 2\phi, x \sin 2\phi - y \cos 2\phi).$$

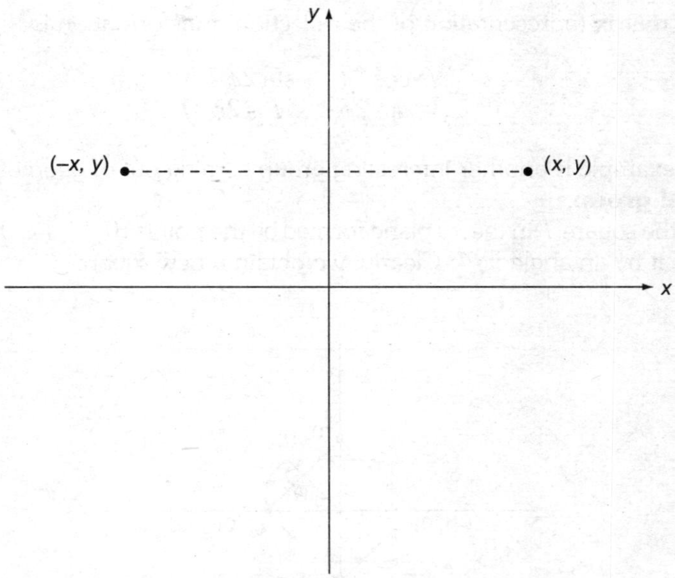

Figure 8.4 Reflection on the y axis

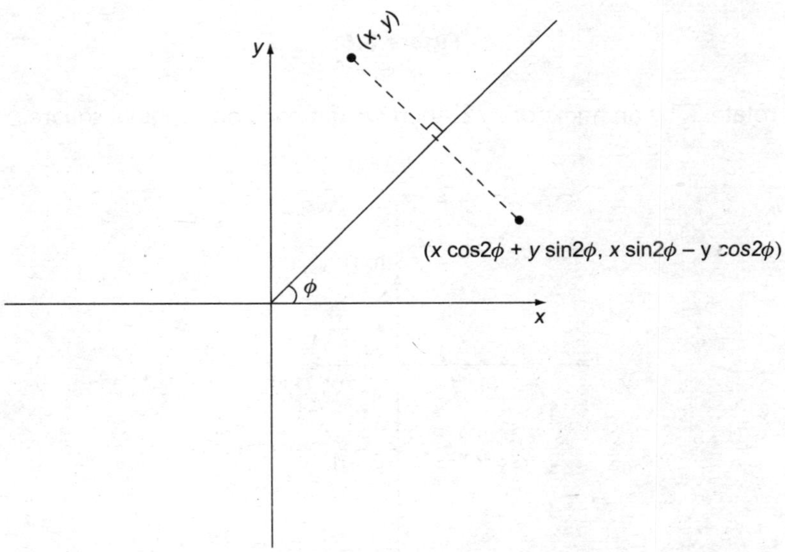

Figure 8.5 Reflection on a line through the origin

We write the matrix form as

$$\begin{pmatrix} \cos 2\phi & \sin 2\phi \\ \sin 2\phi & -\cos 2\phi \end{pmatrix} \begin{pmatrix} x \\ y \end{pmatrix} = \begin{pmatrix} x\cos 2\phi + y\sin 2\phi \\ x\sin 2\phi - y\cos 2\phi \end{pmatrix}.$$

Thus, the matrix representation of the reflection transformation is

$$S_\phi = \begin{pmatrix} \cos 2\phi & \sin 2\phi \\ \sin 2\phi & -\cos 2\phi \end{pmatrix}.$$

The next example is another interesting group coming out of geometry known as the **Dihedral group**.

Consider the square T in the xy plane formed by the points $(0, 1), (1, 0), (0, -1), (-1, 0)$. Let us rotate it by an angle $\pi/4$. Clearly, we obtain a new square.

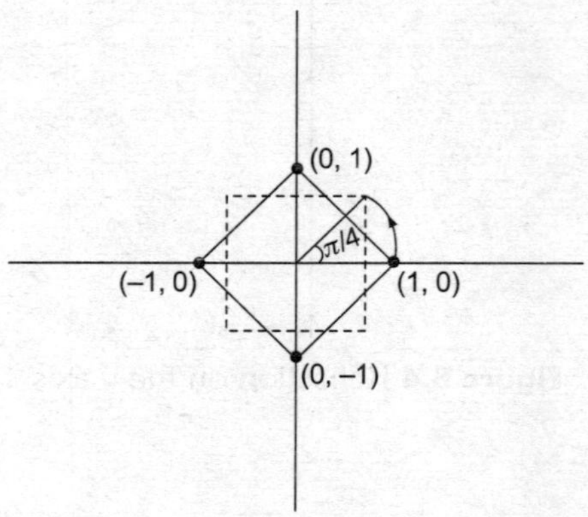

Figure 8.6

If we rotate T by an angle of $\pi/2$, then we get back our original square.

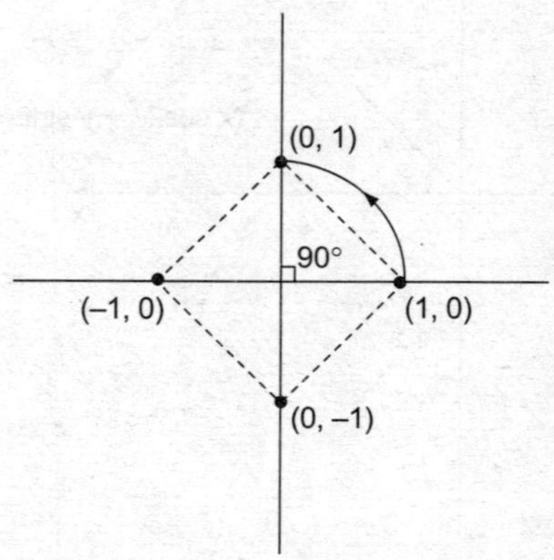

Figure 8.7

Thus $R_{\frac{\pi}{4}}$ is a transformation that does not alter T. Similarly, R_π, $R_{\frac{3\pi}{4}}$ and $R_{2\pi}$ $= R_0 = I$ are all rotations that do not alter T. The interesting fact is that the set of all rotational transformations that do not alter T given by $\left\{R_0, R_{\frac{\pi}{2}}, R_\pi . R_{\frac{3\pi}{2}}\right\}$ is a group. Can you think of geometric proof of this fact?

Next, we ask ourselves whether there is any other geometric transformation that does not alter T.

It is clear that the transformations of reflections on the x-axis, the y-axis, the lines $y = x$ and $y = -x$ also do not alter T. These are transformations given by the matrices S_0, S_π, $S_{\frac{\pi}{2}}$ and $S_{\frac{3\pi}{2}}$.

The fact that the set $\left\{S_0, S_\pi, S_{\frac{\pi}{2}}, S_{\frac{3\pi}{2}}\right\}$ does not form a group easily follows from the fact that the identity is not an element of this set.

The interesting result is that the set of all rotations and reflections that do not alter T form a group. This can be proved algebraically. But let us look at the geometric ideas behind the proofs.

Let P be a point in the xy plane, l be the line making an angle ϕ with the x-axis, and Q be the reflection of P on l. Let B be the point where l meets the line segment PQ. We take an arbitrary point A on the x-axis and O is the origin.

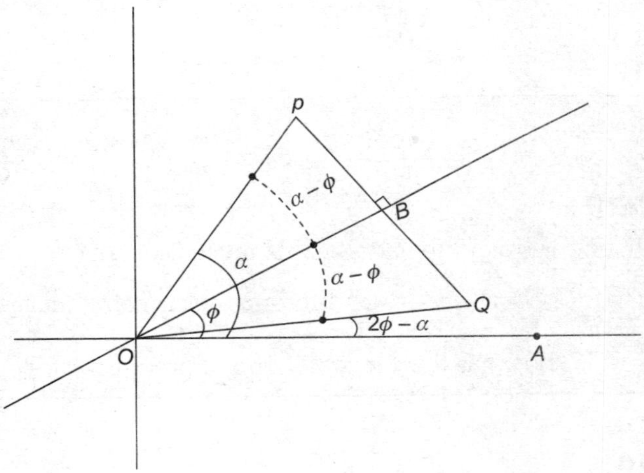

Figure 8.8 (a)

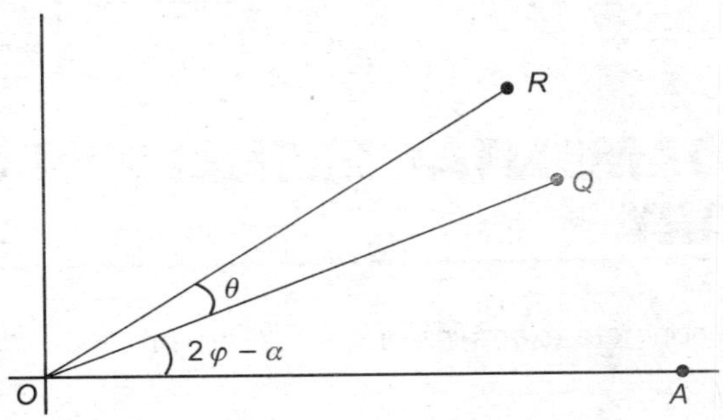

Figure 8.8 (b)

Let us denote $\angle AOP = \alpha$. From the fact that the line segments OP and OQ make the same angle with l, and $PB = QB$, we obtain the following angles: $\angle BOP = \alpha - \phi = \angle QOB$, $\angle AOQ = \phi - (\alpha - \phi) = 2\phi - \alpha$.

OP is called the radius vector of the point P. Now, after reflecting P to get Q, we would like to rotate OQ about the origin by an angle θ. If R is the tip of this rotated line segment, then $\angle AOR = 2\phi - \alpha + \theta = 2(\phi + \theta/2) - \alpha$.

In the previous paragraph, we started with the assumption that Q is the reflection of P on the line OB. Conversely, we can use Figure 8.8(a), to show that if OQ = OP and $\angle AOQ = 2\phi - \alpha$, then Q is the reflection of P on OB. So, since $\angle AOR = 2\left(\phi + \frac{\theta}{2}\right) - \alpha$, R becomes the reflection of P on a line passing through the origin, making an angle $\frac{\theta}{2} + \phi$ with the x-axis,

This tells us that the effect of rotation followed by reflection is another reflection. The line on which this new reflection takes place makes an angle $\phi + \theta/2$ with the x-axis. This is the geometric proof of the simple algebraic fact that

$$R_\theta S_\phi = S_{\frac{\theta}{2}+\phi}.$$

PRACTICE 8.17

Show the following:

(i) $R_\theta = R_{\theta+2\pi}$.
(ii) $R_{\theta+\phi} = R_\theta R_\phi$.
(iii) $\left\{R_0, R_{\frac{\pi}{2}}, R_\pi R_{\frac{3\pi}{2}}\right\}$ is a group under matrix multiplication.
(iv) $\left\{R_0, R_{\frac{\pi}{2}}, R_\pi . R_{\frac{3\pi}{2}}, S_0, S_\pi, S_{\frac{\pi}{2}}, S_{\frac{3\pi}{2}}\right\}$ is group under matrix multiplication.

The group in (iv) is known as the dihedral group of order 8.

PRACTICE 8.18

Consider a regular polygon T_n with n sides one of whose vertices is $(1, 0)$. Identify the rotations and reflections that do not alter T_n. Show that this set is a group. This group is known as the dihedral group of order $2n$.

8.8 FUNDAMENTAL RESULTS IN GROUP THEORY

Having seen lot of concrete examples of groups, we prove interesting results for a general group.

Proposition 8.2

The identity element of a group is unique.

Proof
Let, e_1 and e_2 be two identities for a group $(G, *)$. Then, we have

$$e_1 * e_2 = e_2 \quad \text{since } e_1 \quad \text{is an identity.}$$
$$e_1 * e_2 = e_1 \quad \text{since } e_2 \quad \text{is an identity.}$$

From the above two equations,we get $e_1 = e_2$. Thus, the identity element is unique. ■

Proposition 8.3

Every element of a group has a unique inverse.

Proof
Let, b and c be two inverses for a in a group $(G, *)$. Then, by the definition of inverse, we have

$$a * b = b * a = e = a * c = c * a. \tag{8.1}$$

So,

$$\begin{aligned} b &= b * e \\ &= b * (a * c) \quad \text{(From (8.1))} \\ &= (b * a) * c \quad \text{(Associativity)} \\ &= e * c \quad \text{(From (8.1))} \\ &= c. \end{aligned}$$

So, $b = c$ and hence the inverse of a is unique. ■

Example 8.34

Let x, a, b be real numbers. Show that $x + a = x + b \rightarrow a = b$.

SOLUTION

$$\begin{aligned} x + a &= x + b \\ -x + x + a &= -x + x + b \quad \text{(We add } \; -x \text{ both sides)} \\ 0 + a &= 0 + b \\ a &= b. \end{aligned}$$

Observing that $0 \times 3 = 0 \times 4$ but $3 \neq 4$, the inference $xa = xb \rightarrow a = b$ is not valid. But we have the following inference:

$$(x \neq 0 \wedge xa = xb) \rightarrow a = b. \tag{8.2}$$

■

PRACTICE 8.19

Prove equation (8.2).

Example 8.34 and Practice 8.19 are special cases of Proposition 8.4.

Proposition 8.4

Let $(G, *)$ be a group and $x, a, b \in G$. Then,

(i) If $a * x = b * x$, then $a = b$. (This is known as right cancellation law.)
(ii) If $x * a = x * b$, then $a = b$. (This is known as left cancellation law.)

Proof
Let $a * x = b * x$. Since $x \in G$, there is $x^{-1} \in G$

$$a * x = b * x$$
$$\Rightarrow (a * x) * x^{-1} = (b * x) * x^{-1}$$
$$\Rightarrow a * (x * x^{-1}) = b^{*}(x * x^{-1})$$
$$\Rightarrow a * e = b * e$$
$$\Rightarrow a = b.$$

■

PRACTICE 8.20

Supply the reasons for each step in the proof of the right cancelation law. Prove the left cancelation law.

PRACTICE 8.21

Let G be a non-abelian group and let $x * a = b * x$. Can we conclude $a = b$?

Proposition 8.5

In any group $(a * b)^{-1} = b^{-1} * a^{-1}$.

Proof
To show that y is an inverse of x we must show that $x * y = y * x = e$. Now, consider

$$\begin{aligned}(a * b) * (b^{-1} * a^{-1}) &= [(a * b) * b^{-1}] * a^{-1} \quad \textit{(Associativity)}\\ &= [a * (b * b^{-1})] * a^{-1} \quad \textit{(Associativity)}\\ &= (a * e) * a^{-1} \quad \textit{(Property of Inverse)}\\ &= a * a^{-1} \quad \textit{(Property of Identity)}\\ &= e \quad \textit{(Property of Inverse)}.\end{aligned}$$

Next,

$$\begin{aligned}(b^{-1} * a^{-1}) * (a * b) &= [(b^{-1} * a^{-1}) * a] * b \quad (\textit{Associativity})\\ &= [b^{-1} * (a^{-1} * a)] * b \quad (\textit{Associativity})\\ &= (b^{-1} * e) * b \quad (\textit{Property of Inverse})\\ &= b^{-1} * b \quad (\textit{Property of Identity})\\ &= e \quad (\textit{Property of Inverse}).\end{aligned}$$

Therefore, $(a*b)^{-1} = b^{-1}a^{-1}$. ■

Example 8.35

If $(G, \oplus)$ is a group, we have proved that the cancelation laws hold. Existence of the inverse plays a crucial role in the proof of the cancelation laws. Now, if $(G, \oplus)$ is a monoid in which cancelation laws hold, can we show the existence of inverse elements? In other words, if cancelation laws hold in a monoid, is it a group? The answer is **no**. Consider $(N, \times)$, which is a monoid and if $a \times b = a \times c$, then $b = c$. The other cancelation law is also valid. But $(N, \times)$ is not a group. This makes Theorem 8.1 an interesting result. ■

Theorem 8.1

Let $(G, \oplus)$ be a *finite* semigroup in which both the cancelation laws hold. Then $(G, \oplus)$ is a group.

Proof

Let $G = \{x_1, x_2, ..., x_n\}$. Consider the subset of G defined as

$$\{x_1 \oplus x_1, x_1 \oplus x_2, ..., x_1 \oplus x_n\}.$$

This subset contains n elements, because if it were to contain less than n elements, then for some i and j with $i \neq j$, we would have $x_1 \oplus x_i = x_1 \oplus x_j$. But by the cancellation law, we obtain $x_i = x_j$. Now, it follows that

$$G = \{x_1 \oplus x_1, x_1 \oplus x_2, ..., x_1 \oplus x_n\}.$$

In particular,

$$x_1 = x_1 \oplus x_i \tag{8.3}$$

for some i. Now, we claim that x_i is an identity element in this semigroup. That is, we claim that $x_j = x_j \oplus x_i$ for all j. From equation (8.3), we obtain

$$\begin{aligned}x_1 \oplus x_j &= (x_1 \oplus x_i) \oplus x_j\\ x_1 \oplus x_j &= x_1 \oplus (x_i \oplus x_j)\\ x_j &= x_i \oplus x_j.\end{aligned}$$

Note that in the above, we have used the left cancelation law. It is easy to prove that

$$x_j = x_j \oplus x_i.$$

Thus x_i is an identity.

Now, we must prove the existence of an inverse for each x_j.

As before, we can prove that

$$\{x_j \oplus x_1, x_j \oplus x_2, ..., x_j \oplus x_n\} = G.$$

Hence, there must be some k such that $x_i = x_j \oplus x_k$. Once we prove that $x_i = x_k \oplus x_j$, x_k is an inverse of x_j and the theorem follows. We leave the proof of $x_i = x_k \oplus x_j$ as an exercise. ■

Proposition 8.6

Let n be a positive integer. Show that the set of all positive integers less than n and relatively prime to n forms a group under multiplication (mod n).

Proof

Let a, b, c be numbers less than n and relatively prime to n. Let $ab \cong ac \mod n$ and without loss of generality $b > c$. $a(b-c) \cong 0 \pmod{n}$. Since a is relatively prime to n, $(b-c)$ is a multiple of n. But since $b-c < n$, $b-c$ has to be equal to 0. Thus, the left cancelation law holds and since multiplication (mod n) is commutative, the right cancelation law is also true. Hence, the result follows from Theorem 8.1. ■

8.9 HOMOMORPHISM AND ISOMORPHISM OF GROUPS

Roughly speaking, if there is a *homomorphism* between two groups, then they have similar algebraic structures. First, let us look at some examples.

Example 8.36

Let $G_1 = \{0, 1, 2\}$ be the group under addition (modulo 3) and $G_2 = \{1, \omega, \omega^2\}$ be the group of cube roots of unity under multiplication. We give the tables of these two groups:

$\oplus_3$	0	1	2
0	0	1	2
1	1	2	0
2	2	0	1

$\times$	1	ω	ω^2
1	1	ω	ω^2
ω	ω	ω^2	1
ω^2	ω^2	1	ω

By observing the tables we find that the structures of these two groups are essentially the same. By 'renaming' 0 as 1, 1 as ω and 2 as ω^2, G_1 becomes G_2. We shall soon learn about isomorphism of groups, which is a precise way of saying that two groups are 'essentially same'. ■

Definition 8.9: Let $(G_1, \oplus)$ and $(G_2, \odot)$ be two groups $f : G_1 \longrightarrow G_2$ a function. We say that f is a homomorphism if

$$f(x \oplus y) = f(x) \odot f(y).$$

Example 8.37

This is a counter example rather than an example. $(\mathbf{Z}, +)$ and $(\mathbb{R}, +)$
Define $f : Z \longrightarrow R$ as $f(x) = x^2$. Now,

$$f(x + y) = (x + y)^2 = x^2 + y^2 + 2xy.$$

But

$$f(x) + f(y) = x^2 + y^2.$$

So,

$$f(x + y) = f(x) + f(y)$$

is **not true**. Therefore, f is **not** a homomorphism. ■

Example 8.38

Define $f : (Q^*, \times) \longrightarrow (\mathbb{R}^*, \times)$ as $f(x) = \frac{1}{x}$. Show that f is a homomorphism.

SOLUTION

$$f(xy) = \frac{1}{xy} = \frac{1}{x}\frac{1}{y} = f(x)f(y).$$

Thus, f is a homomorphism. ■

Example 8.39

Define $f : (Z, +) \longrightarrow (Z, +)$ as

$$f(n) = 3n.$$

Then,

$$f(n + m) = 3(n + m) = 3n + 3m = f(n) + f(m).$$

Hence, f is a homomorphism. ■

Example 8.40

Define $f : (\mathbb{R}, +) \longrightarrow (\mathbf{C}^*, \times)$ as

$$f(x) = \cos x + i \sin x.$$

Now

$$f(x+y) = \cos(x+y) + i\sin(x+y) = (\cos x + i\sin x)(\cos y + i\sin x) = f(x)f(y).$$

Hence $f(x)$ is a homomorphism. ■

Example 8.41

Let G be the group of all 2×2 matrices under matrix addition. Let $p, q, r, s \in \mathbb{R}$ be fixed. Define $f(\mathbb{R}, +) \longrightarrow G$ as

$$f(x) = \begin{pmatrix} px & qx \\ rx & sx \end{pmatrix}$$

$$\begin{aligned} f(x+y) &= \begin{pmatrix} p(x+y) & q(x+y) \\ r(x+y) & s(x+y) \end{pmatrix} \\ &= \begin{pmatrix} px & qx \\ rx & sx \end{pmatrix} + \begin{pmatrix} py & qy \\ ry & sy \end{pmatrix} \\ &= f(x) + f(y). \end{aligned}$$

Hence f is a homomorphism. ■

Example 8.42

Let $f : (\mathbb{R}^*, \times) \longrightarrow GL_2(R)$ be defined as

$$f(x) = \begin{pmatrix} 3x & x \\ x & x \end{pmatrix}.$$

Is this a homomorphism?

$$\begin{aligned} f(xy) &= \begin{pmatrix} 3xy & xy \\ xy & xy \end{pmatrix} \\ f(x)f(y) &= \begin{pmatrix} 3x & x \\ x & x \end{pmatrix} \begin{pmatrix} 3y & y \\ y & y \end{pmatrix} \\ &= \begin{pmatrix} 10xy & 4xy \\ 4xy & 2xy \end{pmatrix} \neq f(xy). \end{aligned}$$

Hence f is *not* a homomorphism. ■

PRACTICE 8.22

Check whether the following functions are group homomorphisms:

(1) $f : (Z, +) \longrightarrow (Z, +)$ defined as $f(n) = 5n + 1$.
(2) $f(Z, +) \longrightarrow (Q^*, \times)$ defined as $f(n) = 3n$.
(3) $f(Z, +) \longrightarrow (\mathbb{R}^*, \times)$ defined as $f(n) = 2^n$.
(4) $f(2Z, +) \longrightarrow (Z, +)$ as $f(n) = \frac{n}{2}$.
(5) $f(\mathbb{R}, +) \longrightarrow (\mathbb{R}^*, \times)$ as $f(x) = e^x$.

PRACTICE 8.23

Let $G = \left\{ \begin{pmatrix} a & 0 \\ 0 & a \end{pmatrix} : a \neq 0 \right\}$. Show that G is an abelian group under matrix multiplication. Let $f : (\mathbb{R}^*, \times) \longrightarrow G$ be defined as

$$f(x) = \begin{pmatrix} x & 0 \\ 0 & x \end{pmatrix}.$$

Show that f is a homomorphism.

PRACTICE 8.24

Let $G = \left\{ \begin{pmatrix} x & x \\ x & x \end{pmatrix} : x \in \mathbb{R}^* \right\}$ with a binary operation of matrix multiplication. Define $f : R^* \longrightarrow G$ as $f(x) = \begin{pmatrix} x & x \\ x & x \end{pmatrix}$. Show that f is **not** a homomorphism.

PRACTICE 8.25

Define $g : \mathbb{R}^* \longrightarrow G$ as $g(x) = \begin{pmatrix} \frac{x}{2} & \frac{x}{2} \\ \frac{x}{2} & \frac{x}{2} \end{pmatrix}$. Show that g is a homomorphism.

An isomorphism is a special case of homomorphism. If two groups are isomorphic, then they are essentially the same. Now, we define the notion of an isomorphism.

Definition 8.10: A one-to-one and onto homomorphism is called an isomorphism.

Example 8.43

Show that $f : (Z, +) \longrightarrow (Z, +)$ defined as $f(x) = 7x$ is a homomorphism but not an isomorphism.

SOLUTION

$$f(x + y) = 7(x + y) = 7x + 7y = f(x) + f(y).$$

So, f is a homomorphism. Further, $f(x) = f(y) \Rightarrow 7x = 7y \Rightarrow x = y$. Hence f is one-to-one. But for $1 \in Z$, there is no $x \in Z$ such that $7x = 1$. Hence this is not an isomorphism. ■

Example 8.44

Let GL_n be the group of all $n \times n$ non-singular matrices under matrix multiplication. Define $f : GL_n \longrightarrow R^*$ as $f(A) = |A|$. Show that f is an onto homomorphism, which is not one-to-one. Here the binary operation on R^* is multiplication.

SOLUTION

$$f(AB) = |AB| = |A||B| = f(A)f(B).$$

Thus, f is a homomorphism. Now let $p \in R^*$. Consider the matrix $C = \begin{pmatrix} p & 0 \\ 0 & 1 \end{pmatrix}$. Now $f(C) = |C| = p$. Hence, there is $C \in GL_n$ such that $f(C) = p$. So, f is onto.

But f is not one-to-one. Consider $A = \begin{pmatrix} 1 & 0 \\ 0 & 6 \end{pmatrix}$ and $B = \begin{pmatrix} 2 & 0 \\ 0 & 3 \end{pmatrix}$.

Now,

$$f(A) = |A| = 6 \quad \text{and} \quad f(B) = |B| = 6.$$

But $A \neq B$. Hence f is not one-to-one. ■

Example 8.45

Show that $f : (\mathbb{R}, +) \longrightarrow (\mathbb{R}, +)$ defined as $f(x) = 7x$ is an isomorphism.

SOLUTION

$$f(x + y) = 7(x + y) = 7x + 7y = f(x) + f(y).$$

So, f is a homomorphism. Next, to show that f is onto, consider $y \in R$. Defining $x = \frac{y}{7}$, $f(x) = y$. Thus, f is onto.

Next, to prove that f is one-to-one consider $f(x) = f(y)$. This means that $7x = 7y$ and hence $x = y$. Thus, f is one-to-one.

Therefore, f is an isomorphism. ■

Example 8.46

Let G_1 be the group of real numbers with respect to addition, and G_2 be the group of complex numbers with unit modulus with respect to multiplication. Let $f : G_1 \longrightarrow G_2$ be defined as $f(x) = \cos x + i \sin x$. Show that f is an onto homomorphism but not an isomorphism.

SOLUTION

Let $x, y \in G_1 = \mathbb{R}$

$$f(x + y) = \cos(x + y) + i \sin(x + y) = (\cos x + i \sin x)(\cos y + i \sin y) = f(x)f(y).$$

Hence, f is a homomorphism.

Now, $f(0) = f(2\pi) = 1$. Hence f is **not** one-to-one. Therefore, f is not isomorphism.
Now, let $z \in G_2$, we have $|z| = 1$ and letting θ to be its argument, we have $z = \cos\theta + i\sin\theta$. Thus, $f(\theta) = z$. Thus, f is onto. ■

PRACTICE 8.26

Let $G_1 = (Z_3, \oplus_3)$ and G_2 be the group of third roots of unity under multiplication. Show that the map $f : G_1 \longrightarrow G_2$ defined as $f(0) = 1, f(1) = \omega, f(2) = \omega^2$ is an isomorphism.

PRACTICE 8.27

Show that the set G_1 of all strictly positive real numbers is a group under multiplication. Let G_2 be the group of all real numbers under addition. Define $f(x) = \log x$. Show that $f : G_1 \longrightarrow G_2$ is an isomophism.

PRACTICE 8.28

Consider the group

$$G = \left\{1, \omega, \omega^2, \ldots, \omega^{n-1}\right\}$$

consisting of complex nth roots of unity under multiplication. Show that Z_n and G are isomorphic under a suitably defined isomorphism.

PRACTICE 8.29

Show that $G = \left\{5^n : n \in Z\right\}$ is a group under multiplication. Define $f : Z \longrightarrow G$ as $f(n) = 5^n$. Show that f is an isomorphism.

PRACTICE 8.30

Show that $f : G \longrightarrow G$ defined as $f(a) = a^{-1}$ is an isomorphism, if G is an abelian group.

PRACTICE 8.31

Let $G_1 = (\mathbb{R}, +)$ and $G_2 = (\mathbb{R}^*, \times)$. Define $f : G_1 \longrightarrow G_2$ as $f(x) = e^x$. Check if it exhibits homomorphism or isomorphism.

Example 8.47

Let

$$f : \mathbb{R}^* \longrightarrow \left\{ \begin{pmatrix} \cos\theta & -\sin\theta \\ \sin\theta & \cos\theta \end{pmatrix} : \theta \in \mathbb{R} \right\}$$

be defined as

$$f(x) = \begin{pmatrix} \cos x & -\sin x \\ \sin x & \cos x \end{pmatrix}$$

which is an onto homomorphism, which is not one-to-one.

SOLUTION

Let $x, y \in R^*$.

$$\begin{aligned} f(x+y) &= \begin{pmatrix} \cos(x+y) & -\sin(x+y) \\ \sin(x+y) & \cos(x+y) \end{pmatrix} \\ &= \begin{pmatrix} \cos x \cos y - \sin x \sin y & -\sin x \cos y - \cos x \sin y \\ \sin x \cos y + \cos x \sin y & \cos x \cos y - \sin x \sin y \end{pmatrix} \\ &= \begin{pmatrix} \cos x & -\sin x \\ \sin x & \cos x \end{pmatrix} \begin{pmatrix} \cos y & -\sin y \\ \sin y & \cos y \end{pmatrix} \\ &= f(x)f(y). \end{aligned}$$

Hence f is a homomorphism. By the very definition it is onto. Since $f(0) = f(2\pi)$, f is not an isomorphism. ∎

Example 8.48

Show that if f is a homomorphism, then $f(a^{-1}) = f(a)^{-1}$.

SOLUTION

Let $a \in G_1$.

$$\begin{aligned} e_1 &= aa^{-1} \\ \Rightarrow f(e_1) &= f(aa^{-1}) = f(a)f(a^{-1}) \\ \Rightarrow e_2 &= f(a)f(a^{-1})e_1 = a^{-1}a \\ \Rightarrow f(e_1) &= f(a^{-1}a) = f(a^{-1})f(a) \\ \Rightarrow e_2 &= f(a)f(a^{-1}). \end{aligned}$$

So, we have $f(a)f(a^{-1}) = f(a^{-1})f(a) = e_2$. Thus, $f(a^{-1})$ is the inverse of $f(a)$. ∎

8.10 SUBGROUPS

Example 8.49

Consider the group $(\mathbf{Z}, +)$. Consider the following subsets of Z:

(i) $\{2n+1 : n \in \mathbf{Z}\}$.
(ii) $\{2n : n \in \mathbf{Z}\}$.
(iii) $\{0\}$.
(iv) Fix $n \in \mathbf{Z}$. $n\mathbf{Z} = \{nk : k \in Z\}$.
(v) Set of all prime integers.
(vi) $\{..., -4, -1, 2, 5, 8, 11\}$.
(vii) Set of all integers, which leaves a remainder of 1 when divided by 3.

Note that (ii), (iii) and (iv) are also groups under addition but the remaining are not groups. ■

Definition 8.11: Let $(G, *)$ be a group. A subset H is said to be a group if $(H, *)$ is a group.

The proof of the following proposition follows from the definition.

Proposition 8.7

A subset H is a subgroup if

(i) H is closed with respect to $*$.
(ii) $e \in H$.
(iii) For every $a \in H$, $a^{-1} \in H$.

Proposition 8.8

A non-empty subset H of G is a subgroup of G if and only if for every $a, b \in H$, $a^{-1} * b \in H$.

Proof

First, assume that for all $a, b \in H$, $a^{-1} * b \in H$. Since H is non-empty there is at least one $x \in H$. Now, $x^{-1} * x \in H$ and hence $e \in H$. Now, let $a \in H$. Since $e \in H$, $a^{-1} * e \in H$. That is $a^{-1} \in H$.

Next we shall prove that H is closed with respect to the binary operation. Let $a, b \in H$. $a^{-1} \in H$ and hence $(a^{-1})^{-1} * b \in H$.

That is $a * b \in H$. So, we have proved that H is a subgroup.

Next, assume that H is a subgroup and let $a, b \in H$. Since H is a subgroup $a^{-1} \in H$ and since H is closed $a^{-1} * b \in H$. The proof is complete. ■

Example 8.50

Fix $k \in \mathbf{Z}$. Show that the set

$$k\mathbf{Z} = \{kn : n \in \mathbf{Z}\} = \{0, \pm k, \pm 2k, \pm 3k, ...\}$$

which is the set of all multiples of k, is a subgroup of $\mathbf{Z}$.

SOLUTION

Let kn and km be two elements of kZ. Consider $(kn)^{-1} + km = -kn + km = k(m-n)$. Thus, $(kn)^{-1} + km$ is an element of kZ. So, kZ is a subgroup. ■

PRACTICE 8.32

Show that

$$SL_n(\mathbb{R}) = \left\{ \begin{pmatrix} a & b \\ c & d \end{pmatrix} : a, b, c, d \in \mathbb{R}, ad - bc = 1 \right\}$$

is a subgroup of $GL_2(\mathbb{R})$, the group of non-singular 2×2 matrices under matrix multiplication.

Example 8.51

Show that $(\mathbb{R}^3, +)$ is a group where $+$ is defined as $(x, y, z) + (p, q, r) = (x + p, y + q, z + r)$. Also show that the xy plane which is the subset defined as $H = \{(x, y, 0) : x, y \in \mathbb{R}\}$ is a subgroup.

SOLUTION

By its very definition $+$ is a binary operation on $\mathbb{R}^3$ and its associativity is easily proved. Clearly, $(0, 0, 0)$ is the identity element and the inverse of (x, y, z) is $(-x, -y, -z)$. Hence $(\mathbb{R}^3, +)$ is a group.

To show that H is a subgroup, take $\alpha = (x, y, 0)$ and $\beta = (p, q, 0) \in H$. Now $\alpha + \beta^{-1} = (x, y, 0) + (-p, -q, 0) = (x - p, y - q, 0) \in H$. Thus, H is a subgroup. ■

PRACTICE 8.33

Let H be the set of all strictly positive real numbers. Show that H is a subgroup of $(\mathbb{R}^*, \times)$.

Example 8.52

Show that $H = \{x + 4ix : x \in \mathbb{R}\}$ is a subgroup of $\mathbf{C}$ under addition.

SOLUTION

Consider $a, b \in H$. Let $a = x + 4ix$ and $b = y + 4iy$ where x, y are real numbers. Now $a^{-1} + b = -x - 4ix + y + 4iy = (-x + y) + 4i(-x + y)$, which is an element of H. Thus, H is a subgroup of $(\mathbf{C}, +)$. ■

Remark 8.1

Instead of writing $a * b$, we shall write ab for simplicity. You must interpret ab as the result of applying the binary operation of the group on (a, b).

Proposition 8.9

Let G be a group and H be a subgroup. Then the set

$$H_1 = \{x \in G : xh = hx \quad \text{for all} \quad h \in H\}$$

is a subgroup.

Proof

Let $a, b \in H_1$. Then, for all $h \in H$, $ah = ha$ and $bh = hb$. To show $a^{-1}b \in H$, we must prove that $a^{-1}bh = ha^{-1}b$ for all $h \in H$.

$ah = ha \Rightarrow aha^{-1} = h$ and $bh = hb \Rightarrow bhb^{-1} = h$. Thus, $bhb^{-1} = aha^{-1}$.

$$\begin{aligned} &bhb^{-1} = aha^{-1} \\ &\Rightarrow a^{-1}bhb^{-1} = ha^{-1} \\ &\Rightarrow a^{-1}bh = ha^{-1}b \end{aligned}$$

and the proof is complete. ■

Example 8.53

Show that $H = \{x + 4ix : x \in R^*\}$ is not a subgroup of $(C^*, \times)$.

SOLUTION

Let $a = x+4ix$ and $b = y+4iy$ be elements of H. Now, $ab = xy-16xy+i(8xy) \notin H$. So, H is not closed and H is **not** a subgroup of G. ■

PRACTICE 8.34

Show that $H = \{(x, mx) : x \in \mathbb{R}\}$ is a subgroup of $\mathbb{R}^2$.

PRACTICE 8.35

Show that $H = \left\{(x, y) \in \mathbb{R}^2 : ax + by = 0\right\}$ is a subgroup of $\mathbb{R}^2$.

PRACTICE 8.36

Show that $H = \left\{(x, y, z) \in \mathbb{R}^3 : ax + by + cz = 0\right\}$ is a subgroup of $\mathbb{R}^3$.

Example 8.54

Show that the set of all symmetric matrices form a subgroup of M_n under addition.

SOLUTION I

Let H be the set of all symmetric matrices.

(i) Let $A, B \in H$. Then $A^T = A$ and $B^T = B$. Now, $(A+B)^T = A^T + B^T = A+B$. Thus, $A+B \in H$ and hence H is closed with respect to matrix addition.

(ii) Matrix addition is associative.

(iii) Let $\mathbf{0}$ be the zero matrix, which is the identity element of M_n. Clearly, $\mathbf{0}$ is symmetric. Thus, the identity element belongs to H.

(iv) If A is symmetric, $-A$ is also symmetric. Thus, for every element of H, its inverse is also in H.

So, H is a subgroup.

SOLUTION II

We shall use the criteria for subgroups. Let $A, B \in H$. Consider $(-A+B)^T = (-A)^T + B^T = -(A^T) + B = -A + B$. Thus, $-A + B \in H$. Hence, H is a subgroup. ∎

Proposition 8.10

Let $f : G_1 \longrightarrow G_2$ be a homomorphism. Then $f(G_1)$ is a subgroup of G_2.

Proof

Let $a, b \in f(G_1)$. Then, we have $x, y \in G_1$ with $f(x) = a$ and $f(y) = b$. Since G_1 is a subgroup, $x^{-1}y \in G_1$. But this means that $f(x^{-1}y) \in G_2$. Using the properties of homomorphism, we get $f(x)^{-1}f(y) = a^{-1}b \in G_2$. Thus, $f(G_2)$ is a subgroup. ∎

Theorem 8.2

Let $(G, *)$ be a group. Then G is isomorphic to a subgroup of $S(G)$.

Proof

Recall that $S(G)$ is the group of one-to-one and onto maps on G. Let $a \in G$. Using this element a we want to define a one-to-one and onto map on G. Define $f_a : G \longrightarrow G$ as $f_a(x) = a * x$. Now, we show that f_a is in fact a one-to-one and onto map.

Let $f_a(x) = f_a(y)$. Then $a * x = a * y$. By the left cancelation law, $x = y$. So, f_a is one-to-one. Next, to prove f_a is onto, take $y \in G$. We have to find $x \in G$ such that $f_a(x) = y$. That is, to find x such that $a * x = y$. It is obvious that if we take $x = a^{-1} * y$, then $a * x = y$. Thus, f_a is onto. Now, define $F : G \longrightarrow S(G)$ as $F(a) = f_a$. Now, we show that F is a homomorphism. The binary operation in G is $*$ and the binary operation in $S(G)$ is the composition $\circ$. Thus, we need to prove $F(a * b) = F(a) \circ F(b)$. To show that these two functions are equal, we must show that their values are equal for all $x \in G$.

$$\begin{aligned} F(a*b)(x) &= f_{a*b}(x) \\ &= (a*b)*x \\ &= a*(b*x) \\ &= a*f_b(x) \\ &= f_a(f_b(x)) \\ &= (f_a \circ f_b)(x). \end{aligned}$$

Thus, F is a homomorphism. Next, we must show that F is one-to-one. Let $F(a) = F(b)$. That is $f_a = f_b$. This means that $f_a(x) = f_b(x)$ for all $x \in X$. So. $a*x = b*x$ for all $x \in G$. By the right cancelation law, this gives us $a = b$. So, F is one-one. Now $F(G)$ is a subgroup of $S(G)$. $F : G \longrightarrow F(G)$ is one-to-one and onto and the proof is complete. ■

Example 8.55

Let $f : (\mathbb{R}^2, +) \longrightarrow (\mathbb{R}, +)$ be defined as $f(x, y) = 2x+3y$. We can easily show that this is a homomorphism. Now, consider the straight line L given by the equation $2x + 3y = 0$. Now

$$L = \{(x, y) : f(x, y) = 0\}$$

can be easily shown to be a subgroup of $(\mathbb{R}^2, +)$. ■

This example provides a motivation for Definition 8.12 and the following proposition.

Definition 8.12: Let $f : G_1 \longrightarrow G_2$ be two groups. Then the set $\{x \in G_1 : f(x) = e_2\}$ where e_2 is the identity of G_2 is called the kernel of the homomorphism f.

Proposition 8.11

The kernel of a homomorphism is a subgroup of the domain group.

Proof

Let $a, b \in \{x \in G_2 : f(x) = e_2\}$. Then, $f(a) = e_2$ and $f(b) = e_2$.
Now, $f(a^{-1}b) = f(a^{-1})f(b) = f(a)^{-1}e_2 = e_2^{-1}e_2 = e_2e_2 = e_2$. Thus, $a^{-1}b$ belongs to the kernel and hence the kernel is a subgroup. ■

PRACTICE 8.37

(i) Is $\{10, 100, 1000, ...\}$ a subgroup of **Z**?

(ii) Is $\left\{...\frac{1}{100}, \frac{1}{10}, 1, 10, 100, ...\right\}$ a subgroup of $(\mathbb{R}^*, \times)$?

(iii) Show that $\left\{a + b\sqrt{3} : a \in Q, b \in Q, ab \neq 0\right\}$ is a subgroup of $\mathbb{R}^*$.

(iv) Let H_1 be the set of all non-singular 2×2 matrices with integer entries. Is this a subgroup of $GL_2(\mathbb{R})$? Let H_2 be the set of all non-singular 2×2 matrices with determinant 1. Is this a subgroup of $GL_2(\mathbb{R})$?

8.10.1 Lagrange Theorem

Lagrange Theorem is an interesting theorem for finite groups. It says that the number of elements of a subgroup is a factor of the number of elements of G. For a finite group G, its order $|G|$ is defined as the number of elements in G.

Lagrange Theorem: Let G be a finite group and H be a subgroup. Then the order of H divides the order of G. The proof of the following lemma can be carried out by the reader.

PRACTICE 8.38

Define a relation R on G as $a\ R\ b$ if $ab^{-1} \in H$. Then show that R is an equivalence relation.

Proof of Lagrange Theorem: Let $Ha = \{ha : h \in H\}$. First, we show that Ha is nothing but the equivalence class of a under the relation R as defined in Practice 8.38.

$$\begin{aligned} & x \in [a] \\ & \Leftrightarrow xRa \\ & \Leftrightarrow xa^{-1} \in H \\ & \Leftrightarrow xa^{-1} = h \text{ for some } h \in H. \\ & \Leftrightarrow x = ha \text{ for some } h \in H. \\ & \Leftrightarrow x \in Ha. \end{aligned}$$

Hence, $\{Ha : a \in G\}$ is a partition of G. The subset Ha is called a right coset of H. Next, we show that all right cosets have the same number of elements. This is shown by defining a one-to-one and onto map from Ha to Hb. Let $x \in Ha$. Then there is $h \in H$ such that $x = ha$. We define $f(x) = hb \in bH$. To show that f is one-to-one, let $f(x) = f(y)$. There is $h_1 \in H$ such that $x = h_1a$ and there is $h_2 \in H$ such that $y = h_2a$. Now, $f(x) = h_1b$ and $f(y) = h_2b$. Since $f(x) = f(y)$, we have $h_1b = h_2b$.

By the right cancellation law, $h_1 = h_2$. This implies that $x = y$. Hence the function is one-to-one.

To show that f is onto:

Let $y \in Hb$. That is, $y = bh$ for some $h \in H$. Define $x = ha$. It is clear that $f(x) = y$ and hence f is onto.

Thus, all cosets have the same number of elements. The number of elements in each coset is $|H|$ since $H = He$ is itself a coset. If the number of distinct cosets is k, then $k|H| = |G|$. So, $|H|$ divides $|G|$.

8.10.2 Cyclic Subgroups

Let us observe that $i^1 = i$, $i^2 = -1$, $i^3 = -i$, $i^4 = 1$, $i^5 = i$, $i^6 = -1$, $i^7 = -i$, $i^8 = 1$, etc. It is clear that higher and higher powers of i would produce the numbers $i, -1, -i, 1$ again and again. We have already seen that $\{i, -i, 1, -1\}$ is a group and this is an example of a cyclic group. The word cyclic refers to things happening again and again. But by a cyclic group we mean a group where every element is a power of the same element.

Let $(G, \oplus)$ be a group. For simplicity, we denote $x \oplus y$ by xy and in tune with this simplification, a^n is $a \oplus a \oplus \ldots n$ times. The formal definition of a cyclic group is given below.

Definition 8.13: Let G be a group and $a \in G$.

(i) The subgroup generated by a is defined as $\{a^n : n \in Z\}$.

(ii) A group G is said to be cyclic if $G = \{a^n : n \in Z\}$ for some $a \in G$. In this case, We say that a is the generator of G and G is generated by a.

Proposition 8.12

G is a cyclic group if and only if there exists $a \in G$ such that every element of G is a power of a.

The proof of this proposition can easily be worked out.

Example 8.56

For any $n \in N$, Z_n is a cyclic group. Let $k \in Z_n$. Then, $k = 1 + 1 + \ldots$ k times and hence Z_n is generated by 1. ■

Example 8.57

Consider the group $(\mathbf{Z}_5^*, \otimes_5)$ under multiplication (mod 5). Show that it is cyclic and find its generator.

SOLUTION

$[1]$ is not a generator because $[1]^n = [1]$ for any n. The cyclic subgroup generated by 1 is the trivial subgroup $\{[1]\}$

$$[3]^0 = [1]$$

$$[3]^1 = [3]$$

$$[3]^2 = [4]$$
$$[3]^3 = [2].$$

Thus, 3 is a generator to G. ■

Example 8.58

Show that $(\mathbb{R}, +)$ is not cyclic and find the cyclic subgroup generated by 3.

SOLUTION

We prove that $(\mathbb{R}, +)$ is not cyclic by contradiction. Let us suppose that $(\mathbb{R}, +)$ is cyclic and hence it has a generator $a \neq 0$. This implies that any $x \in \mathbb{R}$ can be written as $x = a + a + \ldots n$ times $= na$. In particular $\frac{a}{2} = na$. Since $a \neq 0$, we get $\frac{1}{2} = n$ an integer. This is false. Thus, $(\mathbb{R}, +)$ is not cyclic.

The cyclic subgroup generated by 3 is

$$\{3n : n \in Z\} = \{\ldots, -9, -6 - 3, 0, 3, 6, 9, \ldots\}.$$ ■

PRACTICE 8.39

(i) Show that [2] and [4] also are generators of the group $\{[1], [2], [3], [4]\}$ under multiplication modulo 5.
(ii) Show that $(Z, +)$ is a cyclic group and it is generated by 1.
(iii) Show that $(\mathbb{R}^*, \times)$ is not cyclic and find the cyclic subgroup generated by 3.

8.11 NORMAL SUBGROUPS

Let G be a group and let H be a subgroup. Fix $a \in H$. The left coset corresponding to a is defined as

$$aH = \{ah : h \in H\}$$

and the right coset is defined as

$$Ha = \{ha : h \in H\}.$$

Example 8.59

Let $G = (Z, +)$, $H = 2Z$ and $a = 3$. Compute the coset $a + H$.

SOLUTION

Since this group is abelian, the left and right cosets are the same.

$$aH = \{ah : h \in H\} = \{3 + n : n \in 2Z\}$$
$$[1ex] = \{3 + 2m : m \in Z\}$$
$$[1ex] = 3 + 2Z = 2Z + 3$$
$$[1ex] = \{..., -6 + 3, -4 + 3, -2 + 30 + 3, 2 + 3, 4 + 3, 6 + 3, ...\}$$
$$[1ex] = \{..., -3, -1, 1, 3, 5, 7, 9, ...\}$$

the set of odd integers. ■

Example 8.60

Let us recall the following facts from Sections 8.3 and 8.4:

(i) $(Z, +)$ is a group.

(ii) If $m \in Z$, then $H = \{0, \pm m, \pm 2m, \pm 3m, ...\}$ is a subgroup of Z.

(iii) R defined as xRy if $x - y \in H$ is an equivalence relation.

(iv) $Z_m = \{[0], [1], ..., [m-1]\}$ is the set of all equivalence classes.

(v) The binary operation $\oplus_m$ defined as Z_m as $[x] \oplus_m [y] = [x + y]$ makes Z_m a group.

Now let G be a group and let H be a subgroup. Define R on G as aRb if $ab^{-1} \in H$. In Section 8.9.1, we have already noted the following:

(i) R is an equivalence relation.

(ii) The equivalence class of a corresponding to the relation R is given by $[a] = Ha$

Imitating Z_m, we would like to define a binary operation on the set of all right cosets as follows:

$$HaHb = Hab. \tag{8.4}$$

Further, we would like to know if the set of right cosets themselves form a group under this binary operation. Consider equation (8.2). If $Ha = Hx$ and $Hb = Hy$, then we must have $HaHb = Hab = Hxy$. $ax^{-1} \in H$ and $ay^{-1} \in H$ should imply that $ab(xy)^{-1} \in H$. This is not true for all subgroups. Subgroups for which this is true are known as normal subgroups. ■

Definition 8.14: A subgroup H is said to be a normal subgroup if $aH = Ha$ for all $a \in H$.

Proposition 8.13

A subgroup is normal if and only if for $a \in G$ and $h \in H$, $aha^{-1} \in H$.

Proof

Let H be a normal subgroup. Let $a \in H$. By hypothesis, $aH = Ha$. Now, for $h \in H$, $ah \in aH = Ha$. Therefore, there is $h_1 \in H$ such that $ah = h_1 a$. This implies that $h_1 = aha^{-1}$. Since $h_1 \in G$, $aha^{-1} \in G$.

Next, assume that for $a \in G$ and $h \in H$, $aha^{-1} \in H$. Let us prove that $aH = Ha$. Let $x \in aH$. That is, $x = ah$ for some $h \in H$. Now, $xa^{-1} = aha^{-1} \in H$. Denoting $aha^{-1} = h_1$, $xa^{-1} = h_1$ and so $x = h_1 a \in Ha$. Thus, $aH \subseteq Ha$.

Clearly, $a^{-1}ha \in H$ for every $a \in G$ and $h \in H$. Using this, we can prove that $Ha \subseteq aH$. Thus, $aH = Ha$. The proof is complete. ∎

Theorem 8.3

If H is a normal subgroup, then the set of all right cosets, which is denoted by G/H, forms a group under the binary operation defined by equation (8.2).

Proof

Let us observe that $H = He$ is a right coset. Now $HaHe = Hae = Ha$ and hence H is the identity in G/H. Next, from $HaHa^{-1} = Haa^{-1} = H$, it follows that Ha^{-1} is the inverse of Ha in G/H. Thus, G/H is a group. ∎

Definition 8.15: The index of a subgroup H is the number of distinct left or right cosets of H.

Proposition 8.14

A subgroup of index two is normal.

Proof

Let H be a subgroup. As there are only two left cosets, the left cosets are aH and H where $a \notin H$. G is the disjoint union of aH and H. Since $a \notin H$, $Ha \neq H$. As the number of right cosets is also two, G is the disjoint union of Ha and H. So, $aH = Ha$ and H is normal. ∎

Example 8.61

Let G be the group of all $n \times n$ non-singular matrices under matrix multiplication and $G_1 = \{A \in G : |A| = 1\}$. Show that G_1 is a normal subgroup.

SOLUTION

First, let us prove that G_1 is a subgroup. Let A and B be in G_1. So, $|A| = |B| = 1$.

$$|A^{-1}B| = |A^{-1}||B| = \frac{1}{|A|}|B| = 1.$$

Hence $A^{-1}B \in G_1$ and G_1 is a subgroup. Next, let $A \in G$ and $H \in G_1$.

$$|A^{-1}HA| = \frac{1}{|A|}|H||A| = 1.$$

So, $A^{-1}HA \in G_1$ and G_1 is normal. ∎

PRACTICE 8.40

Show that the kernel of a homomorphism is a normal subgroup of the domain group.

PRACTICE 8.41

Let $f : \{1, 2, ..., n\}$ be a function such that there are $1 \leq p, q \leq n$ such that $f(p) = q$ and $f(q) = p$ and $f(x) = x$ for $x \neq p$ and $x \neq q$. A permutation corresponding to a function of the type f is called a transposition. Show that any permutation is a product of transpositions. A permutation is said to be even if it is a product of an even number of transpositions. Show that the set of all even permutations is a normal subgroup of the set of all permutations.

Example 8.62

Show that $H = \{(x, 0, 0) : x \in \mathbb{R}\}$ is a normal subgroup of $\mathbb{R}^3$. Find the cosets of H.

SOLUTION

The proof of H being a subgroup is easy. Its normality follows from the fact that for an abelian group, every subgroup is normal. Now if $\alpha = (p, q, r) \in \mathbb{R}^3$, the coset is given by $\alpha + H = \{(x + p, q, r) : x \in \mathbb{R}\}$. Since $x \in \mathbb{R}$ is arbitrary, $\alpha + H = \{(p, q, r) : q, r \in \mathbb{R}\}$. This is nothing but the line passing through the point $(0, q, r)$ and parallel to the x-axis.

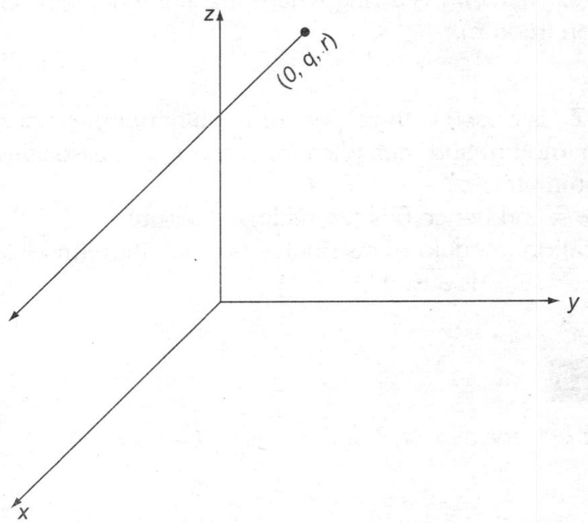

Figure 8.9

8.12 RINGS

Definition 8.16: Let R be a set and let $(\oplus, \odot)$ be two binary operations on R. We say that $(R, \oplus, \odot)$ is a ring if the following hold:

(i) $\oplus$ and $\odot$ are associative.
(ii) $a \oplus b = b \oplus a$ for all $a, b \in R$.
(iii) There exists an element in R, which is denoted by 0 with the property that $a \oplus 0 = a$.
(iv) For every element $a \in \mathbb{R}$, there is an element $-a \in \mathbb{R}$ such that $a \oplus -a = 0$.
(v) The following distributive properties hold:

$$a \odot (b \oplus c) = (a \odot b) \oplus (a \odot c)$$

$$(b \oplus c) \odot a = (b \odot a) \oplus (c \odot a).$$

The above definition can otherwise be summarized as follows: Let R be a set and $\oplus, \odot$ be two associative binary operations on R. We say that $(R, \oplus, \odot)$ is a ring if the following hold:

(i) $(R, \oplus)$ is an *abelian* group.
(ii) The distributive properties hold :

$$a \odot (b \oplus c) = (a \odot b) \oplus (a \odot c)$$

$$(b \oplus c) \odot a = (b \odot a) \oplus (c \odot a).$$

Most of the times, we denote $x \oplus y$ by $x + y$ and $x \odot y$ by xy.

Example 8.63

Show that $(\mathbf{Z}_n, \oplus_n, \odot_n)$ is a ring where $\oplus_n$ is the addition (mod n) and $\odot_n$ is the multiplication (mod n).

SOLUTION

(i) The set $\mathbf{Z}_n$ is closed with respect to the binary operations $\oplus_n$ and $\odot_n$.
(ii) Addition (mod n) and multiplication (mod n) are associative.
(iii) $\oplus_n$ is commutative.
(iv) $a \oplus_n 0 = a$ and hence 0 is the additive identity.
(v) Multiplication (modulo n) distributes over addition (modulo n).

Hence $(\mathbf{Z}_n, \oplus_n, \odot_n)$ is a ring. ■

Proposition 8.15

In a ring R, $0x = 0$ for all $x \in R$ and $(-x)y = x(-y) = -xy$.

Proof

The assertions of this proposition might seem to be trivial but they need to be proved. Here R is any ring and 0 is the additive identity of R, which need not be the usual 0. Further, x and y may not be numbers and even if they are, xy may not be the usual product of x and y. Here $-x$ is the additive inverse of x. The expansion $-x = (-1)x$ may not be meaningful in a general ring.

Now, let us prove $0x = 0$. It is clear that $0+0=0$ since 0 is the additive identity. Using the distributive law,

$$0x = (0+0)x$$

$$0 + 0x = 0x + 0x \quad \textit{Property of the identity and Distributive Property}$$

$$0 = 0x \quad \textit{Cancelation Law}$$

To show that $(-x)y = -xy$ we need to show that the additive inverse of $(-x)y$ is xy. Now, $(-x)y + xy = (-x+x)y = 0y = 0$.

Similarly, $x(-y) + xy = x(-y+y) = x0 = 0x = 0$ and the result is proved. ■

Example 8.64

Let R be the set of all polynomial functions. Show that R is a ring under the usual addition and multiplication of polynomials.

A polynomial p is a function that is defined as

$$p(x) = a_0x^n + a_1x^{n-1} + \ldots + a_n$$

where a_i's are real numbers and $n \in N$.

The sum and product of polynomial functions is again a polynomial and hence R is closed with respect to addition. Polynomial addition is commutative.

The function θ that is identically equal to zero can be written as $\theta(x) = 0x^n + 0x^{n-1} + \ldots. + 0$ and hence, it is a polynomial function. For any polynomial p, $p + \theta = p$ and hence θ is the additive identity.

Let $p(x) = a_0x^n + a_1x^{n-1} + \ldots + a_n$. The function $-p$ defined as

$$-p(x) = -a_0x^n - a_1x^{n-1} + \ldots - a_n$$

is also a polynomial and $p(x) + -p(x) = 0$. So $-p$ is the additive inverse of p.

For polynomials p, q and r,

$$p(x)[q(x) + r(x)] = p(x)q(x) + p(x)r(x)$$

and

$$[q(x) + r(x)]p(x) = q(x)p(x) + r(x)p(x).$$

Thus, the distributive properties hold. Hence R is a ring. ■

PRACTICE 8.42

(i) Let Q_n be the set of all polynomials of degree n. Show that Q_n is *not* a group under polynomial addition.
(ii) Let P_n be the set of all polynomials of degree less than or equal to n. Show that P_n is a group under polynomial addition.
(iii) Is P_n a ring?
(iv) Show that the set of all polynomials with integer coefficients is a ring.

8.12.1 Subrings, Ideals and Quotients

Example 8.65

The set M_2 of all 2×2 matrices is a ring with respect to matrix addition and multiplication. Now the set S of all 2×2 matrices with integer entries is also a ring with respect to matrix addition and multiplication. Further, $S \subseteq M_2$. We say that S is a subring of M_2. ■

Definition 8.17: If $(R, +, \times)$ is a ring then we say that R_1 is a subring if R_1 is closed with respect to $+$ and $\times$ and $(R_1, +, \times)$ is a ring.

PRACTICE 8.43

Exhibit the two subrings of $(R, +\times)$.

Example 8.66

Consider the ring of integers $(\mathbf{Z}, +, \times)$ and $(2\mathbf{Z}, +)$, which is a subgroup of $(\mathbf{Z}, +)$. For any arbitrary $n \in \mathbf{Z}$ and $m \in 2\mathbf{Z}$, $nm \in \mathbf{Z}$. In other words, the set $2\mathbf{Z}$ is such that any element outside it multiplied with an inside element produces an inside element. $2\mathbf{Z}$ is an example of an ideal. ■

We now formally define an ideal of a ring.

Definition 8.18: A subset I of a ring $(R, +, \times)$ is called an ideal if $(I, +)$ is a subgroup of $(R, +)$ and for every $x \in R$ and $y \in I$, $xy \in I$.

PRACTICE 8.44

Let I be an ideal of the ring of integers. Let $k \in I$ be the least positive element of I. Show that $I = k\mathbf{Z}$ and hence find all the ideals of $\mathbf{Z}$.

PRACTICE 8.45

Show that the set of all multiples of a fixed polynomial p is an ideal in the ring of all polynomials with real coefficients.

Example 8.63 tells us that $\mathbf{Z}_n$ is a ring. Recalling that $\mathbf{Z}_n$ is nothing but the collection of the equivalence classes with respect to the relation R defined as $x\ R\ y$ if $x - y \in n\mathbf{Z}$ and the fact that $n\mathbf{Z}$ is an ideal, we may like to generalize the construction of a ring analogous to that of $\mathbf{Z}_n$ in a general ring. The following theorem achieves this.

Theorem 8.4

Let $(R, +, \times)$ be a ring and let I be an ideal of the ring. Define a relation on R as $a \equiv b$ if $a - b \in I$. Let $R/I = \{[x] : x \in R\}$ where $[x]$ is the set $\{y \in R : x - y \in I\}$.

Define $+$ and $\times$ on R/I as follows: $[x]+[y]=[x+y]$, $[x]\times[y]=[x\times y]$. R/I is a ring that is known as the quotient ring with respect to the ideal I.

Sketch of the Proof: First, we shall have to show that $\equiv$ is an equivalence relation and hence we can define the equivalence classes $[x]$.

To show that the new binary operations are well defined, we have to show that

$$(a \in [x] \wedge b \in [y]) \rightarrow (a+b \in [x+y] \wedge a \times b \in [x \times y]).$$

We have to show that the equivalence class of 0 is I and further $[x]+I=[x]$. That is, we have to show that I is the additive identity. Further, to get the additive inverse of elements of R/I, we have to show that $[x]+[-x]=I$. Further, the distributive properties have to be proved. ■

PRACTICE 8.46

Complete the proof of Theorem 8.4.

Example 8.67

Defining $f:\mathbf{Z} \longrightarrow Z_n$ as $f(k)=[k]$, by the very definition $f(x+y)=f(x)+f(y)$ and $f(xy)=f(x)f(y)$. This is an example of a ring homomorphism.

Let us now define a ring homomorphism. ■

Definition 8.19: Let $f:(R_1,+,\times) \longrightarrow (R_2,\oplus,\odot)$ be a function such that $f(x+y) = f(x)\oplus f(y)$ and $f(x\times y)=f(x)\odot f(y)$. Then f is known as a ring homomorphism.

PRACTICE 8.47

Let $R=R_1\times R_2$ be the cartesian product of the rings in which we define $(x_1,y_1) + (x_2,y_2)=(x_1+x_2,y_1+y_2)$ and $(x_1,y_1)(x_2,y_2)=(x_1x_2,y_1y_2)$. Show that R is a ring. Define $f:R \longrightarrow R_1$, $g:R_2 \longrightarrow R$ and $f(x,y)=x$ and $g(y)=(0,y)$. Show that both f and g are ring homomorphisms.

Example 8.68

Let $f:R_1 \longrightarrow R_2$ be a ring homomorphism. Let $R_3=\{x\in R_1 : f(x)=0\}$. Show that R_3 is a subring.

Proof

The fact that R_3 is an additive subgroup of R_1 follows from proposition 8.11. Now, we have to show that R_3 is closed with respect to multiplication.

Let $x,y\in R_3$. Then $f(x)=0$ and $f(y)=0$. Now $f(xy)=f(x)f(y)=0\times 0=0$. Hence $xy\in R_3$. The result is proved. ■

PRACTICE 8.48

Examine whether R_3 is an ideal of R_1.

Definition 8.20: A ring is said to be an integral domain if $xy = 0$ implies that either $x = 0$ or $y = 0$.

PRACTICE 8.49

Show that $(Z_4, \oplus_4, \odot_4)$ is not an integral domain. Further, show that $(Z_5, \oplus_5, \odot_5)$ is an integral domian.

8.13 FIELDS

Definition 8.21: Let F be a set and let $(\oplus, \odot)$ be two binary operations on F. We say that $(F, \oplus, \odot)$ is a *field* if the following hold:

(i) $\oplus$ and $\odot$ are associative.
(ii) $a \oplus b = b \oplus a$ for all $a, b \in F$.
(iii) There exists an element in F, which is denoted by 0 with the property that $a \oplus 0 = a$ for all $a \in F$.
(iv) For every element $a \in F$, there is an element $-a \in F$ such that $a \oplus -a = 0$.
(v) $a \odot b = b \odot a$ for all $a, b \in F$.
(vi) There exists an element in F, which is denoted by 1 with the property that $a \odot 1 = a$ for all $a \in F$.
(vii) For every element $a \neq 0$ in F, there is an element $a^{-1} \in F$ such that $a \oplus a^{-1} = 1$.
(viii) The distributive properties:

$$a \odot (b \oplus c) = (a \odot b) \oplus (a \odot c)$$

$$(b \oplus c) \odot a = (b \odot a) \oplus (c \odot a).$$

Remark 8.2

The above definition can be summarized as follows: Let F be a set and let $(\oplus , \odot)$ be two associative binary operations on F. We say that $(F, \oplus, \odot)$ is a field if the following hold:

(i) $(F, \oplus)$ is an *abelian* group.
(ii) $(F - \{0\} , \odot)$ is an *abelian* group.
(iii) The distributive properties hold:

$$a \odot (b \oplus c) = (a \odot b) \oplus (a \odot c)$$

$$(b \oplus c) \odot a = (b \odot a) \oplus (c \odot a).$$

PRACTICE 8.50

Show that $(\mathbb{R}, +, \times)$ is a field.

Example 8.69

The set F of all real numbers of the type $a + \sqrt{2}b$ where a and b are rational is a field under addition and multiplication.

SOLUTION

The addition and multiplication of real numbers is commutative and associative.

Consider two elements $a+\sqrt{b}$ and $p+\sqrt{q}$, of F. a, b, p, q are rational numbers.

$$(a + \sqrt{2}b) + (p + \sqrt{2}q = (a + p) + \sqrt{2}(b + q).$$

Since $a + p$ and $b + q$ are rational numbers, the above sum is an element of F. Thus, F is closed with respect to addition.

$$(a + \sqrt{2}b)(p + \sqrt{2}q) = ap + \sqrt{2}aq + \sqrt{2}bp + 2bq = (ap + 2bq) + \sqrt{2}(aq + bp).$$

Since $ap + 2bq$ and $aq + bp$ are rational, the above product is in F. Hence, F is closed with respect to multiplication.

$0 = 0 + \sqrt{2}(0) \in F$ and thus 0 is the additive identity.

$1 = 1 + \sqrt{2}(0) \in F$ and thus 1 is the multiplicative identity.

When a and b are rational $-a$ and $-b$ are also rational. Thus $-a+\sqrt{2}(-b) \in F$. $a+\sqrt{2}+-a+\sqrt{2}(-b) = 0$ and hence every element in F has an additive inverse.

Now let $a + \sqrt{2}b \in F$ be non-zero. That is, at least one of a and b is non-zero. We have

$$\begin{aligned}\frac{1}{a+\sqrt{2}b} &= \frac{a-\sqrt{2}b}{(a+\sqrt{2}b)(a-\sqrt{2}b)} \\ &= \frac{a-\sqrt{2}b}{a^2-2b^2} \\ &= \frac{a}{a^2-2b^2} - \sqrt{2}\frac{b}{a^2-2b^2}.\end{aligned}$$

Clearly, $\frac{a}{a^2-2b^2} - \sqrt{2}\frac{b}{a^2-2b^2} \in F$. Let us note that if a and b are rational numbers $a^2 - 2b^2 \neq 0$. For, if $a^2 - 2b^2 = 0$, $\frac{a^2}{b^2} = 2$, which contradicts the irrationality of $\sqrt{2}$. Thus, every non-zero element in F has an inverse. Thus, F is a field. ■

Example 8.70

Show that a finite ring such that $xy = 0 \to x = 0 \vee y = 0$ is a field.

SOLUTION

Let $(R, +, \times)$ be the ring We have to show that $(R^*, \times)$ is a group. Now let $xy = xz$ with $x, y, z \in R^*$ By distributivity and Proposition 8.15, we get $x(y + (-z)) = 0$. Since $x \neq 0$, we must have $y + -z = 0$ or that $y = z$. Thus, we have proved the left cancelation law and the right cancelation law can also be proved. The result now follows from Theorem 8.1. ■

SUMMARY

- $(G, \odot)$ is a semigroup if $\odot$ is associative.
- A monoid is a semigroup if it has an identity.
- A group is a monoid where every element has an inverse.
- A subgroup is a subset of the group, which is also a group with respect to the binary operation on the group.
- The number of elements of a subgroup is a divisor of the number of elements of a finite group.
- A subgroup gives rise to an equivalence relation on the group whose equivalence classes are known as cosets.
- A normal subgroup is a subgroup such that every left coset is also a right coset.
- A normal subgroup is used to define the quotient group.
- $(R, \oplus, \odot)$ is a ring if $(R, \oplus)$ is an abelian group, $(R, \odot)$ is a semigroup and $\odot$ distributes over $\oplus$.
- A subring of a ring is a subset of the ring, which is also a ring with respect to the binary operations on the ring.
- A subring $I \subseteq R$ of a ring is an ideal if $x \in R$, $y \in I$ imply that $xy, yx \in I$.
- An ideal of a ring gives rise to its quotient ring.
- $(F, \oplus, \odot)$ is a field if $(R, \oplus)$ is an abelian group, $(F^*, \odot)$ is an abelian group and $\odot$ distributes over $\oplus$.

PROBLEMS

1. Which of the following are all semigroups/monoids:

(i) $G =$ The set of all odd integers $= \{2n + 1 : n \in Z\} = \{\pm 1, \pm 3, \pm 5, \ldots\}$ under the operation of addition.

(ii) $2Z =$ The set of all even integers $= \{2n : n \in Z\} = \{0, \pm 2, \pm 4, \pm 6, \ldots\}$ under the operation of addition.

(iii) $\mathbb{R}$ under the binary operation defined as $a * b = a + 2b$.

(iv) Q^* under division.

(v)

$$G = \left\{ \begin{pmatrix} a & b \\ b & c \end{pmatrix} : a, b, c \in \mathbb{R} \right\}$$

under the operation of matrix multiplication.

(vi)

$$G = \left\{ \begin{pmatrix} a & b \\ c & d \end{pmatrix} : a, b, c, d \in \mathbb{N} \right\}$$

under the operation of matrix multiplication

(vii)

$$G = \left\{ \begin{pmatrix} a & b \\ c & d \end{pmatrix} : a, b, c, d \in \mathbf{C} \right\}$$

under the operation of matrix addition.

(viii)

$$G = \left\{ \begin{pmatrix} a & b \\ c & d \end{pmatrix} : a, b, c, d \in \mathbf{C} |a| = |b| = |c| = |d| = 1 \right\}$$

under the operation of matrix addition.

(ix) $2Z$ = The set of all even integers $= \{2n : n \in Z\} = \{0, \pm 2, \pm 4, \pm 6, \ldots\}$ under the operation of addition.

(x) $2Z$ = The set of all even integers $= \{2n : n \in Z\} = \{0, \pm 2, \pm 4, \pm 6, \ldots\}$ under the operation of multiplication.

2. Show that the set $\{1, 3, 7, 9\}$ forms a group under multiplication modulo 10.
3. Show that a monoid $(G, \odot)$ is a group if every element has an inverse.
4. Show that the set

$$\left\{ \begin{pmatrix} -\frac{1}{2} & -\frac{\sqrt{3}}{2} \\ \frac{\sqrt{3}}{2} & -\frac{1}{2} \end{pmatrix}, \begin{pmatrix} -\frac{1}{2} & \frac{\sqrt{3}}{2} \\ -\frac{\sqrt{3}}{2} & -\frac{1}{2} \end{pmatrix}, \begin{pmatrix} 1 & 0 \\ 0 & 1 \end{pmatrix} \right\}$$

is a group under matrix multiplication.

5. Show that the set

$$\left\{ \begin{pmatrix} 0 & -1 \\ 1 & 0 \end{pmatrix}, \begin{pmatrix} -1 & 0 \\ 0 & -1 \end{pmatrix}, \begin{pmatrix} 0 & 1 \\ -1 & 0 \end{pmatrix}, \begin{pmatrix} 1 & 0 \\ 0 & 1 \end{pmatrix} \right\}$$

is a group under matrix multiplication.

6. Show that the set $SL_n(\mathbb{R})$ of all $n \times n$ matrices whose determinant value is 1 is a group under matrix multiplication.
7. A matrix is said to be orthogonal if $AA^t = A^tA = I$. Show that the set of all orthogonal $n \times n$ matrices form a group under matrix multiplication.
8. Show that the set

$$\left\{ \begin{pmatrix} \cos\theta & -\sin\theta \\ \sin\theta & \cos\theta \end{pmatrix} : \theta \in \mathbb{R} \right\}$$

is a group under matrix multiplication.

Remark: An element of this set is called a rotational matrix. It represents a rotation by θ in the plane.

9. Give an example of a monoid, which is not a group.
10. Give an example of a semigroup, which is not a monoid.
11. Give an example of a non-abelian group.
12. Show that $(Z_m, \otimes_m)$ is a monoid.
13. Find the solutions of $x \otimes_m x = 1$ in $\mathbf{Z_m}$ for $m = 5$, $m = 6$ and $m = 7$.
14. Show that the subset of Z_m containing only $[a]$ such that $(a, m) = 1$ is a group under $\otimes_m$.
15. Define functions f_1, f_2, f_3, f_4 from $\mathbb{R}^*$ into $\mathbb{R}^*$ as $f_1(x) = x$, $f_2(x) = -x$, $f_3(x) = \frac{1}{x}$, $f_4(x) = \frac{-1}{x}$. Show that the set $\{f_1, f_2, f_3, f_4\}$ forms a group under composition.
16. Show that $(\mathbb{R}^n, +)$ is a group where $+$ is defined as $(x_1, x_2, \ldots, x_n) + (y_1, y_2, \ldots, y_n) = (x_1 + y_1, x_2 + y_2, \ldots, x_n + y_n)$
17. An element of a monoid $(G, *)$ is said to be an idempotent if $x * x = x$. Show by an example that a monoid may contain more than one idempotent.
18. In a group, show that the identity is the only idempotent element.

19. Show that the group $\{i, -i, 1, -1\}$ under complex multiplication is a cyclic group. Find its generator. Does -1 generate this group?
20. Show that $\{[1], [2], ..., [10]\}$ is a cyclic group with respect to $\otimes_{11}$.
21. Let n be a multiple of 4. Show that $H = \{1, -1, i, -i\}$ is a subgroup of the multiplicative group of all nth roots of unity. Find all the cosets of H.
22. Let $f : (\mathbb{R}^*, \times) \longrightarrow GL_2(\mathbb{R})$ be defined as

$$f(\theta) = \begin{pmatrix} \cos\theta & \sin\theta \\ -\sin\theta & \cos\theta \end{pmatrix}.$$

Show that f is a homomorphism but it is neither one-one nor onto.
23. Let l, m and n be fixed real numbers. Show that $L = \{(lt, mt, nt) : t \in R\}$ is a subgroup of $(\mathbb{R}^3, +)$. L is a line in space passing through the origin.
24. Let $H = \{(x_1, x_2, ..., x_k, 0, 0, ..., 0) : x_1, x_2, ..., x_k \in \mathbb{R}\}$ be the subset of $\mathbb{R}^n$ containing only those points whose $(k+1)$th $(k+2)$th,...,nth coordinates are zero. Show that H is a normal subgroup of $(\mathbb{R}^n, +)$. Show that $\mathbb{R}^n/H$ is isomorphic to the group $(\mathbb{R}^{n-k}, +)$.
25. Let G_1 be the group $(\mathbb{R}^2, +)$ and let $(G_2, +)$ be the set of diagonal 2×2 diagonal matrices. Define $F : G_1 \longrightarrow G_2$ as

$$F(x, y) = \begin{pmatrix} x & 0 \\ 0 & y \end{pmatrix}.$$

Show that F is an isomorphism. Examine the nature of F if G_2 is taken as the set of all 2×2 matrices.
26. Check whether $f : (Z, +) \longrightarrow (\mathbb{R}, +)$ defined as $f(x) = 7x$ is an isomorphism.
27. Let $k \in \mathbf{Z}$ and define $f : \mathbf{Z} \longrightarrow \mathbf{Z}$ as $f(n) = kn$. Examine whether f is a ring homomorphism.
28. Show that $(\mathbf{C}, +, \times)$ is a field.
29. Show that $\mathbf{Z_m}$ is a field if and only if m is a prime.
30. Let $\mathbb{R}$ be a ring with a multiplicative identity 1. Further, assume that $xy = 0$ implies that at least one of x and y is zero. Define an equivalence relation on $\mathbb{R} \times \mathbb{R} - \{0\}$ as follows: $(x, y) \equiv (a, b)$ if and only if there is $0 \neq p \in \mathbb{R}$ such that $a = px$ and $b = py$. Show that $\equiv$ is an equivalence relation. Let $[(x, y)]$ denote the equivalence class of (x, y). $F = \{[(x, y)] : x, y \in \mathbb{R}, y \neq 0\}$. Define $[(x, y)] + [(a, b)] = [(bx + ay, yb)]$ and $[(x, y)][(a, b)] = [(ax, by)]$.

    (i) Show that these binary operations are well defined. That is, when $(x_1, y_1) \in [(x, y)]$ and $(a_1, b_1) \in [(a, b)]$, we have $[(b_1x_1 + a_1y_1, y_1b_1)] = [(bx + ay, yb)]$ and $[(a_1x_1, b_1y_1)] = [(ax, by)]$.
    (ii) Show that $[(0, 1)]$ is the additive identity and $[(1, 1)]$ is the multiplicative identity of F.
    (iii) Show that $[(x, y)]$ has the additive inverse as $[-x, -y)]$ and the multiplicative inverse as $[(y, x)]$ if $x \neq 0$.
    (iv) Show that F is a field.

FURTHER READING

Tremblay, J. P. and Manohar, R. (1997), *Discrete Mathematical structures with applications to computer science,* Tata McGraw-Hill, New Delhi.

Herstein, I. N. (1999), *Topics in Algebra*, John Wiley & Sons, New York.

Fraleigh, J. B. (1982), *A First Course in Abstract Algebra*, Narosa Publishing House, New Delhi.

CHAPTER

9

Graph Theory

Learning Objectives

On completing this chapter, you should be able to:

- define the notions of a graph, directed graph and a simple graph
- define and calculate the degree of a vertex in a graph
- define the concept of a connected graph
- define the notions of walks and paths connecting two vertices of a given graph
- identify connected graphs
- identify the connected components of a graph, which is not connected
- define the notion of bipartite graphs
- define the notion of a tree
- prove simple results on connected graphs
- prove simple results on trees
- define the notion of an Euler graph
- prove the characterization of an Euler graph

9.1 INTRODUCTION AND MOTIVATION

Example 9.1

Graphs can be used to show the inter-connectivity among servers, PCs, hubs, etc. in a network. There are different types of networks and the word topology refers to the manner in which various elements of the network are connected to each other. We illustrate ring topology, star topology and tree topology using graphs in Figure 9.1.

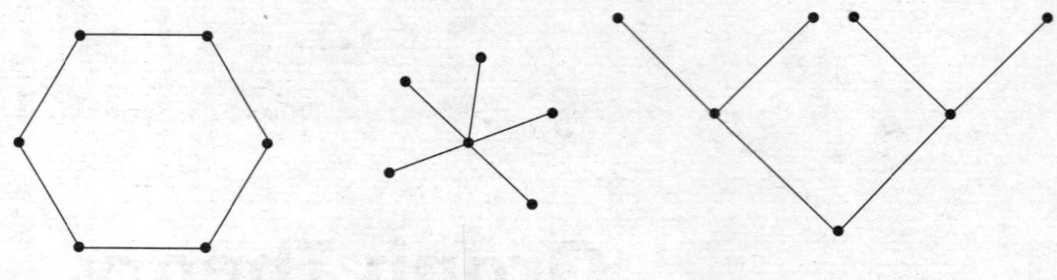

Figure 9.1

Example 9.2

You are aware that Web pages have links to other Web pages. It is possible that you start the day with a Web page on coffee and end it with a Web page on Saint Theresa by following up the links. It is unlikely that a Web page on *rice* would lead you to a Web page on *The Theory of Relativity* but one cannot be absolutely sure. We may represent the Web pages by dots A, B, C, etc., and connect the dots by an arrow if the corresponding Web sites are linked. To be precise, if there is a link to the Web page C from the Web page A, we draw an arrow from A to C.

Figure 9.2 shows an example of a directed graph.

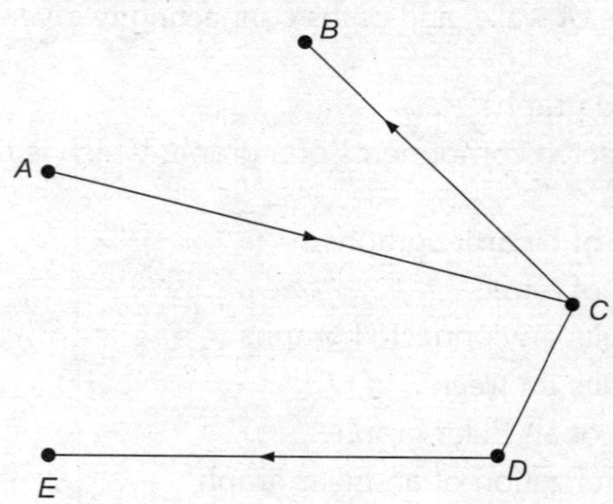

Figure 9.2

Loosely speaking, by a graph we mean a set of points and lines/arrows connecting some of these points. The points are called vertices and the lines/arrows are called edges.

Let us look at the graphs in Figures 9.3 and 9.4.

If you want to describe this graph to your friend over the telephone what would you do? Naturally, you will first name all the vertices. For every edge e_i, your friend must know which vertices are connected by it. In a mathematical description of a graph, we have two sets V and E. V is called the set of vertices and E is called the set of edges. For each of the edges e, we must know the pair of vertices that is connected by e. This is specified by a function whose domain is the set of edges and whose co-domain is the set of all non-empty subsets of V containing not more than two elements.

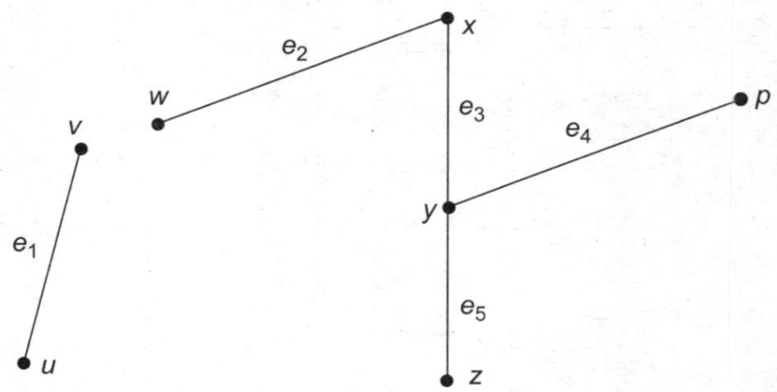

Figure 9.3

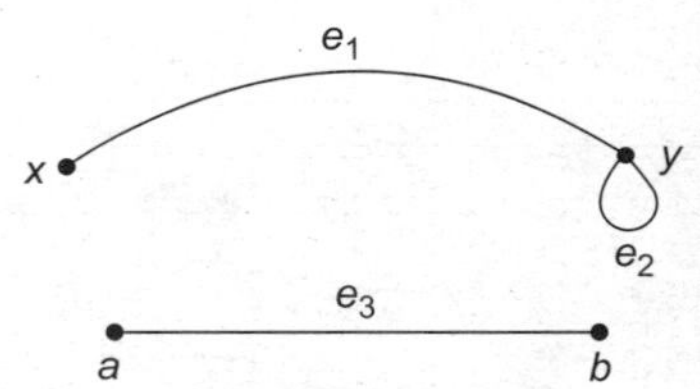

Figure 9.4

Definition 9.1: (i) A graph $G = (V, E, \phi)$ consists of V a set of vertices, E a set of edges and a function $\phi : E \longrightarrow \{\{u, v\} \,|u, v \in V\}$.

(ii) A directed graph $G = (V, E, \phi)$ consists of V a set of vertices, E a set of edges and a function $\phi : E \longrightarrow \{(u, v) : u, v \in V\}$.

The reader must carefully note that in (i), the set $\{u, v\}$ is used but in (ii) the ordered pair (u, v) is used.

In (ii), if $\phi(e) = (u, v)$, then e is directed from u to v.

Note that in the above definition we do not demand that $u \neq v$. If $u = v$, then the set $\{u, v\}$ is simply $\{u\}$. Hence, an edge might connect a vertex with itself. Such an edge is called a loop.

Example 9.3

Consider the graph in Figure 9.5:

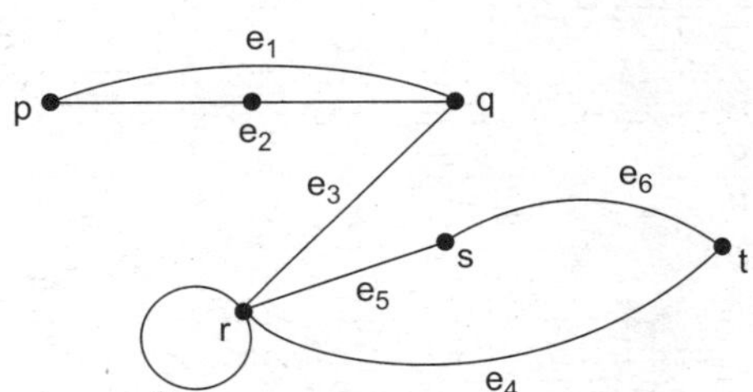

Figure 9.5

$V = \{p, q, r, s, t\}$, $E = \{e_1, e_2, e_3, e_4, e_5, e_6\}$ and ϕ is given by

$$\phi(e_1) = \{p, q\}$$
$$\phi(e_2) = \{p, q\}$$
$$\phi(e_3) = \{q, r\}$$
$$\phi(e_4) = \{r, t\}$$
$$\phi(e_5) = \{r, s\}$$
$$\phi(e_6) = \{t, s\}$$
$$\phi(e_7) = \{r\}$$

For example, $\phi(e_5) = \{r, s\}$ means that edge e_5 connects vertices r and s. $\phi(e_7) = \{r\}$ means that edge e_7 is a loop joining vertex r to itself. ■

PRACTICE 9.1

Draw the graph (V, E, φ), where $V = \{a, b, c, d\}$, $E = \{e_1, e_2, e_3\}$ and φ is given by

$$\phi(e_1) = \{a, b\}$$
$$\phi(e_2) = \{a, d\}$$
$$\phi(e_3) = \{a, e\}.$$

We say that u and v are adjacent if there is an edge connecting them. The vertices connected by an edge are called the ends of the vertices. If u is an end of e then we say that u is incident with e and e is incident with u. The formal definition is given below:

Definition 9.2: Let (G, V, ϕ) be a graph, $u, v \in V$ and $\phi(e) = \{u, v\}$. Then we say that

(i) The edge e connects the vertices u and v.
(ii) u is an end or end vertex of e.
(iii) v is an end or end vertex of e.
(iv) u is incident with e.
(v) v is incident with e.
(vi) e is incident with u.
(vii) e is incident with v.
(viii) u and v are adjacent.

Example 9.4

In the graph shown in Figure 9.6, we find that the ends of both e_2 and e_4 are p and a. Such edges are called parallel.

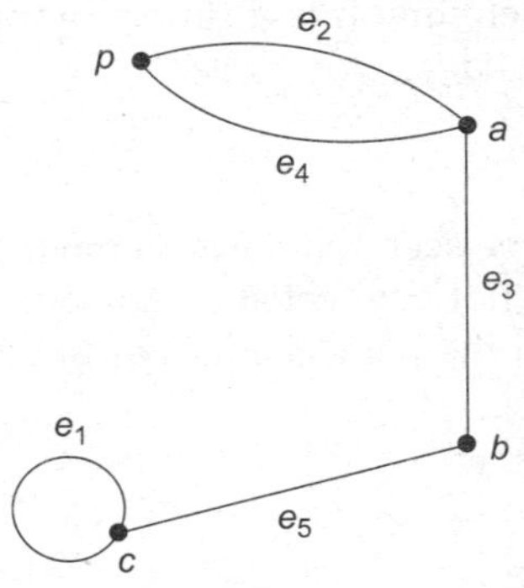

Figure 9.6

e_1 is an example of a loop. ■

Definition 9.3: An edge e is called a loop if $\phi(e)$ is a singleton. Two edges e_1 and e_2 are called parallel if $\phi(e_1) = \phi(e_2)$.

Definition 9.4: A graph is called simple, if there are no loops and no parallel edges. The formal definition of parallel edges and loop follows.

PRACTICE 9.2

Identify the simple graphs among the following graphs:

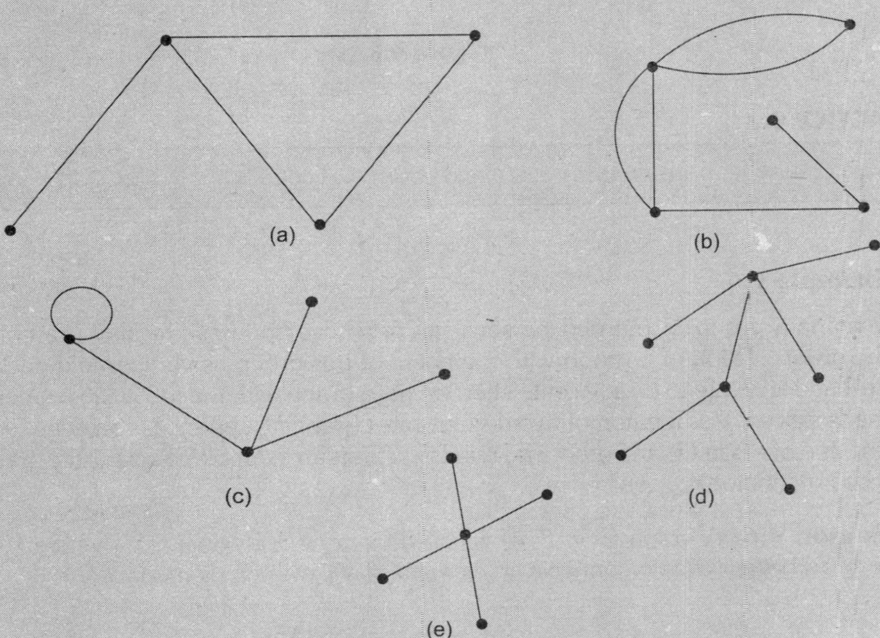

Figure 9.7

Definition 9.5: A complete graph is a simple graph where every pair of distinct vertices is connected by an edge.

Example 9.5

From Figure 9.8, it can be seen that (a) is a complete graph; (b) is not a complete graph since q and s are not connected by an edge; (c) is a complete graph; (d) is not complete because of the parallel edges e_1 and e_2.

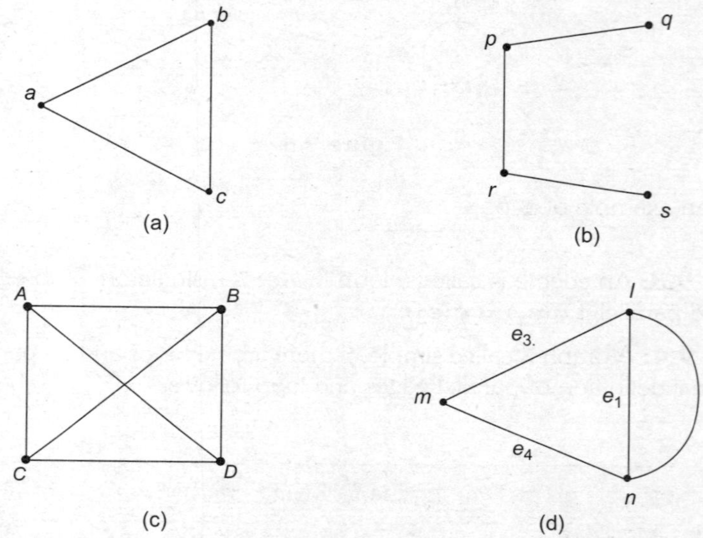

Figure 9.8 ■

PRACTICE 9.3

Draw complete graphs with five, six and seven vertices.

Example 9.6

Imagine a group of married persons, all of whose spouses are also present in the group. Think of a graph with members of this group as vertices and declare husband and wife to be adjacent. Then we get a graph with the following property: the vertex set V is a union of two disjoint sets U_1 and U_2, and if e is an edge, then one end of e is in U_1 and other end is in U_2. This is an example of a bipartite graph, whose definition is given below. ■

Definition 9.6: A graph $G(V, E, \varphi)$ is called bipartite if $U = V_1 \cup V_2$ where $V_1 \cap V_2 = \phi$ and every edge e connects an element of V_1 with an element of V_2.

Example 9.7

Among the following graphs in Figure 9.9, (a) and (b) are bipartite, whereas (c) is not bipartite.

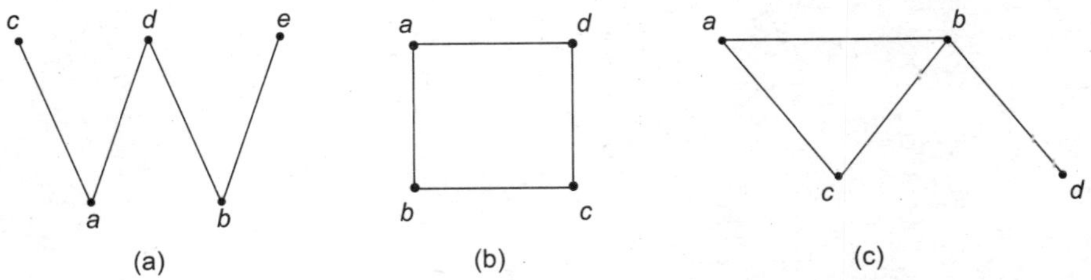

Figure 9.9

For the first graph, it is easy to identify the vertex subsets V_1 and V_2, respectively, as $V_1 = \{c, d, e\}$ and $V_2 = \{a, b\}$. But for the second graph, they are not obvious from the figure. But one can easily verify that $V_1 = \{a, c\}$ and $V_2 = \{b, d\}$.

Now, we show that c is not bipartite. Note that vertices b and c are connected to a each by an edge. They are also connected to each other by an edge. Thus, if $a \in V_1$, then $b \in V_2$ and $c \in V_2$. But since $b \in V_2$ and $c \in V_1$ also, it is a contradiction to $V_1 \cap V_2 = \phi$. Hence the graph (c) cannot be bipartite. ■

PRACTICE 9.4

Identify the bipartite graphs among the following in Figure 9.10:

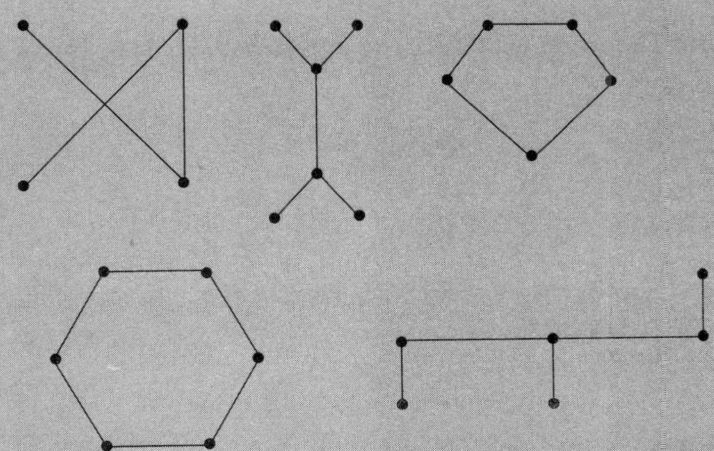

Figure 9.10

Definition 9.7: The degree of a vertex in a graph is the number of edges incident on it. A loop contributes two to the degree.

Example 9.8

Find the degrees of each vertex in the following graph in Figure 9.11:

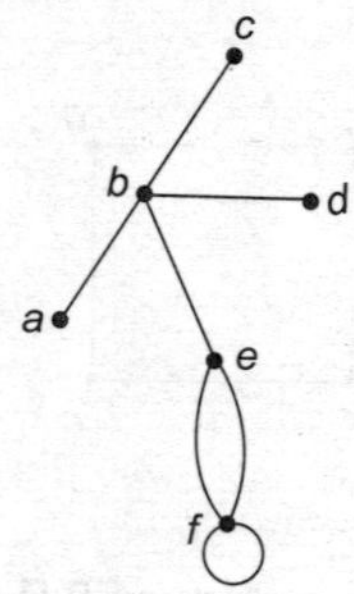

Figure 9.11

SOLUTION

The degree of a is 1
The degree of b is 4
The degree of c is 1
The degree of d is 1
The degree of e is 3
The degree of f is $1 + 1 + 2 = 4$
(A loop contributes 2 to the degree) ■

Handshaking Theorem: Let $G(V, E)$ be a graph with E edges. Then

$$2e = \sum_{v \in V} deg(v).$$

Proof
Every edge corresponds to two vertices. So, the sum of the degrees of the vertices is twice the number of degrees. ■

Example 9.9

A graph consists of four vertices each of degree three and an isolated vertex. Find the number of edges.

SOLUTION

The sum of the degrees $= 4 \times 3 + 0$. Hence, the number of edges $= \frac{12}{2} = 6$. ■

PRACTICE 9.5

Is it possible for a graph with vertices a, b and c to have the following data? $deg(a) = 3$, $deg(b) = 2$, $deg(c) = 2$? Why?

Theorem 9.1

A graph has an even number of vertices of odd degree.

Proof

Divide the set of vertices into two subsets V_1 and V_2 where V_1 is the set of all vertices of even degree and V_2, the set of vertices of odd degree. Now,

$$2e = \sum_{v \in V_1} deg(v) + \sum_{v \in V_2} deg(v).$$

Clearly, $2e$ is even. In the first summation, each $deg(v)$ is even. Thus, the first term on the RHS is also even. So, the second term is even. Now as each of the term in the second sum is odd, the number of terms has to be even. ■

PRACTICE 9.6

A group of persons can be treated as the vertices of a graph. We say that there is an edge connecting u and v, if they shake hands with each other. Using this graph theoretic interpretation, show that in any group of persons, the number of people who shake hands with an odd number of others is even.

Example 9.10

Let A and B be managers of two different sections in an office. D is the secretary to A and C is the secretary to B. The LAN connections between the PCs of these four people are as given in the graph in Figure 9.12.

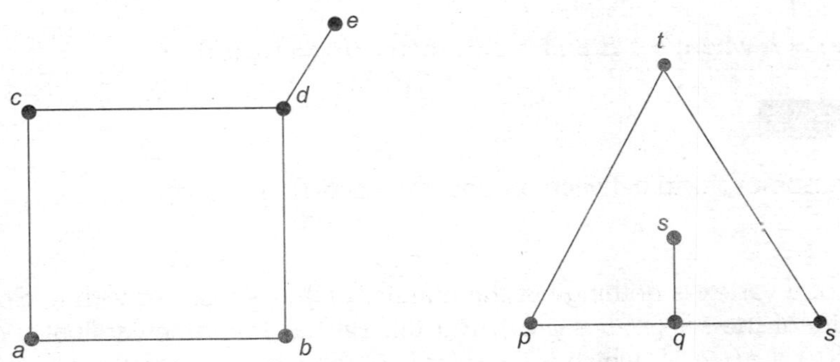

Figure 9.12

Imagine that the office where C sits is near a noisy highway and due to some illness C can no longer occupy her present office. D kindly agreed to switch her office with C, and their respective bosses have approved this change. Naturally, C and D do not want to switch their PCs. Their PCs would also be shifted to their respective new offices. Now, the original LAN connection needs to be maintained. Thus, the new LAN connection is as shown in Figure 9.13.

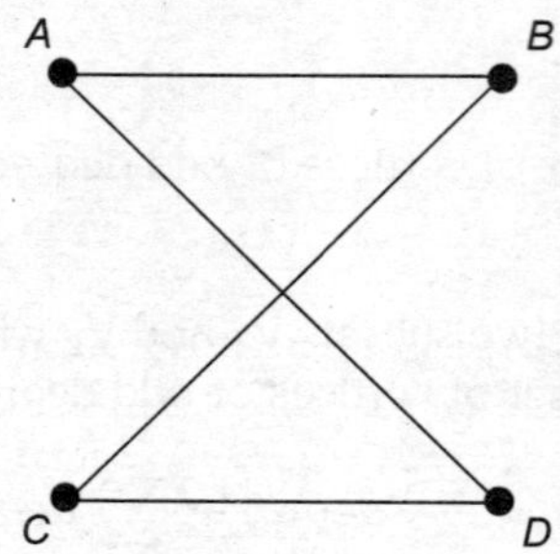

Figure 9.13

It is clear that the new arrangement would require a longer length of cable to be used. Although such physical changes have taken place, the original connectivity has not changed. This is an example of graph isomorphism.

If two graphs are isomorphic, they have the same number of vertices and edges. Further, every vertex and edge of one graph corresponds to a vertex and an edge of the second graph in such a way that the incidence relation is preserved. In other words, if u and v are vertices in one graph corresponding to vertices A and B in the other graph and if e is an edge connecting u and v, then the corresponding edge in the second graph connects A and B. We now formally define an isomorphism between two graphs. ■

Definition 9.8: The graphs $G(U, E, \varphi)$ and $G(V, F, \phi)$ are said to be isomorphic if there is a one-to-one and onto function $f : U \cup E \longrightarrow V \cup F$ such that

(i) $f(U) = V$.
(ii) $f(E) = F$.
(iii) If $f(e) = h$ where $e \in E$ and $h \in F$, then $\phi(h) = f(\varphi(e))$.

Theorem 9.2

If f is an isomorphism between graphs, then $\deg(f(u)) = \deg(u)$.

Proof

Degree of a vertex is nothing but the number of edges incident with u. So, it is the cardinality of the set $\{e : u \in \varphi(e)\}$. But this set has the same cardinality of the set $\{f(e) : f(u) \in \phi(f(e))\}$, which is the degree of the vertex $f(e)$ in the second graph.

Theorem 9.2 is useful to determine whether two given graphs are isomorphic or not. If isomorphic, the isomorphism can also be found. But these methods would take a lot of time if the number of vertices and edges is very big. Let us look at some examples. ■

Example 9.11

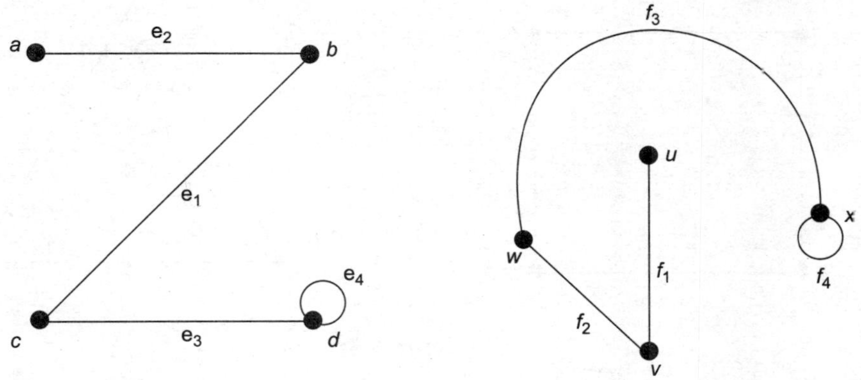

Figure 9.14

It is obvious that $h : \{a, b, c, d, e_1, e_2, e_3, e_4\} \longrightarrow \{u, v, w, x, f_1, f_2, f_3, f_4\}$ defined as $h(a) = u$, $h(b) = v$, $h(c) = w$, $h(d) = x$, $h(e_i) = h(f_i)$ for $i = 1, 2, 3, 4$ is an isomorphism and hence the two graphs are isomorphic (Fig. 9.14). ■

Example 9.12

Examine whether the following graphs shown in Figure 9.15 are isomorphic:

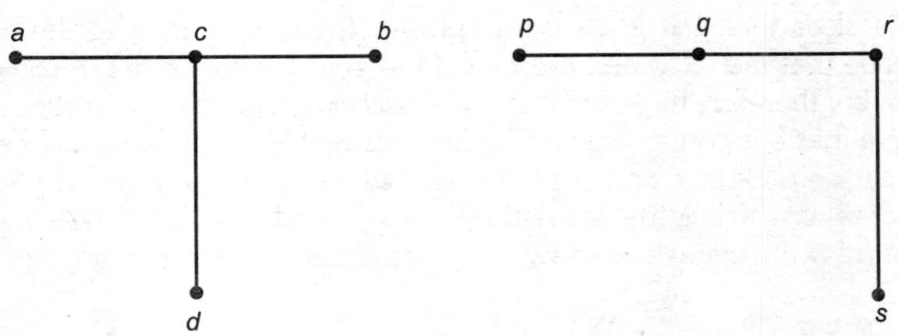

Figure 9.15

SOLUTION

We find that the degree of c is three and no vertex in the second graph has degree three. Hence, by Theorem 9.2, these graphs cannot be isomorphic. ■

Example 9.13

Examine whether the following graphs shown in Figure 9.16 are isomorphic and if so find an isomorphism.

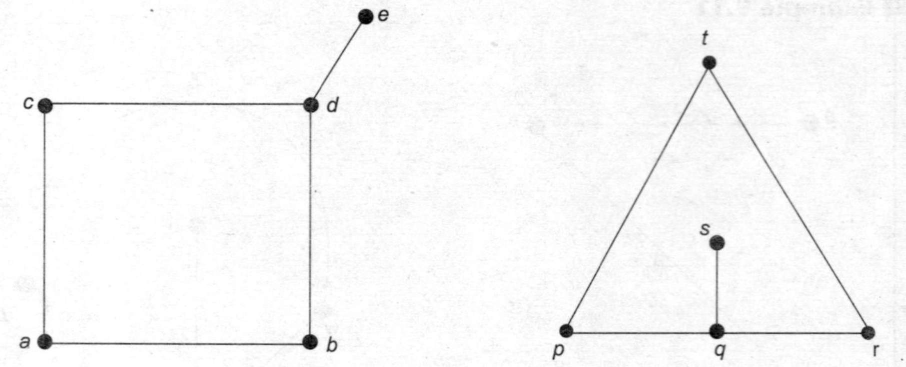

Figure 9.16

SOLUTION

We prepare a table of degrees of vertices for both the graphs.

Vertex	Degree	Vertex	Degree
a	2	p	2
b	2	q	3
c	2	r	2
d	3	s	1
e	1	t	2

It is clear that if h is an isomorphism, then by looking at the degrees, we conclude that $h(d) = q$ and $h(e) = s$. How shall we define $h(b)$? The degree of b is two, but there are three vertices in the second graph whose degrees are two. We observe that b is connected to d by an edge and hence $h(b)$ should be connected to $h(d) = q$. Both r and p are connected to q. We choose $h(b) = r$. Similar considerations tell us that we define $h(c) = p$ and $h(a) = t$. Now, one can easily define h for the edges and show that h is indeed an isomorphism. ■

PRACTICE 9.7

By modifying h in Example 9.13 as $h(b) = p$ and $h(c) = r$, show that different isomorphisms are possible between the same graphs.

Example 9.14

Show that any two complete graphs with the same number of vertices are isomorphic.

SOLUTION

Let $G_1 = (U_1, E_1)$ and let $G_2 = (U_2, E_2)$ be the complete graphs. We take $U_1 = \{u_1, u_2, \ldots, u_n\}$ and $U_2 = \{v_1, v_2, \ldots, v_n\}$. Define $F : U_1 \cup E_1 \longrightarrow U_2 \cup E_2$

as follows: $F(u_i) = v_i$ and the edge connecting u_i and u_j is mapped to the edge connecting v_i and v_j. Since there is a unique edge connecting any two vertices of each of the graphs, F is an isomorphism. ■

PRACTICE 9.8

Show that a graph with 10 vertices and 20 edges cannot be a complete graph.

9.1.1 Subgraphs

Example 9.15

Consider the graphs shown in Figure 9.17.

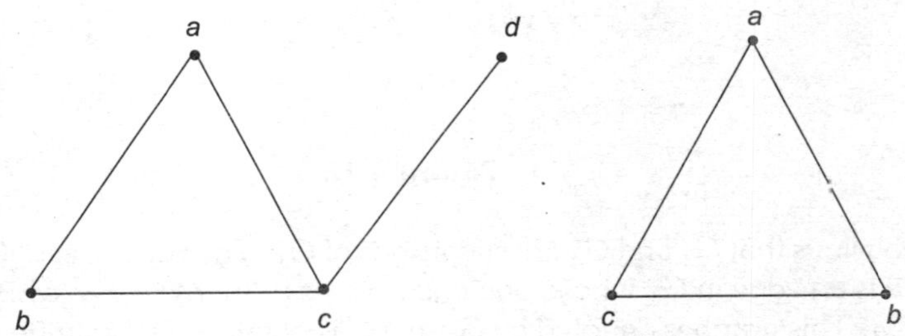

Figure 9.17

It is clear that the second graph is a part of the first graph. We say that the second graph is a subgraph of the first.

If G_2 is a subgraph of G_1, then the vertex set of G_2 is a subset of the vertex set of G_1 and the edge set of G_2 is a subset of the edge set of G_1.

Further, if $u \in V_2$ and $v \in V_2$ are the end vertices of an edge $e \in E_2$, they themselves should be the end points of e in G_1 also. Let us now define the idea of a subgraph. ■

Definition 9.9: Let $G_1 = (V_1, E_1, \phi)$ and let $G_2 = (V_2, E_2, \varphi)$ be graphs. We say that G_2 is a subgraph of G_1 if $V_2 \subseteq V_1$, $E_2 \subseteq E_1$ and if for $e \in E_2$, $\phi(e) = \{u, v\} \rightarrow \varphi(e) = \{u, v\}$. Any graph H that is isomorphic to G_2 is also called a subgraph of G_1.

Now, imagine a computer network in which computers are connected by cables. We treat the computers as vertices and the cables as the edges of a graph. Now, if a cable becomes faulty, connectivity is affected but the computers connected to the cable can still function. But if a computer itself stops functioning, then any cable connected to it is useless. We may obtain a subgraph of a graph by omitting some vertices and some edges, in a manner similar to the one described above.

If e is an edge connecting u and v in G_1, and $u, v \in V_2$, we do not demand that $e \in E_2$. That is, we may remove an edge without removing the ends of the edge. But when a vertex is removed, all the edges incident with it are removed.

Example 9.16

Consider the graphs shown in Figure 9.18.

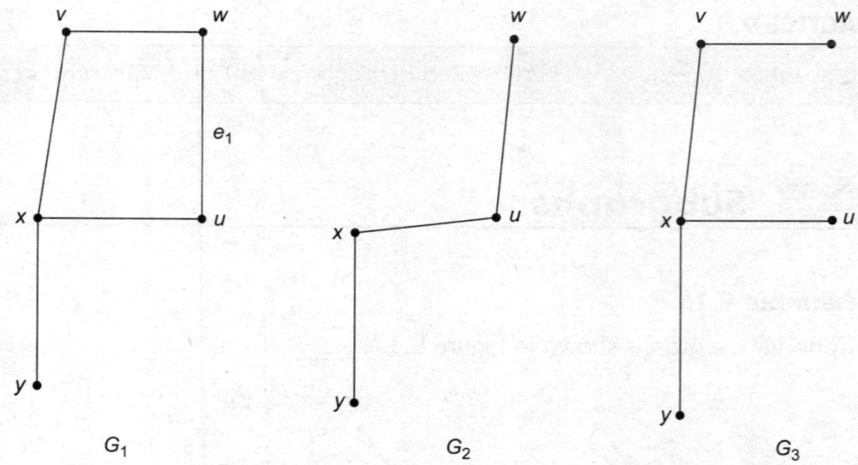

Figure 9.18

It is obvious that G_2 and G_3 are subgraphs of G_1. The vertex set of G_2 is $V_1-\{v\}$. And if e is an edge in G_1 whose one end point is u, then $e \notin E_2$, which is the edge set of G_2. This graph is denoted by $G-v$. G_3 is obtained by omitting edge e_1 from G. Note that we have not omitted the vertices incident with e_1. G_3 is denoted by $G-e$. ■

Definition 9.10: Let $G(V, E, \varphi)$ be a graph and $U \subseteq V$ and $F \subseteq E$. Then,

(i) the graph $G-U = (V_1, E_1, \varphi)$ where $V_1 = V - U$, $E_1 = \{e \in E : V \cap \varphi(e) = \phi\}$ and $\varphi_1 : E_1 \longrightarrow \{\{u, v\} : u, v \in V_1\}$ is defined as $\varphi_1(e) = \phi(e)$.

(ii) the graph $G - F = (V, E - F, \varphi)$, $\varphi : E - F \longrightarrow \{\{u, v\} : u, v \in V\}$ is defined as $\varphi(e) = \phi(e)$.

(iii) $G - \{u\}$ and $G - \{e\}$ are simply denoted by $G - u$ and $G - e$, respectively.

9.2 PATHS

Example 9.17

Consider the graph shown in Figure 9.19.

Let us imagine that the vertices are towns and the edges are roads connecting them. It is clear that we can travel from A to B by road but there is no such roadway between D and H. To reach B from A, we take the following route:

(i) A to E via e_3.
(ii) E to D via e_4.
(iii) D to C via e_5.
(iv) C to B via e_6.

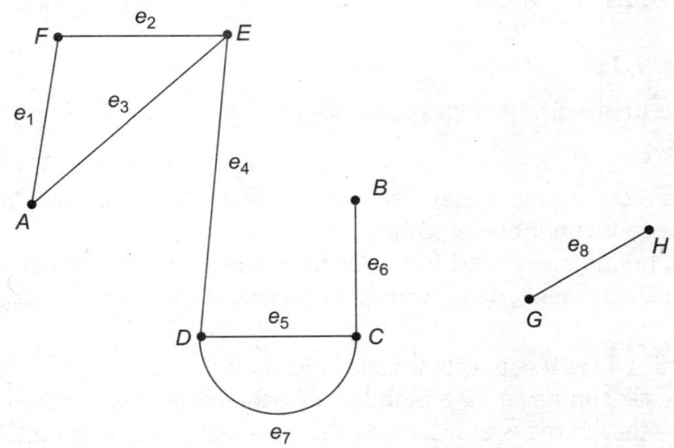

Figure 9.19

Naturally, we may represent the above route as $Ae_3Ee_4De_5Ce_6B$. Further, $Ae_3Ee_4De_7Ce_6B$ is also a route between A and B. $Ae_1Fe_2Ee_4De_5Ce_6B$ is also a route, perhaps roundabout, between A and B. Let us also note that Ee_6C is not a meaningful route at all, for E is not an end of e_6. We formalize the idea of a route, called a *walk*, in the following definition. ■

Definition 9.11: Let G be a graph and u, v be two vertices. A $u - v$ walk in G is defined as a finite sequence

$$W = u_0e_1u_1e_2...u_{k-1}e_ku_k$$

where $u_0 = u$, $u_k = v$ and e_i is an edge with ends u_{i-1} and u_i. The number of edges in W, namely k is called the *length* of the walk.

PRACTICE 9.9

Find the length of the walks in Example 9.17.

Definition 9.12: For any vertex v of G, $W = v$ is called a *trivial walk*.
It is obvious that the length of a trivial walk is 0.

Example 9.18

In Example 9.17, we mentioned that $Ae_1Fe_2Ee_4De_5Ce_6B$ is a roundabout route between A and B. But still one may travel by that route perhaps to meet someone living in F. Let us now consider another roundabout route between A and B in the same example, namely $Ae_3Ee_4De_5Ce_7De_5Ce_6B$. In this route the towns C and D are visited twice. This is an example of a walk, which is not a path. In a path, defined below, we demand that a vertex appears only once. ■

Definition 9.13: A walk is called a *path* if its vertices are distinct.

Example 9.19

Show that in a path the edges are also distinct.

SOLUTION

If an edge e occurs more than once in a walk, then its ends also occur more than once. Hence it cannot be a path.

During holidays, we start from our hometown, visit many places and return to our hometown. This type of a route gives rise to the idea of a closed walk. ■

Definition 9.14: A $u-v$ walk is called closed if $u = v$.

A closed walk can never be a path because the initial and the final vertices are the same. If the other vertices are distinct, then we call it a closed path. But we should keep in mind that a closed path is not a path.

Definition 9.15: A closed path is a non-trivial walk

$$W = v_0 e_1 v_1 e_2 \ldots v_{n-1} e_n v_0$$

in which v_0, v_1,...,v_n are all distinct vertices and e_1, e_2,...,e_n are all distinct edges.

Example 9.20

Consider the graph shown in Figure 9.20.

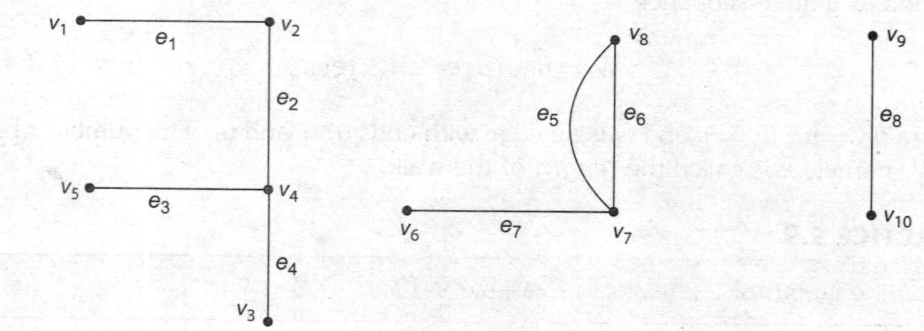

Figure 9.20

$v_5e_3v_4e_4v_3$ is a $v_5 - v_3$ walk of length 2. $v_1e_1v_2e_2v_4e_3v_5$ is a $v_1 - v_5$ walk of length 3. Both these walks are paths. $v_6e_7v_7e_6v_8e_5v_7$ is a $v_6 - v_7$ walk, which is not a path.

Now, it is clear that there cannot be any walk between the following pairs: $v_1 - v_8$, $v_2 - v_9$ and $v_7 - v_{10}$. Treating the vertices as towns and the edges as roads, we find that from v_4 we can travel to v_1, v_2, v_5, v_3. We say that v_4 is *connected* to these vertices. We find that this graph can be divided into three graphs such that in each of these graphs every vertex is connected to every other vertex. ■

Definition 9.16: A vertex u is connected to the vertex v, if there is a $u - v$ walk in the graph. A graph is said to be connected if for any distinct pair of vertices u, v there is a $u - v$ walk.

PRACTICE 9.10

Let G be a graph. Define a relation R on the set of vertices as uRv if there exists a $u - v$ walk in the graph. Show that R is an equivalence relation.

Definition 9.17: Let G be a graph and let v be a vertex. Let $C(v)$ be the set of all vertices connected to v. The subgraph induced by $C(v)$ is called *the connected component* of G containing v.

$C(v)$ is the equivalence class containing v under the equivalence relation defined in Practice 9.10.

Example 9.21

Consider Example 9.20. We have $C(v_1)=\{v_1, v_2, v_3, v_4, v_5\}$, $C(v_6)=\{v_6, v_7, v_8\}$ and $C(v_9) = \{v_9, v_{10}\}$. ■

PRACTICE 9.11

Show that if u and v are in the same connected component of a subgraph of G, then they are in the same connected component of G. Show by an example that the converse is not true.

PRACTICE 9.12

Show that a complete graph is connected but not vice versa.

Example 9.22

Show that a simple graph with four vertices cannot have seven edges.

SOLUTION

A simple graph has only one edge connecting each pair of vertices. Hence, for a graph with four vertices, the maximum number of edges is $\frac{4\times3}{2} = 6$.

Now let $ue_1Ae_2Be_3Ce_4Ae_5De_6v$ be a walk connecting u and v in the graph shown in Figure 9.21.

In this walk, A occurs twice. We can avoid this by taking the route $ue_1Ae_5De_6v$. This idea is used in the solution of Example 9.23.

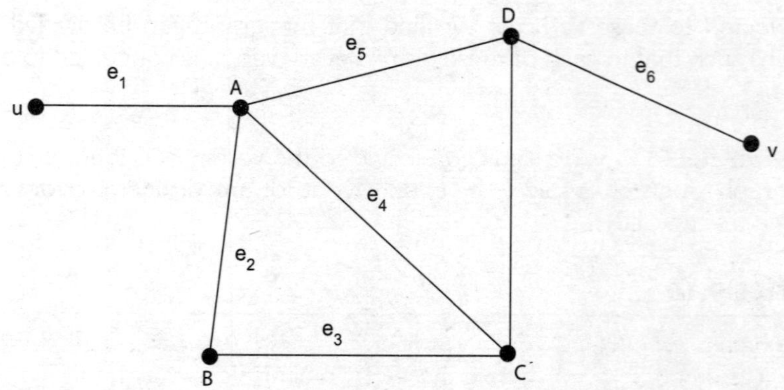

Figure 9.21

Example 9.23

Show that a graph is connected if for any two distinct vertices u and v, there is a path connecting them.

SOLUTION

Let us denote $u = u_0$ and $v = u_n$. Let $u_0e_1u_1e_2....e_nu_n$ be a walk. If w occurs twice as say u_i and u_j, then the walk looks like $u_0e_1u_1e_2...e_iw....e_jw...e_nu_n$. Now removing the segment $e_{i+1}u_{i+1}....e_j$ from the walk, we still have a walk connecting u and v but in which the number of occurrences of w has been reduced by 1. We continue this process till we get a walk containing every vertex only once. ■

Definition 9.18: The length of a walk is defined as the number edges in it where counting is done with repetitions.

PRACTICE 9.13

If a walk covers m vertices (counting with repetitions) show that its length is $m-1$.

9.3 MATRIX REPRESENTATION OF GRAPHS

Graphs can be represented and effectively stored in a computer using matrices. In the following, we discuss two such types of representations, namely, adjacency matrix and incidence matrix.

Definition 9.19: The adjacency matrix of a graph with n vertices $v_1, v_2,...,v_n$ is an $n \times n$ matrix such that $a_{i,j}$ is the number of edges joining vertices v_i and v_j. Now, $a_{i,i}$ is the number of loops the vertex u_i has. If u_i has no loop, then $a_{i,i} = 0$.

We illustrate Definition 9.19 in the following example:

Example 9.24

Consider the following matrices:

$$\begin{pmatrix} 0 & 1 & 0 \\ 1 & 0 & 1 \\ 0 & 1 & 0 \end{pmatrix} \quad \begin{pmatrix} 1 & 1 & 0 & 1 \\ 1 & 0 & 1 & 0 \\ 0 & 1 & 0 & 1 \\ 1 & 0 & 1 & 2 \end{pmatrix} \quad \begin{pmatrix} 1 & 3 & 0 \\ 3 & 0 & 2 \\ 0 & 2 & 0 \end{pmatrix} \quad \begin{pmatrix} 1 & 0 & 0 \\ 0 & 2 & 0 \\ 0 & 0 & 0 \end{pmatrix}$$

The corresponding graphs are shown in Figure 9.22.

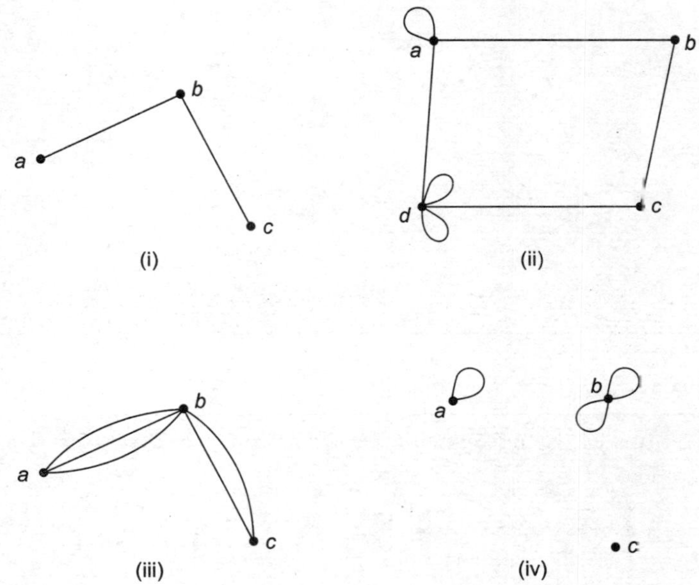

Figure 9.22

The following proposition easily follows from the definition. ■

Proposition 9.1

A graph contains no loops if and only if all the diagonal entries are zero. A graph contains no parallel edges if and only if each entry is either 0 or 1 in the adjacency matrix. ■

PRACTICE 9.14

Show that a graph is totally disconnected if and only if its adjacency matrix is a zero matrix.

PRACTICE 9.15

Write down the adjacency matrix for the graphs shown in Figure 9.23.

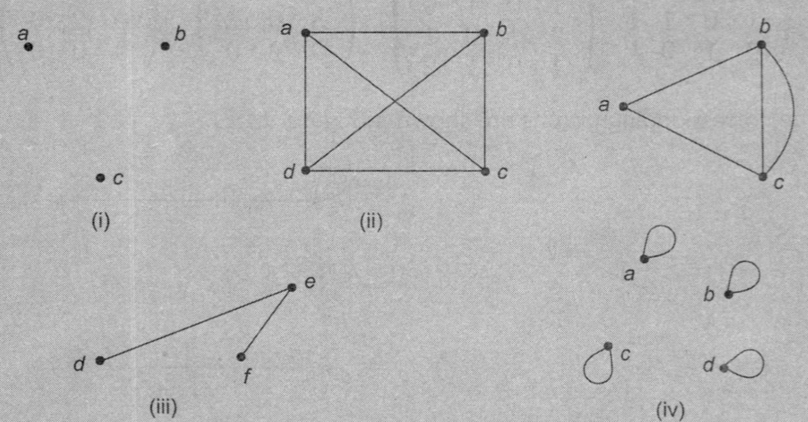

Figure 9.23

PRACTICE 9.16

Draw the graphs whose adjacency matrices are given below:

$$(3)\quad \begin{pmatrix} 0 & 2 & 1 & 0 & 0 \\ 2 & 0 & 1 & 0 & 0 \\ 1 & 1 & 0 & 0 & 0 \\ 0 & 0 & 0 & 1 & 2 \\ 0 & 0 & 0 & 2 & 0 \end{pmatrix} \quad \begin{pmatrix} 0 & 1 & 1 \\ 1 & 0 & 1 \\ 1 & 1 & 0 \end{pmatrix} \quad \begin{pmatrix} 1 & 0 & 0 \\ 0 & 2 & 0 \\ 0 & 0 & 0 \end{pmatrix}$$

Example 9.25

Identify the connected components of the graph whose adjacency matrix is given by

$$\begin{pmatrix} 0 & 1 & 2 & 0 & 0 & 0 \\ 1 & 0 & 3 & 0 & 0 & 0 \\ 2 & 1 & 1 & 0 & 0 & 0 \\ 0 & 0 & 0 & 1 & 2 & 1 \\ 0 & 0 & 0 & 2 & 0 & 1 \\ 0 & 0 & 0 & 1 & 1 & 2 \end{pmatrix}$$

without drawing the graph.

SOLUTION

If we denote the vertices by $u_1, u_2, \ldots, u_6$, from the matrix it is clear the connected components are $\{u_1, u_2, u_3\}$ and $\{u_4, u_5, u_6\}$. ■

Definition 9.20: The incidence matrix of a graph with n vertices and m edges is a matrix such that $a_{i,j} = 0$ if the vertex v_i is not incident on the edge e_j and $a_{i,j} = 1$ if the vertex v_i is incident on e_j.

Example 9.26

The incidence matrix of the graph below is shown in Figure 9.24.

$$\begin{matrix} & a & b & c & d \\ e_1 & 1 & 0 & 1 & 0 \\ e_2 & 1 & 0 & 1 & 0 \\ e_3 & 1 & 0 & 1 & 0 \end{matrix}$$

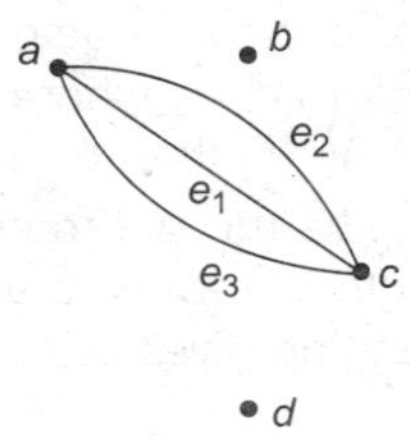

Figure 9.24

PRACTICE 9.17

Give a suitable definition of adjacency and incidence matrices when the graph is a directed graph.

PRACTICE 9.18

Without drawing the graph whose adjacency matrix is given below

$$\begin{pmatrix} 0 & 0 & 0 & 1 & 2 & 1 \\ 0 & 0 & 0 & 1 & 1 & 1 \\ 0 & 0 & 0 & 2 & 1 & 1 \\ 1 & 2 & 1 & 0 & 0 & 0 \\ 2 & 1 & 1 & 0 & 0 & 0 \\ 1 & 1 & 1 & 0 & 0 & 0 \end{pmatrix}$$

show that the graph is bipartite.

Theorem 9.3 is an interesting result, which is an application of the matrix representation of graphs.

If $a_{i,k}a_{k,j} \neq 0$ implies that there is an edge connecting v_i with v_k and further if there is an edge connecting v_k and v_j, then there is a path of length 2 connecting v_i and v_j. The summation $\sum_{k=1}^{n} a_{i,k}a_{k,j}$ contains as many ones as the number of walks of length 2. This is the basic idea behind the proof of Theorem 9.3.

Let us emphasize that an edge may appear more than once in a walk. For example, if e_1 is the edge connecting u and v, then ue_1v is a $u - v$ walk of length 1. ue_1ve_1u is a $u - u$ walk of length 2, etc.

Theorem 9.3

Let G be a graph with n vertices v_1, ..., v_n and let A be the adjacency matrix of G with respect to this ordering of the vertices. Let k be any positive integer. The (i, j)th entry of A^k is the number of different $v_i - v_j$ walks in G of length k.

Proof

We prove this by mathematical induction on k. For $k = 1$, this result is obvious from the definition of adjacency matrix since an edge gives rise to a walk of length 1.

Assume that the result is true for A^{k-1}, where k is some integer greater than 1. Thus, if $b_{i,j}$ is the (i, j)th entry of A^{k-1}, then $b_{i,j}$ is the number of different walks of length $k - 1$ from v_i to v_j. Let $c_{i,j}$ be the (i, j)th entry of A^k. We want to show that $c_{i,j}$ is the number of walks of length k from v_i to v_j. Let $a_{i,j}$ be the (i, j)th element of A. The matrix multiplication definition gives us

$$c_{i,j} = \sum_{l=1}^{n} b_{i,l} \times a_{l,j}. \tag{9.1}$$

Now, every $v_i - v_j$ walk of length k consists of a v_i–v_l walk of length $k - 1$, where v_l is adjacent to v_j, followed by an edge $v_l - v_j$. Since there are $b_{i,l}$ such walks of length $k - 1$ and $a_{l,j}$ edges for each vertex v_l, the total number of all v_i–v_j walks is $\sum_{l=1}^{n} b_{i,l} \times a_{t,l}$. This completes the proof. ■

We can use the above result to determine whether or not a graph is connected.

Theorem 9.4

Let G be a graph with n vertices $v_1, ..., v_n$ and let A denote the adjacency matrix of G with respect to this listing of the vertices. Let $B = (b_{i,j})$ be the matrix defined as

$$B = A + A^2 + \cdots + A^{n-1}.$$

Then G is a connected graph if and only if for $i \neq j$ we have $b_{i,j} \neq 0$.

Proof

Let $a_{i,j}(k)$ be the (i, j)th entry of the matrix A^k. Then

$$b_{i,j} = a_{i,j}(1) + a_{i,j}(2) + \cdots + a_{i,j}(n - 1).$$

By the previous theorem $a_{i,j}(k)$ is the number of distinct walks of length k from v_i to v_j. Therefore, we note that

$$\begin{aligned} b_{i,j} &= \text{number of different } v_i\text{–}v_j \text{ walks of length } 1 \\ &= +\text{number of different } v_i\text{–}v_j \text{ walks of length } 2 \\ &\ldots\ldots \\ &+ \text{number of different } v_i\text{–}v_j \text{ walks of length } n-1, \end{aligned}$$

i.e., $b_{i,j}$ is the number of different u_i–u_j walks of length less than n.

Now suppose that G is connected. Then for every pair of distinct indices i, j there is a path from u_i to u_j. Since G has only n vertices this path goes through at most n vertices and so it has length less than n, i.e., there is at least 1 path from u_i to u_j of length less than n. Hence $b_{i,j}$ (0 for each i, j with i) j, as required.

Conversely, suppose that for each distinct pair i, j we have $b_{ij} = 0$. Then, from above, there is at least 1 walk (of length less than n) from u_i to u_j. In particular, u_i is connected to u_j. Thus, G is a connected graph as required, since i and j were an arbitrary pair of distinct vertices.

As an illustration, we use Theorem 1.6 to find whether a particular graph G is connected or not using its adjacency matrix A (G). Suppose that A(G) is the following matrix A.

$$A = \begin{pmatrix} 0 & 1 & 0 \\ 1 & 0 & 0 \\ 0 & 0 & 0 \end{pmatrix}$$

Then $n = 3$

$$A^2 = \begin{pmatrix} 1 & 0 & 0 \\ 0 & 1 & 0 \\ 0 & 0 & 0 \end{pmatrix} \quad A + A^2 = \begin{pmatrix} 1 & 1 & 0 \\ 1 & 1 & 0 \\ 0 & 0 & 0 \end{pmatrix}$$

Since $b_{1,3} = 0$, G is not connected.

9.4 CUT VERTICES AND CONNECTIVITY

In this section, we shall study the effects of deleting a vertex or an edge from a graph.

Example 9.27

In a network of ring topology, the central vertex is usually a hub and the other vertices may be computers (Fig. 9.25).

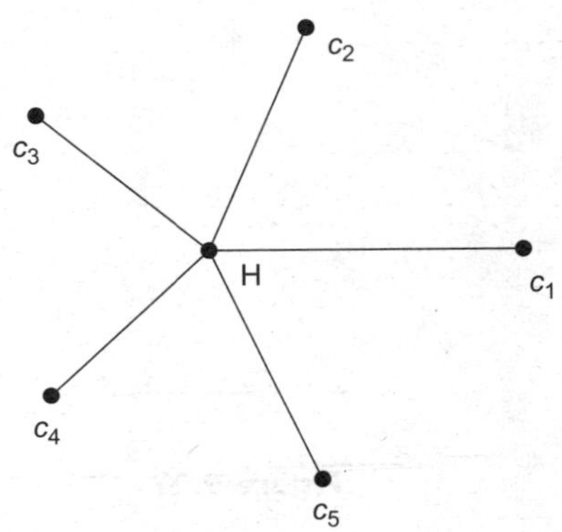

Figure 9.25

If the hub is faulty, then no machine is connected to any other machine. We have the following situation (Fig. 9.26).

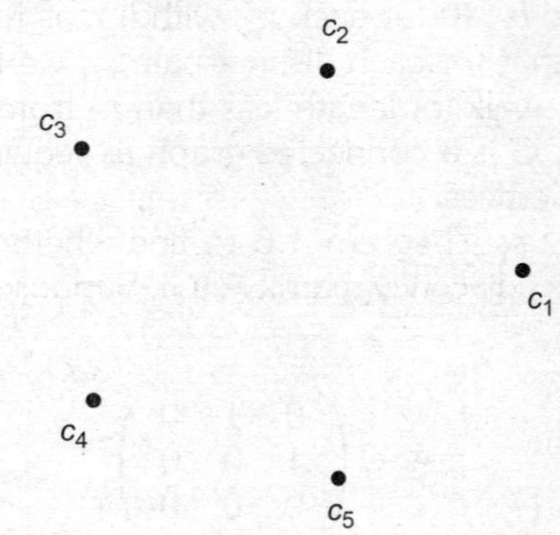

Figure 9.26

As far as connectivity is concerned, the hub is vital. If a particular machine is faulty, it may not affect the connectivity of the other machines. For example, in Figure 9.25 even if machine c_1 is faulty, the connection between the remaining machines is not affected. H is an example of a cut vertex and c_1 is not a cut vertex. ■

Example 9.28

Consider the graph shown in Figure 9.27 and vertices u and v.

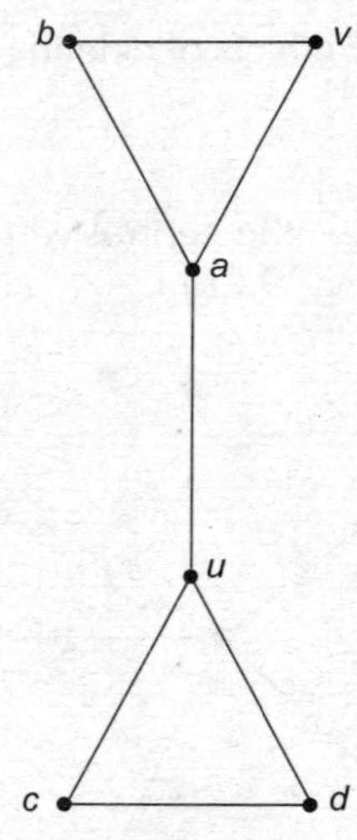

Figure 9.27

Let us examine the effects of removing u or v. The removal of u results in the graph (a) shown below and the removal of v results in graph (b).

We find that $w(G - u) = w(G)$ is connected while $w(G - v) > w(G)$ is not connected. (Fig. 9.28).

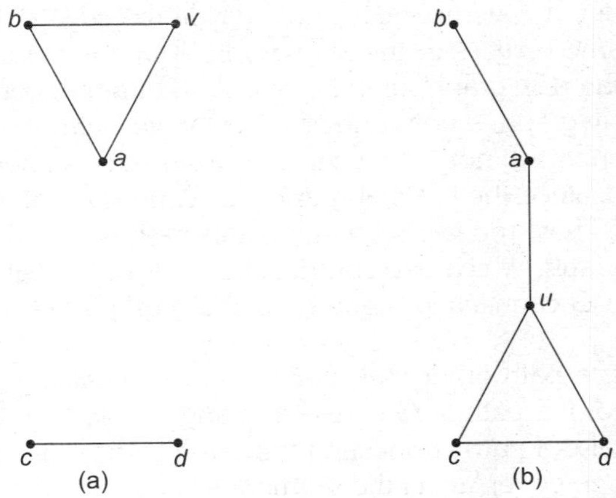

Figure 9.28 ■

This motivates the following definition

Definition 9.21:
A vertex v is called a cut vertex of G if $w(G - v) > w(G)$.

Theorem 9.5

Let v be a vertex of the connected graph G. Then v is a cut vertex of G if and only if there are two vertices u and w of G, $v \neq u$, $w \neq v$ such that every $u - w$ path in G contains v.

Proof
Let v be a cut vertex of G. Then $G - v$ is not a connected graph and hence $G - v$ would have two distinct connected components G_1 and G_2. Let u be a vertex of G_1 and let w be a vertex of G_2. If there is a path in G from u to w without v, then $G - v$ has a path from u to w, which contradicts the fact that u and v are in distinct connected components.

To prove the converse, let u and w be two vertices of G, other than v, such that every path in G from u to w contains v. Then there cannot exist a path from u to w in $G - v$. Thus, $G - v$ is not connected. Therefore, v is a cut vertex.

A vertex of a graph may be incident with any number of edges and hence its removal can badly affect the connectivity of a graph. But an edge is incident with precisely two vertices and hence removal of an edge may not be very bad. The following theorem describes the effect of removing an edge from a graph. ■

Theorem 9.6

When an edge is removed from a graph, the number of connected components of the graph gets decreased by at most 1.

Proof

Let us restate the theorem as $w(G) \leq w(G-e) \leq w(G)+1$. First, we assume that the graph is connected. Let u and v be the ends of the edge getting deleted. If a path in G connecting any two vertices does not include the edge e, then it is also a path connecting the same vertices in the new graph. If uev is the only path connecting u and v, then in the new graph, u and v belong to different connected components and hence the new graph is not connected. Now we claim that there are only two connected components. If not, let x be a vertex, which is neither in $C(u)$ nor in $C(v)$ in the new graph. Since the original graph was connected, there is an $x-u$ path. It must include e. Thus, the last segment of this path is veu and, there is an $x-v$ path in the new graph, which is a contradiction. Thus, we have proved that if uev is the only path in G connecting u and v, then the graph $G-e$ has two connected components.

Now, if there is a path other than uev in G not containing e, we show that the $G-e$ is connected. If a path in G connecting any two vertices does not include the edge e, then it is also a path connecting the same vertices in the new graph. If the path includes e, then by replacing the segment uev by the $u-v$ path not containing e, we get a path connecting the vertices in the new graph. Hence $G-e$ is connected. ■

Example 9.29

Consider the graphs shown in Figure 9.29.

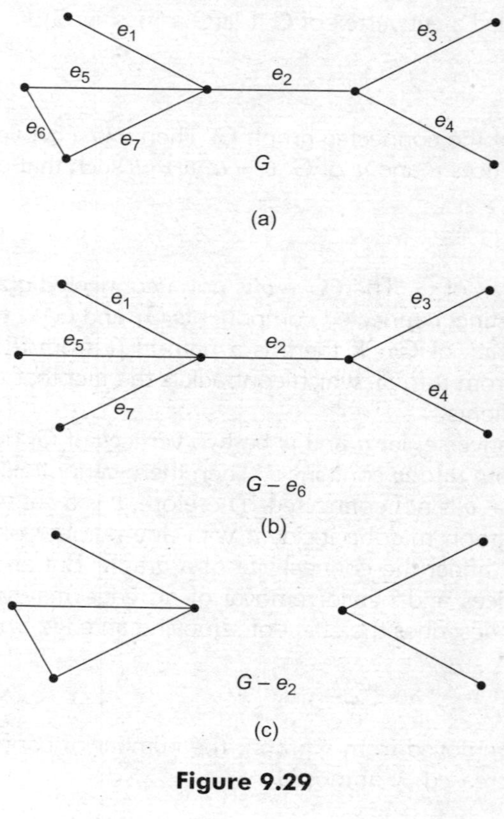

Figure 9.29

Note that $G-e_6$ is connected but $G-e_2$ has two connected components. e_2 is an example of a bridge. Let us note that e_6, which is not a bridge, is part of a closed path in the graph. ■

Definition 9.22: If $w(G-e) = w(G)+1$, then the edge e is called a bridge.

PRACTICE 9.19

Show that e is a bridge if and only if it is a part of a closed path.

PRACTICE 9.20

Show that a complete graph does not contain a cut vertex.

All vertices in a graph cannot be cut vertices. This follows from the next theorem.

Theorem 9.7

Let G be a graph with n vertices, with $n > 2$. Then G has at least two vertices, which are not cut vertices.

Proof

It is enough to show that a connected component with more than two vertices has two vertices that are not cut vertices. Thus, we assume that G is a connected graph.

We prove the result by contradiction. Let us assume that there is at most one vertex in G, which is not a cut vertex. Now let u, v be vertices in G such that the distance $d(u, v) = diam(G)$. Since G is connected and has at least two vertices, $u \neq v$. Because of our assumption, one of these two vertices must be a cut vertex, say v. Then $G - v$ is not connected. So, there must be a vertex w in G, which does not belong to the same component of u in the graph $G - v$. We conclude that every $u - w$ path in G contains the vertex v.

So, we have found a path whose length is longer than $d(u, v)$, which is a contradiction. The proof is complete. ■

9.5 TREES

Look at the graphs in Figure 9.30. They all look like trees. But can we call (b) and (c) trees?

If we think of branches of a tree as the edges of a graph, then the degree of the tip of a branch is one. For (b), the lowest vertex has degree one. The top edges do not have *tips*. In a real tree, two branches after they branch off will not meet again. They also would not give rise to branches that would meet again. This observation is the essence of the following definition of a tree.

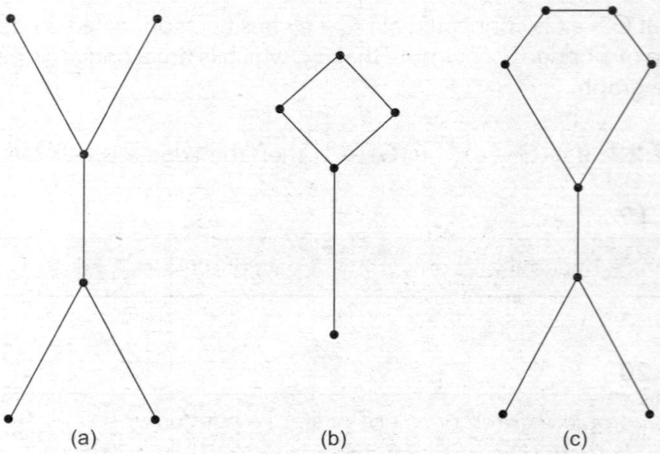

Figure 9.30

Definition 9.23: A graph G is called a tree if it is a connected graph without any closed path.

PRACTICE 9.21

Identify trees among the graphs shown in Figure 9.31.

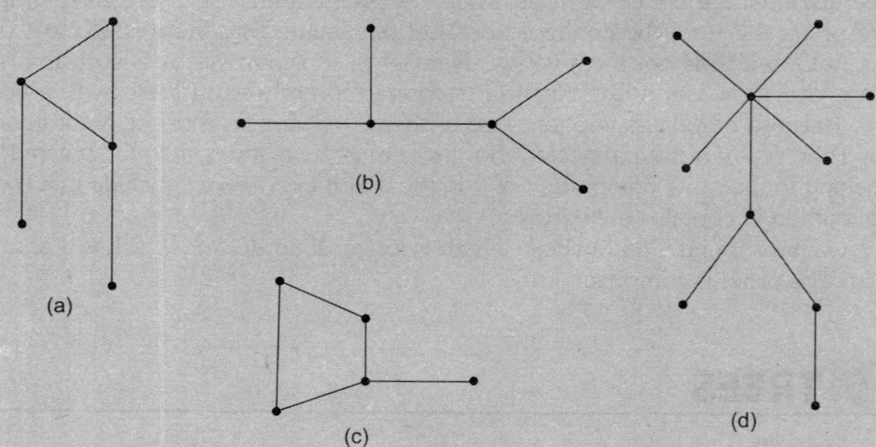

Figure 9.31

Since a tree is a connected graph, there is a path connecting any two vertices. In a general connected graph, there may be more than one path connecting two vertices. But such a thing will not happen in a tree. The following example illustrates this.

Example 9.30

Consider the two different paths from A to D in Figure 9.32.

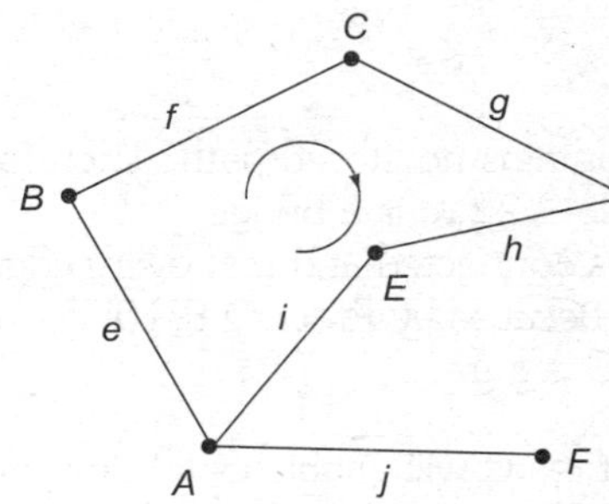

Figure 9.32

$AeBfCgD$ and $AiEhD$. Now we follow the first route from A to D. Next, starting from D, we follow the second route in the reverse direction. That is, we follow the route $DhEiA$. Thus, we have travelled the closed route $AeBfCgDhEiA$. ■

This example tells us that if there are two distinct paths connecting vertices u and v, then we can construct a uv closed path. We use this idea in the proof of the next theorem.

Theorem 9.8

(i) Let u and v be distinct vertices of a tree T. Then there is a unique path from u to v.

(ii) Let G be a graph without any loop. If for every pair of distinct vertices u and v of G there is a unique path from u to v, then G is a tree.

Proof

(i) If the result were to be false, there are two different paths from u to v, say $P_1 = ue_1v_1e_2...e_mv$ and $P_2 = uf_1w_1f_2w_2...f_nv$. Then

$$vf_n...f_2w_1f_1ue_1v_1e_2...e_mv$$

is a uv closed path.

The proof of (ii) is given as Practice 9.22. ■

PRACTICE 9.22

Prove (ii) of Theorem 9.8.

Theorem 9.9

Let G be a connected graph. Then G is a tree if and only if every edge of G is a bridge.

Proof

Let G be a tree. Then G contains no closed path. Therefore, no edge of G belongs to a closed path. By Practice 9.22, e is a bridge.

Next we assume that G is connected and that every edge e of G is a bridge. Then G can have no closed path because by Practice 9.10, an edge belonging to a closed path is not a bridge. Thus G is a tree. ■

The corollary of the next result will enable us to give yet another characterization of trees.

Theorem 9.10

Let G be a graph with n vertices and m edges and let $w(G)$ denote the number of connected components of G. Then $m \geq (n - w(G))$.

Proof

Consider the case $m = 0$. This means that there are no edges. Hence, each vertex is a connected component. Thus, $w(G) = n$ and $m = 0 \geq n - w(G) = n - n = 0$.

We prove the result by induction. Let us assume that the result holds for a graph with m edges. We shall have to prove that the result is true for a graph with $m + 1$ edges. Let G be a graph with n vertices and $m + 1$ edges. Then $G - e$ is a graph with n vertices and $m + 1$ edges. Now $n - w(G - e) \leq m$ by induction hypothesis. Now by Theorem 9.6, $w(G - e) \leq w(G) + 1$. Hence, $n - [w(G) + 1] \leq n - w(G - e) \leq m$, which implies that $n - w(g) \leq m + 1$ and the result is proved.

Example 9.31

Consider the graphs in Figure 9.33.

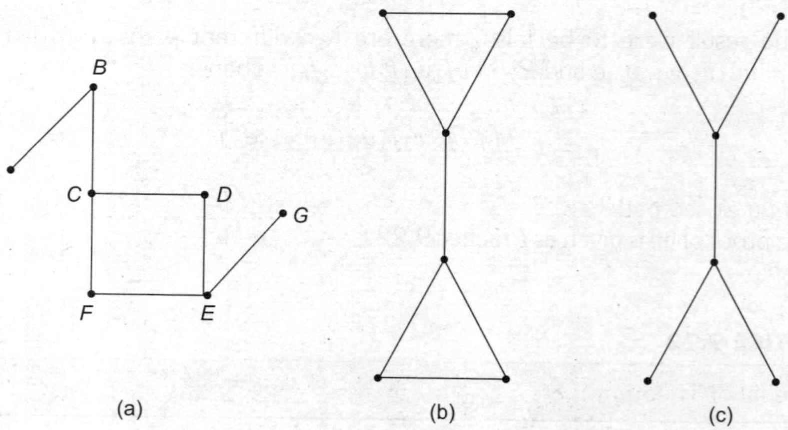

Figure 9.33

(a) and (b) are not trees. (c) is a tree. Further, let us observe that (c) is a subgraph of (b) having the same set of vertices as that of (b). We say that (c) is a spanning tree of (b). ■

Definition 9.24: *A spanning tree* of graph G is a subgraph H of G, which is a tree and which contains all the vertices of G.

PRACTICE 9.23

Find two different spanning trees for (a) of Figure 9.29.

Next, we state a result without proof.

Theorem 9.11

A graph G is connected if and only if it has a spanning tree.

Example 9.32

Show that a tree with at least two vertices is a bipartite graph.

SOLUTION

We first look at an example. The graph shown in Figure 9.34

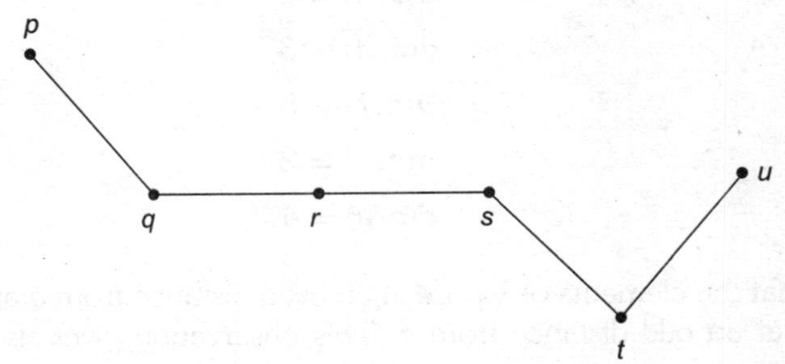

Figure 9.34

is a tree. This picture does not make it obvious that the graph is bipartite. Hence we redraw it as follows. To identify V_1 and V_2, we shall try to arrange the vertices of V_1 horizontally and the vertices of V_2 also in the same manner, but below V_1. We mark p. Let us say that $p \in V_1$. Then as q is connected to p by an edge, $q \in V_2$. Hence q is marked below p. Next, we find $r \in V_1$ and so it is marked next to p above. This process is continued till all the vertices are exhausted. We show the various stages of this process in Figures 9.35(a) – 9.35(d).

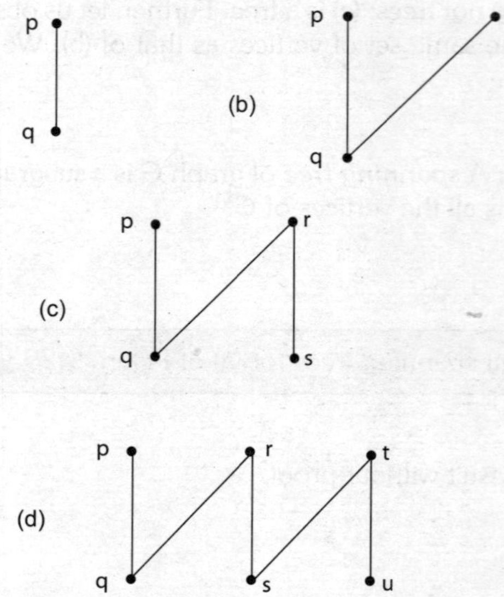

Figure 9.35

Now in a connected tree, there is a unique path connecting two vertices. The number of edges in the path can be thought of as distance between the vertices. Now this distance being a positive integer is either odd or even.

In the above example, we take $V_1 = \{a, c, e\}$ and $V_2 = \{b, d, f\}$. We calculate the distances between a and other elements as

$$d(a, b) = 1$$
$$d(a, d) = 3$$
$$d(a, f) = 5$$
$$d(a, c) = 2$$
$$d(a, e) = 4.$$

We note that the elements of V_1 are at an even distance from a and the elements of V_2 are at an odd distance from a. This observation gives us the idea of the proof.

Let $T(V, E, \phi)$ be a tree. Define a relation on V as follows: $u\ |R|\ v$ if the distance between u and v is even. This relation is well defined because there is a unique path connecting any two distinct vertices in T. Further, $d(u, u) = 0$ for any u in a tree, for there are no loops in a tree. It is easy to show that this relation is an equivalence relation. There are only two equivalence classes. If $[u]$ is the equivalence of u, then $U - [u]$ contains all vertices that are at an odd distance from u and the elements of $U - [u]$ are at even distances from each other. Thus, the two equivalence classes are $[u]$ and $U - [u]$. We take $V_1 = [u]$ and $V_2 = U - [u]$. Now it is obvious that T is bipartite. ■

9.6 EULER PATHS

Example 9.33

An Euler graph is a graph that can be drawn without lifting the pencil from the paper and without tracing an edge more than once, and the vertex we end up with should be the first vertex we started with. Consider the graphs in Figure 9.36.

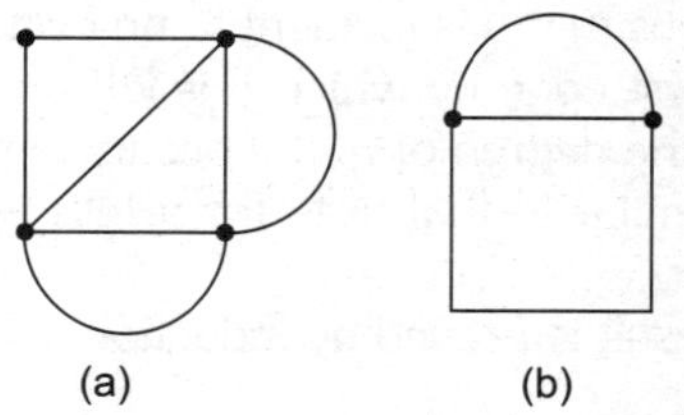

Figure 9.36

The first graph can be easily drawn as described above but the second one cannot be. ■

Definition 9.25: An Euler closed path in a graph is a closed walk containing every *edge* of *G* exactly once.

PRACTICE 9.24

Identify an Euler closed path for graph (a) in Figure 9.36.

If we try to find an Euler circuit for graph (b) in Figure 9.36, we would fail. We can prove this using the famous Theorem 9.12 due to Euler.

Let us think of the edges as roads and the vertices as cities. We are allowed to travel by road only once and we must visit each of the cities at least once. If u is the city from where we start our journey, then finally we have to come back to u. If we had left u by a road, then it cannot be the road by which we come back to it. This tells us that u is concurrent with at least two edges. Let us call these roads e_1 and e_2. If there is a third road e_3 connected to u, then we should have travelled by that road. If we had left u by this third road, then we must have entered u by a road that is different from all the three roads we have mentioned so far. If we had entered u by e_3 we must have left it before coming back to u by e_2. Continuation of this argument tells us that u has an even number of roads connected to it. If some other city v has been visited only once, then there are two roads connected to it, one for entering it and another for leaving it. An argument similar to the previous argument tells us that this city v should also be connected to an even number of roads. This idea is formalized in the proof of Theorem 9.12.

Theorem 9.12

A connected graph has an Euler closed path if and only if each of its vertices has even degree.

Proof

First we assume that G is Euler and let there be a vv Euler closed path for the graph. Let $u \neq v$ be a vertex in G. The vv path must include u. This is because the graph is connected and every edge incident to u should be in the path. Every time u is visited, it is reached and left by a different edge as an Euler closed path should have an edge used only once. This tells us that u is incident to an even number of edges and hence its degree is even. The first edge by which v is left and the last edge by which v is entered contributes 2 to the degree of v. If v occurs in places other than the ends of the path, an argument similar to that used for u tells us that the degree of v should also be even.

The converse of the result is proved by induction and is left as Practice 9.25. ■

PRACTICE 9.25

Show that if the degree of each vertex in a graph is even, then G contains a closed path. Using this result and induction, prove the converse part of Theorem 9.12.

SUMMARY

- A graph is given by a set of vertices and a set of edges where each edge connects a pair of vertices, in which both the vertices may be the same.
- If an edge connects a vertex to itself it is called a loop.
- Two edges connecting the same pair of vertices are called parallel.
- The degree of a vertex is the number of edges incident on it.
- A graph is said to be connected if there is a walk connecting a given pair of distinct vertices.
- $w(G)$ is the number of connected components of a graph.
- A tree is a connected graph without any closed path.
- An Euler graph is a graph such that there is a closed path containing all edges precisely once.
- A graph is an Euler graph if and only if every vertex has an even degree.

PROBLEMS

1. Identify the parallel edges of the graphs shown in Figure 9.37.

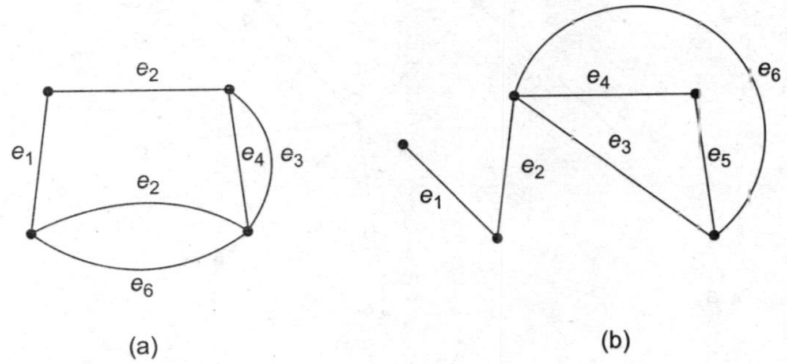

Figure 9.37

2. Identify the loops of the graphs shown in Figure 9.38.

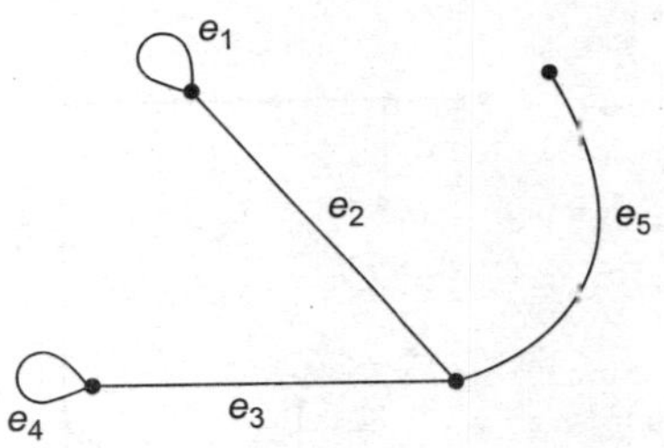

Figure 9.38

3. Identify the bipartite graphs among those shown in Figure 9.39.

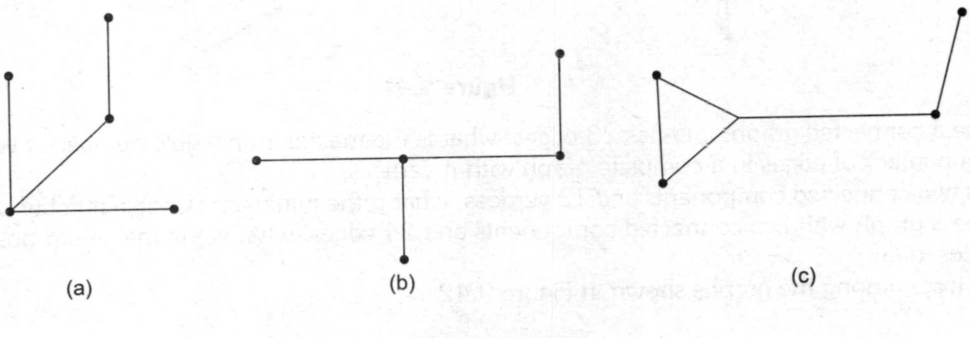

Figure 9.39

4. Find the degrees of all the vertices of the graph shown in Figure 9.40.

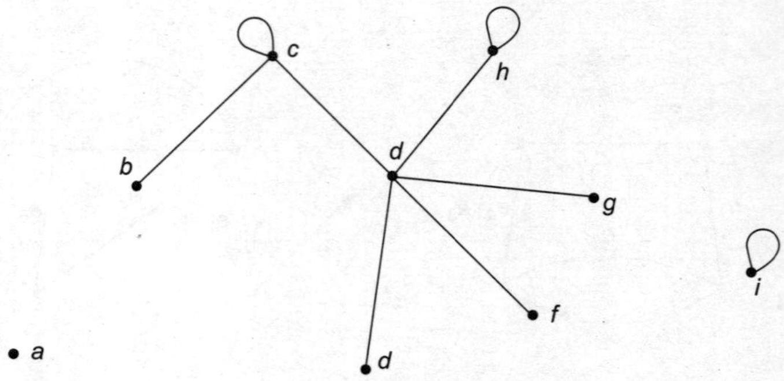

Figure 9.40

5. Identify and group isomorphic graphs in those shown in Figure 9.41.

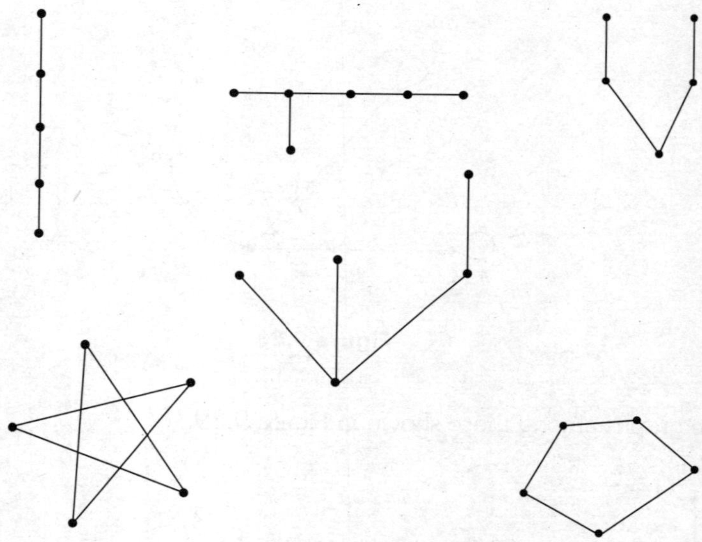

Figure 9.41

6. Let G be a connected graph. If G has 23 edges, what is the maximum possible number of vertices in G?
7. Find the number of edges in a complete graph with n vertices.
8. If G has two connected components and 12 vertices, what is the minimum possible number of edges in G?
9. Let G be a graph with five connected components and 24 edges, what is the maximum possible number of vertices in G?
10. Identify trees among the graphs shown in Figure 9.42.

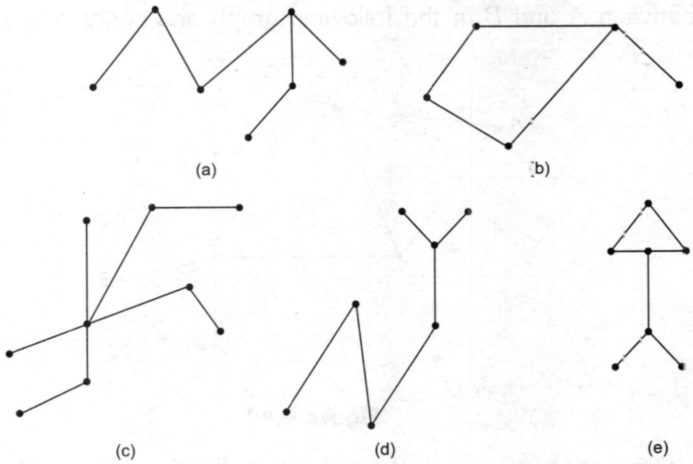

Figure 9.42

11. Describe how you may obtain the matrix representation of the graph $G - U$ where U is a subset of V.
12. Show by an example that a graph can be isomorphic to itself under a function other than the identity function.
13. The matrix representation of a simple graph is such that after deleting the row and column corresponding to a vertex, zero matrix is obtained. What can you say about the vertex?
14. Draw the graphs whose matrix representation is given by

$$\begin{pmatrix} 1 & 0 & 1 \\ 0 & 1 & 1 \\ 1 & 1 & 1 \end{pmatrix} \begin{pmatrix} 1 & 0 & 0 & 1 \\ 0 & 1 & 1 & 1 \\ 0 & 1 & 1 & 0 \\ 1 & 1 & 0 & 1 \end{pmatrix} \begin{pmatrix} 1 & 1 & 1 & 1 \\ 1 & 1 & 1 & 1 \\ 1 & 1 & 1 & 1 \\ 1 & 1 & 1 & 1 \end{pmatrix}.$$

15. Let $M = p + q$ and $N = k + l$. Let A be a $p \times k$ matrix and let B be a $q \times l$ matrix. Let M be the block matrix

$$\begin{pmatrix} A & \mathbf{0}_1 \\ \mathbf{0}_2 & B \end{pmatrix}$$

where $\mathbf{0_1}$ is the zero matrix of order $p \times l$ and $\mathbf{0_2}$ is the zero matrix of order $q \times k$. What can you say about the graph whose adjacency matrix is M?
16. Describe the block structure of the adjacency matrix of a bipartite graph which, is obtained after an appropriate ordering of the vertices.
17. Redraw the graphs shown in Figure 9.43 by clearly showing them as bipartite graphs.

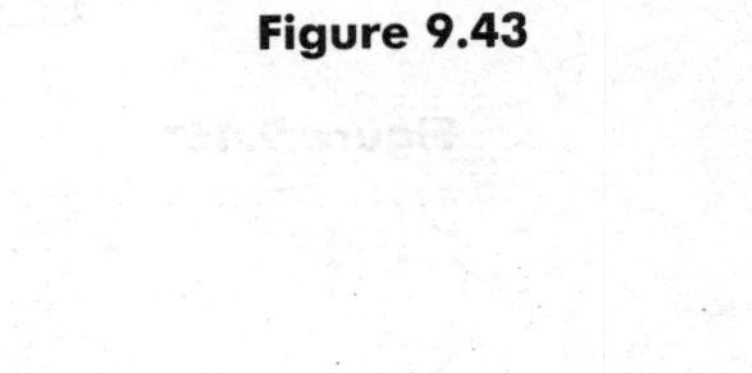

Figure 9.43

18. Find all the walks between A and B in the following graph and verify your result by using the matrix representation.

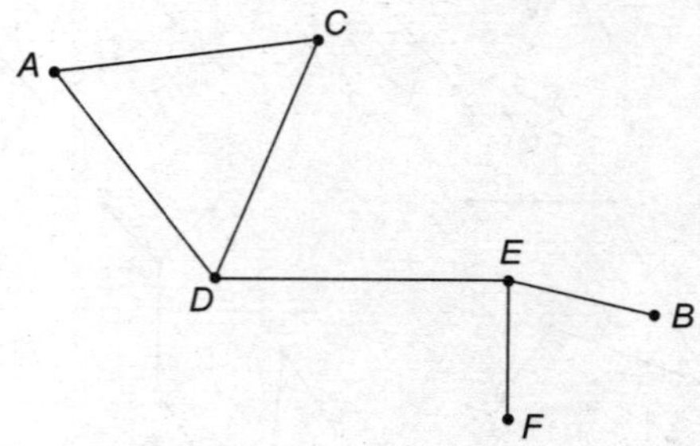

Figure 9.44

19. Show that a graph with n vertices is connected if and only if all the non-diagonal entries of $A+A^2+\ldots A^{n-1}$ are non-zero.
20. Examine which of the following graphs have an Euler Closed Path and find the Euler Paths when they exist.

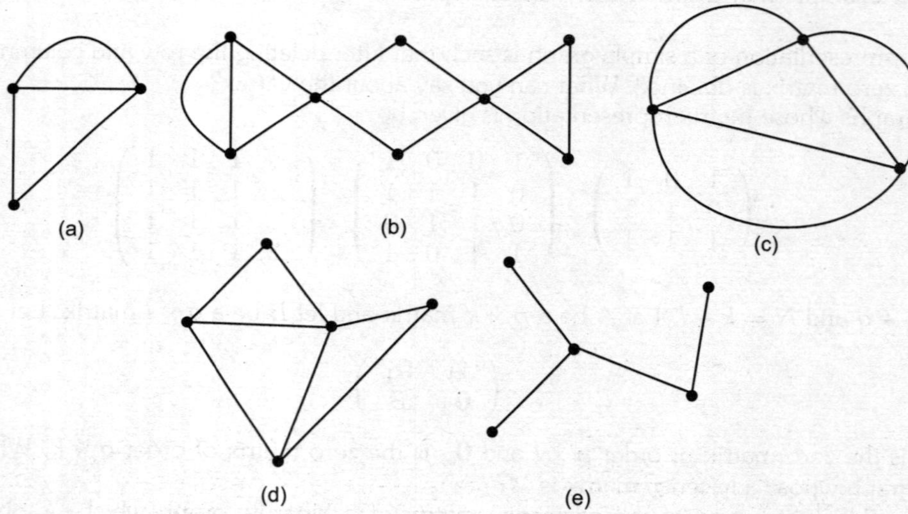

Figure 9.45

21. Give a list of all spanning trees of the graph in Figure 9.45.

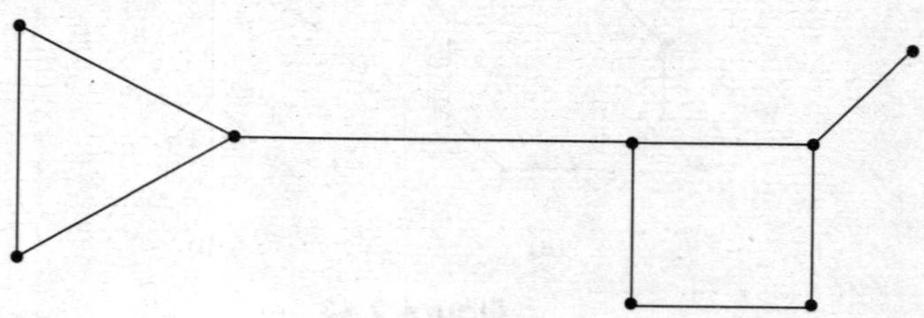

Figure 9.46

FURTHER READING

Clark. J. and Holton. D. A. (1991), *A First Look at Graph Theory*, Allied Publishers Ltd, New Delhi.

CHAPTER

10

Posets, Lattices and Boolean Algebras

Learning Objectives

On completing this chapter, you should be able to:

- state the algebraic definition of a Boolean algebra
- solve problems using the algebraic properties of the elements of a Boolean algebra
- define a poset and find the maximum and minimum elements of subsets of posets when they exist
- find the supremum and infimum of subsets of posets when they exist
- define a lattice and identify lattices among posets
- define and verify distributive property for lattices
- define a sublattice
- state the definitions of a bounded lattice and a complete lattice
- identify bounded and complete lattices
- prove or disprove the existence of a compliment of an element in simple lattices
- state the lattice/order theoretic definition of a Boolean algebra
- prove the equivalence of the two definitions of a Boolean algebra
- identify Boolean algebras
- recognize isomorphic lattices and Boolean algebra
- prove identities in Boolean algebras

10.1 INTRODUCTION AND MOTIVATION

In Chapter 8, we learnt about groups that are algebraic structures equipped with a single binary operation. We also learnt about fields that are structures with two binary operations. Now, let us consider an example of an algebraic structure with two binary operations, but whose properties differ considerably from that of a field or a ring.

Example 10.1

Consider the power set $P(A)$ of a set A. Let $X, Y \in P(A)$. This means that X, Y are subsets of A. Hence, $X \cap Y$ and $X \cup Y$ are also subsets of A. Thus, $X \cap Y$, $X \cup Y \in P(A)$. So, $\cup$ and $\cap$ are binary operations on $P(A)$. Let us recall that the empty set ϕ and the full set A are both elements of $P(A)$. Now, let us list the properties of these binary operations:

(a) $X \cup (Y \cup Z) = (X \cup Y) \cup Z$, $X \cap (Y \cap Z) = (X \cap Y) \cap Z$. Associative laws

(b) $X \cup Y = Y \cup X$, $X \cap Y = Y \cap X$. Commutative laws

(c) $X \cup (Y \cap Z) = (X \cup Y) \cap (X \cup Z)$, $X \cap (Y \cup Z) = (X \cap Y) \cup (X \cap Z)$. Distributive laws

(d) $X \cup \phi = X$, $X \cap A = X$. Identity laws

(e) $X \cup X^c = A$, $X \cap X^c = \phi$. Compliment laws

$(P(A), \cup, \cap)$ is an important example of a Boolean algebra about which we are going to learn in this chapter. Now, we give the algebraic definition of a general Boolean algebra. ■

Boolean Algebra: A Boolean algebra is an ordered triplet $(L, \oplus, \odot)$ where L is a non-empty set, $\oplus$ and $\odot$ are binary operations such that for elements of L

(i) $(x \oplus y) \oplus z = x \oplus (y \oplus z)$, $(x \odot y) \odot z = x \odot (y \odot z)$.

(ii) $x \oplus y = y \oplus x$, $x \odot y = y \odot x$.

(iii) There exist elements $0, 1 \in L$ such that
$x \oplus 0 = x$.
$x \odot 1 = x$.

(iv) $x \odot (y \oplus z) = (x \odot y) \oplus (x \odot z)$, $x \oplus (y \odot z) = (x \oplus y) \odot (x \oplus z)$.

(v) For every element $x \in L$, there is $x^c \in L$ with the following property:
$x \oplus x^c = 1$,
$x \odot x^c = 0$,

where x^c is called the compliment of x.

The following proposition tells us that the Boolean algebra is indeed peculiar. To remember the results easily, think of $\oplus$ as union and $\odot$ as intersection.

Proposition 10.1

In a Boolean algebra, we have

(i) $a \oplus 1 = 1$.

(ii) $a \odot 0 = 0$.

(iii) $a \odot a = a$.
(iv) $a \oplus (a \odot b) = a$.

Proof

(i)
$$\begin{aligned} a \oplus 1 &= (a \oplus 1) \odot 1 \\ &= (a \oplus 1) \odot (a \oplus a^c) \\ &= a \oplus (1 \odot a^c) \\ &= a \oplus a^c \\ &= 1 \end{aligned}$$

(ii)
$$\begin{aligned} a \odot 0 &= (a \odot 0) \oplus 0 \\ &= (a \odot 0) \oplus (a \odot a^c) \\ &= a \odot (0 \oplus a^c) \\ &= a \odot a^c \\ &= 0 \end{aligned}$$

(iii)
$$\begin{aligned} a \odot a &= (a \odot a) \oplus 0 \\ &= (a \odot a) \oplus (a \odot a^c) \\ &= a \odot (a + a^c) \\ &= a \odot 1 \\ &= a \end{aligned}$$

(iv)
$$\begin{aligned} a \oplus (a \odot b) &= (a \odot 1) \oplus (a \odot b) \\ &= a \odot (1 \oplus b) \\ &= a \odot 1 \\ &= a \end{aligned}$$

In Example 10.1, we know that $X \cup A = A$. Since X is a subset of A, while we take the union $X \cup A$, X gets absorbed in A. Since (i) of Proposition 10.1 is similar to $X \cup A = A$, this property is referred to as *absorbtion*.

We have remarked that the idea of a Boolean algebra is a generalization of the power set of a set. We have learnt the set theoretic De Morgan laws in Section 2.1. It is interesting to note that we can prove De Morgan laws in a general Boolean algebra as well. We prove one of the De Morgan laws in the next example, and the other De Morgan law is given as a problem. ■

Example 10.2

Show that in a Boolean algebra,

$$(x \oplus y)^c = x^c \odot y^c$$

SOLUTION

To show that an element b is a compliment of a, we need to prove $a \oplus b = 1$ and $a \odot b = 0$. So, we first consider $(x \oplus y) \oplus (x^c \odot y^c)$.

$$\begin{aligned}
(x \oplus y) \oplus (x^c \odot y^c) &= ((x \oplus y) \oplus x^c) \odot ((x \oplus y) \oplus y^c) \\
&\quad \text{Distributive law} \\
&= (x^c \oplus (x \oplus y)) \odot (x \oplus (y \oplus y^c)) \\
&\quad \text{Commutativity and associativity} \\
&= ((x^c \oplus x) \oplus y) \odot (x \oplus 1) \\
&\quad \text{Associativity and compliment} \\
&= (1 \oplus y) \odot 1 \\
&\quad \text{Compliment and absorbtion} \\
&= 1 \odot 1 \\
&\quad \text{Absorbtion} \\
&= 1
\end{aligned}$$

Next, we consider $(x \oplus y) \odot (x^c \odot y^c)$.

$$\begin{aligned}
(x \oplus y) \odot (x^c \odot y^c) &= (x \odot (x^c \odot y^c)) \oplus (y \odot (x^c \odot y^c)) \\
&\quad \text{Distributive law} \\
&= ((x \odot x^c) \odot y^c)) \oplus ((x^c \odot y^c) \odot y) \\
&\quad \text{Commutativity and associativity} \\
&= (0 \odot y) \oplus (x^c \odot (y^c \odot y)) \\
&\quad \text{Associativity and compliment} \\
&- 0 \oplus (x^c \odot 0) \\
&\quad \text{Compliment and Proposition 10.1} \\
&= 0 \oplus 0 \\
&\quad \text{Absorbtion} \\
&= 0
\end{aligned}$$

■

PRACTICE 10.1

Show that $a \oplus a = a$ and $a \odot (a \oplus b) = a$.

For real numbers, we know that $x(y + z) = xy + xz$. But it is **not** true that $x + yz = (x + y)(x + z)$. In other words, in a field, multiplication distributes over addition, but *addition does not distribute over multiplication*. In contrast, each of ∪ and

$\cap$ distributes over the other in $P(A)$. Further, we have the interesting properties, $X \cap X = X$ and $X \cup X = X$ in $P(A)$. Thus, we may say that the algebraic laws for a Boolean algebra are a bit peculiar. To understand this peculiarity, we have to understand the underlying *order structure* of a Boolean algebra. So, we begin the next section with the definition of a *partial order* and we later define Boolean algebra from the point of view of a partial order. We have already defined the concept of a partial order in Section 2.3 of Chapter 2.

10.2 POSETS

Example 10.3

There is a university with three departments A, B and C. There are two affiliated colleges each having the same departments. There is a vice chancellor for the university, principals for each college and nine heads of the department. Each of these individuals has a PC of their own. The local area network has been designed in the following manner: The vice chancellor can read any file from any of these computers. The principal of a particular college can read any file from any of the three computers of the heads of the same college. The head of a particular department at the university can read the files from the computers of the corresponding two heads of the department in the college. Let us now define a relation on the set of all PCs described above. Let P and Q be two PCs. We say that a $P\ R\ Q$ if the user of Q has access to read the files from P.

Vice chancellor	V
Head of university department A	U_A
Head of university department B	U_B
Head of university department C	U_C
Principal of college one	M_1
Principal of college two	M_2
Head, department A, college one	H_A
Head, department B, college one	H_B
Head, department C, college one	H_C
Head, department A, college two	h_A
Head, department B, college two	h_B
Head, department C, college two	h_C

Figure 10.1 symbolically represents the above configuration.

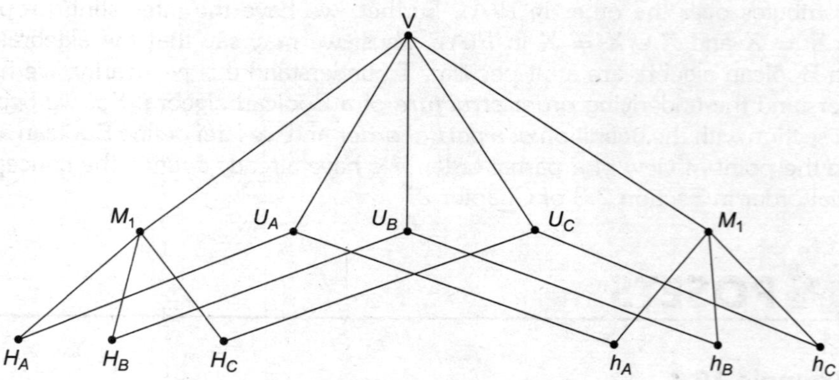

Figure 10.1

We assume that a user of a computer can read files from his own computer. This tells us that R is a reflexive relation. We also find that if the user of a computer P can read the files of the computer Q and if the user of a computer Q can read the files of a computer S , then the user of a computer P can read the files of the computer S. Thus, R is transitive. Further, we note that for no pair of computers P and Q each user can read the other's files. This implies that R is anti-symmetric. Thus, we conclude that R is a partial order. ■

This example motivates the following definition of a poset and its Hasse diagram.

Definition 10.1: Let $\leq$ be a partial order relation on a set S. Then $(S, \leq)$ is called a poset.

Hasse Diagram for a Poset: If $(S \leq)$ is a poset, then its Hasse diagram is drawn as follows:

(1) The elements of S are represented by dots.

(2) If $a \leq b$ and $a \neq b$, then the dot corresponding to a is *below* the dot for b and there is a sequence of lines connecting them.

(3) The dots for a and b are connected by a single line, if $a \leq b$, $a \neq b$ and there is no element c such that $a \leq c \leq b$.

Example 10.4

Show that the set $S = \{1, 2, 3, ..., 10\}$ is a poset with respect to the relation defined as $m\ R\ n$ if n is a multiple of m, and draw its Hasse diagram.

SOLUTION

(i) Every integer is a multiple of itself. Hence $n\ R\ n$ for all $n \in S$. Thus R is reflexive.

(ii) Let $m R n$ and $n R m$. That is, m is a multiple of n and n is a multiple of m. Since both m and n are positive, $m = n$. So, R is anti-symmetric.

(iii) If m is a multiple of l and n is a multiple of m, then n is a multiple of l. So, R is transitive.

The given relation is reflexive, anti-symmetric and transitive So, it is a partial order. Thus (S, R) is a poset. Its Hasse diagram is shown in Figure 10.2.

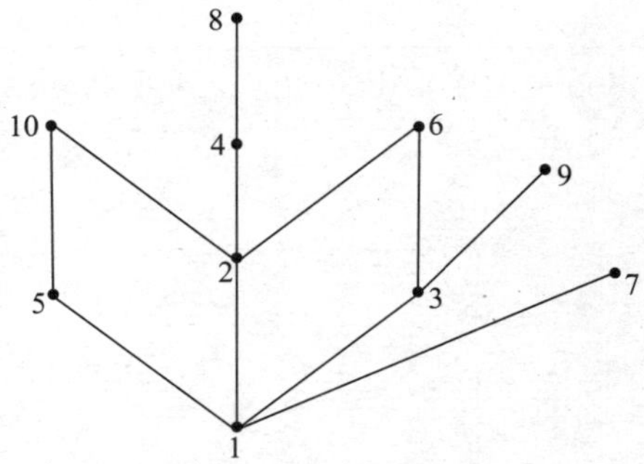

Figure 10.2

Example 10.5

Show that the set of all positive divisors of 42 form a poset under the relation $\leq$ defined as $a \leq b$ if a is a divisor of b. Draw its Hasse diagram.

SOLUTION

Let $S = \{1, 2, 3, 6, 7, 14, 21, 42\}$.

(i) Every element a is a divisor of itself and hence $a R a$. The relation is reflexive.

(ii) If $a R b$ and $b R a$, then we have 'a is a divisor of b' and 'b is a divisor of a'. Since S has only positive elements, $a = b$. Thus, R is anti-symmetric.

(iii) If $a R b$ and $b R c$, then a is a divisor of b and b is a divisor of c. Then, a is a divisor of c, and hence $a R c$.Thus, R is transitive.

As the given relation is reflexive, anti-symmetric and transitive, R is a partial order, and S is a poset with respect to R. Now,

$$\begin{aligned} R = \; & \{(1,1), (1,3), (1,3), (1,6), (1,7), (1,14), (1,21), (1,42) \\ & (2,2), (2,4), (2,6), (2,14), (2,42) \\ & (3,3), (3,6), (3,21), (3,42) \\ & (4,4), (4,42) \\ & (6,6), (6,42) \\ & (7,14), (7,21), (7,42) \\ & (14,14), (14,42) \\ & (21,21), (21,42) \\ & (42,42)\} \end{aligned}$$

Its Hasse diagram is shown in Figure 10.3.

Figure 10.3

PRACTICE 10.2

For the diagram in Figure 10.4, write the partial order.

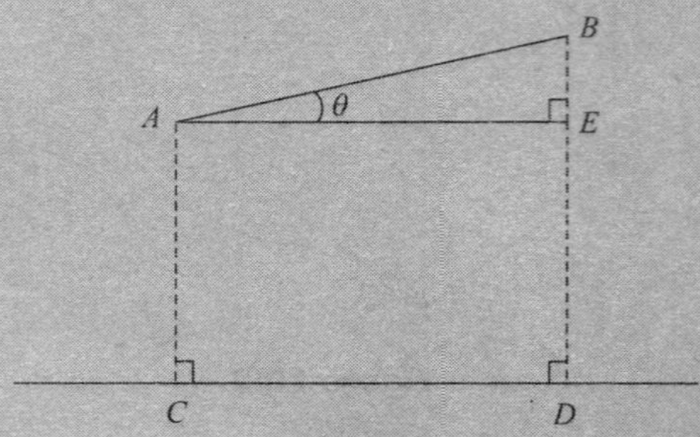

Figure 10.4

Example 10.6

Define $\leq$ on the set of all complex numbers as $a\,R\,b$ if $|a| \leq |b|$. Show that this is **not** a partial order.

SOLUTION

This relation is not anti-symmetric. Consider $a = 1$ and $b = i$. Since $|1| = |i|$, $a\,R\,b$ and $b\,R\,a$. But $a \neq b$. Thus $\leq$ is not a partial order. ■

Example 10.7

Let $S = \{0, 1, 1, +i, -3i, 7\}$. The partial order on S is defined as $a\,R\,b$ if $|a| \leq |b|$. Show that (S, R) is a poset.

SOLUTION

(i) $a\,R\,a$ for all $a \in S$, since $|a| = |a|$ for all $a \in S$ and hence R is reflexive.

(ii) Now if $a\,R\,b$ and $b\,R\,a$, then $|a| = |b|$. Noting that *all the elements of S have distinct modulus*, we find that $|a| = |b| \Rightarrow a = b$. Hence the relation is anti-symmetric.

(iii) $|a| \leq |b|$ and $|b| \leq |c| \Rightarrow |a| \leq |c|$ and, hence the relation is transitive.

The Hasse diagram for this poset is shown in Figure 10.5.

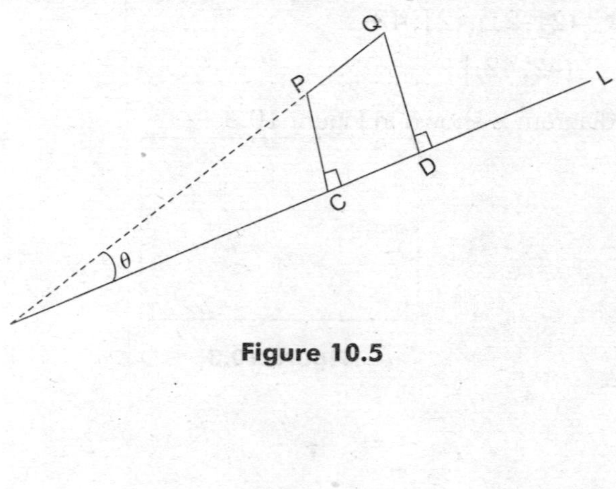

Figure 10.5 ■

PRACTICE 10.3

Draw the Hasse diagram for the following partial order relation on $\{a, b, c, d, e, f\}$ defined as

$$\begin{aligned} R = \{&(a,a), (a,b)(a,c), (a,d), (a,e)(a,f) \\ &(b,b), (b,d)(b,e), (b,f)(d,d)(d,f), (e,e), (e,f), (f,f)\} \end{aligned}$$

Consider Example 10.5 and its Hasse diagram. Consider the set $A = \{1, 3, 7, 21\}$. The elements 3 and 7 are not comparable. That is, neither of the statements $3 \leq 7$ and $7 \leq 3$ are true.

Let us consider a real-life example. It is clear that within a state, the chief minister is more powerful than other ministers in the cabinet. However, the ministers themselves are not more powerful than each other. For example, the health minister is no more powerful than the education minister and vice versa. The health minister and the education minister are like the elements 3 and 7, which cannot be compared with each other. Now, the element 21 is like the chief minister. Precisely, under the partial order of Example 10.7, $a \leq 21$ for all $a \in A$. We say that 21 is the maximum element of A.

Definition 10.2: Let A be a subset of a poset S. An element $b \in A$ is said to be a maximum of A if $a \leq y$ for all $a \in A$.

PRACTICE 10.4

Define the minimum element of A.

Example 10.8

Consider the poset $(\mathbb{R}, \leq)$ where $\leq$ is the usual order relation 'less than or equal to'. Let $A = \mathbb{N}$. Then, $\mathbb{N}$ does not have any maximum.
The set $(0, 3) = \{x \in R : 0 < x < 3\}$ also does not have a maximum.
The set $(0, 3]$ has the maximum element 3 but does not have any minimum element. ■

Example 10.9

Let $S = \mathbb{N}$. Consider the partial order on S defined as $a \leq b$ if b is a divisor of a. For a subset A of this poset, the maximum element of A is an element that is a divisor of all other elements in A.

Let $A = \{2, 3, 6\}$. Its Hasse diagram is shown in Figure 10.6.

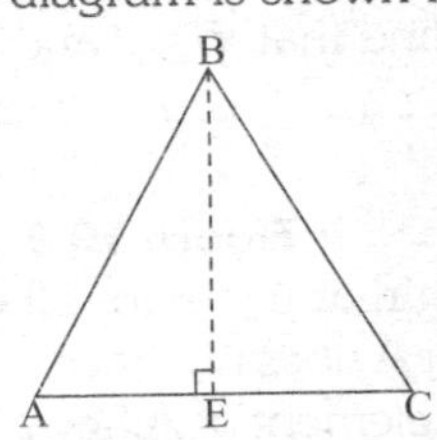

Figure 10.6

Hence, this set does not have any maximum. Its minimum element is 6, which is a multiple of all elements of A. ■

Example 10.10

Consider the poset represented by the Hasse diagram in Figure 10.7.

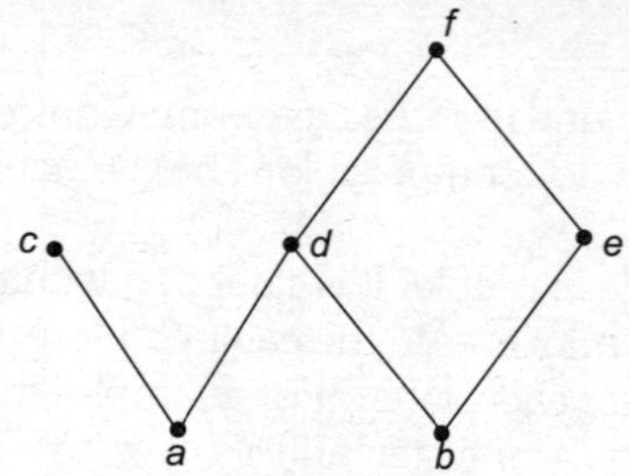

Figure 10.7

Examine which of the following subsets have maximum elements and find them when they exist

(i) $\{a, c, d\}$.
(ii) $\{a, b, d\}$.
(iii) $\{b, d, e\}$.
(iv) $\{f, d, e\}$.
(v) $\{a, b, c, d, e, f\}$.

SOLUTION

From the diagram, it is clear that the set $\{a, c, d\}$ does not have a maximum element. But we shall prove this statement rigorously. Since $a \leq c$, a cannot be the maximum element. c is not maximum because $d \leq c$ is false. Further, d is not maximum because $c \leq d$ is false. So, we conclude that $\{a, c, d\}$ does not have a maximum element.

For the remaining sets, while we give a solution based on the diagram, the readers must supply the proof themselves.

The maximum element of $\{a, b, d\}$ is d. The set $\{b, d, e\}$ does not have a maximum. ■

Example 10.11

Consider R with the usual order $\leq$ as a poset. Let $A = (-infty, 3)$. If $x \in A$, then $x < 3$. Taking $y = \frac{x+3}{2}$, we find that $x \leq y$ and $y \in A$ (Fig. 10.8).

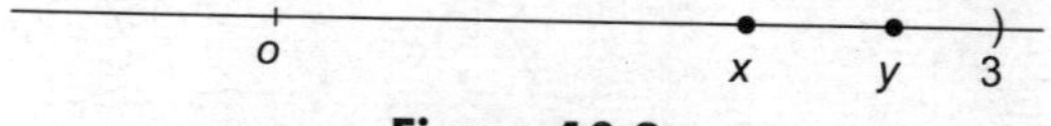

Figure 10.8

So, there is no $x \in A$ such that $y \leq x$ for all $y \in A$.

From this we conclude that A does not have a maximum. Indeed $y \leq 3$ for all $y \in A$, but 3 is not the maximum element of A because $3 \notin A$. Motivated by this example, we make the following definition. ■

Definition 10.3: Let A be a subset of a poset $(S, \leq)$. An element $b \in S$ is said to be an upper bound of A if $a \leq b$ for all $a \in A$.

Consider Example 10.11 again. 3 is an upper bound for A. But we also note that 3.001, 4, 234 are all upper bounds for A. In fact, any $b \geq 3$ is an upper bound for the set A. Among the upper bounds, 3 has a special status in the sense that it is the smallest upper bound. This motivates the following definition.

Definition 10.4: An element y is said to be a least upper bound of A if y is an upper bound and for any other upper bound z, $y \leq z$.

The least upper bound is also known as the supremum.

Example 10.12

Consider the poset $(\mathbb{R}, \leq)$ where $\leq$ is the usual order relation 'less than or equal to'. Let $A = \mathbb{N}$. Then, $\mathbb{N}$ does not have any upper bound. The upper bound for set $(0, 3)$ is any real number greater than or equal to 3. Its least upper bound is 3. ■

Example 10.13

Consider Example 10.10 and Figure 10.7. It may be noted that the set $\{b, d, e\}$ does not have a maximum. But f is an upper bound of this set. Being the only upper bound, it is also the supremum of $\{b, d, e\}$. ■

Example 10.14

In the following poset represented by its Hasse diagram (Fig. 10.9), find all the upper bounds of $\{a, b, c\}$ and its supremum.

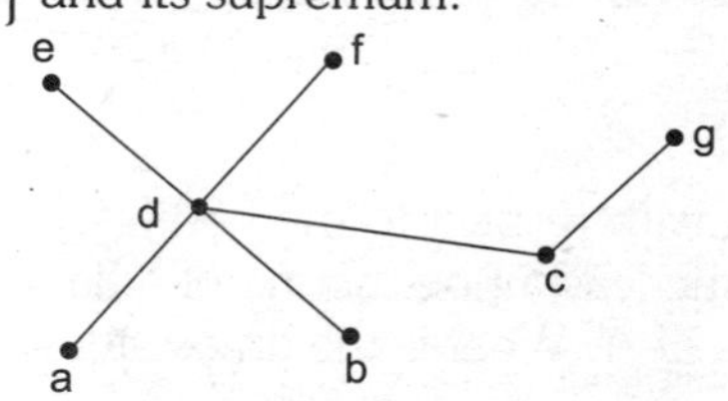

Figure 10.9

SOLUTION

From the Hasse diagram, it is clear that d, e, f are the upper bounds of the given set. Further d is the supremum. ■

Example 10.15

If a is the maximum element of A, show that a is also the supremum of A.

SOLUTION

We have $x \leq a$ for all $x \in A$. So, a is an upper bound of A. Now, let b be any other upper bound of A. We have $x \leq b$ for all $x \in A$. Since $a \in A$, we obtain $a \leq b$. Thus, a is the supremum of A. ■

PRACTICE 10.5

Define lower bound and the greatest lower bound. The greatest lower bound is also known as the infimum.

PRACTICE 10.6

Consider the poset represented by the following Hasse diagram (Fig. 10.10).

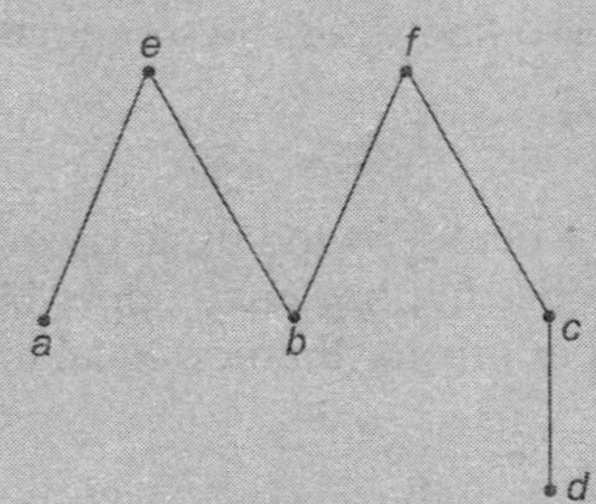

Figure 10.10

Examine which of the following subsets have minimum elements and find them when they exist. Also find the infimums if they exist.

(i) $\{e, a, b\}$.
(ii) $\{e, b, f\}$.
(iii) $\{f, c, d\}$.
(iv) $\{f, d, e\}$.
(v) $\{b, f, c, d\}$.

Example 10.16

Consider $\mathbb{N}$ as a poset with respect to the partial order $\leq$ defined as $a \leq b$ if b is a multiple of a. Then the least upper bound of a finite subset A is nothing but the lcm of all the elements of A. What is the greatest lower bound for A?

PRACTICE 10.7

Consider $\mathbb{N}$ as a poset with respect to the partial order $\leq$ defined as $a \leq b$ if b is a divisor of a. Express the least upper bound and the greatest lower bound of a finite set.

Definition 10.5: If the subset $\{a, b\}$ has a supremum, it is denoted by $a \vee b$ and the infimum of the set is denoted by $a \wedge b$, if it exists.

Example 10.17

Consider the poset $P(S)$ of a set of all subsets of some set S. $A \vee B = A \cup B$ and $A \wedge B = A \cap B$.

Example 10.18

Let S be a poset and $A \subseteq S$ be such that for all $x, y \in A$, either $x \leq y$ or $y \leq x$. That is, every pair of elements of A is comparable. Such a set is called *totally ordered.* Show that a finite totally ordered subset of a poset has a minimum element and a maximum element. ■

Proof

Let $A = \{x_1, x_2, x_3, ..., x_n\}$.

Consider the elements x_i and x_j of A. Since either $x_i \leq x_j$ or $x_j \leq x_i$, $x_i \vee z_j$ exists and is an element of A, for $x_i \vee x_j$ is either x_i or x_j. Now, define $y_1 = x_1$, $y_2 = y_1 \vee x_2$, $y_3 = y_2 \vee x_3$ and so on upto $y_n = y_{n-1} \vee x_n$. It is clear that $y_1 \leq \cdot \cdot \leq y_n$ and hence y_n is the maximum element. ■

PRACTICE 10.8

Complete the solution to Example 10.18 by showing that A has a minimum element.

Example 10.19

Show that if a maximum element of a subset of a exists, then it s unique. ■

Proof

Let A be a subset of a poset and let $x \in A$ be a maximum element of A. If $y \in A$ is also a maximum element, then we have to show that $x = y$. From the definition, it follows that $a \leq x$ for all $a \in A$ and $a \leq y$ for all $a \in A$. In particular, we have $x \leq y$ and $y \leq x$. By anti-symmetry, we get $x = y$. ■

PRACTICE 10.9

In a small tuition center, the managing director m can teach his two employees p and q, and the students x, y, m, n, r. p and q can teach all the students. The students x and y who belong to a higher class than that of m, n and r can teach the students m, n and r. Further, p can teach q and m can teach his classmates. Consider the set $\{m, n, r, x, y, p, q, m\}$. We say that $a \leq b$ if b can teach a. Draw the Hasse diagram for this poset and find the maximum elements of S and $\{m, n, r\}$. Does the set $\{x, y\}$ have a maximum element? What are the upper bounds of $\{x, y\}$? Does it have a least upper bound? Does S have a minimum element?

Example 10.20

Let $S_{30} = \{1, 2, 3, , 5, 6, 15, 30\}$ be the set of all divisors of 30. We define a partial order $\leq$ on S_{30} as follows: $a \leq b$ if b is a divisor of a.

Let us note that, in this poset, $3 \leq 1$. Its Hasse diagram is given in Figure 10.11.

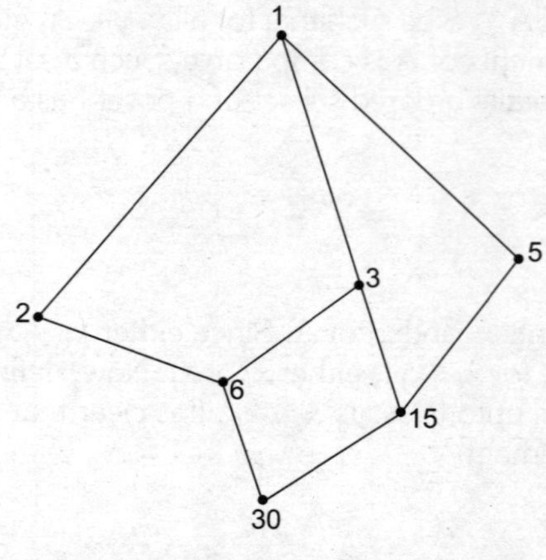

Figure 10.11

The set $\{2, 3, 15\}$ neither has a maximum nor a minimum. Its supremum is 1 and infimum is 30. Further, $3 \wedge 5 = 15$ and $3 \vee 5 = 1$. We also have $2 \wedge 15 = 30$ and $2 \vee 15 = 1$.

For the set $\{1, 2, 3\}$, 6 and 30 are lower bounds and 6 is the greatest lower bound, that is, the infimum (here, $30 \leq 6$). ■

10.3 LATTICES

Definition 10.6: A lattice is a poset L such that for $a, b \in L$, elements sup $\{a, b\} = lub\,\{a, b\} = a \vee b$ and inf $\{a, b\} = glb\,\{a, b\} = a \wedge b$ exist.

Example 10.21

Consider a poset whose Hasse diagram is given in Figure 10.12.

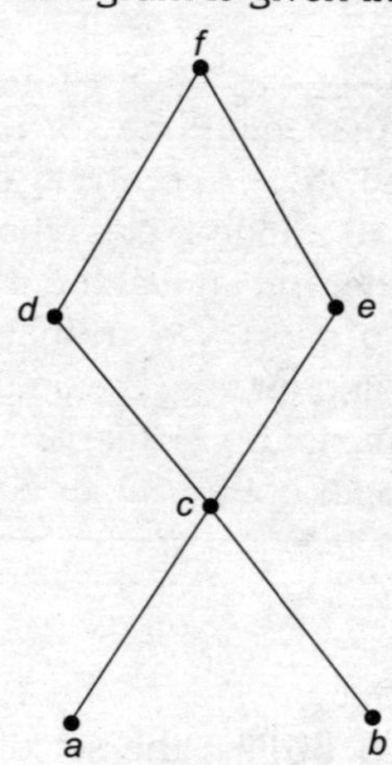

Figure 10.12

For this figure: $a \vee b = c$, $e \vee f = f$, $d \vee e = f$, $d \wedge e = c$, $f \wedge e = e$. $a \wedge b$ does not exist. Hence this is not a lattice. ■

Example 10.22

Let L be the set of all subsets of $\{1, 2, 3\}$ and let $\subseteq$ be the partial order on L. Is this a lattice?

SOLUTION

From Example 10.17, we know that $X \vee Y = X \cup Y$ and $X \wedge Y = X \cap Y$. Hence L is a lattice. The Hasse diagram for this lattice is given in Figure 10.13.

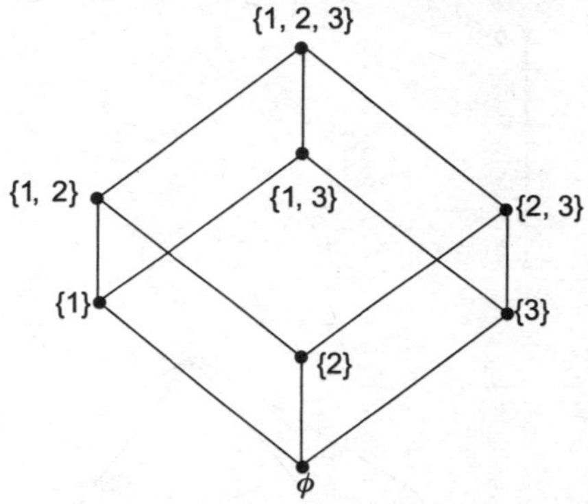

Figure 10.13 ■

Example 10.23

Consider the set $P = \{\phi, \{a\}, \{b\}\}$. Find $\{a\} \wedge \{b\}$. Is it a lattice with respect to the partial order of set inclusion?

SOLUTION

The Hasse diagram for this poset is show in Figure 10.14.

Figure 10.14

It is clear that $\{a\} \wedge \{b\}$ does not exist. Thus P is not a lattice. ■

Example 10.24

Consider the following Hasse diagrams of certain posets (Fig. 10.15). Which of them are lattices? Why?

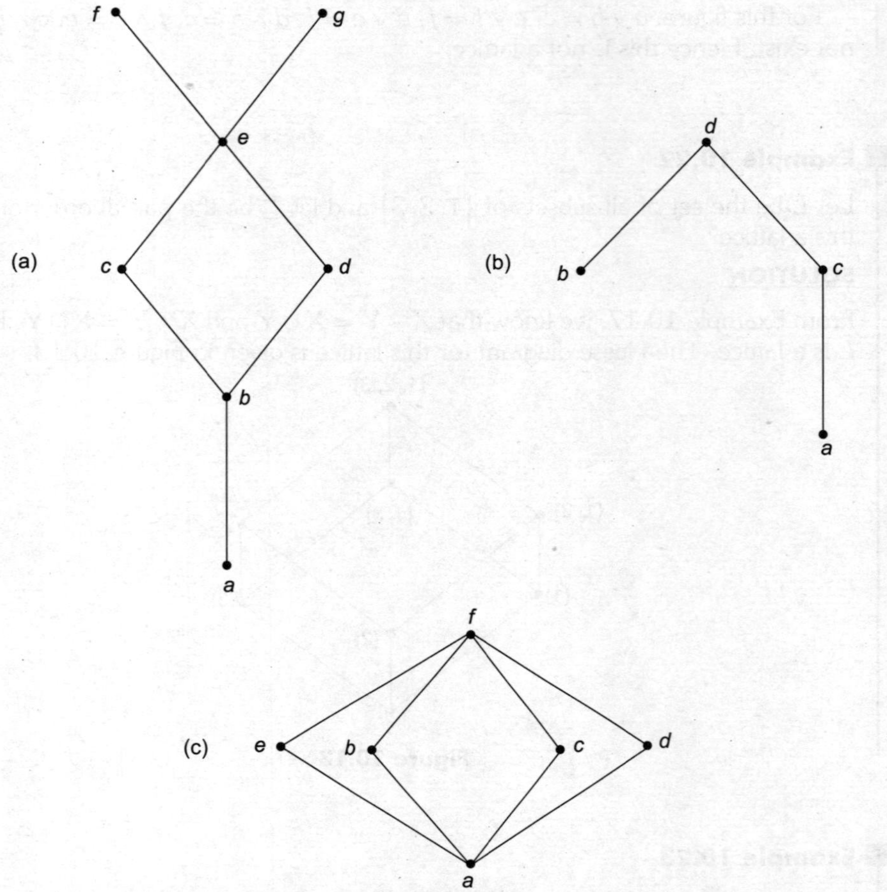

Figure 10.15

SOLUTION

(i) Consider (a). $f \vee g$ does not exist. Hence this poset is not a lattice.

(ii) Consider (b). $b \wedge c$ does not exist. Hence this poset is not a lattice.

(iii) Consider (c). In a poset if $p \leq q$, then $p \vee q = q$ and $p \wedge q = q$. So, we consider only two-element subsets $\{a, b\}$ with a and b not comparable. From the lattice diagram,

$$b \vee c = f; \quad b \wedge c = a$$

$$b \vee e = f; \quad b \wedge e = a$$

$$d \vee e = f; \quad d \wedge e = a$$

Thus, we have shown that every two-element subset has an infimum and a supremum. Thus, the given poset is a lattice. ■

PRACTICE 10.10

Show that S_{30} of Example 10.20 is a lattice.

Note that in a lattice every finite set has an infimum and a supremum. But infinite subsets may not have an infimum and a supremum.

Example 10.25

Consider the poset $(\mathbb{N}, \leq)$ where $\leq$ is the usual 'less than or equal to' relation. Now, for any $a, b \in \mathbb{N}$, $a \vee b = \max\{a, b\}$ and $a \wedge b = \min\{a, b\}$. Thus, $(\mathbb{N}, \leq)$ is a lattice. But $\mathbb{N}$ does not have a maximum. It has a minimum element, namely 1. Similarly, the lattice $\{-1, -2, -3, \ldots\}$ has no minimum but has the maximum element -1. ■

Example 10.26

Let $L = \{1, 2, 4, 7, 14, 28\}$ and on L the partial order $\leq$ is defined as $a \leq b$ if a is a divisor of b. Show that L is a lattice.

SOLUTION

Note that L is the set of all positive integral divisors of 28. We have already seen that $\leq$ is a partial order. Now, we can easily show that for $a, b \in L$, $a \wedge b$ must be the gcd of a and b. Since a and b are factors of 28, their gcd is also a factor. Hence $a \wedge b$ exists in L. Similarly, $a \vee b$ is the lcm and hence L is a lattice. ■

PRACTICE 10.11

Let $N \in \mathbb{N}$ and D_N be the set of all positive integral factors of N. Define $\leq$ as $a \leq b$ if a is a factor of b. Show that D_N is a lattice.

Example 10.27

Let A be any set and let $P(A)$ be the power set of S. Show that it is a lattice under the partial order defined as $X \leq Y \leftrightarrow X \subseteq Y$.

SOLUTION

We have already seen that the relation $\subseteq$ is a partial order on $P(A)$. Consider $X, Y \in P(S)$. That is $X, Y \subseteq S$. Now, we have to show that sup $\{X, Y\} = X \vee Y$ exists. By definition, $X \vee Y$ must contain both X and Y. Further, if any other set contains X and Y, it must contain $X \vee Y$. Thus, it is clear that $X \vee Y$ has to be $X \cup Y$. Similarly, $X \wedge Y = \inf\{X, Y\} = X \cap Y$. Thus $P(A)$ is a lattice. ■

Definition 10.7: A poset $(S, \leq)$ is said to be totally ordered if for $a, b \in S$ either $a \leq b$ or $b \leq a$. A subset C of a poset S is said to be a chain if for $a, b \in C$ either $a \leq b$ or $b \leq a$.

Example 10.28

D_{15} is not totally ordered. We note that neither 3 is a factor of 5 nor is 5 a factor of 3. ■

Example 10.29

Consider the given Hasse diagram in Figure 10.16.

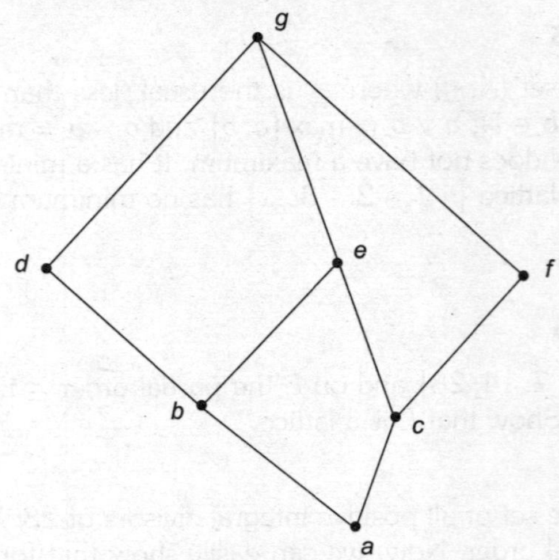

Figure 10.16

The sets $\{a, b, d, g\}$, $\{a, b, e, g\}$, $\{a, c, e, g\}$ are all chains in this poset. ■

PRACTICE 10.12

Let A contain at least two elements. Show that $P(A)$ is not totally ordered.

Example 10.30

$\mathbb{N}$ with the usual order *less than or equal to* is totally ordered. ■

Proposition 10.2

A totally ordered set L is a lattice.

Proof

Let $a, b \in L$. Without loss of generality, let $a \leq b$. Then, $a \vee b = b$ and $a \wedge b = a$. Hence, L is a lattice.

If L is a lattice, then $\vee$ and $\wedge$ give rise to binary operations on L. If $S \subset L$, then S is a poset with respect to the order relation on L. But S need not be a lattice. ■

Example 10.31

Let $L = \{a, b, c, d\}$ and $S = \{a, b, c\}$ be posets represented by the Hasse diagrams, given in Figure 10.17.

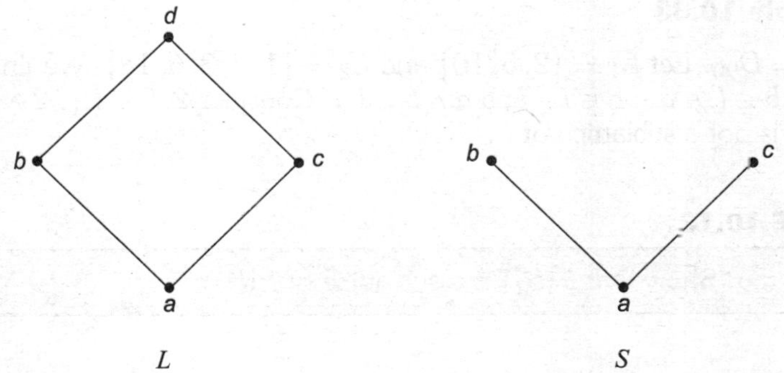

Figure 10.17

It is clear that L is a lattice but S is not a lattice since $b \vee c$ does not exist in S.

Next, even if $S \subseteq L$ is a lattice, it may so happen that the supremum of a set $\{a, b\}$ in S could be different from the supremum of $\{a, b\}$ in L. In the next example, $a \vee_S b$ denotes the supremum of the set $\{a, b\}$ in S. ■

Example 10.32

Let $L = \{a, b, c, d, e\}$ and $S = \{a, b, d, e\}$ be posets represented by the Hasse diagrams given in Figure 10.18.

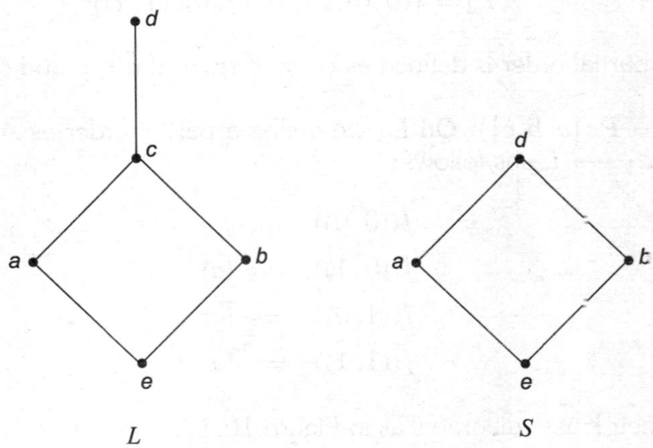

Figure 10.18

It is clear that both L and S are lattices. Further, the partial order on S is the restriction of the partial order on L. Now, $a \vee b = c$ but $a \vee_S b = d$.

In this example, L is a lattice, $S \subseteq L$, S is a lattice with respect to the same partial order, but S is not a sublattice of L.

Now, we formally define the idea of a sublattice. ■

Definition 10.8: Let L be a lattice and $S \subset L$. If for all $a, b \in S$, $a \vee b \in S$ and $a \wedge b \in S$, then S is called a sublattice of L.

Example 10.33

Let $L = D_{90}$. Let $L_1 = \{2, 5, 10\}$ and $L_2 = \{1, 2, 3, 6, 18\}$. We find that whenever $a, b \in L_2$, $a \vee b \in L_2$ and $a \wedge b \in L_2$. Consider $2, 5 \in L_1$; $2 \wedge 5 = 1 \notin L_1$. So, L_1 is not a sublattice of L. ■

PRACTICE 10.13

Let $S_1 \subset S_2$. Show that $P(S_1)$ is a sublattice of $P(S_2)$.

PRACTICE 10.14

Let $L = D_{32}$ and $L_1 = \{2, 4, 8\}$. On both the sets, the partial order is defined as $a \le b$ if and only if a is a factor of b. Show that L_1 is a lattice, but that it is not a sublattice of L.

Example 10.34

Consider the lattice

$$L_1 = \{(0,0), (0,1), (1,0), (1,1)\}$$

where the partial order is defined as $(x, y) \le (p, q)$ if $a \le p$ and (b, q).

Let $L_2 = P(\{a, b, c\})$. On L_2 we define a partial order as $A \le B$ if $A \subseteq B$. Define $f : L_1 \longrightarrow L_2$ as follows:

$$\begin{aligned} f((0,0)) &= \phi \\ f((0,1)) &= \{a\} \\ f((1,0)) &= \{c\} \\ f((1,1)) &= \{a, c\} \end{aligned}$$

We draw their Hasse diagrams as in Figure 10.19.

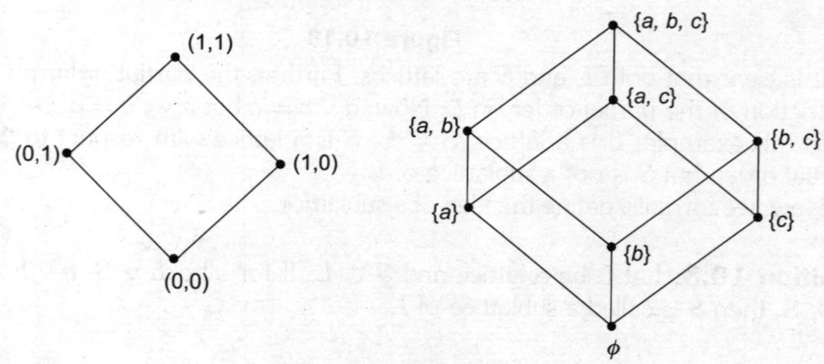

Figure 10.19

We note that $f(x \vee y) = f(x) \vee f(y)$ and $f(x \wedge y) = f(x) \wedge f(y)$ for every $x, y \in L_1$. This is an example of lattice homomorphism. ■

Definition 10.9: Let $(L_1, \leq)$ and $(L_2, \leq)$ be lattices. A function $f : L_1 \longrightarrow L_2$ is a lattice homomorphism if $f(a \vee b) = f(a) \vee f(b)$ and $f(a \wedge b) = f(a) \wedge f(b)$.

Proposition 10.3

A lattice homomorphism is order preserving. That is, if $f : L_1 \longrightarrow L_2$ is a lattice homomorphism, then

$$a \leq b \rightarrow f(a) \leq f(b)$$

Proof

Let $a \leq b$. Then, $a \wedge b = a$ and $a \vee b = b$. Thus, from the definition of lattice homomorphism, $f(a \wedge b) = f(a) \wedge f(b) = f(a)$ and $f(a \vee b) = f(a) \vee f(b) = f(b)$. Since $f(a) \wedge f(b) \leq f(a) \vee f(b)$, we obtain $f(a) \leq f(b)$. ■

Definition 10.10: A lattice L is said to be complete if for every non-empty subset $S \subseteq L$, $\sup S$ and $\inf S$ exist.

A lattice is said to be bounded, if there is $1 \in L$ such that $a \leq 1$ for all $a \in L$ and if there is $0 \in L$ with $0 \leq a$ for all $a \in L$.

Example 10.35

Let $S = [0, 1) \cup (1, 2]$ and the partial order on S is taken as the usual *less than or equal to* relation. This lattice is bounded since $x \leq 2$ for all $x \in S$ and $0 \leq x$ for all $x \in S$. But *this lattice is not complete* since for the subset $A = [0, 1)$, $\sup A$ does not exist. ■

PRACTICE 10.15

In the previous example, find $B \subseteq S$ such that $\inf B$ does not exist.

Proposition 10.4

Every complete lattice L is bounded.

Proof

Every non-empty subset has a supremum and an infimum and so, in particular, L does. Let $1 = \sup L$ and $0 = \inf L$. ■

Example 10.36

The set $L = \{1, 2, 3, 4, 5\}$ with the usual order relation of *less than or equal to* is a complete lattice.

SOLUTION

Note that L is a finite set. Let A be any non-empty subset of L. $\sup A$ is the greatest element in A and $\inf A$ is the least element in A. Thus, L is complete. ■

Example 10.37

The lattice whose Hasse diagram is given in Figure 10.20 is a bounded lattice.

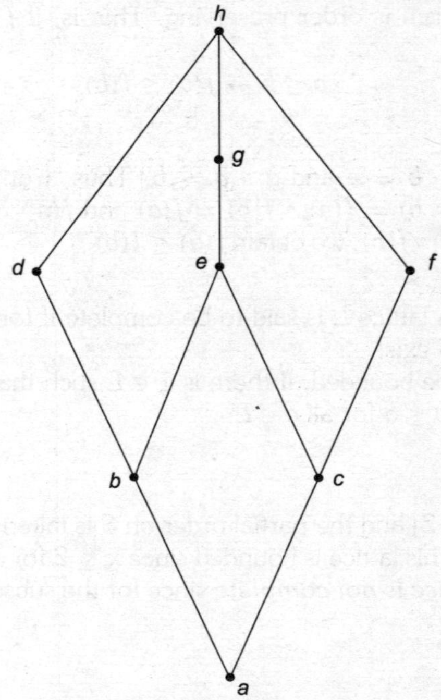

Figure 10.20

From the diagram, we find that for all $x \in L$, $x \leq h$ and $a \leq x$. So we may denote h by 1 and a by 0. ■

Example 10.38

The poset $2\mathbb{N} = \{2, 4, 6, 8, \ldots\}$ is a lattice since the gcd and lcm of two even numbers are even and hence are in $2\mathbb{N}$. But this is not a complete lattice, because there is no element in $2\mathbb{N}$, which is a multiple of all elements in $2\mathbb{N}$. Note, however, that there is a minimum element, namely 2. ■

PRACTICE 10.16

Let $L = \{1, 3, 5, 15\}$ and consider the partial order on L defined as $a \leq b$ if b is a multiple of a. Show that L is a complete lattice. What are the maximum and minimum elements for this lattice?

Proposition 10.5

Show that every finite lattice L is complete.

Proof

Let S be a non-empty subset of L. Since L is finite, S is also finite. Let $S = \{a_1, a_2, a_3, \ldots, a_n\}$. Define $b = a_1 \wedge a_2 \wedge \cdots \wedge a_n$ and $c = a_1 \vee a_2 \vee \cdots \vee a_n$. Clearly, $b = \inf S$ and $c = \sup S$. ■

Example 10.39

Show that $P(S)$ is a complete lattice. ■

Proof

Let $\Gamma \subset P(S)$. Define A to be the union of all elements in Γ. It is easy to show that $A = \sup \Gamma$. Let B be defined as the intersection of all elements in Γ. Then $B = \inf \Gamma$. Hence $P(S)$ is a complete lattice. ■

Definition 10.11: A lattice is said to be distributive if the following properties hold:

(i) $a \vee (b \wedge c) = (a \vee b) \wedge (a \vee c)$.

(ii) $a \wedge (b \vee c) = (a \wedge b) \vee (a \wedge c)$.

Example 10.40

Let S be any set. Show that $P(S)$, the set of all subsets of S, is a distributive lattice under the partial order $A \leq B$ if $A \subseteq B$. ■

Proof

We have already seen that $P(S)$ is a lattice.

Now, $A \vee B$ is nothing but the union $A \cup B$, and $A \wedge B$ is nothing but the intersection $A \cap B$. We know that the following distributive laws are valid:

$A \cap (B \cup C) = (A \cap B) \cup (A \cap C)$.

$A \cup (B \cap C) = (A \cup B) \cap (A \cup C)$.

Thus, $P(S)$ is a distributive lattice. ■

Example 10.41

Show that the lattice in Figure 10.21 is not distributive.

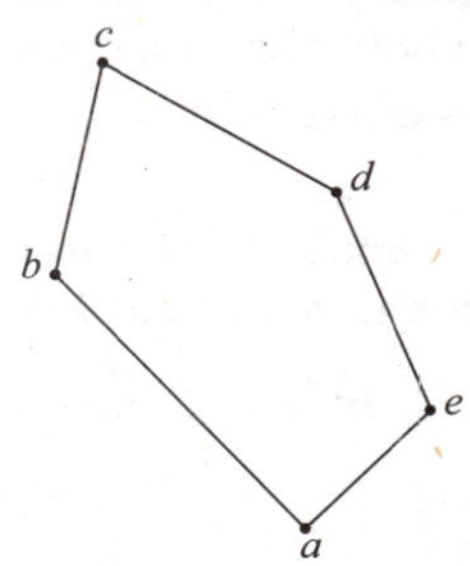

Figure 10.21

Consider $d \wedge (b \vee e) = d \wedge c = d$. Next, $(d \wedge b) \vee (d \wedge e) = a \vee e = e$. Hence, $d \wedge (b \vee e) \neq (d \wedge b) \vee (d \wedge e)$. So, this lattice is not distributive. ■

PRACTICE 10.17

Show that the lattice in Figure 10.22 is not distributive.

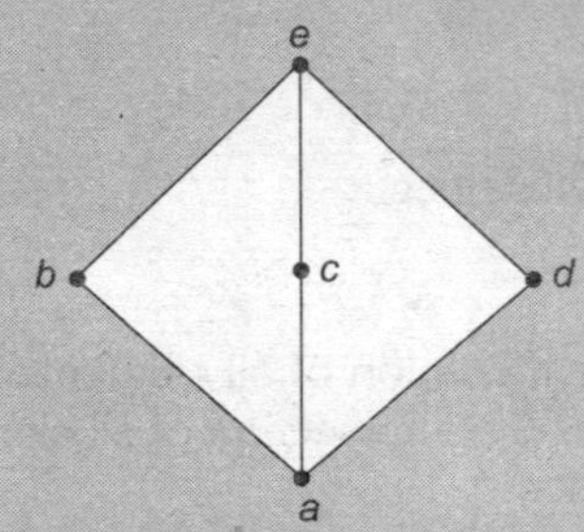

Figure 10.22

Example 10.42

Consider the lattices D_{35} and $P(A)$ where $A = \{a, b\}$. Their Hasse diagrams are shown in Figure 10.23.

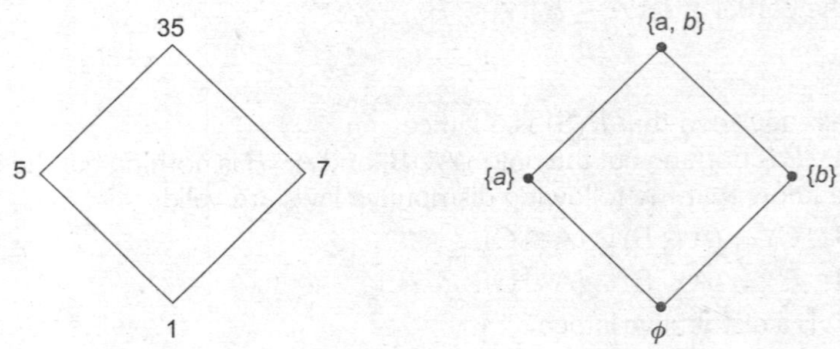

Figure 10.23

It is clear that these two lattices are essentially the same. We have already seen an example of a lattice homomorphism. Here we have an example of a lattice isomorphism whose definition is given below. ■

Definition 10.12: Let $(L_1, \leq)$ and $(L_2, \leq)$ be two lattices. These are said to be isomorphic if there is a one-one and onto map $f : L_1 \to L_2$ such that

$$f(a \vee b) = f(a) \vee f(b)$$
$$f(a \wedge b) = f(a) \wedge f(b)$$

Let us recall that a one-one and onto map is called a *bijection*. Using this terminology, a lattice isomorphism is a bijective lattice homomorphism.

Next, we state an important theorem on distributive lattices without any proof.

Theorem 10.1

A lattice is distributive if and only if it contains no sublattice isomorphic to any of the lattices in Example 10.41 and Practice 10.17.

It is easy to see that if there is a sublattice isomorphic to the lattices given in the theorem, then the lattice is not distributive. The converse is proved in an advanced course on lattice theory.

Definition 10.13: Let L be a bounded lattice. We say that $b \in L$ is a compliment of a if $a \vee b = 1$ and $a \wedge 1 = 0$. It is denoted by a^c.

Example 10.43

Let L be a bounded lattice. Then, $1^c = 0$. ■

Proof

It is clear that $1 \vee 0 = 1$ and $1 \wedge 0 = 0$. Hence, they are compliments of each other. ■

PRACTICE 10.18

If b is a compliment of a, then show that a is a compliment of b.

Example 10.44

Let $L = D_{30}$. Find the compliment of 3.

SOLUTION

$D_{30} = \{1, 2, 3, 5, 6, 10, 15, 30\}$. Let x be the compliment of 3. We must have $3 \wedge x = 1$. This means that $\gcd(3, x) = 1$. The possibilities are 1, 2, 5 and 10. We should also have $\text{lcm}(3, x) = 30$. This implies that $x = 10$ and so the compliment of 3 is 10. ■

Example 10.45

Let $L = D_{12}$. Show that 6 does not have a compliment.

SOLUTION

$D_{12} = \{1, 2, 3, 4, 6, 12\}$. Let x be the compliment of 6. We must have $6 \wedge x = 1$. This means that $\gcd(6, x) = 1$. The only possibility is 1. We should also have $\text{lcm}(6, 1) = 12$, which is not true. This implies that 6 does not have a compliment. ■

Example 10.46

Consider the lattice given in Figure 10.24.

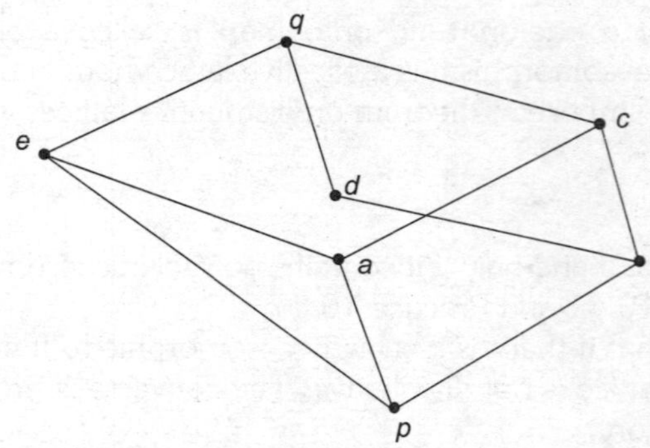

Figure 10.24

Show that d has two compliments and that c does not have any compliment.

SOLUTION

From the Hasse diagram, it is clear that q is the maximum and p is the minimum. Further, $d \vee a = q$ and $d \wedge a = p$. Hence a is a compliment of d. One can also verify that e is also a compliment of d.

Next, we shall show that c does not have a compliment. If $x = c'$, then $c \wedge x = p$. Clearly p is the only possible element with this property. Now, $c \vee p = c \neq q$. Thus, c cannot have a compliment. ■

In Example 10.46, we noted that the compliment of an element need not be unique. But in a distributive lattice, the compliment of an element is unique. The following proposition tells us more.

Proposition 10.6

Let $(L, \leq)$ be a distributive lattice. Show that, if for some $a \in L$, $a \wedge x = a \wedge y$ and $a \vee x = a \vee y$, then $x = y$.

Proof

Expanding $y \vee (a \wedge x)$ we obtain

$$\begin{aligned} y \vee (a \wedge x) &= (y \vee a) \wedge (y \vee x) && \text{Distributivity} \\ &= (x \vee a) \wedge (x \vee y) && \text{Given hypothesis and commutativity} \\ &= x \vee (a \wedge y) && \text{Distributivity} \\ &= x \vee (a \wedge x) && \text{Given hypothesis} \\ &= x && \text{Absorption} \end{aligned}$$

Thus, we get $x = y \vee (a \wedge x)$. Since $y \leq y \vee (a \wedge x)$, we get $y \leq x$. Similarly, we can show that $x \leq y$. Hence $x = y$. ■

Definition 10.14: A bounded lattice is said to be a *complimented lattice* if every element has at least one compliment.

PRACTICE 10.19

Show that D_{30} is a complimented lattice.

Example 10.47

Consider the set S of all strings of zeros and ones of length 4. Define $\leq$ as $xyuv \leq pqrs$ if and only if $x \leq p$, $y \leq q$, $u \leq r$ and $v \leq s$. Show that S is a complimented lattice.

SOLUTION

Consider the set $B = \{0, 1\}$ as a poset under the usual ordering $\leq$. It is clear that this set is a totally ordered set and so we have $1 \vee 1 = 1 \vee 0 = 0 \vee 1 = 1$. Similarly, $0 \vee 0 = 0$, $1 \wedge 1 = 1$, $0 \wedge 1 = 1 \wedge 0 = 0 \wedge 0 = 0$. Further, $1^c = 0$ and $0^c = 1$. Now using this notation, we can easily show that $xyuv \vee pqrs$ is nothing but the string $(x \vee p)(y \vee q)(u \vee r)(v \vee s)$. For example, $1001 \vee 0010 = 1011$. Similarly,

$$xyuv \wedge pqrs = (x \wedge p)(y \wedge q)(u \wedge r)(v \wedge s)$$

Thus, we have shown that S is a lattice. Obviously, 1111 is the maximum of S and 0000 is the minimum of S. So, S is a bounded lattice. Now, the compliment of $xyuv$ is just the string $x^c y^c u^c v^c$. ■

The next example is an important example.

Example 10.48

Show that the set of all positive divisors of a positive integer N denoted by D_N is a complete, distributive lattice under the partial order defined as $a \leq b$ if a is a divisor of b. Also, find the maximum and minimum element of this lattice.

SOLUTION

Every positive integer is a divisor of itself. Thus, $a \leq a$. Hence, $\leq$ is *reflexive*.

Now, let $a \leq b$ and $b \leq a$. This means that a divides b and b divides a. As D_N consists only of positive integers, we must have $a = b$. Hence, this relation is *anti-symmetric*.

Next, let $a \leq b$ and $b \leq c$. That is, a is a divisor of b and b is a divisor of c. But this implies that a is a divisor of c. So, $a \leq c$ and hence $\leq$ is transitive.

We have proved that $(D_N, \leq)$ is a poset. Next, we have to show that for any a and b in D_N, $a \vee b$ and $a \wedge$ is also in D_N.

Now, as per the definition, $a \vee b$ must be a common multiple of a and b, and if k is any other common multiple, then $a \vee b$ should be a divisor of k. Clearly, $a \vee b$ is the lcm of a and b. Since if a and b are divisors of N, their lcm is also a divisor of N, $a \vee b \in D_N$. Similarly, $a \wedge b$ is the gcd of a and b, which is also a divisor of N.

Thus, we have shown that every two-element subset of D_N has a supremum and an infimum. Thus, D_N is a lattice.

Now, we know from number theory that

$$\text{lcm}(a, \text{gcd}(b, c)) = \text{gcd}(\text{lcm}(a, b)\ \text{lcm}(a, c))$$

Thus,

$$a \wedge (b \vee c) = (a \wedge b) \vee (a \wedge c)$$

Likewise,

$$\text{gcd}(a, \text{gcd}(b, c)) = \text{lcm}(\text{gcd}(a, b)\ \text{gcd}(a, c))$$

Thus,

$$a \vee (b \wedge c) = (a \vee b) \wedge (a \vee c)$$

So, the distributive properties hold.

D_N is a distributive lattice. Next, we have to show that this lattice is complete. This follows from the fact that D_N is finite.

Since 1 is a divisor of N, $1 \in S_N$. Further, 1 is a divisor of every element in S_N. Hence $1 \leq a$ for all $a \in D_N$. Hence 1 is the minimum element.

Since N is a divisor of N, $N \in D_N$. Further, every $a \in N$ is a divisor of D_N. Hence $a \leq N$ for all $a \in D_N$. Hence N is the maximum element. ■

Example 10.49

Show that the compliment of 12 in D_{60} is 5.

SOLUTION

It is clear that $12 \vee 5 = 60$ and $12 \wedge 5 = 1$ and hence $12^c = 5$. ■

PRACTICE 10.20

Show that the compliment of m in D_N is $\frac{N}{m}$, if $\frac{N}{m}$ and m do not have any common factor.

Example 10.50

Show that D_{27} is a totally ordered.

SOLUTION

The Hasse diagram of D_{27} is given in Figure 10.25, from which the proof is obvious.

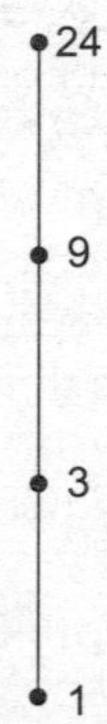

■

Figure 10.25

PRACTICE 10.21

Under what conditions on N is D_N totally ordered?

Example 10.51

Show that the lattices $\{000, 001, 010, 100, 011, 101, 110, 111\}$ and $P(A)$ where $A = \{a, b, c\}$ are isomorphic.

SOLUTION

$$\begin{aligned} f(000) &= \phi. \\ f(001) &= \{c\} \\ f(010) &= \{b\} \\ f(100) &= \{a\} \\ f(011) &= \{b, c\} \\ f(101) &= \{a, c\} \\ f(110) &= \{a, b\} \\ f(111) &= \{a, b, c\} \end{aligned}$$

■

Example 10.52

Let $(L_1, \leq)$ and $(L_2, \leq)$ be two isomorphic lattices. Show that a set $S \subseteq L_1$ has a maximum if and only if $f(S) \subseteq L_2$ has a maximum. Further, if a is the maximum of S then $f(a)$ is the maximum of $f(S)$.

SOLUTION

Let $a \in S$ be such that

$$x \leq a, \quad \forall x \in S \tag{10.1}$$

Clearly, $f(a) \in f(S)$. Now, let $y \in f(S)$. Then, there is $x \in S$ such that $y = f(x)$. Now from equation (10.3) and Proposition 10.3, we get $y = f(x) \leq f(a)$. Thus, $f(a)$ is the maximum of $f(S)$.

To prove the converse, if $b \in f(S)$ is the maximum, then we can show that $f^{-1}(b) \in S$ is the maximum of S. ■

PRACTICE 10.22

Let $(L_1, \leq)$ and $(L_2, \leq)$ be two isomorphic lattices. Show that L_1 is bounded if and only if L_2 is bounded.

10.4 BOOLEAN ALGEBRA

In Section 10.1, we saw the algebraic definition of a Boolean algebra. In this section, we shall define Boolean algebra using the partial order structure. Of course, we have to prove that the two definitions are equivalent.

Consider D_N, which is a complete, distributed lattice. Comparing D_N with the Boolean algebra $P(A)$, we find that in D_N an element may not have a compliment. This motivates the definition of a Boolean algebra.

Definition 10.15: A Boolean algebra is a complimented distributive lattice.

Example 10.53

Show that D_{165} is a Boolean algebra. Is D_{12} a Boolean algebra?

SOLUTION

By Example 10.48, both D_{165} and D_{12} are complete distributive lattices. Next, we have to examine the existence of compliments of the elements.

Consider the Hasse diagram for D_{165}, which is given Figure 10.26.

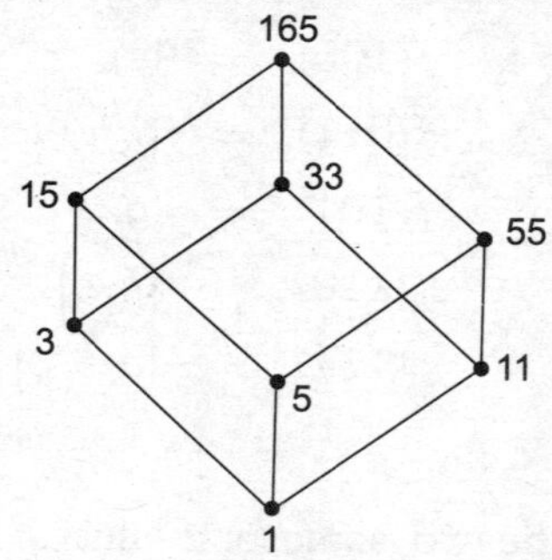

Figure 10.26

Note that $3 \vee 55 = 165$ and $3 \wedge 55 = 1$. Thus, 55 is the compliment of 3 and vice versa. Similarly, 5 and 33 are compliments of each other; 15 and 11 are compliments of each other.

Since D_{165} is a distributive lattice and every element in D_{165} has a compliment, D_{165} is a Boolean algebra.

Next, consider the Hasse diagram for D_{12}, which is given in Figure 10.27.

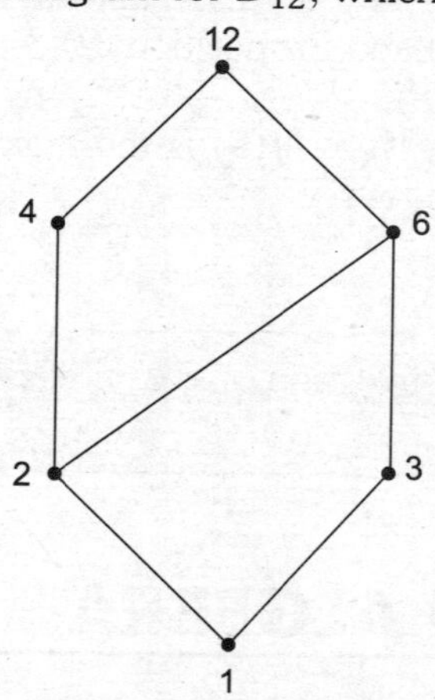

Figure 10.27

If a is the compliment of 2, then we must have $2 \wedge a = 1$. The only such elements are 1 or 3. But $1 \vee 2 = 2$ and $3 \vee 2 = 6$. Both of them are not equal to

12, the greatest element. Thus, 2 does not have a compliment. Thus, D_{12} is not a Boolean algebra. ■

PRACTICE 10.23

Show that for any set A, $(P(A), \subseteq)$ is a Boolean algebra.

Example 10.54

Show that the set $\{1, 2, 3\}$ with the usual order $\leq$ is not a Boolean algebra.

SOLUTION

In this lattice, the maximum element is 3 and the minimum element is 1. We show that 2 does not have a compliment. If 1 is the compliment of 2, then we must have $1 \vee 2 = 3$, which is false. If 3 is the compliment, then we must have $3 \wedge 2 = 1$, which is also false. Obviously, 2 is not its own compliment. The proof is complete. ■

Theorem 10.2

The algebraic and the lattice theoretic definition of a Boolean algebra are equivalent.

Sketch of the Proof: Let $(L, \leq)$ be a complete distributive lattice. We have to define the binary operations $\odot$ and $\oplus$ on L. We define $a \odot b = a \wedge b$ and $a \oplus b = a \vee b$. Now the algebraic properties are easily verified.

Next, let $(L, \oplus, \odot)$ be an algebraically defined Boolean algebra. Now, we must define the partial order on L. We define $a \leq b$ if $a \oplus b = b$. Now, the lattice theoretic properties can be verified.

Example 10.55

Let $L = \{a_1, a_2, a_3, a_4, a_5, a_6, a_7, a_8\}$ and define a relation R on L as

$$\{(a_8, a_5), (a_8, a_6), (a_8, a_7), (a_8, a_2), (a_8, a_3), (a_8, a_4), (a_8, a_1)$$
$$(a_5, a_2), (a_5, a_3), (a_5, a_1)$$
$$(a_6, a_2), (a_6, a_4), (a_6, a_1)$$
$$(a_7, a_4), (a_7, a_3), (a_7, a_1)$$
$$(a_2, a_1)$$
$$(a_3, a_1)$$
$$(a_4, a_1)$$
$$(a_8, a_8), (a_7, a_7), (a_6, a_6), (a_5, a_5), (a_4, a_4), (a_3, a_3), (a_2, a_2),$$
$$(a_1, a_1)\}$$

Its Hasse diagram is given in Figure 10.28.

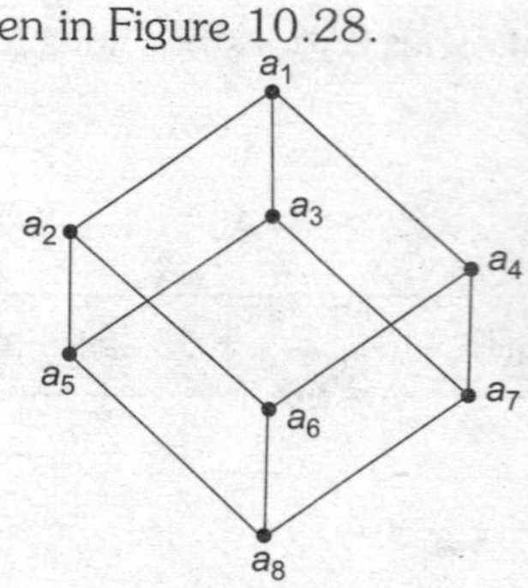

Figure 10.28

It is clear that L is a bounded lattice with the greatest element being a_1 and the least element being a_8. From the Hasse diagram it is clear that L is distributive. Now, from the Hasse diagram we find that

$$a_5 \wedge a_4 = a_8; \quad a_5 \vee a_4 = a_1$$

Thus, a_5 and a_4 are compliments of each other.

$$a_6 \wedge a_3 = a_8; \quad a_6 \vee a_3 = a_1$$

Thus, a_6 and a_3 are compliments of each other.

$$a_2 \wedge a_7 = a_8; \quad a_2 \vee a_7 = a_1$$

Thus, a_2 and a_7 are compliments of each other.

Clearly a_1 and a_8 are compliments of each other. Hence L is a Boolean algebra. ■

Example 10.56

Let L be the set of all statement formulas that can be formed with the three atomic statements P, Q and R. Two statements X and Y are considered equal if they are logically equivalent. Show that $(L, \vee, \wedge)$ is a Boolean algebra.

SOLUTION

It is clear that **T** is the identity for $\wedge$ since $P \wedge T = P$ and **F** is the identity for $\vee$ as $P \vee \mathbf{F} = P$.

How many elements does this Boolean algebra have? 2^8. ■

Example 10.57

Let $a \in A$. Let $L = \{X \in P(A) : a \notin X\}$. Show that $(L, \subseteq)$ is a Boolean algebra.

SOLUTION

If $a \notin X$ and $a \notin Y$, then $a \notin X \cup Y$ and $a \notin X \cap Y$. Thus, L is a lattice. Distributive laws are obviously true. Now, let $\phi \neq S \subseteq L$. We have to show that supremum S exists. Define M as

$$M = \bigcup_{X \in S} X$$

Now, clearly $a \notin M$ and hence $M \in L$. Further, for any $X \in S$, $X \subseteq M$. Thus, M is an upper bound of S. If $N \in L$ is such that $X \subseteq N$ for all $X \in S$, then $M = \bigcup_{X \in S} X \subseteq N$. Thus, M is the supremum. Similarly, we can show that $P = \bigcap_{X \in S} X$ is the infimum. We have shown that any non-empty subset of L has a supremum and an infimum. This makes L a complete lattice. To complete the proof, we have to show that every element has a compliment. Now, it is obvious that in L, $1 = A - \{a\}$ and $0 = \phi$.

At this stage, we need to modify the notation for the compliment of an element in a lattice. For the Boolean algebra $P(A)$, X^c denotes both the set theoretic compliment $A - X$ and the lattice theoretic compliment. But in this problem, we will have to distinguish between these two types of compliment. For $X \subseteq A$, X^c denotes the usual set theoretic compliment $A - X$. In the lattice L, we use $\overline{X}$ to denote the lattice theoretic compliment of X. For $X \in L$, take $\overline{X} = X^c - \{a\}$. One can easily show that

$$\begin{aligned} X \cup \overline{X} &= A - \{a\} = 1 \\ X \cap \overline{X} &= \phi = 0 \end{aligned}$$

Now, we have shown that every element of L has a compliment and, hence, L is a Boolean algebra. ■

PRACTICE 10.24

Let $a \in A$. Let $L = \{X \in P(A) : a \in X\}$. Show that $(L, \subseteq)$ is a Boolean algebra.

Example 10.58

Show that in a Boolean algebra, $(a \oplus b) \odot (a^c \oplus c) = (a \odot c) \oplus (a^c \odot b) \oplus (b \odot c)$.

SOLUTION

$$\begin{aligned} (a \oplus b) \odot (a^c \oplus +c) &= [(a \oplus b) \odot a^c] \oplus [(a \oplus b) \odot c] \\ &= [(a \odot a^c) \oplus (b \odot a^c)] \oplus [(a \odot c) \oplus (b \odot c)] \\ &= [0 \oplus (b \odot a^c)] \oplus [(a \odot c) \oplus (b \odot c)] \\ &= (a^c \odot b) \oplus (a \odot c) \oplus (b \odot c) \end{aligned}$$ ■

Example 10.59

Show that in a Boolean algebra $a = b \leftrightarrow a \odot b^c \oplus a^c \odot b = 0$.

SOLUTION

We solve this problem using the order structure of a Boolean algebra. If $a = b$, then $a \odot b^c \oplus a^c \odot b = a \odot a^c \oplus a^c \odot a = 0 \oplus 0 = 0$.

Now, let $a \odot b^c \leq a \odot b^c \oplus a^c \odot b = 0$. So, $a \odot b^c \leq 0$.
Since 0 is the minimum, we have

$$a \odot b^c = 0 \tag{10.2}$$

Similarly, we can show that

$$a^c \odot b = 0 \tag{10.3}$$

From equation (10.2), we obtain

$$a^c \oplus b = 1 \tag{10.4}$$

From equations (10.3) and (10.4), we obtain $b = (a^c)^c = a$. ■

If we reexamine all the finite Boolean algebras we have considered so far, we shall find that their cardinality would be 2^n for some positive integer n. Here, we say that the boolean algebra $(L, \oplus, \odot)$ is finite if L is a finite set.

We state the following important result for finite Boolean algebras.

Theorem 10.3

Every finite Boolean algebra has a cardinality of 2^n for some positive integer n. Further, two finite Boolean algebras having the same cardinality are isomorphic.

SUMMARY

- A Boolean algebra is an algebraic structure $(L, \oplus, \odot)$ such that $(L, \oplus)$ and $(L, \odot)$ are monoids and each binary operation distributes over the other.
- A set L with a relation $\leq$ defined on it is a poset if $\leq$ is reflexive, anti-symmetric and transitive.
- An element $a \in L$ is an upper bound of $S \subseteq L$ if $b \leq a$ for all $b \in S$.
- An upper bound a is the supremum of a set if $a \leq b$ for every upper bound b of S.
- An element $a \in L$ is a lower bound of $S \subseteq L$ if $a \leq b$ for all $b \in S$.
- An upper bound a is the infimum of a set if $b \leq a$ for every lower bound b of S.
- A poset is called a lattice if for every set containing two elements there exists a supremum and an infimum.
- A lattice is said to be bounded if it has a maximum and a minimum element.
- An element b is called a compliment of a if $a \vee b = 1$ and $a \wedge b = 0$.
- A Boolean algebra is a complete, bounded complimented lattice.

PROBLEMS

1. Let $(L, \oplus, \odot)$ be a Boolean algebra as defined in Section 10.1. Prove the De Morgan law

$$(x \odot y)^c = x^c \oplus y^c$$

2. Prove the following in a Boolean algebra as defined in Section 10.1
   (a) $a \oplus (a^c \odot b) = a \oplus b$.
   (b) $(a \odot b \odot c) \oplus (ab)$.
   (c) $(a \oplus b^c) \odot (b \oplus c^c) \odot (c \oplus a^c) = (a^c \oplus b) \odot (b^c \oplus c) \odot (c^c \oplus a)$.
3. Define the relation $\leq$ on $N \times N$ as $(m, n) \leq (p, q)$ if $m + n \leq p + q$. Is $\leq$ a partial order?
4. Let $S = \{a, b, c, d\}$, $R = \{(a, c), (b, c), (d, b), (d, c), (a, a), (b, b), (c, c), (d, d)\}$. Draw the Hasse diagram for R.
5. Let $S = \{\phi, \{a\}, \{b\}, \{a, b\}\}$ and $T = \{\phi, \{a\}, \{a, b\}\}$. Define $\leq$ on both S and T as follows:

$$A \leq B \leftrightarrow |A| \leq |B|$$

   (Here $|A|$ is the number of elements of A. In this problem, the symbol $\leq$ has two different meanings.) Show that $(S, \leq)$ is not a poset, whereas $(T, \leq)$ is a poset. Draw the Hasse diagram of T.
6. Let (S, R_1) and (T, R_2) be posets. Define a relation R on $S \times T$ as $(s_1, t_1)\, R\, (s_2, t_2)$ if $s_1 R_1 s_2$ and $t_1 R_2 t_2$. Show that under this relation $S \times T$ is a poset.
7. Let S be a set of statements. Declare $P \leq Q$ when $P \to Q$. Treating logically equivalent statements as identical, show that $(S, \leq)$ is a poset.
8. Examine the following Hasse diagrams (Fig. 10.29) and determine which of them represent lattices:

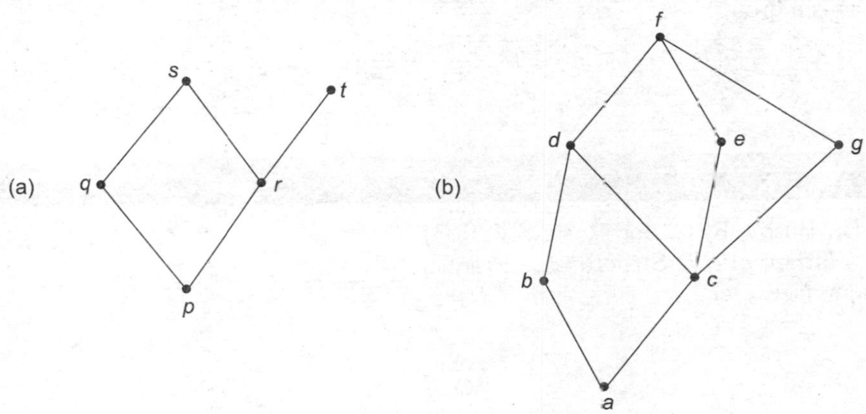

Figure 10.29

9. Let $(Q, \leq)$ be the poset of rational numbers where $\leq$ is the usual order. Show that the set $S = \{x \in Q : x^2 \leq 2\}$ has many upper bounds but no least upper bound.
10. Show that the minimum element of a subset of a poset is unique when it exists. Repeat this exercise for the supremum and the infimum.

11. A collection of disjoint non-empty subsets of S such that their union is S is called the partition of S. Let $\Pi(S)$ be the set of all partitions of S. Define $\leq$ on $\Pi(S)$ as follows: $\alpha, \beta \in \Pi(S)$, $\alpha \leq \beta$ if every element of α is a subset of some element of β. Show that $\Pi(S)$ is a lattice.
12. Show that D_{32} is a chain.
13. Let $S = \{00, 10, 20, 01, 11, 21, 02, 12, 22\}$. Consider S as a set of strings of length 2 made of 0, 1 and 2. Define the relation $\leq$ on S as follows: $xy \leq pq$ if $x \leq p$ and $y \leq q$. Draw the Hasse diagram for S and identify all the chains of S.
14. Let B be a non-empty subset of A. Let $L = \{X \in P(A) : X \cap B = \phi\}$. Show that $(L, \subseteq)$ is a Boolean algebra.
15. Find the compliment of 35 in D_{105}.
16. Find an element in D_{100} without a compliment.
17. Find all the sublattices of D_{14}.
18. Show that the lattices of Example 10.47 and D_{2002} are isomorphic.
19. Let A be a set containing n elements and let L be a set of all strings made of zeros and ones and of length n. Define a partial order on L such that the lattice $(L, \leq)$ is isomorphic to $P(A)$.
20. Show that in a Boolean algebra
    (a) $a = 0 \leftrightarrow (a \odot b^c) \oplus (a^c \odot b) = b$.
    (b) $a \leq b \rightarrow a \oplus (b \odot c) = b \odot (a \oplus c)$.
21. Exhibit an isomorphism between the lattices D_{385} and the power set of $\{0, 1, 2\}$.
22. Show that D_N is a Boolean algebra if and only if N is a product of distinct primes.
23. Let $(L_1, \leq)$ and $(L_2, \leq)$ be two isomorphic lattices and let $a \in L_1$. If $a \in L_1$, then show that $(f(a))^c = f(a^c)$, where $f : L_1 \rightarrow L_2$ is an isomorphism.
24. Let $(L_1, \leq)$ and $(L_2, \leq)$ be two isomorphic lattices. Show that L_1 is complete if and only if L_2 is complete. Further, show that $(L_1, \leq)$ is a Boolean algebra if and only if $(L_2 \leq)$ is a Boolean algebra.
25. Let $(L, \leq)$ be a Boolean algebra. Define the binary operation $*$ on L as $a * b = (a \odot b^c) \oplus (a^c \odot b)$. Show that $(L, *)$ is a group.

FURTHER READING

Kolman, B., Busby, R. C. and Ross, S. (2001), *Discrete Mathematical Structures*, Prentice Hall of India, New Delhi.

CHAPTER

11

Formal Languages and Language Acceptors

Learning Objectives

On completing this chapter, you should be able to:

- define the concept of a language
- define a finite automaton
- define the notion of a regular language
- construct a finite automaton that accepts a given regular language
- identify the regular language accepted by a finite automaton
- write down regular expressions for a given regular language
- identify the regular langauge represented by a given regular expression
- define a pushdown automaton
- define the notion of a context-free language
- construct pushdown automata for simple context-free languages
- define a Turing machine
- define a linear bounded automaton
- define the notion of a context-sensitive language
- construct a linear bounded automaton for simple context-sensitive languages
- define the notions of recursive and recursively enumerable languages
- define the notions of regular grammar, context-free grammar, context-sensitive grammar and unrestricted grammar
- construct a regular grammar for a given regular language
- construct a context-free grammar for a given context-free language
- construct a context-sensitive grammar for simple context-sensitive languages

11.1 ALPHABET, STRINGS AND LANGUAGES

Example 11.1

Consider the specification of an *integer constant* in the C language. We know that 234 and 82905 are valid integer constants, whereas $7,325$, 67.45, 034 and 45 23 are not valid integer constants. $7,325$ is illegal because of the presence of a *comma*, the punctuation mark. 67.45 is illegal because of the presence of the decimal point. 034 is illegal because an integer constant must not begin with a zero. 45 23 is illegal because of the presence of the blank space. Now, the set of all legal integer constants is an infinite set. The computer does not understand the concepts of integer, rational numbers, etc. A legal integer constant is just a string made up of the symbols $0, 1, 2, 3, 4, 5, 6, 7, 8$ and 9 such that the first symbol is different from 0. This set of all legal integer constants is an example of a language over the set of symbols $\{0, 1, 2, 3, 4, 5, 6, 7, 8, 9\}$. If I is used to denote this set, then we have $234 \in I$ and $02905 \notin I$. ■

Example 11.2

Though $7, 325$ is a valid representation of an integer in our day-to-day life, such a representation is not used for programming with the C language. Now, the set of all usual integer representations is an example of a language over the set of symbols $\{0, 1, 2, 3, 4, 5, 6, 7, 8, 9, ,\}$. In this set of symbols, the last *comma* is not a punctuation mark but a symbol on its own right. If this language is denoted by I_1, then $7, 235 \in I_1$ but $72, 35 \notin I_1$. ■

Example 11.3

Consider the following mathematical formulae:

(a) $2[(x - y)^2 + z^2] = 2x^2 + 2y^2 + 2z^2 + 4xy$.

(b) $2[(x - y)^2 + z^2 = 2x^2 + 2y^2 + 2z^2 - 4xy$.

(a) is a wrong formula. It is wrong because the correct formula has $-4xy$ instead of $+4xy$. Is (b) a correct formula? We cannot decide about its correctness, as it is given. Note that the left bracket [does not have a corresponding right bracket. Now, consider

(c) $2[(x - y)^2 + z^2] = 2x^2 + 2y^2 + 2z^2 - 4xy$.

(d) $2[(x - y)^2] + z^2 = 2x^2 + 2y^2 + 2z^2 - 4xy$.

(c) is correct. (d) is not correct, but its meaning is clear unlike that of (b). Since (d) expresses clearly what it wants to express, we can declare it to be wrong. We say that (b) has a syntax error but (d) has a factual error.

Let us take the sentences:

(e) Yesterday there was a snake in the garden.

(f) Yesterday there is a snake in the garden.

We do not know whether (e) is factually correct because we do not know the context in which this statement is made. We do know that it is grammatically correct unlike (f), which is grammatically wrong.

Now imagine some student who is writing a computer programme to find the roots of a quadratic equation. If there is a syntax error in the programme, the compiler would detect it. But if he has used the wrong formula

$$\frac{b \pm \sqrt{b^2 - 4ac}}{2a}$$

the compiler would not correct it. ■

In this chapter we are going to learn about mathematical models of compilers. Consider the statement (f). How do we detect the grammatical error in it? We note the word *yesterday* and store it in our mind. We expect to see verbs like *was* or *were* in the sentence. As soon as we find *is* we know that it is a mistake. If the sentence had been '*Yesterday there were a snake in the garden*', we would not have rejected it before encountering the article *a* in it. The moment we register the word *were*, we expect it to be followed by a *plural* form, and hence reject it when we see *a*.

A compiler scans the given programme symbol by symbol, registers it in some manner and would continue to scan till it encounters an incompatible symbol. At this stage, it stops and gives an error message. If there is no incompatibility, the compilation is successful.

The language acceptors are abstract mathematical models of compilers. The acceptors can be in one of *many internal states*. To understand the meaning of the internal states let us imagine a computer. This computer consists of a single programme that calculates the square of an integer. If the input is 3, the output is 9. Now, let us imagine that the program to determine the square has been deleted, and instead a program that computes the cube is stored. If the input now is 3, then the output is 27. In some sense, when we write a program, we alter the internal state of the computer and the *output depends on this internal state as well as the input*.

Before getting into mathematical models of a compiler, we should know what a language is. Unlike natural languages that are very complicated, mathematical languages are simple to describe. Based on Examples 1 and 2, we have the following ingredients for a language:

(i) A set of symbols.

(ii) Rules for the formation of strings with the symbols.

A string like *kattikkuttazq* has no meaning in English, but it could very well be a word in a formal language. For example, L can be the set of strings formed by using symbols from the English alphabet, whose first letter is k and the last letter is q. In simple words, a string can be defined as an ordered sequence of symbols from a predetermined set.

Definition 11.1: Let Σ be a set of symbols. A string of symbols from Σ is a finite sequence of symbols. Σ^* is the set of all strings formed by using the symbols of Σ including the empty string. Σ^+ is the set of all non-empty strings formed by the symbols of Σ.

Example 11.4

Let $\Sigma^* = \{a, b, c, d, ..., x, y, z\}$ be the set of all English letters. Then every word in English is an element of Σ^*. Further, even meaningless strings like *pygssskioll*, *iiikjhg* are elements of Σ^*. But $\alpha pq \notin \Sigma^*$, $222aab \notin \Sigma^*$. ■

Example 11.5

Let $\Sigma = \{a, b, c, d, ..., x, y, z\} \cup \{0, 1, 2, 3, 4, 5, 6, 7, 8, 9\}$. Then, $14325 \in \Sigma^*$, $7abfh648iolk \in \Sigma^*$, $ybvcshj \in \Sigma^*$. ■

Example 11.6

Show that Σ^* is an infinite set if $\Sigma \neq \phi$.

SOLUTION

Let $x \in \Sigma$. It is clear that the set $M = \{x, xx, xxx, xxxx...\}$ is a countably infinite set (refer Chapter 2). Since M is an infinite subset of Σ^*, Σ is an infinite set. ■

PRACTICE 11.1

Investigate whether Σ^* is countable or not.

In the theory of languages, a set of symbols Σ is called an alphabet.

Definition 11.2: Let Σ be an alphabet. A language L over Σ is defined as a subset of Σ^*.

Let L_1 and L_2 be languages over Σ. It is clear that $L_1 \cup L_2$ and $L_1 \cup L_2$ are also languages over Σ. Now, we are going to learn yet another method of getting a new language out of old languages.

Concatenation of Strings: If u and v are strings, then uv is the string formed by placing the string v next to the string u. uv is called the concatenation of u and v.

Note that concatenation defines a binary operation on Σ^* (see Chapter 8). It is clear that the concatenation of u with the empty string is u itself. Further, concatenation is associative and hence Σ^* is a monoid under concatenation where the empty string is the identity. Let us also note that concatenation is not commutative.

Example 11.7

If $u = ary$ and $v = pplm$, then $uv = arypplm$ $vu = pplmary$. ■

Concatenation of Languages: If L_1 and L_2 are languages, then L_1L_2 is defined as $L_1L_2 = \{uv : u \in L_1 \text{ and } v \in L_2\}$.

If L_1 is a language over Σ_1 and L_2 is a language over Σ_2, then $L_1 \cup L_2$, $L_1 \cap L_2$ and L_1L_2 are all languages over $(\Sigma_1 \cup \Sigma_2)^*$.

Example 11.8

If

$$\begin{aligned} L_1 &= \{pp, pqp, qqpr\} \text{ and } L_2 = \{12221, 2121\} \\ L_1L_2 &= \{pp12221, pp2121, pqp2121, pqp12221, qqpr2121, qqpr12221\} \\ L_2L_1 &= \{12221pp, 2121pp, 2121pqp, 12221pqp, 2121qqpr, 12221qqpr\} \end{aligned}$$ ■

Example 11.9

Consider an octal integer constant in the C language. An octal integer constant must begin with a zero and it only consists of the digits 0, 1, 2, 3, 4, 5, 6, 7. Thus, the set of all octal integer constants I_8 is a language over the alphabet $\{0, 1, 2, 3, 4, 5, 6, 7\}$. We have the following:

(i) $0734 \in I_8$.
(ii) $0458 \notin I_8$.
(iii) $070.76 \notin I_8$.

It is convenient to have the following notation: If $u \in \Sigma^*$, then u^0 is the empty string, $u^1 = u$, $u^2 = uu$, etc. For example, if $u = cbbabbb$, then $u^3 = abbabbbabbabbbabbabbb$ and $a^5 = aaaaa$. ■

Example 11.10

Let L be the set of strings in $\{a, b\}^*$ containing only b's. That is, the only members of L are the empty string, b, bb, bbb, etc.

Therefore $L = \{b^n : n \in N \cup \{0\}\}$. ■

Example 11.11

Let $L \subseteq \{a, b\}^*$ be the set of all strings of a's and b's containing at least one a.Then, $a \in L$. $b \notin L$, $abbba \in L$, $bbb \notin L$.

An element of L has at least one a. Choose any one of the a's. Let u be the part of the string occurring before the chosen a and let v be the part of the string occurring after it. Under this notation, this member can be written as uav. For example, if $w = bbababbaa$ and if we choose the second a, we take $u = bbab$ and $v = bbaa$.

If $w = abbb$, then u is the empty string and $v = bbb$. From these examples, it is clear that

$$L = \{uav : u, v \in \{a, b\}^*\}$$ ■

Example 11.12

Let $L = \{aubvbw : u, , v, w \in \{a, b\}^*\}$. Show that L is the set of all strings containing at least two b's.

SOLUTION

If $x = aubvbw$, then it is obvious that the first symbol of x is a and that x must have two b's. Since the strings u, v and w may or may not contain b, we can say that x contains at least two b's. Thus, if $x \in L$, then x starts with a and contains at least two b's. Next, we prove the converse.

Let x begin with a and let there be at least two b's. For definiteness, let us consider the first two b's of x. Take u to be the part of x between a and the first b, v to be the part of x between the first b and the second b, and w to be the part of x after the second b. We can write $x = aubvbw$. Thus $x \in L$. Now, we have proved that L is the set of all strings starting with a and containing at least two b's. ■

PRACTICE 11.2

Let L be the set of strings over $\{a, b\}$ ending with two b's. Write down the set theoretic description of L.

Example 11.13

Let $L \subseteq \{a, b\}^*$ be a language consisting of strings having exactly one a. Then $L = \{b^n a b^m : n, m \in N \cup \{0\}\}$. ■

PRACTICE 11.3

Write down the set theoretic description of the language $L \subset \{a, b\}^*$ of strings containing at least one a and at least one b.

PRACTICE 11.4

Consider $L = \left\{a^m (ba)^n a^k b^l : m, n, k \geq 0,\ l \leq 1\right\}$. How would you describe L verbally?

Example 11.14

Let $w = abbbaab$ and $u = baabbba$. How are these two strings related to each other? u is nothing but the string obtained by writing the symbols of w in reverse order. We use the notation $u = w^R$ to denote this. The next definition formally defines w^R for any w. ■

Definition 11.3: Let $w \in \Sigma^*$. w^R is the string in which the symbols of w are written in the reverse order.

Example 11.15

Let $w = kig54$. Then, $w^R = 45gik$. ■

PRACTICE 11.5

Show that $(w^R)^R = w$.

PRACTICE 11.6

Let $w = malayalam$. What is w^R?

In Practice 11.6, we have an example of w such that $w = w^R$. For such a string, when we write its symbols in reverse order, we get back the original string. These strings have a special name, which is given by the next definition.

Definition 11.4: A word w is said to be a palindrome if $w = w^R$.

PRACTICE 11.7

Describe $L_1 = \{uavawa : u, v, w \in \{a, b\}^*\}$ and $L_2 = \{uavawax : u, v, w, x \in \{a, b\}^*\}$ verbally. Find which of these two languages contain the other and prove your result.

Proposition 11.1

$u \in \Sigma^*$ is a palindrome if and only if $u = ww^R$ for some $w \in \Sigma^*$ or $u = wxw^R$ for some $w \in \Sigma^*$ and $x \in \{a, b\}$.

Proof

Let $w \in \Sigma^*$ and $u = wxw^R$. Then, $u^R = (w^R)^R x w^R = wxw^R = u$ and hence u is a palindrome. Now, let u be a palindrome. If u contains an even number of symbols, take w to be the first half of u, we have $u = ww^R$.

If u contains an odd number of symbols, then let x be the symbol in the middle and let w be the part of u that precedes x. We have $u = wxw^R$. ■

Definition 11.5: Let $w \in L$. The number of symbols (counted with repetition) is denoted by $n(w)$. For the empty string, $n(\epsilon) = 0$. For $a \in \Sigma$, $n_a(w)$ is the number of a's in w.

Example 11.16

Let $\Sigma = \{1, 2\}$. Then, $n(12211121221) = 11$ and $n_1(12211121221) = 6$ and $n_2(12211121221) = 5$. ■

Example 11.17

For $w \in \{a, b, c\}^*$, $n(w) = n_a(w) + n_b(w) + n_c(w)$. ■

Example 11.18

Let L be the language of strings over $\{a, b\}$ of length less than or equal to 5 containing no consecutive b's. Find the number of strings in L_1 using the ideas from Chapter 4. ■

Example 11.19

Let $L_1 = \{w \in \{a, b\}^* : n_a(w) = n_b(w)\}$ and $L_2 = \{a^m b^m : m \in N \cup \{0\}\}$. Show that L_2 is a proper subset of L_1.

Proof

Let $w \in L_2$. Then $w = a^m b^m$ for some $m \in N \cup \{0\}$. Since $n_a(a^m b^m) = m = n_b(a^m b^m)$, $w \in L_1$. Hence $L_2 \subseteq L_1$.

Now, take $w = abba \in L_1$. It is clear that $w \notin L_2$. Hence $L_2 \neq L_1$ and hence L_2 is a proper subset of L_1. ■

PRACTICE 11.8

Let $L_1 = \{w \in \{a, b\}^* : n_a(w) = 2n_b(w)\}$ and $L_2 = \{a^{2m} b^m : m \in N \cup \{0\}\}$. Show that L_2 is a proper subset of L_1.

11.2 FINITE AUTOMATA AND REGULAR LANGUAGES

We employ finite automata as language acceptors. Roughly speaking, a finite automaton reads a given string starting from the leftmost symbol and moves to the right by scanning each symbol in turn. Depending on the symbol being scanned, the automaton changes into a new state. After scanning the last symbol, if the automaton reaches a final state, we say that the string has been accepted. If not, the string has been rejected.

11.2.1 Finite Automata

Definition 11.6: A finite automaton is a system $(Q, \Sigma, \delta, q_0, F)$ where

(a) Q is a finite set of **internal states**.

(b) Σ is a finite set of symbols called the **input alphabet**.

(c) $\delta : Q \times \Sigma \longrightarrow Q$ is the **transition function**.

(d) $q_0 \in Q$ is the initial *state*.

(e) $F \subseteq Q$ is a set of **final states**.

Example 11.20

A finite automaton that accepts all strings with at least one a:

$$\begin{aligned}\delta(q_0, a) &= q_1\\ \delta(q_0, b) &= q_0\\ \delta(q_1, a) &= q_1\\ \delta(q_1, b) &= q_1\\ F &= \{q_1\}\end{aligned}$$

This finite automaton, while scanning the given string, is looking for the symbol a. If the string contains only b's then the automaton will not reach a final state. But once it scans an 'a', it goes to a final state and it stops after scanning the last symbol. ■

Example 11.21

A finite automaton that accepts all strings with exactly one a:

$$\begin{aligned}\delta(q_0, a) &= q_1\\ \delta(q_0, b) &= q_0\\ \delta(q_1, a) &= q_0\\ \delta(q_1, b) &= q_1\\ F &= \{q_1\}\end{aligned}$$

Scanning b will not alter the state of this automaton. When it scans the first a it reaches a final state. As far as this automaton is concerned, a second a is not welcome, and so if it is in a final state when it scans a second a, it goes back to the initial state. If it does not encounter any further a, it stops in the final state after scanning the last symbol. ■

Example 11.22

A finite automata that accepts all strings with no more than three a's:

$$\begin{aligned}\delta(q_0, a) &= q_1\\ \delta(q_0, b) &= q_0\\ \delta(q_1, a) &= q_2\\ \delta(q_1, b) &= q_1\\ \delta(q_2, a) &= q_3\\ \delta(q_2, b) &= q_2\end{aligned}$$

$$\begin{aligned}
\delta(q_3, a) &= q_4 \\
\delta(q_3, b) &= q_3 \\
\delta(q_4, a) &= q_4 \\
\delta(q_4, b) &= q_4 \\
F &= \{q_0, q_1, q_2, q_3\}
\end{aligned}$$

This automaton has four final states. The state q_i with $i = 0, 1, 2, 3$ corresponds to the number of a's scanned so far. The symbol b will not change any state. Now as a fourth a is not welcome, if the automaton encounters a while in the state q_3, it reaches a non-final state, which is q_4.

The abbreviation FA stands for a finite automation. ■

Example 11.23

A finite automaton that accepts all strings over $\{x, y\}$ with an even number of y's:

$$\begin{aligned}
\delta(q_0, x) &= q_0 \\
\delta(q_0, y) &= q_1 \\
\delta(q_1, x) &= q_1 \\
\delta(q_1, y) &= q_0 \\
F &= \{q_0\}
\end{aligned}$$

Consider the string $xxyxyyx$. This string has three y's and hence has to be rejected. Let us now record the changes of state undergone by the finite automaton as it scans the given string. The first coloumn represents the part of the string scanned so far and the second coloumn gives the state of the FA into which it goes after the current scan.

Table 11.1

	q_0
x	q_0
xx	q_0
xxy	q_1
$xxyx$	q_1
$xxyxy$	q_0
$xxyxyy$	q_1
$xxyxyyx$	q_1

From the table, it is clear that the state would be q_0 if the number of y's scanned so far is even and the state would be q_1 if the number of y's scanned so far is odd. Since q_0 is the only final state, a string containing an odd number of y's would be rejected and a string containing an even number of y's would be accepted. ■

PRACTICE 11.9

Construct a finite automaton that accepts all strings of zeros and ones with a number of ones being a multiple of 3.

PRACTICE 11.10

Construct a finite automaton that accepts strings over $\{0, 1\}$ with exactly one zero and not more than two ones.

Example 11.24

Consider Figure 11.1. Identify the finite automaton it represents and also the language accepted by it.

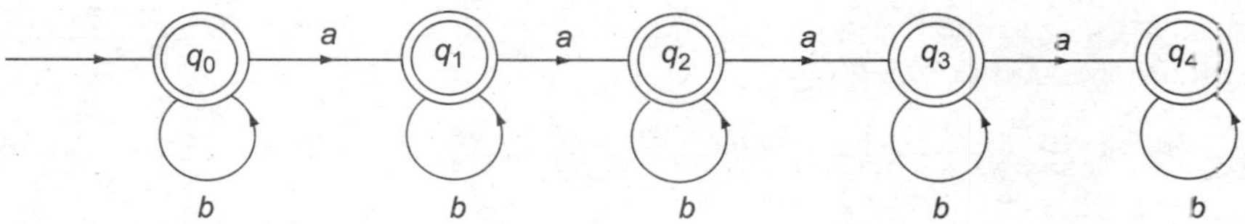

Figure 11.1

Definition 11.7: A string is said to be accepted by a finite automaton if the finite automaton reaches a final state after scanning the last symbol of the string. A string is said to be rejected by a finite automaton if the finite automaton reaches a non-final state after scanning the last symbol of the string.

Definition 11.8: Let L be a language over Σ. We say that an FA, whose input alphabet is Σ, accepts a language L if every string in L gets accepted by it and every string in L^c is rejected by it. Here L^c is the compliment of L in Σ^*. A language is said to be regular if there exists a finite automaton that accepts it.

PRACTICE 11.11

Show that $L = \{w \in \{a, b\}^* : n(w) \geq 6\}$ is a regular language.
Hint: Construct a finite automaton that accepts L.

PRACTICE 11.12

Show that the language L over $\{a, b\}$ consisting of all strings containing an odd number of a's and an even number of b's is a regular language.

Example 11.25

Identify the language L over $\{a, b\}$, which is accepted by the finite automaton whose transition function is given by

$$\delta(q_0, a) = q_1$$
$$\delta(q_0, b) = q_0$$
$$F = \{q_1\}$$

SOLUTION

Since q_0 is not an element of F, we find that if b is the last symbol scanned, then the string would be rejected. We also find that if a is the last symbol that is scanned, then the string would be accepted. Thus, this finite automaton accepts L over $\{a\}$ consisting of strings, which end with a. ■

PRACTICE 11.13

Identify the language over $\{a\}$ accepted by the finite automaton whose transition function is given by

$$\delta(q_0, a) = q_0$$
$$\delta(q_0, b) = q_1$$
$$\delta(q_1, a) = q_1$$
$$\delta(q_1, b) = q_1$$
$$F = \{q_1\}$$

PRACTICE 11.14

Construct a finite automaton accepting the language of strings over $\{a, b\}$ whose first symbol is a and the last symbol is b.

The following example is a good illustration of how the abstract set theory is useful in a highly practical field such as computer science.

Example 11.26

If L_1 and L_2 are regular languages over Σ, show that $L_1 \cap L_2$ is regular.

SOLUTION

Since L_1 and L_2 are regular, there exist finite automata that accept them. So, let $(Q_1, \Sigma, \delta_1, q_1, F_1)$ and $(Q_2, \Sigma, \delta_2, q_2, F_2)$ be, respectively, the finite automata accepting L_1 and L_2. Now, to show that $L_1 \cap L_2$ is regular, we construct a finite automaton that accepts it. Let $Q = Q_1 \times Q_2$, $q_0 = (q_1, q_2)$ and $F = F_1 \times F_2$.

Let us now define the transition function $\delta : Q \times \Sigma \to Q$.

$$\delta((u, v), x) = (\delta_1(u, x), \delta_2(v, x))$$

If a string is in L_1, then after the scanning of the last symbol, this new FA would be in the state (u, v) where $u \in F_1$. If a string is in L_2, then after the scanning of

the last symbol, this new FA would be in the state (u, v) where $v \in F_2$.
Since $(u, v) \in F$ if and only if $u \in F_1$ and $v \in F_2$, the string would be accepted if and only if it is in both $L_1 \cap L_2$.
Thus, we have constructed an FA, which accepts $L_1 \cap L_2$, and hence it is regular. ■

PRACTICE 11.15

Consider the finite automaton of Example 11.22. Define a new finite automaton in which the set of final states is changed to $F^c = \{q_4\}$. Identify the language accepted by this new finite automaton.

11.2.2 Regular Expressions for Regular Languages

A regular expression is a way of expressing a regular language. Before giving the formal definition, let us see some examples to understand the idea.

Example 11.27

Let $\Sigma = \{a, b\}$. The expression $(aab + b)^*$ represents a language $L \subseteq \Sigma^*$. Consider $\Sigma_1 = \{aab, b\}$ as a new set of symbols. Now, $L = \Sigma_1^*$. L consists of all words that can be formed by the symbols aab and b. It is clear that $bbaabb \in L$, $aba \notin L$, $aabaabbbbaab \in L$, $baaab \notin L$, etc. L is nothing but strings made of a's and b's such that every a in the string, if any, must be followed by ab. ■

Example 11.28

Give a regular expression for the following language:

$$L = \{w \in \{1, 0\}^* : n(w) \geq 2\}$$

SOLUTION

L is the set of all strings of length at least 2.

We note that a string in L must contain any one of aa, ab, ba or bb as its substring. The symbols occurring before or after them are irrelevant. Hence its regular expression is

$$(a + b)^*aa(a + b)^* + (a + b)^*ab(a + b)^* + (a + b)^*ba(a + b)^*$$
$$+(a + b)^*bb(a + b)* = (a + b)^*(aa + ab + ba + bb)(a + b)^*$$ ■

Example 11.29

The regular expression for the language consisting of integer constants in C is

$$(1 + 2 + 3 + 4 + 5 + 6 + 7 + 8 + 9)$$
$$(0 + 1 + 2 + 3 + 4 + 5 + 6 + 7 + 8 + 9)^*$$

Now, we formally define languages described by a regular expression: ■

Definition 11.9: The language $L(r)$ denoted by any regular expression r is defined using the following rules.

(i) ϕ is a regular expression denoting the empty set.
(ii) ϵ is a regular expression denoting $\{\epsilon\}$.
(iii) **a** is the regular expression for $\{a\}$.
(iv) $L(r_1 + r_2) = L(r_1) \cup L(r_2)$.
(v) $L(r_1 r_2) = L(r_1)L(r_2)$.
(vi) $L((r_1)) = L(r_1)$.
(vii) $L(r_1^*) = (L(r_1))^*$.

Example 11.30

Give a regular expression for the following language:

$$L = \{w \in \{1, 0\}^* : n(w) \geq 3\}$$

SOLUTION

L is the set of all strings of length at least 3. Its regular expression is

$$(a + b)^*(aaa + aab + aba + baa + bab + bba + abb + bbb)(a + b)^*$$ ■

Example 11.31

Give regular expressions for the following languages:

(a) All strings in $\{0, 1, 2\}^*$containing exactly two twos.
(b) All strings that contain at least one occurence of each symbol in $\Sigma = \{a, b, c\}$.
(c) All strings in $\{0, 1\}^*$ with no two consecutive zeros.
(d) Strings in $\{0, 1\}$ with an even number of ones.

SOLUTION

(a) $(0 + 1)^*2(0 + 1)^*2(0 + 1)^*$.
(b) $(a + b + c)^*a(a + b + c)^*b(a + b + c)^*c(a + b + c)^*$.
(c) $(1 + 01)^*(0 + \epsilon)$.
(d) $(0^*10^*10^*)^* + 0^*$. ■

Example 11.32

Find the regular expressions for $L = \{a^n b^m, n \geq 4, m \leq 3\}$ and its complement.

SOLUTION

The regular expression for L is $aaaa(a^*)(\epsilon + b + bb + bbb)$. Next, we find the compliment of L in Σ^*. Note that L is a subset of L_1 whose regular expression is given by a^*b^*. The compliment of L in L_1 is easily written as $\{a^n b^m : n < 4\} \cup \{a^n b^m : m > 3\}$.

The regular expression for this is $(\epsilon + a + aa + aaa)b^* + a^*b^*bbbb$.

But $L^c = (L_1 - L) \cup L_1^c$. L_1^c is nothing but the set of strings in which a b must be followed by an a whose regular expression is $(a+b)^*ba(a+b)^*$. Hence the required regular expression is

$$(\epsilon + a + aa + aaa)b^* + a^*b^*bbbb + (a+b)^*ba(a+b)^* \qquad \blacksquare$$

PRACTICE 11.16

Find the regular expression for the language consisting of octal integer constants in C.

Example 11.33

A hexadecimal integer constant in the C language is represented by a string that begins with $0X$ and that is made up of the symbols $0, 1, 2, 3, 4, 5, 6, 7, 8, 9, A, B, C, D, E, F$. Let us recall that A represents 10, B represents 11 and so on up to F representing 15. The set of all hexadecimal integer constants is a language over the alphabet

$$\{0, 1, 2, 3, 4, 5, 6, 7, 8, 9, X, A, B, C, D, E, F\}$$

Its regular expression is

$$0X(0+1+2+3+4+5+6+7+8+9+A+B+C+D+E+F)^* \qquad \blacksquare$$

Example 11.34

Show that the language consisting of all strings with an even number of a's and an even number of b's is a regular language. Write down a regular expression for it.

SOLUTION

At every stage of scanning, for the number of a's and b's scanned so far, we have the following possibilities:

Even a's and even b's.

Odd a's and even b's.

Even a's and odd b's.

Odd a's and odd b's.

We can represent the above situation by the states q_0, q_1, q_2 and q_3, respectively. If we are in state q_2, and we scan a, then we move to q_- and we can similarly find the other moves. Now,the transition function δ can be defined based on these above moves:

$$\begin{aligned}
\delta(q_0, a) &= q_1 \\
\delta(q_0, b) &= q_2 \\
\delta(q_1, a) &= q_0 \\
\delta(q_1, b) &= q_3 \\
\delta(q_2, a) &= q_3 \\
\delta(q_2, b) &= q_0 \\
\delta(q_3, a) &= q_2 \\
\delta(q_3, b) &= q_1 \\
F &= \{q_0\}
\end{aligned}$$

Its regular expression is $(aa + bb + (ba + ab)(aa + bb)^*(ba + ab))^*$ ■

Example 11.35

(a) Write the regular expression for strings beginning with 101 and construct a finite automaton that accepts it.

(b) Write the regular expression for strings ending with 101 and construct a finite automaton that accepts it.

SOLUTION

(a) The regular expression is given by $101(1 + 0)^*$. The transition function for the finite automaton is as follows:

$$\begin{aligned}
\delta(q_0, 0) &= q_0 \\
\delta(q_0, 1) &= q_1 \\
\delta(q_1, 0) &= q_2 \\
\delta(q_2, 1) &= q_3 \\
\delta(q_3, 0) &= q_3 \\
\delta(q_3, 1) &= q_3 \\
\delta(q_1, 1) &= q_1 \\
\delta(q_2, 0) &= q_2 \\
F &= \{q_3\}
\end{aligned}$$

(b) The regular expression is given by $(1 + 0)^*101$. The transition function for the finite automaton is as follows:

$$\begin{aligned}
\delta(q_0, 0) &= q_0 \\
\delta(q_0, 1) &= q_1 \\
\delta(q_1, 1) &= q_1
\end{aligned}$$

$$\begin{aligned}
\delta(q_1, 0) &= q_2 \\
\delta(q_2, 1) &= q_3 \\
\delta(q_2, 0) &= q_2 \\
\delta(q_3, 0) &= q_0 \\
\delta(q_3, 1) &= q_0 \\
F &= \{q_3\}
\end{aligned}$$

■

Example 11.36

Write the regular expression for $L = \{w \in \{a, b\}^* : n_a(w) \leq 2\}$.

L is the language over $\{a, b\}$ consisting of not more than two a's. Its regular expression is $b^* + b^*ab^* + b^*ab^*ab^*$. ■

Example 11.37

Write the regular expression for $L = \{w \in \{a, b\}^* : n(w) \equiv 2 \bmod 3\}$ and construct a finite automaton that accepts it.

SOLUTION

Its regular expression is $(a + b)((a + b)(a + b)(a + b))^*(a + b)$:

$$\begin{aligned}
\delta(q_0, a) &= q_1 \\
\delta(q_0, b) &= q_1 \\
\delta(q_1, a) &= q_2 \\
\delta(q_1, b) &= q_2 \\
\delta(q_2, a) &= q_0 \\
\delta(q_2, b) &= q_0 \\
F &= \{q_2\}
\end{aligned}$$

■

PRACTICE 11.17

Write the regular expression for $L = \{w \in \{a, b\}^* : n_a(w) \equiv 2 \bmod 3\}$ and construct a finite automaton that accepts it.

PRACTICE 11.18

Write the regular expression for $L = \{w \in \{a, b\}^* : n(w) \equiv 1 \bmod 4\}$ and construct a finite automaton accepting it.

Example 11.38

Write the regular expression for $L = \{w \in \{a, b\}^* : n_b(w) \equiv 2 \bmod 3\}$ and construct a finite automaton that accepts it.

SOLUTION

$(a^*ba^*ba^*ba^*)^*bb + b(a^*ba^*ba^*ba^*)^*b + bb(a^*ba^*ba^*ba^*)^*$.

The transition function for the finite automaton is given by:

$$\begin{aligned} \delta(q_0, a) &= q_0 \\ \delta(q_0, b) &= q_1 \\ \delta(q_1, a) &= q_1 \\ \delta(q_1, b) &= q_2 \\ \delta(q_2, a) &= q_2 \\ \delta(q_2, b) &= q_0 \\ F &= \{q_2\} \end{aligned}$$

PRACTICE 11.19

Write the regular expression for $L = \{w \in \{a, b, c\}^* : n_a(w) \leq 2\}$. Also, construct a finite automaton that accepts it.

We now state a theorem whose proof can be found in the list of references.

Theorem 11.1

Any language that can be described by a regular expression is a regular language. That is, there is a finite automaton accepting it.

Example 11.39

A Language that is not Regular: Consider the language $\{a^n b^n : n \geq 1\}$. Can it be described by a regular expression? No. To create a string in this language, we have to simultaneously work at both the ends, and this cannot be achieved by a regular expression. As per the above theorem, there can be no finite automaton that accepts this language.

11.3 PUSHDOWN AUTOMATA AND CONTEXT-FREE LANGUAGES

Recall the expression (b) of Example 11.3. We observed that the first square bracket [did not have a matching left bracket]. To simplify the discussion, let us forget about the variables and the arithmetic operations and consider only the brackets. Now,

[[[]]] is a valid arrangement, whereas [[[]] is not. Now, we can think of a language over the alphabet {], [}, which contains the correctly matched strings [], [[]], [[[]]], etc. The question we ask now is whether there is a finite automaton that accepts these correctly matched strings and that rejects strings like [[], etc. It is clear that the language under discussion is simply the language $\{a^n b^n : n \geq 1\}$ in a different form.

In the previous section, we noted that the language $\{a^n b^n : n \geq 1\}$ is not a regular language. This language is an example of a *context-free language*. Context-free languages are more complicated than the regular languages and they require a language acceptor more advanced than that of a finite automaton.

After a finite automaton scans a symbol and changes its state, it does not remember the symbol it has scanned. It is clear that a machine that must accept the language $\{a^n b^n : n \geq 1\}$ must have some memory device to keep track of the number of a's it has scanned. A pushdown automaton has such a device.

Definition 11.10: A non-deterministic pushdown automaton (NPDA) is a system $(Q, \Sigma, \Gamma, \delta, q_0, z, F)$ where

(a) Q is a finite set of **internal states**.

(b) Σ is a finite set of symbols called the **input alphabet**.

(c) Γ is a finite set of symbols called the **stack alphabet**.

(d) $\delta : Q \times (\Sigma \cup \{\epsilon\}) \times \Gamma \longrightarrow$ finite subsets of $(Q \times \Gamma^*)$ is the **transition function**.

(e) $q_0 \in Q$ is the initial state.

(f) $z \in \Gamma$ is the stack start symbol.

(g) $F \subseteq Q$ is a set of **final states**.

The pushdown automata, in addition to an 'input file', has a memory device called 'stack'. The input file is shown as a horizontal tape in Figure 11.2. The vertical strip represents the stack. We can only change the topmost object of the stack. When the topmost object is removed, the object immediately below the removed object comes up. When we place a new object on the stack, the earlier topmost object moves down and we can only observe the new topmost object.

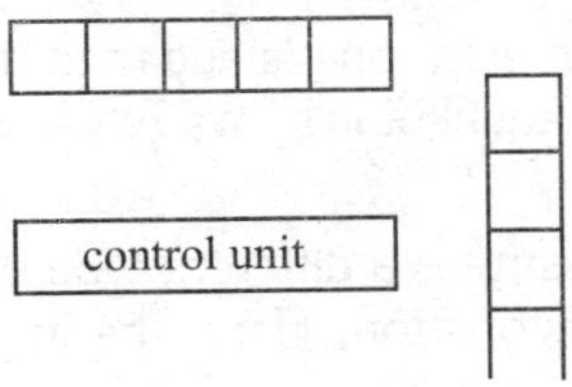

Figure 11.2

The stack is used as a memory device to record the scanned symbols. Depending on the symbol being scanned, a new stack symbol is added or removed. The NPDA starts with z in its topmost stack.

Unlike the finite automaton of the previous section, the transition function is *multi-valued*. For example, if $\delta(q, x, \alpha) = \{(p_1, \beta\alpha), (p_2, \gamma\alpha), (p_3, z)\}$ then when the symbol x is scanned in state q one of the following things occur:

(a) The state changes to p_1 and a β is added to the stack.

(b) The state changes to p_2 and a γ is added to the stack.

(c) The state changes to p_3 and the symbol α is removed from the stack.

Just as in the case of an FA, if the NPDA reaches a final state after scanning the last symbol of the input string, we say that the string has been accepted. But there is also another mode of acceptance, which is defined below.

Definition 11.11: A language is said to be accepted by an NPDA as an *empty stack* if it is a set of all inputs for which some sequence of moves causes it to empty the stack.

Before describing the pushdown automaton that accepts the language $\{a^n b^n :\}$, we give the example of a pushdown automaton that accepts a bigger but less complicated language. This illustrates the use of the stack.

Example 11.40

An NPDA that accepts the language $\left\{w \in \{a, b\}^+ n_a(w) = n_b(w)\right\}$

$$\begin{aligned}
\delta q_0, a, z &= \{(q_0, 0z\} \\
\delta q_0, b, z &= \{q_0, 1z\} \\
\delta q_0, a, 1 &= \{(q_0, \lambda\} \\
\delta q_0, b, 1 &= \{q_0, 11\} \\
\delta q_0, a, 0 &= \{(q_0, 00\} \\
\delta q_0, b, 0 &= \{q_0, \lambda\}
\end{aligned}$$

Whenever an a is scanned, if the top of the stack is empty, then a zero is placed there. Whenever a b is scanned, if the top of the stack is empty, then a one is placed there. Whenever an a is scanned, if the top of the stack contains zero, then an extra zero is added to the stack. Whenever a b is scanned, if the top of the stack contains zero, it is removed. Whenever an a is scanned, if the top of the stack contains one, then it is removed. Whenever a b is scanned, if the top of the stack contains one then an extra one is added to the stack. Thus, only when the number of a's and b's are equal, finally, we reach an empty stack. ■

The following example illustrates a different type of language acceptance, which is similar to that of a finite automaton. Here, the input string causes the NPDA to reach a final state.

Example 11.41

An NPDA that accepts the language $\{a^n b^n :\} \cup \{a\}$.

$$\begin{aligned}
Q &= \{q_0, q_1, q_2, q_3\} \\
\Sigma &= \{a, b\} \\
\Gamma &= \{0, 1\} \\
z &= 0 \\
F &= \{q_3\}
\end{aligned}$$

We now describe the transition function. Here λ is the empty string corresponding to the stack symbols.

$$\begin{aligned}
\delta(q_0, a, 0) &= \{(q_1, 10), (q_3, \lambda)\} \\
\delta(q_0, \lambda, 0) &= \{(q_3, \lambda)\} \\
\delta(q_1, a, 1) &= \{(q_1, 11)\} \\
\delta(q_1, b, 1) &= \{(q_2, \lambda)\} \\
\delta(q_2, b, 1) &= \{(q_2, \lambda)\} \\
\delta(q_2, \lambda, 0) &= \{(q_3, \lambda)\}
\end{aligned}$$

■

PRACTICE 11.20

Modify the above NPDA to make it accept $\{a^n b^n : n \geq 1\}$.

PRACTICE 11.21

Construct an NPDA that accepts the language $L = \left\{ww^R : w \in \{a, b\}^+ :\right\}$, which is the set of even palindromes in a, b.

Definition 11.12: A language is said to be context-free if there is a non-deterministic pushdown automaton that accepts it.

We have noted that an FA is not suitable to recognize context-sensitive languages. They require pushdown automata. So, the next natural question is to ask if there are languages that cannot be recognized by a pushdown automaton. The answer is yes.

Example 11.42

An example of a language that is not context-free:

The language $\{a^n b^n c^n : n \geq 1\}$ is not context free and hence there is no NPDA automaton that accepts it. ■

The above language is an example of context-sensitive language.

11.4 TURING MACHINES

In a strict logical order, a linear bounded automaton must be discussed before the Turing machine. But it is easier to understand the Turing machine first and the linear bounded automaton later.

11.4.1 Introduction and Motivation

Imagine a row of houses near Satyamangalam forest area. A determined and happy Veerappan is standing in front of the first house. He enters the house and finds that the house does not have anyone who could be kidnapped; but he continues to be happy and he moves to the next house. In the representation given below, B denotes an empty house, S denotes the fact that the house is occupied by a policeman of STF, T denotes the presence of a tribal person, and H denotes the state that Veerappan is happy. A denotes the fact that he is afraid.

B	S	B	T	T	S	T	S	T	T
H									

Now after entering the second house, he finds that it is occupied by a sleepy policeman of the special task force (STF). He comes out a bit afraid and decides to move to the next house.

B	S	G	T	T	S	T	S	T	T
	H								

B	S	G	T	T	S	T	S	T	T
		A							

Now, in the third house, he finds one of his own gangmen and so comes out happily and moves to the second house again to tackle the policeman.

B	S	G	T	T	S	T	S	T	T
	H								

Now, let T represent the tribal people of the forest. You can continue the story in whichever way you like.
The important points to be noted are the following:

(i) Veerapan's behaviour depends not only on who is there in the house, but also on his mental state.

(ii) His movement to the left or to the right also depends on the inmate of the house he examines.

(iii) Depending on who is present in the house, his mental state also changes.

Table 11.2

	0	B
q_0	$(q_0, 0, R)$	$(q_1, 0, R)$
q_1		(q_2, B, R)
q_2		

This idea illustrates the idea of a Turing machine.

Informally, a Turing machine consists of a one-way infinite tape with some inputs on the cells. A control keeps moving to the left or to the right. Depending on the input and what is known as an internal state of the control, it may change the input, go into a different state and move either to the left or to the right. To understand the meaning of the internal states, let us imagine a computer. This computer consists of a single program that calculates the square of an integer. If the input is 3, the output is 9. Now let us imagine that this program that calculates the square has been deleted, and that instead a program that computes the cube is stored. If the input now is 3, then the output is 27. In some sense, when we write a program, we alter the internal state of the computer and the output depends on this state as well as the input.

Before we give the formal definition of a Turing machine, consider a one-way infinite tape and a control that can exist in the states q_0, q_1 and q_2. The tape cells are either blank or contain zeros. A blank cell is supposed to contain the symbol B. The state q_2 is a final state. That is, when the control attains this state, it halts.

This control, when it encounters 0 in q_0, moves to the right without changing the input and without changing its state. When the control encounters a B in state q_0, it replaces the latter by 0, changes to state q_1 and moves to the right. When the control encounters a B in state q_1, it changes its state to q_2 and moves to the right. These details can be summarized in a tabular form as follows.

Now, let us start with an infinite tape consisting of three zeros in the first three cells and blanks everywhere else. Let the control be placed at the left-end cell. Let us trace the changes the tape undergoes.

0	0	0	B	B	B
q_0					

0	0	0	B	B	B
	q_0				

0	0	0	B	B	B
		q_0			

0	0	0	B	B	B
			q_0		

0	0	0	0	B	B
				q_1	

0	0	0	0	B	B
					q_2

The control has stopped after increasing the number of zeros by 1. Now, it is clear that if we start with the initial tape configuration as

0	0	0	0	B	B
					q_2

then the final configuration will be

0	0	0	0	0	B	B
					q_2	

If there are n zeros in the tape initially, then when the Turing machine stops, there would be $n+1$ zeros in the tape. Hence, the above Turing machine can be thought of as a machine that computes the function $f : N \longrightarrow N$ defined as $f(n) = n+1$. In Section 2 of Chapter 12, we shall study Turing machines as machines that would compute functions. In Section 6 of Chapter 11, we shall study the Turing machines as language acceptors.

Just as in the case of finite automata, the transition function δ is the heart of a Turing machine. The transition function δ for the above Turing machine is defined as follows: $\delta(q_0, 0) = (q_0, 0, R)$, $\delta(q_0, 0) = (q_1, 0, R)$, $\delta(q_1, B) = (q_2, B, R)$.

You may note that in the previous example $\delta(q_2, 0)$ and $\delta(q_2, 1)$ are not defined. Thus, δ is only a *partial function*.

If $\delta(q_2, 0)$ were to be defined, it would mean that the Turing machine would carry out a further task even after reaching a final state. This is like a computer still doing some calculations after producing the output. Forgetting parallel processing and such advanced notions, we would like to know when the calculations come to an end. Thus, it is desirable that $\delta(q_2, 0)$ is left undefined.

The concept of a function defined in Chapter 2 is a bit too rigid to accommodate the transition functions. Thus, we make the following definition.

Definition 11.13: Let $B \subseteq A$ and $f : B \longrightarrow C$ be a function. We say that f is a partial function from A into C.

The reason we have to work with the idea of a partial function is that in many practical situations connected with language acceptors, we know what A is but we we do not know what B would be, in advance. If $A_1, A_2, \cdots, A_n$ are sets, then an element of A_1 times A_2 times A_3 times $\cdots$ A_n is called an n-tuple. Now, we give the formal definition of a Turing machine.

Definition 11.14: A Turing machine is a 7-tuple $M = (Q, \Sigma, \Gamma, \delta, q_0, B, F)$ where

(1) Q is the finite set of states.

(2) Γ is the finite set of allowable tape symbols.

(3) B is a symbol of Γ, which is the blank.

(4) Σ, a subset of not including B, is the set of input symbols.

(5) δ is the next move function, which is a partial function from $Q \times \Gamma \longrightarrow Q \times \Gamma \times \{L, R\}$.
(6) q_0 is the start state.
(7) $F \subset Q$ is the set of final states.

Definition 11.15: We say that the Turing machine halts if the control is in the state q, the input symbol that is being scanned is x and $\delta(q, x)$ is not defined.

Note that if $q \in F$, by the very definition of final state, $\delta(q, x)$ is not defined for any $x \in \Gamma$.

Definition 11.16: When the control reaches a final state, we say that the machine halts in an accepting state and if $q \notin F$ but $\delta(q, x)$ is not defined, we say that the machine halts in a non-accepting state.

We have already seen that finite automata can be used as langauge acceptors. In this section, we shall see how a Turing machine can be used to accept some languages.

Definition 11.17: The language accepted by a Turing machine M denoted by $L(M)$ is the set of those words in Σ^* that causes M to enter a final state when placed justified at the left, on the tape of M, with M in state q_0, and the tape head of M at the leftmost cell.

Remark 11.1

Note that for a string not in Σ^*, the Turing machine may either stop at a non-final state or may not stop at all.

Example 11.43

A Turing machine that accepts a string containing three consecutive zeros:

$$\begin{aligned}
\delta(q_0, 0) &= (q_1, R, 0)\\
\delta(q_0, 1) &= (q_0, R, 1)\\
\delta(q_1, 0) &= (q_2, R, 0)\\
\delta(q_1, 1) &= (q_0, R, 1)\\
\delta(q_2, 0) &= (q_f, R, 0)\\
\delta(q_2, 1) &= (q_0, R, 1)
\end{aligned}$$

■

We shall learn more about languages accepted by Turing machines in Section 11.6.

11.5 LINEAR BOUNDED AUTOMATA AND CONTEXT-SENSITIVE LANGUAGES

We have seen that the language $\{a^n b^n c^n : n \geq 1\}$ is not context-free and hence there is no pushdown automaton to accept it. We have also stated that it is an example of a context-sensitive language. The language acceptors that accept the class of context-sensitive languages are known as *Linear Bounded Automata*.

Informally, a linear bounded automaton is a non-deterministic Turing machine whose head moves only within the cells occupied by the input string. Are there languages whose acceptance requires scanning of cells beyond those occupied by the input string? There are such languages, but they cannot be constructed in an explicit manner. We learn about these languages in the next section.

Now, we define a linear bounded automaton. There are two special symbols denoted by [and]. If $w \in \Sigma^*$ is the word to be tested, the input string is taken as $[w]$. When the symbol [is scanned, only a right move is allowed and when the symbol] is scanned, only a left move is allowed.

Definition 11.18: A linear bounded automaton is a 7-tuple $M = (Q, \Sigma, \Gamma, \delta, q_0, B, F)$ where

(1) Q is the finite set of states.

(2) The two special symbols [,] must belong to Q.

(3) Γ is the finite set of allowable tape symbols.

(4) B is a symbol of Γ, which is the blank.

(5) Σ, a subset not including B, is the set of input symbols.

(6) $\delta : Q \times \Gamma \longrightarrow 2^{Q\times\Gamma\times\{L,R\}}$ is the next move function such that $\delta(q_i, [)$ can contain only elements of the form $(q_j, [, R)$, and $\delta(q_i,])$ can contain only elements of the form $(q_j,], L)$.

(7) q_0 is the start state.

(8) $F \subset Q$ is the set of final states.

Example 11.44

A linear bounded automaton that accepts the language $\{a^n b^n c^n : n \geq 1\}$.

We define the transition function for this linear bounded automata as follows:

$$\begin{aligned}
\delta(q_0, [) &= (q_0, [, R)\\
\delta(q_0, a) &= (q_1, X, R)\\
\delta(q_1, a) &= (q_1, X, R)\\
\delta(q_1, Y) &= (q_1, Y, R)\\
\delta(q_1, b) &= (q_2, Y, R)\\
\delta(q_2, Y) &= (q_2, Y, R)\\
\delta(q_2, Z) &= (q_2, Z, R)\\
\delta(q_2, c) &= (q_3, Z, R)\\
\delta(q_3,]) &= (q_5,], L)\\
\delta(q_3, c) &= (q_4, c, L)\\
\delta(q_4, Z) &= (q_4, Z, L)\\
\delta(q_4, Y) &= (q_4, Y, L)
\end{aligned}$$

$$\begin{aligned}
\delta(q_4, a) &= (q_4, a, L) \\
\delta(q_3, X) &= (q_4, X, R) \\
\delta(q_4, a) &= (q_1, X, R) \\
\delta(q_5, Z) &= (q_5, Z, L) \\
\delta(q_5, Y) &= (q_5, Y, L) \\
\delta(q_5, X) &= (q_f, X, R)
\end{aligned}$$

After changing the first a into X, the head looks for b and changes it into Y; then it goes to look for c to change it into Z. The head moves to the left, and finds the next a. ■

PRACTICE 11.22

Consider the strings *aaabbc* and *aabacc*. Trace the action of the above linear bounded automaton on these strings.

PRACTICE 11.23

Construct a linear bounded automaton that accepts $\{ww : w \in \{a, b\}^+\}$.
Hint: First, we have to locate the middle of the string. The leftmost symbol is changed into X or Y depending on whether it is a or b. Then rightmost symbol is changed in a similar fashion. After changing the rightmost symbol, the leftmost *unchanged a or b* is changed and so on. We give the transition function as follows:

$$\begin{aligned}
\delta(q_0, [) &= ([, q_0, R) \\
\delta(q_0, a) &= (X, q_1, R) \\
\delta(q_0, b) &= (Y, q_1, R) \\
\delta(q_1, a) &= (a, q_1, R) \\
\delta(q_1, b) &= (b, q_1, R) \\
\delta(q_1,]) &= (], q_2, L) \\
\delta(q_1, X) &= (X, q_2, L) \\
\delta(q_1, Y) &= (Y, q_2, L) \\
\delta(q_2, a) &= (X, q_3, L) \\
\delta(q_2, b) &= (Y, q_3, L) \\
\delta(q_3, X) &= (q_4, P, L) \\
\delta(q_3, Y) &= (q_4, Q, L)
\end{aligned}$$

Let us explain what happens if the length of the string is odd with an example. If the length is 7, then after changing the fourth symbol, the machine would be in state q_2 and would be looking for a's or b's, which it cannot find and the machine would be stopped in a non-accepting state. If the number of symbols is even, after changing the last symbol of the left half it would change the first symbol of the right half and immediately detect X or Y on its left. At this stage, the machine would be in state q_4. The last symbol of the first half is marked by P or Q.

Next, the symbols have to be matched to find if the given string is ww. The left-most symbol of the first half has to be matched with the left-most symbol of the second half and so on. Further states need to be defined and δ has to be extended.

11.6 TURING MACHINES AND RECURSIVELY ENUMERABLE LANGUAGES

A Turing machine differs from other language acceptors in an essential way. A Turing machine need not halt on some inputs. This gives rise to two classes of languages associated with a Turing machine. These are known as *recursively enumerable* and *recursive* languages. We have already defined what we mean by a language being accepted by a Turing machine in Section 11.4. Now, we define the above-mentioned classes of languages.

Definition 11.19: A language is said to be *recursively enumerable* if there exists a Turing machine that accepts it.

Now, let L be a recursive language $w \notin L$ and let M be a Turing machine accepting L. Then, there are two possibilities for the input w. That is, either the machine would halt in a non-accepting state or it will not halt at all. This gives rise to a subclass of recursive languages, which is defined below.

Definition 11.20: A language is said to be *recursive* if there exists a Turing machine, which halts on every input in Σ^+ that accepts it.

Almost any formal language one can think of is context-free. There exist languages that are not context-free but recursively enumerable. But we cannot describe such a language in an explicit manner.The reader may refer to the books suggested for further reading.

The language of the following example is already known to us. There is an NPDA accepting it. But as Turing machines have extra features, it is easy to construct a Turing machine that accepts this language.

Example 11.45

A Turing machine that accepts a strings of zeroes and ones with an equal number of them:

$$\delta(q_0, 0) = (q_1, R, B)$$
$$\delta(q_0, 1) = (q_2, R, B)$$

$$
\begin{aligned}
\delta(q_1, 0) &= (q_1, R, 0) \\
\delta(q_2, 1) &= (q_2, R, 1) \\
\delta(q_2, 0) &= (q_b, L, B) \\
\delta(q_1, 1) &= (q_b, L, B) \\
\delta(q_b, 0) &= (q_b, L, , 0) \\
\delta(q_b, 1) &= (q_b, L, 1) \\
\delta(q_b, B) &= (q_0, R, B) \\
\delta(q_2, 0) &= (q_0, R, B) \\
\delta(q_0, B) &= (q_f, R, B)
\end{aligned}
$$

■

In Chapter 12, we shall see how Turing machines can be used as models for computing.

11.7 FORMAL LANGUAGES AND GRAMMAR

11.7.1 Introduction

In this section, first we present an introduction to grammars. Then, we describe some formal languages that are simple models of programming languages. Next, we discuss how these languages can be generated by grammars. Then, we specify how these grammars are related to the machines described in the previous sections.

A native speaker of Malayalam can easily identify whether a particular sentence written in Malayalam is grammatically correct, even if he does not know the rules of grammar. But for someone who is learning Malayalam anew, a knowledge of grammar would be useful to identify mistakes. The grammars of natural languages are very complicated and designing a computer programme to find grammatical errors in a passage is not easy. But for programming languages, this process is easier. Informally, the grammar for a language is a set of rules that tell us how to construct correct sentences. Before giving the definition of a grammar, we discuss a simple sentence formation. This discussion would help us understand the definition of a grammar.

Consider the following sentence:

Yesterday, Muthu wrote a letter.

Now, to get a grammatically correct sentence from the above sentence, the name Muthu can be replaced by any other name.

We record the above observation in the following manner:

Yesterday, (name) wrote a letter.

name → Muthu
name → Ram
name → Fatima

The arrow indicates the allowed replacements.
Next, we consider further modifications:
Yesterday, (name) wrote a @@@@

@@@@ → poem
@@@@ → story
@@@@ → complaint

The above 'replacements' give us grammatically correct sentences like 'Yesterday, Ram wrote a story' and 'Yesterday, Fathima wrote a complaint'. We shall now define various types of grammar.

Definition 11.21: An unrestricted grammar is a quadruple (V, T, S, P) where V is a finite set of objects called variables, T is a finite set of objects called terminal symbols, $S \in V$ is called the start variable and P is a finite set of productions.

A production is of the form $x \to y$, where $x \in (V \cup T)^+$ and $y \in (VUT)^*$.

Definition 11.22: A grammar $G = (V, T, S, P)$ is said to be context-free if all productions in P have the form $A \to x$, where $A \in V$ and $x \in (V \cup T)^*$.

Definition 11.23:

(i) A grammar $G = (V, T, S, P)$ is said to be *right-linear* if all productions in P have the form $A \to xB$ or $A \to x$.

(ii) A grammar $G = (V, T, S, P)$ is said to be *left-linear* if all productions in P have the form $A \to Bx$ or $A \to x$.

(iii) A grammar $G = (V, T, S, P)$ is said to be *regular* if it is either left-linear or right-linear.

A grammar generates strings in a language over T. A variable plays the role of words like 'noun' and 'verb'. Ultimately, a grammatically correct string must contain only the terminals.

11.7.2 Regular Languages and Regular Grammars

We have already learnt about regular languages. In this section, we shall see how these languages are generated by regular grammars.

(i) A grammar $G = (V, T, S, P)$ is said to be *right-linear*, if all productions in P have the form $A \to xB$ or $A \to x$.

(ii) A grammar $G = (V, T, S, P)$ is said to be *left-linear*, if all productions in P have the form $A \to Bx$ or $A \to x$.

(iii) A grammar $G = (V, T, S, P)$ is said to be *regular* if it is either left-linear or right-linear.

Example 11.46

Let $G = (V, T, S.P)$, $V = \{S, S_1\}$, $T = \{x, y, z\}$ and the productions be given by

$$\begin{aligned} S &\rightarrow xS \\ S &\rightarrow yS_1 \\ S_1 &\rightarrow yS_1 \\ S_1 &\rightarrow z \end{aligned}$$

The first line indicates that S can be replaced by xS and so on. The final string should not involve any variables and should involve only terminals. Consider the following examples of formation of a string using this grammar:

(a) $S \rightarrow xS \rightarrow xxS \rightarrow xxyS_1 \rightarrow xxyz$

(b) $S \rightarrow xS \rightarrow xyS_1 \rightarrow xyS_1 \rightarrow xyyS_1 \rightarrow xyyz$

Now, $xxyz$ and $xyyz$ are words in the language generated by this grammar.

The first two productions tell us that a string in this language can have any number of x. The second and third productions tell us that the string of x should be followed by a string of y's. The last production tells us that the string must end with z. Thus, the language is $\{x^m y^n z, n \geq 1, m \geq 1\}$. ■

Note that the regular expression for the above language is xx^*yy^*z.
In Practice 11.24, we consider $L(x^*y^*z) = \{x^m y^n z, n \geq 0, m \geq 0\}$.

PRACTICE 11.24

Find right-linear grammar for the regular expression x^*y^*z.

Example 11.47

Consider the regular expression $0(01)^*$. Find a right-linear regular grammar that generates this language

SOLUTION

Let $V = \{S, A\}$, $T = \{x, y, z\}$. The productions are given by

$$\begin{aligned} S &\rightarrow 0A \\ A &\rightarrow 10A \\ A &\rightarrow \epsilon \end{aligned}$$

■

Example 11.48

Consider the regular expression $a^*ba^*b(a + b)^*$. A right-linear regular grammar that generates this language is

$$
\begin{aligned}
S &\rightarrow aS \\
S &\rightarrow bA \\
A &\rightarrow aA \\
A &\rightarrow bB \\
B &\rightarrow aC \\
B &\rightarrow bC \\
A &\rightarrow b \\
C &\rightarrow a \\
C &\rightarrow b
\end{aligned}
$$

■

Example 11.49

Construct right-linear and left-linear grammars for the language

$$L = \{a^n b^m : n \geq 2, m \geq 3\}$$

SOLUTION

$$
\begin{aligned}
S &\rightarrow Bbbb \\
B &\rightarrow Bb \\
B &\rightarrow Aaa \\
A &\rightarrow Aa \\
A &\rightarrow \epsilon
\end{aligned}
$$

■

Example 11.50

Find a right-linear regular grammar that generates the language $L(ba^*(a + bb)^*)$.

SOLUTION

$$
\begin{aligned}
S &\rightarrow bA \\
A &\rightarrow aA \\
A &\rightarrow \epsilon \\
A &\rightarrow bbB \\
B &\rightarrow aB \\
B &\rightarrow bbB \\
B &\rightarrow \epsilon
\end{aligned}
$$

■

PRACTICE 11.25

(a) Find a left-linear regular grammar that generates the language $L(ba^*(a+bb)^*)$.
(b) Find a regular grammar that generates the language $L(bb^*(ba+b)^*)$.

Example 11.51

Find right-linear and left-linear grammars for strings in $\{0,1\}^*$
(a) starting with 101
(b) ending with 101

(a) Starting with 101

Right linear	Left linear
$S \to 101$	$S \to A0$
$S \to 101A$	$S \to A1$
$A \to 0A$	$A \to A0$
$A \to 1A$	$A \to A1$
$A \to \epsilon$	$A \to B1$
$A \to 0$	$B \to C0$
$A \to 1$	$C \to 1$

(b) Ending with 101

Right linear	Left linear
$S \to A101$	$S \to B1$
$A \to A0$	$B \to C0$
$A \longrightarrow A1$	$C \to D1$
$A \to \epsilon$	$D \to D0$
$A \to 0$	$D \to D1$
$A \to 1$	$D \longrightarrow \epsilon$

■

Example 11.52

Let $a(a+b)^*$ be the regular expression for L_1 and let $(a+b)^*b$ be the regular language expression for L_2. Show that $L_1 \cup L_2$ is a regular language.

SOLUTION

Since $a(a+b)^* + (a+b)^*b$ is the regular expression for $L_1 \cup L_2$, it is trivial to see that $L_1 \cup L_2$ is regular. But we want to give a proof using the idea of grammar. Let us write down two right-linear grammars for L_1 and L_2. The grammar for L_1 is as follows:

$V = \{S_1, A\}$, $T = \{a, b\}$

$$S_1 \to aA$$
$$A \to aA$$

$$A \rightarrow bA$$
$$A \rightarrow \epsilon$$

The grammar for L_2 is as follows:
$V = \{S_2, B\}$, $T = \{a, b\}$

$$S_2 \rightarrow aS_2$$
$$S_2 \rightarrow bS_2$$
$$S_2 \rightarrow b$$

Now, to show that $L_1 \cup L_2$ is regular, it is enough to show that there is a right-linear grammar generating it. We construct such a grammar by introducing a new start symbol S and combining the productions of both the above grammars.

A right-linear grammar for $L_1 \cup L_2$ is as follows: $V = \{S, S_1, S_2, A, B\}$, $T = \{a, b\}$ and the productions are given by

$$S \rightarrow S_1$$
$$S \rightarrow S_2$$
$$S_1 \rightarrow aA$$
$$A \rightarrow aA$$
$$A \rightarrow bA$$
$$A \rightarrow \epsilon$$
$$S_2 \rightarrow aS_2$$
$$S_2 \rightarrow bS_2$$
$$S_2 \rightarrow bs$$

It is clear that any string generated by this new grammar can also be generated by one of the given grammars and vice versa. Thus, the new grammar is a grammar for $L_1 \cup L_2$. ■

PRACTICE 11.26

Let L_1 and L_2 be regular languages. Describe how one can obtain a regular grammar for $L_1 \cup L_2$.

The grammar of a formal language tells us how the strings of that language are formed. An acceptor tells us whether a given string belongs to the language or not. We have already defined a regular language as a language for which there exists a finite automaton that accepts it. Now, the next theorem tells that a regular language is a language that is generated by a left-linear or a right-linear grammar. For the proof, the reader may refer to the list of suggested books.

Theorem 11.2

A language is regular if and only if it is generated by a left-linear grammar or a right-linear grammar.

11.7.3 Context-Free Grammars and Context-Free Languages

We have already learnt about context-free languages. The grammar that generates a context-free language is called a context-free grammar.

Definition 11.24: A grammar $G = (V, T, S, P)$ is said to be context-free if all productions in P have the form $A \to x$, where $A \in V$ and $x \in (V \cup T)^*$.

Example 11.53

Find a context-free grammar for $L = \{a^n b^n : n \geq 0\}$:

SOLUTION

$$\begin{aligned} S &\to aSb \\ S &\to \epsilon \end{aligned}$$

■

Example 11.54

Find a context-free grammar for $L = \left\{ww^R : w \in \{0, 1\}^*\right\}$:

SOLUTION

$$\begin{aligned} S &\to aSa \\ S &\to bSb \\ S &\to \epsilon \end{aligned}$$

■

PRACTICE 11.27

Construct a context-free grammar for $L = \left\{wxw^R : w \in \{0, 1\}^*, x \in \{0, 1\}\right\}$.

Example 11.55

A context-free grammar for $L = \left\{a^n b^{2n} : n \geq 0\right\}$:

$$\begin{aligned} S &\to aSbb \\ S &\to \epsilon \end{aligned}$$

■

Example 11.56

Write down a context-free grammar for strings having an twice as many a's as b's:

$$
\begin{aligned}
S &\to abaA \\
S &\to aabA \\
S &\to baaA \\
A &\to abaA \\
A &\to aabA \\
A &\to baaA \\
A &\to \epsilon \\
S &\to \epsilon
\end{aligned}
$$

PRACTICE 11.28

Write down a context-free grammar for strings having an equal number of a's and b's

Example 11.57

Construct a context-free grammar for $L = \{a^n b^m : n \leq 2m\}$:

SOLUTION

$$
\begin{aligned}
S &\to \epsilon \\
S &\to S_1 b \\
S_1 &\to S_1 b \\
S_1 &\to aS_1 \\
S_1 &\to aaS_1 b \\
S_1 &\to \epsilon
\end{aligned}
$$

PRACTICE 11.29

Find a context-free grammar for $L = \{a^n b^m : n \neq m + 2\}$.

11.7.4 Context-Sensitive Grammars and Context-Sensitive Languages

We have already learnt about context-sensitive languages. The grammar that generates a context-sensitive language is called a context-sensitive grammar.

Definition 11.25: A grammar $G = (V, T, S, P)$ is said to be context-sensitive if all productions in P have the form $x \to y$, where $x, y \in (V \cup T)^+$ and $|x| \le |y|$.

Example 11.58

Find a context-sensitive grammar for $\{a^n b^n c^n : n \ge 1\}$.

SOLUTION

$$\begin{aligned}
S &\to abc \\
S &\to aAbc \\
Ab &\to bA \\
Ac &\to Bbcc \\
bB &\to Bb \\
aB &\to aa \\
aB &\to aaA
\end{aligned}$$

Example 11.59

Find a context-sensitive grammar for $\{a^n b^m c^n d^m : n \ge 1, m \ge 1\}$.

SOLUTION

$$\begin{aligned}
S &\to aAB \\
S &\to aB \\
A &\to aAX \\
A &\to aX \\
B &\to bBD \\
B &\to bYD \\
Xb &\to bX \\
XY &\to Yc \\
Y &\to c
\end{aligned}$$

PRACTICE 11.30

Discuss the following attempt to create a context-sensitive grammar for $\{ww : w \in \{a, b\}^*\}$:

$$\begin{aligned}
S &\to E_1 A E_2 \\
Aa &\to aA
\end{aligned}$$

$$
\begin{aligned}
Ab &\rightarrow bA \\
AE_2 &\rightarrow E_2A \\
E_2A &\rightarrow A_1aE_2 \\
bA_1 &\rightarrow A_1b \\
aA_1 &\rightarrow A_1a \\
E_1A_1 &\rightarrow aE_1A \\
E_1A_1 &\rightarrow aE_1B \\
S &\rightarrow E_1BE_2 \\
Ba &\rightarrow aB \\
Bb &\rightarrow bB \\
BE_2 &\rightarrow E_2B \\
E_2B &\rightarrow B_1bE_2 \\
bB_1 &\rightarrow B_1b \\
aB_1 &\rightarrow B_1a \\
E_1B_1 &\rightarrow bE_1A \\
E_1B_1 &\rightarrow bE_1B
\end{aligned}
$$

SUMMARY

- A language L over Σ is a set of strings made of symbols in Σ.
- A regular language is a language for which there exists a finite automaton that accepts it.
- A regular expression is an expression that describes the strings of a regular language.
- A grammar is a set of rules for generating the strings of a language.
- A language is regular if and only if it is generated by a left-linear grammar or a right-linear grammar.
- A context-free language is a language for which there exists a non-deterministic pushdown automaton that accepts it.
- A language is context-free if and only if it is generated by a context-free grammar.
- A context-sensitive language is a language for which there exists a linear bounded automaton that accepts it.
- A language is context-sensitive if and only if it is generated by a context-sensitive grammar.
- A recursively enumerable language is one for which there exists a Turing machine that accepts it.

PROBLEMS

1. Consider the language L over $\{a, b, c\}$ containing strings of length less than or equal to four. Find the number of elements in L, using the ideas from Chapter 4.
2. Construct a finite automaton that accepts strings of length less than or equal to 2 over the alphabet $\{0, 1\}$.
3. Represent the FA's of Example 11.23 and Practices 11.10 and 11.11.
4. Show that the language $L \subseteq \{a, b\}$ consisting of all strings with an even number of a's and an odd number of b's is a regular language. Write down a regular expression for it.
5. Let $(Q, \Sigma, \delta, q_0, F)$ be the finite automaton for L. Show that $(Q, \Sigma, \delta, q_0, F^c)$ is the finite automaton for L^c. Here F^c is the compliment of F in Q and L^c is the compliment of L in Σ^*.
6. Write down the regular expression for the language over $\{0, 1, 2\}$, which contains at least three ones.
7. If the regular expression for L_1 is $a(ba)^*b$ and the regular expression for L_2 is $(ab)^*$, how are L_1 and L_2 related?
8. Consider the finite automaton whose transition function is given below:

$$\begin{aligned}
\delta(q_0, a) &= q_1 \\
\delta(q_0, b) &= q_0 \\
\delta(q_1, b) &= q_2 \\
\delta(q_1, a) &= q_3 \\
\delta(q_2, b) &= q_2 \\
\delta(q_2, a) &= q_1 \\
\delta(q_3, a) &= q_3 \\
\delta(q_3, b) &= q_3 \\
F &= \{q_2\}
\end{aligned}$$

Identify the regular language accepted by the above finite automaton.
9. Construct a finite automaton for the language whose regular expression is given by $a(a + b)^*a$.
10. Consider the finite automaton whose transition function is given below:

$$\begin{aligned}
\delta(q_0, b) &= q_1 \\
\delta(q_0, a) &= q_0 \\
\delta(q_1, a) &= q_2 \\
\delta(q_1, b) &= q_1 \\
\delta(q_2, a) &= q_2 \\
\delta(q_2, b) &= q_1 \\
F &= \{q_2\}
\end{aligned}$$

Identify the regular language accepted by the above finite automaton.
11. Construct an NPDA that accepts $L = \{a^{2n}b^n : n \geq 1\}$.
12. Show that $L\, \{w \in \Sigma^* : n_a(w) = 2n_b(w)\}$ is context-free.

13. Construct a linear bounded automaton that accepts $L\{a^n b^m c^n d^m : n, m \geq 1\}$.
14. Construct a Turing machine that accepts $\{a^n b^{2n} : n \geq 1\}$.
15. Construct a regular grammar for the language represented by the regular expression $(bba + ab)^*$.
16. Identify the language generated by the grammar:

$$S \rightarrow aaSbbb$$
$$S \rightarrow \epsilon$$

17. Construct a finite automaton that accepts the language generated by the grammar

$$S \rightarrow 0A$$
$$A \rightarrow 10A$$
$$A \rightarrow \epsilon$$

18. Let L_1 and L_2 be regular languages. Show that $L_1 \cup L_2$ is regular by proving that any two right-linear grammars generated by L_1 and L_2 can be combined to give a right-linear grammar generating $L_1 \cup L_2$.
19. Examine if $L = \{a^n b^m : n \neq 3m\}$ is a context-free language, and if so, construct a context-free grammar for it.
20. Construct a context-free grammar for $L = \{a^n b^m : n = m + 3\}$
21. If L_1 and L_2 are context-free languages, describe how one may construct a context-free grammar for the concatenation $L_1 L_2$.
22. If L_1 and L_2 are context-sensitive languages, describe how one may construct a context-sensitive grammar for $L_1 \cup L_2$.

FURTHER READING

Martin, J. C. (1998), *Introduction to Languages and the Theory of Computation*, Tata McGraw-Hill, New Delhi.

Hopcroft, J. E. and Ullman, J. D. (1979), *Introduction to Automata Thoery, Languages and Computation*, Narosa Publishing House, New Delhi.

Linz, P. (2002), *An Introduction to Formal Languages and Automata*, Narosa Publishing House, New Delhi.

Morel, B. M. (2002), *The Theory of Computation*, Pearson Education, Singapore.

CHAPTER

12

Turing Machines and Computable Functions

Learning Objectives

On completing this chapter, you should be able to:

- define *initial functions* that are the building blocks of computable functions
- present primitive recursive definitions for commonly used functions
- state the definition of a recursive function
- compute the values of the Ackerman's function for various values of its variables
- state the definition of a computable function
- prove that a given function is computable by constructing a Turing machine, which computes the given function
- give the sketch of the proof of the theorem, which asserts that every recursive function is computable
- recognize that the class of recursive and computable functions are identical

12.1 RECURSIVE AND PRIMITIVE RECURSIVE FUNCTIONS

In this chapter, unlike in previous chapters, the symbol $\mathbb{N}$ denotes the set of all non-negative integers $\{0, 1, 2, 3...\}$.

Let us recall the definition of the number e, which is used as a base for the natural logarithm:

$$e = 1 + 1 + \frac{1}{2!} + \frac{1}{3!} + \frac{1}{4!} + \cdots$$

It is clear that no computer programme can be written to calculate the value of e exactly. Even if we ignore the truncation errors in each of the individual terms above, there are infinitely many terms and no computer can add infinitely many numbers. Let us now define a sequence of numbers $e_0, e_1, e_2, \ldots,$ as

$$\begin{aligned} e_0 &= 1 \\ e_1 &= 1+1 \\ e_2 &= 1+1+\frac{1}{2!} \\ e_3 &= 1+1+\frac{1}{2!}+\frac{1}{3!} \\ &\vdots \\ e_k &= 1+1+\frac{1}{2!}+\frac{1}{3!}+\frac{1}{4!}+\ldots\frac{1}{k!} \end{aligned}$$

and so on.

Depending on the storage capacity of the computer, we can calculate e_k for a large k. For simple calculations, e is approximately 2.718. If k is large, then we can approximate e as 2.71828182845904523536028747135 27. Suppose that we have calculated e_{100} and we want a better approximation in a new problem. How are we going to calculate e_{101}? We need not do all the calculations again, if we note that

$$e_k = e_{k-1} + \frac{1}{k!}$$

$$\begin{aligned} e_1 &= 2.0000 \\ e_2 &= e_1 + 1/2! = 2.5000 \\ e_3 &= e_2 + 1/3! = 2.5000 + 0.1667 = 2.6667 \\ e_4 &= e_3 + 1/4! = 2.6667 + 0.0417 = 2.7084 \\ e_5 &= e_4 + 1/5! = 2.7084 + 0.0083 = 2.7167 \\ e_6 &= e_5 + 1/6! = 2.7167 + 0.0014 = 2.7181 \end{aligned}$$

The above is an example of recursion. Roughly speaking, recursion is a step-by-step calculation. The calculation at the kth stage uses the calculation at the $(k-1)$th stage. Keeping the informality of our discussion, we may say that a recursively defined function is a function whose calculation can be carried out by an algorithm.

Almost all the calculations in a computer deal with rational numbers whose decimal expansions terminate. Most of these computations can be reformulated as operations on non-negative integers. For example, the addition of the decimals $a.a_1a_2\ldots a_k$ and $b.b_1b_2\ldots b_n$ involves operations with the non-negative integers $a, b, a_1, b_1,$ etc. Thus, theoretically, we consider functions of one or many *non-negative integral variables.*

Among the recursive functions, primitive recursive functions are of a simpler type. Before we state what we mean by a function of *many* variables to be defined in a primitive recursive fashion, we discuss an example and give the definitions for functions of one and two variables.

Example 12.1

Consider the function $f : \mathbb{N} \longrightarrow \mathbb{N}$ defined as $f(n) = n!$ We make the following observations

$$\begin{aligned} f(0) &= 1 \\ f(1) &- f(0) \times 1 = 1 \\ f(2) &= f(1) \times 2 = 2 \end{aligned}$$

and so on. In general, we have

$$f(y+1) = f(y) \times y$$

Thus, we find that $f(y+1)$ is a function of $f(y)$ and y.

Further, we cannot calculate $f(0)$ in terms of $f(-1)$ as $-1 \notin \mathbb{N}$. So, $f(0)$ has to be specified. This motivates the following definition. ■

Definition 12.1: A function $f : \mathbb{N} \longrightarrow \mathbb{N}$ is said to be defined as recursively primitive if there exist a constant g_0 and a function $h : \mathbb{N} \times \mathbb{N} \longrightarrow \mathbb{N}$ such that

$$\begin{aligned} f(0) &= g_0 \\ f(y+1) &= h(y, f(y)) \end{aligned}$$

Example 12.2

Let $f : \mathbb{N} \longrightarrow \mathbb{N}$ be defined as $f(y) = y^2$. How do we present a recursive definition of this function? Since $f(0) = 0$, we take $g_0 = 0$. Next, let us examine $f(y+1) = (y+1)^2 = y^2 + 2y + 1$. Thus, if we define $h(y, z) = 1 + 2y + z$, then $h(y, f(y)) = 1 + 2y + y^2 = (y+1)^2 = f(y+1)$. Thus, we obtain the recursive definition of this function as follows:

$$\begin{aligned} f(0) &= 0 \\ f(y+1) &= y + 2y + f(y) \end{aligned}$$ ■

Definition 12.2: A function $f : \mathbb{N} \times \mathbb{N} \longrightarrow \mathbb{N}$ is said to be defined as recursively primitive if there exist functions $g : \mathbb{N} \longrightarrow \mathbb{N}$ and $h : \mathbb{N} \times \mathbb{N} \times \mathbb{N} \longrightarrow \mathbb{N}$ such that

$$\begin{aligned} f(x, 0) &= g(x) \\ f(x, y+1) &= h(x, y, f(x, y)) \end{aligned}$$

Definition 12.3: A function $f : \mathbb{N}^{n+1} \rightarrow \mathbb{N}$ is said to be defined as recursively primitive if there exist functions $g : \mathbb{N}^n \longrightarrow \mathbb{N}$ and $h : \mathbb{N}^{n+2} \longrightarrow \mathbb{N}$ such that

$$\begin{aligned} f(x_1, x_2, ..., x_n, 0) &= g(x_1, x_2, ..., x_n) \\ f(x_1, x_2, ..., x_n, y+1) &= h(x_1, x_2, ..., x_n, y, f(x_1, x_2, ..., x_n, y)) \end{aligned}$$

Example 12.3

What function is obtained by primitive recursion if $g(x) = x$ and $h(x, y, z) = z+2$?

SOLUTION

We have

$$\begin{aligned} f(x, 0) &= x \\ f(x, y+1) &= h(x, y, f(x, y)) \\ &= f(x, y) + 2 \\ f(x, 1) &= f(x, 0) + 2 \\ &= x + 2 \\ f(x, 2) &= x + 2 + 2 \\ &= x + 4 \end{aligned}$$

We can see that $f(x, y) = x + 2y$. ■

Definition 12.4: Define the functions $Z : \mathbb{N} \longrightarrow \mathbb{N}$, $S : \mathbb{N} \longrightarrow \mathbb{N}$ and $U_i^n : \mathbb{N}^n \longrightarrow \mathbb{N}$ as

$$\begin{aligned} Z(x) &= 0 \\ S(x) &= x + 1 \\ U_i^n(x_1, x_2, ..., x_n) &= x_i \end{aligned}$$

These functions called *zero function*, *successor function* and *projection function*, respectively, are collectively known as *initial functions*.

In Definition 12.3, we do not specify the *nature of the functions* g and h. But the following definition demands that g and h come from the class of initial functions. In some sense, the initial functions are the building blocks of primitive recursive functions.

Definition 12.5: A function f is called *primitive recursive* if it can be defined from *the initial functions* by a finite number of operations of composition and primitive recursion.

To show that a function is *primitive recursive*, we have to define it using primitive recursion where the functions g and h are obtained using the initial functions.

We now give examples of primitive recursive functions below.

Example 12.4

Fix $k \in \mathbb{N}$. Let $f : \mathbb{N} \longrightarrow \mathbb{N}$ be the constant function $f(x) = k$. Show that f is primitive recursive.

SOLUTION

We note that $f(0) = k$ and hence we take $g_0 = k$ where g_0 is as in Definition 12.1. Next, we have to express $f(x+1)$ as a function of x and $f(x)$ using the initial functions. Since $f(x+1) = f(x)$, using the projection function U_2^2, we obtain $f(x+1) = U_2^2(x, f(x))$. Thus, the primitive recursive definition of the constant function is as given below:

$$\begin{aligned} f(0) &= k \\ f(x+1) &= U_2^2(x, (f(x)) \end{aligned}$$ ■

Example 12.5

Define $f : \mathbb{N} \longrightarrow \mathbb{N}$ as $f(x) = 2x$. Show that f is primitive recursive.

SOLUTION

$f(0)$ and $f(x+1) = 2(x+1) = 2x + 2f(x) + 2 = S(S(f(x))$. Hence the required definition is

$$\begin{aligned} f(0) &= 0 \\ f(x+1) &= S(S(U_2^2(x, f(x)))) \end{aligned}$$ ■

PRACTICE 12.1

Show that $f : \mathbb{N} \longrightarrow \mathbb{N}$ defined as $f(x) = 3x + 1$ is primitive recursive.

Example 12.6

Show that the functions $f_1, f_2 : \mathbb{N} \times \mathbb{N} \longrightarrow \mathbb{N}$ defined as $f(x, y) = x + 1$ and $g(x, y) = y + 2$ are primitive recursive.

SOLUTION

Recalling Definition 12.2, we find that $f_1(x, 0) = x + 1$. Defining $g_1(x) = f_1(x, 0) = x + 1$, we find that $g_1(x) = S(x)$ where S is the successor function. Next, we have to express $f_1(x, y+1)$ as a function of x, y and $f(x, y)$ using the initial functions. Now, $f_1(x, y+1) = f_1(x, y)$ and hence $f_1(x, y+1) = U_3^3(x, y, f_1(x, y))$.

The final primitive recursive definition of f_1 is as follows:

$$\begin{aligned} f_1(x, 0) &= S(x) \\ f_1(x, y+1) &= U_3^3(x, y, f_1(x, y)) \end{aligned}$$

Now, consider f_2. $f_2(x, 0) = 2$. With reference to Definition 12.2, we take $g_2(x) = f_2(x, 0) = 2$. Now we have to express g_2 in terms of the initial functions. It is easy to see that $g_2(x) = S(S(Z(x)))$. Next, consider $f_2(x, y+1) = y+3 = y+2+1 = f(x, y)+1 = S(f(x, y))$. But as per Definition 12.2, we should express $f_2(x, y+1)$ as a function of x, y and $f_2(x, y)$. As in the previous examples, using the projection function we write $f_2(x, y+1) = S(U_3^3(x, y, f(x, y)))$.

The final primitive recursive definition of f_2 is as follows:

$$\begin{aligned} f_2(x, 0) &= S(S(Z(x))) \\ f_2(x, y+1) &= S(U_3^3(x, y, f_2(x, y))) \end{aligned}$$

■

PRACTICE 12.2

Define $f : \mathbb{N} \times \mathbb{N} \times \mathbb{N} \longrightarrow \mathbb{N}$ as $f(x_1, x_2, x_3) = x_1 + x_3$. Show that f is primitive recursive.

Example 12.7

In this example, we present **the sign function or non-zero test function sg**. $sg : \mathbb{N} \longrightarrow \mathbb{N}$ is defined as

$$\begin{aligned} sg(0) &= 0 \\ sg(x) &= 1, \quad x \neq 0 \end{aligned}$$

Its primitive recursive definition is as follows:

$$\begin{aligned} sg(0) &= 0 \\ sg(y+1) &= SZ(U_2^2(y, sg(y)) \end{aligned}$$

■

Example 12.8

Now we define the **zero test function** $\bar{s}g$. $\bar{s}g : \mathbb{N} \longrightarrow \mathbb{N}$ is defined as

$$\begin{aligned} \bar{s}g(0) &= 1 \\ \bar{s}g(x) &= 0 \quad x \neq 0 \end{aligned}$$

Its primitive recursive definition is as follows:

$$\begin{aligned} \bar{s}g(0) &= 1 \\ \bar{s}g(y+1) &= Z(U_2^2(y, \bar{s}g(y)) \end{aligned}$$

■

Example 12.9

The **predecessor function** $P : \mathbb{N} \longrightarrow \mathbb{N}$ is defined as

$$\begin{aligned} P(0) &= 0 \\ P(y) &= x - 1, \ x \neq 0 \end{aligned}$$

Its primitive recursive definition is as follows:

$$\begin{aligned} P(0) &= 0 \\ P(y+1) &= y = U_1^2(y, P(y)) \end{aligned}$$

■

Example 12.10

The addition function *Add* defined as *Add* $(x, y) = x + y$ is primitive recursive. Its primitive recursive definition is as follows:

$$\begin{aligned} Add\ (x, 0) &= x = U_1^1(x) \\ Add\ (x, y+1) &= x + y + 1 \\ &= Add\ (x, y) + 1 \\ &= SU_2^2(x, y, Add\ (x, y)) \end{aligned}$$

■

Example 12.11

Show that the function $f(x, y) = x + 2y$ is primitive recursive.

SOLUTION

$$\begin{aligned} f(x, 0) &= x \\ f(x, y+1) &= x + 2(y+1) = x + 2y + 2 \\ &= f(x, y) + 2 \\ &= (S(S(U_3^3(x, y, f(x, y))))) \end{aligned}$$

Here we take $g(x) = x$. Note that g is nothing but the initial function $x = U_1^1(x)$.

■

Example 12.12

Show that the function $f(x, y) = 3x + 4y$ is primitive recursive.

SOLUTION

$$\begin{aligned} f(x, 0) &= 3x \\ f(x, y+1) &= 3x + 4(y+1) = 3x + 4y + 4 \\ &= f(x, y) + 4 \end{aligned}$$

$$= S(S(S(S(U_3^3(x, y, f(x, y))))))$$

Now, we have to show that $g(x) = 3x$ can be obtained from initial functions by primitive recursion:

$$\begin{aligned} g(0) &= 0 \\ g(x+1) &= 3x + 3 = S(S(S(U_2^2(x, g(x))))) \end{aligned}$$

This proves that g is primitive recursive.

It is clear that as the functions become more and more complicated, the definition of functions in terms of *initial functions* would also become increasingly complicated. For example, if we take $g(x) = 100x$, then to express $g(x+1)$, the successor function has to be applied 100 times. ■

The next proposition, the proof of which evident, simplifies the procedure.

Proposition 12.1

Initial functions are primitive recursive. Any function that can be obtained by primitive recursion using primitive recursive functions are also primitive recursive.

Remark 12.1

Let

$$\begin{aligned} f(x, 0) &= g(x) \\ f(x, y+1) &= h(x, y, f(x, y)) \end{aligned}$$

It is clear that f is primitive recursive if g and h are *initial functions*. The above proposition tells us that f is primitive recursive if g and h are *primitive recursive*.

The next example illustrates how Proposition 12.1 simplifies the proofs of primitive recursiveness.

Example 12.13

Show that $g : \mathbb{N} \longrightarrow \mathbb{N}$ defined as $g(x) = 100x$ is primitive recursive.

SOLUTION

As we have already pointed out, if we have to use only the initial functions to express $g(x+1)$ in terms of $g(x)$, we have to apply the successor function 100 times. But now, we proceed as follows:

$$g(x+1) = 100(x+1) = 100x + 100 = g(x) + 100 = Add\,(U_2(x, g(x)), f_{100}(x))$$

Here $f_{100}(x)$ is the constant function such that $f_{100}(x) = 100$ for all $x \in \mathbb{N}$ and *Add* is the addition function. By Example 12.10 and Example 12.4, *Add* and f_{100} are primitive recursive. Hence g is primitive recursive. ■

Example 12.14

The **odd and even parity function** Pr:

$$\begin{aligned} Pr(x) &= 0, \quad x \text{ is even} \\ Pr(x) &= 1, \quad x \text{ is odd} \end{aligned}$$

Pr is a primitive recursive function because

$$\begin{aligned} Pr(0) &= 0 \\ Pr(y+1) &= \bar{s}g(U_2^2(y, Pr(y))) \end{aligned}$$

Example 12.15

Proper subtraction function is defined as

$$x \overset{\circ}{-} y = 0, \quad \text{if } x < y$$

$$x \overset{\circ}{-} y = x - y, \quad \text{if } x > y$$

Its primitive recursive nature is proved as follows:

$$\begin{aligned} x \overset{\circ}{-} 0 &= x \\ x \overset{\circ}{-} (y+1) &= P(x \overset{\circ}{-} y) \end{aligned}$$

Example 12.16

Show that $Mult\,(m, n) = mn$ is primitive recursive.

$$\begin{aligned} Mult\,(m, 0) &= 0 \\ Mult\,(m, n+1) &= m(n+1) \\ &= mn + m \\ &= Add\,(U_1^3(m, n, Mult\,(m, n)), U_3^3(m, n, Mult\,(m, n))) \end{aligned}$$

Example 12.17

Show that $f(x, y) = x^y$ is primitive recursive.

$$\begin{aligned} f(x, 0) = 1 &= SZ(x) \\ x^{y+1} &= x^y \times x \\ &= Mult\,(U_1^3(x, y, f(x, y)), U_3^3(x, y, f(x, y))) \end{aligned}$$

PRACTICE 12.3

Show that the function $f(x, y) = x + y^2$ is primitive recursive.

Example 12.18

Let $Q(x, y)$ be the quotient when y is divided by x. Show that $Q(x, y)$ is recursive.

SOLUTION

Now, we have to express $Q(x, y+1)$ in terms of $Q(x, y)$. Let $q = Q(x, y)$. Then, $y - qx$ is the remainder when y is divided by x. Hence $y - qx < x$. If $(y+1) - qx$ is less than x, then we get the same quotient q when $y+1$ is divided by x. But if $y - qx = x$, then the new quotient is $q+1$.

In other words, depending on whether $y - qx$ is strictly greater than zero or not, we either have the same quotient or 1 plus the old quotient. This is where the sign function is useful.

$$\begin{aligned} Q(x, 0) &= 0 \\ Q(x, y+1) &= Q(x, y) + sg(y - qx) \\ &= Q(x, y) + sg(y - Q(x, y)x) \end{aligned}$$ ■

PRACTICE 12.4

Let $R(x, y)$ be the remainder when y is divided by x. Show that $R(x, y)$ is primitive recursive.

Example 12.19

For a real number r, $\lfloor r \rfloor$ denotes the greatest integer smaller than or equal to r. For example, $\lfloor 3.78 \rfloor = \lfloor 3.999 \rfloor = 3$, $\lfloor -4 \rfloor = -4$, $\lfloor 5 \rfloor = 5$, $\lfloor 0.8 \rfloor = 0$, $\lfloor -0.8 \rfloor = -1$, $\lfloor -2.3 \rfloor = -3$, etc.

Define $f : \mathbb{N} \longrightarrow \mathbb{N}$ as $f(x) = \lfloor \frac{x}{2} \rfloor$.

For example, $f(0) = 0$, $f(1) = 0$, $f(2) = 1$, $f(3) = 1$ and $f(4) = 2$ etc.

In this example, we show that f is primitive recursive.

SOLUTION

We note that $f(x+1)$ is either $f(x)$ or $f(x)+1$. To be specific, we find that if x is odd, $f(x+1) = x+1$ and if x is even, $f(x+1) = f(x)$. Hence $f(x+1) = x+Pr(x)$. Here $Pr(x)$ is the odd and even parity function of Example 12.14. This shows that f is primitive recursive. ■

PRACTICE 12.5

Let (x) be the smallest integer greater than or equal to x. Let $g : \mathbb{N} \longrightarrow \mathbb{N}$ be defined as $g(x) = (x/2)$. Show that g is primitive recursive.

Example 12.20

Let $f(x, y) = \max(x, y)$. Show that f is primitive recursive.

SOLUTION

If $x \geq y$, then $\max(x, y) = x$. Further, $y \overset{\circ}{-} x = 0$. Hence $x = x + y \overset{\circ}{-} x$. If $y > x$, then $\max(x, y) = y = x + (y - x) = x + y \overset{\circ}{-} x$.
So

$$\max(x, y) = x + y \overset{\circ}{-} x$$

and hence f is primitive recursive. ■

Example 12.21

Show that $f : \mathbb{N} \longrightarrow \mathbb{N}$ defined as $[\sqrt{x}]$ is primitive recursive.

SOLUTION

If $([\sqrt{x}] + 1)^2 < x + 1$, then $[\sqrt{x+1}] = [\sqrt{x}]$ and if $([\sqrt{x}] + 1)^2 = x + 1$, then $[\sqrt{x+1}] = [\sqrt{x}] + 1$. Thus, to obtain $f(x+1)$ from $f(x)$ we should add 0, if

$$((x + 1) - ([\sqrt{x}] + 1)^2) > 0$$

or we should add 1 if

$$((x + 1) - ([\sqrt{x}] + 1)^2) = 0$$

So, using the zero test function of Example 12.8, we find that

$$\left[\sqrt{x+1}\right] = [\sqrt{x}] + \bar{s}g(x + 1 - ([\sqrt{x}] + 1)^2)$$

From the definition of primitive recursive functions, it is clear that primitive recursive functions are computable. Here, we use the term computable in an intuitive sense. Later, in this section we shall make the definition of a computable function in a precise manner. Now, we ask ourselves the following question:

Is every computable function primitive recursive?

The answer is **NO**. Primitive recursive functions are simple functions, for the computation of which one can give a step-by-step procedure. But there are functions that are more complicated than primitive recursive functions *but that can be defined by giving an algorithm.* Here an algorithm is a set of rules that help us to find the values of the function for a given set of variables. One such example is the Ackerman's function. Below, we define Ackerman's function, which is recursive but not primitive recursive. The proof of this fact is beyond the scope of this book. The definition of a recursive function is given after defining Ackerman's function. ■

Example 12.22

Ackerman's function is defined below:

(i) $A(0, y) = 1$.
(ii) $A(1, 0) = 2$.

(iii) $A(x,0) = x + 2$ for $x \geq 2$.
(iv) $A(x+1, y+1) = A(A(x, y+1), y)$.
(i) and (ii) define $A(x,0)$ for $x = 0$ and $x = 1$. ■

Example 12.23

Find
(i) $A(0,5)$,
(ii) $A(1,0)$,
(iii) $A(4,0)$,
(iv) $A(2,5)$,
where A is Ackerman's function.

SOLUTION

From the definition, it is clear that $A(0,5) = 1$, $A(1,0) = 2$ and $A(4,0) = 4 + 2 = 6$.

Now, to compute $A(3,5)$, we proceed as follows:

$$\begin{aligned} A(2,5) &= A(A(1,5),4) \\ A(1,5) &= A(A(0,5),4)) = A(1,4) = A(A(0,4),3)) = A(1,3) = A(A(0,3),2) \\ &= A(1,2) = A(A(0,2),1) = A(1,1) = A(A(0,1)),0) = A(1,0) = 2 \end{aligned}$$

Now ,

$$\begin{aligned} A(2,5) &= A(A(1,5),4) = A(2,4) = A(A(1,4),3) = A(2,3) = A(A(1,3),2) \\ &= A(2,2) = A(A(1,2),1) = A(2,1) = A(A(1,1),0) = A(2,0) = 4 \end{aligned}$$ ■

PRACTICE 12.6

Compute $A(3,3)$, $A(4,3)$, $A(3,4)$.

As it has been already remarked, primitive recursive functions are *simple* functions which can be computed by a step-by-step procedure. From the previous example and the practice problem, we find that there are functions which are not primitive recursive, but which can be computed by a set of rules. Roughly speaking, recursive functions are functions for which one can give a procedure of computing. Now, we present the formal definition of a recursive function.

Definition 12.6: We say that a partial function is a *recursive function* if it is defined by a finite set of rules that for various arguments specify the function in terms of variables, non-negative integer constants, the successor function, the function itself or an expression built from these by composition functions.

It is clear that all primitive recursive functions are recursive. Ackerman's function disproves the converse.

Note that giving a recursive definition is easier than giving a primitive recursive definition. We have a lot more freedom in the case of recursive definitions. Consider the functions *Add* and *Mult*, which are primitive recursive and hence recursive. But below we directly show that they are recursive.

Example 12.24

Define $f : \mathbb{N} \times \mathbb{N} \longrightarrow \mathbb{N}$ as $f(m, n) = m + n$. Then, show that f is recursive.

Proof

$$\begin{aligned} f(m,0) &= m \\ f(m,n+1) &= S(f(m,n)) \\ f(0,n) &= n \\ f(m+1,n) &= S(f(m,n)) \end{aligned}$$

This function is denoted by *Add*. ■

Example 12.25

Recursive definition of $Mult(m, n) = mn$

$$\begin{aligned} Mult(m,0) &= 0 \\ Mult(0,n) &= 0 \\ Mult(m,n) &= (((Add(m,m),m),m)\ldots) \quad n \text{ times}, \quad n > 0 \end{aligned}$$ ■

Example 12.26

Recursive definition of $n!$

$$\begin{aligned} Fact(0) &= 1 \\ Fact(n) &= n \times Fact(n-1) \end{aligned}$$ ■

12.2 TURING MACHINES AND COMPUTABLE FUNCTIONS

Roughly speaking, *recursive* functions are functions that can be computed by *providing a set of rules. Computable functions* are functions that can be computed with the help of a *computer.* It is intuitively clear that if rules are given, then one can write a computer program to compute it and vice versa. This result, which states that a function is recursive if and only if it is computable, is proved in an advanced computer science course.

We have already given a formal definition of a recursive function. Now, we proceed to define a computable function formally. This is where we use the concept of a Turing machine as a mathematical model of a computer. Earlier, we used Turing machines as language acceptors. In this section, we shall learn how functions can be computed using Turing machines. Further, we give a sketch of the proof that a recursive function is computable.

In Turing machines, the input and output are represented by sequences of zeros under the unary system. 0 is represented by a blank, 1 is represented by 0, 2 is represented by 00, etc. In general, n is represented by 0^n. For functions of many variables, we use 1 to separate the different variables.

The elements $(i_1, i_2, \ldots, i_k) \in N^k$ are represented by $0^{i_1}10^{i_2}1\ldots0^{i_k}$.

Definition 12.7: Let the tape initially contain $0^{i_1}10^{i_2}1\ldots0^{i_k}$. If the TM halts (either in an accepting state or in a non-accepting state) with a tape consisting of 0^m for some m then we say that

$$f(i_1, i_2, \ldots, i_k) = m$$

Definition 12.8: We say that a partial function $f : N^n \longrightarrow N$ is a *computable function* if there exists a Turing machine that scans the leftmost cell when the control is in q_0 and the input is $0^{i_1}10^{i_2}1\ldots10^{i_n}$:

(i) halts with the output 0^m if $f(i_1, i_2, \ldots, i_n)$ is defined and is equal to m,

(ii) does not halt when $f(i_1, i_2, \ldots, i_n)$ is not defined.

This definition says that a partial function f is a computable function, if there is a TM that accepts an x in the domain of f as the input, and produces the output $f(x)$. If x is not in the domain, then the TM must not halt. In other words, a partial function is computable if there exists a Turing machine that computes it.

Next, we present examples of computable functions and Turing machines that compute them.

First, we show that the initial functions are computable by designing Turing machines that compute them. In all the following examples, we describe only the transition function δ.

Example 12.27

A Turing machine for the **zero function:** No matter what the input is, the TM should stop after converting all the tape symbols into blank. We describe the transition function for it as follows:

$$\begin{aligned}
\delta(q_0, 0) &= (q_0, B, R)\\
\delta(q_0, 1) &= (q_0, B, R)\\
\delta(q_0, B) &= (q_1, B, R)\\
F &= \{q_1\}
\end{aligned}$$

■

Successor function: There must be an extra zero when the machine halts. So, the first blank encountered must be changed to 0 and the machine halts:

$$\begin{aligned}\delta(q_0, 0) &= (q_0, 0, R)\\ \delta(q_0, B) &= (q_1, 0, R)\\ F &= \{q_1\}\end{aligned}$$

Example 12.28

A TM for **projection functions:**

We defined $U_i^n : \mathbb{N}^n \to$ as $U(x_1, x_2, ..., x_n) = x_i$.

First, let us consider U_1^n. This function, for example, with an input string given as 000100100001...00*BBB*... should produce the output 000*BBBBB*.... The control should not alter the first three zeros and change everything else into blanks. We define the transition function for the TM:

$$\delta(q_0, 0) = (q_0, 0, R)$$

In the state q_0, the control leaves 0 unchanged, and moves to the right:

$$\delta(q_0, 1) = (q_1, B, R)$$

When the control encounters 1, it replaces the 1 by B, changes to state q_1 and moves to the right. From now on, it will keep erasing everything until it reaches B. Then, it changes into the final state q_2 and halts:

$$\begin{aligned}\delta(q_1, 0) &= (q_1, B, R)\\ \delta(q_1, 1) &= (q_1, B, R)\\ \delta(q_1, B) &= (q_2, B, R)\\ F &= \{q_2\}\end{aligned}$$

Projection function U_i^n for $1 < i \leq n$:
For $k = 0, 1, ..., i - 2$,

$$\begin{aligned}\delta(q_k, 0) &= (q_k, B, R)\\ \delta(q_k, 1) &= (q_{k+1}, B, R)\end{aligned}$$

So far, the *TM* has deleted all entries before the *i*th component. Now, it has to preserve the zeros of the *i*th component, delete the 1 at the end of the *i*th component and delete the remaining entries:

$$\begin{aligned}\delta(q_{i-1}, 0) &= (q_{i-1}, 0, R)\\ \delta(q_{i-1}, 1) &= (q_i, B, R)\\ \delta(q_i, 0) &= (q_i, B, R)\\ \delta(q_i, 1) &= (q_i, B, R)\end{aligned}$$

When the control scans B, it must halt. Thus,

$$\delta(q_i, B) = (q_{i+1}, B, R); \quad F = \{q_{i+1}\}$$

■

Example 12.29

Turing machine for $f(n) = a$: This is a Turing machine with $Q = \{q_0, q_1, ..., q_a\}$, $\Sigma = \{0\}$, $\Gamma = \{B, 0\}$ and $F = \{q_a\}$. The function δ is defined as follows:

$$
\begin{aligned}
\delta(q_0, 0) &= (q_1, 0, R) \\
\delta(q_0, B) &= (q_1, 0, R) \\
\delta(q_i, 0) &= (q_{i+1}, 0, R) \\
\delta(q_i, B) &= (q_{i+1}, 0, R) \\
\cdots &= \cdots \\
\delta(q_{a-1}, 0, R) &= (q_a, 0, R) \\
\delta(q_{a-1}, B, R) &= (q_a, 0, R)
\end{aligned}
$$

■

PRACTICE 12.7

Construct a Turing machine to compute $f : \mathbb{N} \longrightarrow \mathbb{N}$ defined as $f(x) = x + a$ where a is a fixed positive integer.

Example 12.30

In this example, we consider **a Turing machine for the non-zero sign function**.

If the input is *BBBBB*..., the TM halts without changing the input. For the input 0^n, it should halt after producing the output 0*BBBB*....

We describe δ for this function.

$$
\begin{aligned}
\delta(q_0, B) &= (q_1, B, R) \\
\delta(q_0, 0) &= (q_2, 0, R) \\
\delta(q_2, 0) &= (q_2, B, R) \\
\delta(q_2, B) &= (q_1, B, R) \\
F &= \{q_1\}
\end{aligned}
$$

■

PRACTICE 12.8

Construct a Turing machine for the zero sign function.

Example 12.31

In this example, we construct a **Turing machine for the parity function**. If the input is blank, the output must also be blank. For other inputs, how would the machine know that the number of zeros is even or odd? The two states q_1 and q_2

change into each other when 0 is encountered. Thus, if the head has scanned an even number of zeros, it would be in state q_2 and if the head has scanned an odd number of zeros it would be in state q_1. If the head encounters B while in q_2, the input is even and if the head encounters B in q_1, the input is odd. At this stage, the head must go back deleting *all the zeros* or all except one zero depending on whether the input is *even* or *odd*. But when the head keeps moving to the left, there must be a way to know that we have reached the first cell. Thus, if the input is not blank, the first zero is changed to X. If the input is even, this X is finally deleted, and if the input is odd, the X is changed into a zero.

$$\begin{aligned}
\delta(q_0, B) &= (q_5, B, R) \\
\delta(q_0, 0) &= (q_1, X, R) \\
\delta(q_1, B) &= (q_3, B, L) \\
\delta(q_1, 0) &= (q_2, 0, R) \\
\delta(q_2, 0) &= (q_1, 0, R) \\
\delta(q_2, B) &= (q_4, B, L) \\
\delta(q_4, 0) &= (q_4, 0, L) \\
\delta(q_4, X) &= (q_5, B, R) \\
\delta(q_3, 0) &= (q_3, B, L) \\
\delta(q_3, X) &= (q_5, 0, R) \\
F &= \{q_5\}
\end{aligned}$$

■

PRACTICE 12.9

Construct a Turing machine for the function

$$\begin{aligned}
f(n) &= 1 \text{ if } n \leq 3 \\
f(n) &= 0 \text{ if } n > 3
\end{aligned}$$

Example 12.32

In this example, we present a **Turing machine for addition**. This TM converts the input $0^n 1 0^m$ into 0^{n+m}.

The required TM simply replaces the separating 1 by a 0. But at this stage, the output represents $m + n + 1$ and not $m + n$. So, the last 0 must be erased.

$$\begin{aligned}
\delta(q_0, 0) &= (q_0, 0, R) \\
\delta(q_0, 1) &= (q_1, 0, R) \\
\delta(q_1, 0) &= (q_1, 0, R) \\
\delta(q_1, B) &= (q_2, B, L)
\end{aligned}$$

$$\delta(q_2, 0) = (q_4, B, R)$$
$$F = \{q_4\}$$

■

PRACTICE 12.10

Construct a Turing machine for multiplication and use it as a 'subroutine' to construct a Turing machine for calculating $n!$

PRACTICE 12.11

Find the output for some values of the input 0^n and find what the following Turing machine tries to do:

	0	X	Y	B
q_0	(q_1, Y, R)	(q_0, X, R)	(q_0, Y, R)	$(q_7, 0, R)$
q_1	$(q_2, 0, R)$	(q_1, X, R)	(q_1, Y, R)	(q_3, B, R)
q_2	(q_1, X, R)	(q_2, X, R)	(q_2, Y, R)	(q_3, B, R)
q_3	$(q_3, 0, R)$	–	–	$(0, q_4, L)$
q_4	$(q_4, 0, L)$	(q_4, X, L)	(q_2, Y, R)	(q_5, B, L)
q_5	$(q_4, 0, L)$	(q_6, X, L)	(q_7, Y, R)	–
q_6	$(q_4, 0, L)$	(q_7, X, L)	(q_7, Y, R)	–
q_7				

As we have already remarked, the initial functions are the building blocks for recursive functions. We have seen that the initial functions are computable by constructing Turing machines that compute them. Further, we have seen examples of Turing machine constructions for recursive functions. Now, the following theorem asserts that a recursive function is computable.

Theorem 12.1

If a function is recursive, then it is computable. That is, if a function is recursive, then there exists a Turing machine that computes it.

Sketch of the proof of Theorem 12.1: Let us recall that a recursive function is a function whose values are computed by specifying a set of rules. These rules apart from the constant functions and the successor function also involve composition of functions. We have seen the construction of TMs that compute constant functions and a TM that computes the successor function. Since the computation of a recursive function requires the computation of the function compositions, we need to combine Turing machines that compute the individual functions. We now indicate the construction of a TM that computes $f \circ g$ when there are Turing machines computing f and g.

Combining Turing machines: Let $T_1 = (Q_1, \Sigma_1, \Gamma_1, \delta_1, B, \{h\})$ and $T_2 = (Q_2, \Sigma_2, \Gamma_2, \delta_2, B, F_2)$ be the Turing machines. Let the initial state of T_1 be q_1 and let the initial state of T_2 be q_2.

Now, let us construct a Turing machine that behaves as T_1, produces an output and then behaves like T_2 treating the above output as the input for T_2.

We assume that $\Gamma_1 \subseteq \Sigma_2$ and $Q_1 \cap Q_2 = \phi$. This assumption can be easily made true after renaming the symbols, if needed.

Now, for the new Turing machine T, we have the following: $q_0 = q_1$. $Q = Q_1 \cup Q_2$. For any $q \in Q_1$ and $X \in \Gamma_1 \cup \{B\}$, if

$$\delta_1(q, X) = (r, Y, D)$$

then

$$\begin{aligned} \delta(q, X) &= (r, Y, D) \quad \text{if } r \neq h \\ &= (q_2, Y, D) \quad \text{if } r \neq h \end{aligned}$$

For any $q_2 \in Q_2$ and $X \in \Gamma_2 \cup \{B\}$,

$$\delta(q, X) = \delta_2(q, X)$$

Now, from the definition of a recursive function, it is clear that to compute it, we need to combine various TM's to get a TM that computes it.

Remark 12.2

The converse of the theorem is also true. This is proved by using the idea of Godel Numbering.

SUMMARY

- *The initial functions* are the building blocks of computable functions.
- A function $f : \mathbb{N} \times \mathbb{N} \longrightarrow \mathbb{N}$ is said to be defined primitive recursively if there exist functions $g : \mathbb{N} \longrightarrow \mathbb{N}$ and $h : \mathbb{N} \times \mathbb{N} \times \mathbb{N} \longrightarrow \mathbb{N}$ such that

$$\begin{aligned} f(x, 0) &= g(x) \\ f(x, y+1) &= h(x, y, f(x, y)) \end{aligned}$$

- Ackerman's function is recursive but not primitive recursive.
- A computable function is a function for which there exists a Turing machine that computes it.
- The class of recursive and computable functions are identical.

PROBLEMS

1. Give a primitive recursive definition of $f : N \longrightarrow N$ as $f(x) = 9x$ using the initial functions.
2. Give a primitive recursive definition of $f : N \times N \longrightarrow N$ as $f(x, y) = 3x + 4y$ using the initial functions.
3. Give a recursive definition of $f(x) = \frac{x(x+1)}{2}$.
4. Let $f : \mathbb{N} \times \mathbb{N} \longrightarrow \mathbb{N}$ be defined as follows:

$$\begin{aligned} f(x, 0) &= 2x \\ f(x, y+1) &= x + yf(y) \end{aligned}$$

   Find f explicitly.
5. Show that the function $f(x) = x^2$ is primitive recursive.
6. Show that the function $f : \mathbb{N} \times \mathbb{N} \longrightarrow \mathbb{N}$ defined as $f(x, y) = \min(x, y)$ is primitive recursive.
7. Show that the function $f : \mathbb{N} \longrightarrow \mathbb{N}$ defined as $[\sqrt[3]{x}$ is primitive recursive.
8. Let $f, g : \mathbb{N} \longrightarrow \mathbb{N}$ be primitive recursive functions. Show that $F(x, y) = f(x)g(y)$ is primitive recursive.
9. Give an example of a function that is recursive but not primitive recursive.
10. Calculate $A(5, 4)$, where A is Ackerman's function.
11. Examine the truth of the statement $A(x, 1) = 2x$ for all $x \geq 1$. If true, give an inductive proof of the same.
12. Write a computer program to evaluate Ackerman's function for $(x, y) \in \{0, 1, 2, ..., 10\} \times \{0, 1, 2, ..., 10\}$.
13. Construct a Turing machine to compute $f(x) = x^2$.
14. Construct a TM that computes $f(x, y) = \max(x, y)$.
15. Let $f : \mathbb{N} \longrightarrow \mathbb{N}$ be defined as follows:

$$\begin{aligned} f(x) &= 0 \text{ if } x \text{ is a multiple of } 3 \\ &= 1 \text{ if } x \text{ gives a remainder of } 1 \text{ when divided by } 3 \\ &= 2 \text{ if } x \text{ gives a remainder of } 2 \text{ when divided by } 3 \end{aligned}$$

16. Construct a Turing machine that computes $f : \mathbb{N} \longrightarrow \mathbb{N}$ defined as $f(x) = 2x$.
17. Let $f : \mathbb{N} \longrightarrow \mathbb{N}$ be defined as

$$\begin{aligned} f(x) &= 0, \quad x \leq 3 \\ &= 1, \quad x > 3 \end{aligned}$$

FURTHER READING

Hein, J. L. (2004), *Discrete Structures, Logic, and Computability*, Narosa Publishing House, New Delhi.

Hopcroft, J. E. and Ullman, J. D. (1979), *Introduction to Automata Thoery, Languages and Computation*, Narosa Publishing House, New Delhi

Linz, P. (2002), *An Introduction to Formal Languages and Automata*, Narosa Publishing House, New Delhi.

CHAPTER

13

Coding Theory

Learning Objectives

On completing this chapter, you should be able to:

- define an encoding function
- define the minimum distance of a code
- calculate the minimum distance of a given code
- define and identify group codes
- calculate the maximum number of errors that can be detected by a given code
- design codes that would detect a given number or fewer errors
- calculate the maximum number of errors that can be corrected by a given code
- decode strings using maximum likelihood procedure
- design codes that would correct a given number or fewer errors
- decode strings using the concept of a syndrome

13.1 INTRODUCTION AND PRELIMINARIES

Example 13.1

A globe-trotting CEO of a firm has four offices in India at Chennai, Mumbai, Kolkata and Bengaluru. She hurriedly leaves a message to her second-in-command Mr Peter asking him to meet her in India. Her message jotted down by someone (who is not familiar with the names of the Indian cities) reads as follows:

Meet me at Kolota

Mr Peter will have no difficulty in decoding this message as referring to Kolkata. This is a type of maximum likelihood decoding. In some sense, the word Kolkata is the nearest word to the string of symbols written as Kolota.

However, this procedure may not always work. For example, if we include *Bangarpet* in our list of places and if the message contains something like *Banger* we have a difficulty. ■

Now let us look at a more technical example. Our words are strings of zeros and ones.

Example 13.2

If the possible guesses for a word in the message is 0110, 0101,1001, 1110, and the actual word received is 1111, then we may decode it as 1110 as it differs least, when compared to the other words. Let us explain this in detail:

(i) The number of places where 0110 differs from 1111 is 2.

(ii) The number of places where 0101 differs from 1111 is 2.

(iii) The number of places where 1001 differs from 1111 is 2.

(iv) The number of places where 1110 differs from 1111 is 1.

Since 1110 contains the least number of errors, we choose it. ■

Imagine a teenager who has a baby brother. This teenager is a nice girl who takes her baby brother to movies and restaurants. But sometimes she prefers to go to these places either alone or with her teenage friends. If her baby brother comes to know about her visit, he would also like to come. So, this teenager while discussing these visits does not use the words 'cinema' and 'hotel'. She has informed her friends that the word *cinema* would be replaced by *special class* and the word *hotel* would be replaced by the word *hospital*.

The following is a part of a telephonic conversation the baby brother heard:

"I will meet you at the MTR hospital at 11 a.m. Then we shall all go to the special class at the Galaxy".

The above is an example of a very simple coded message. The actual meaning of the above message is

I will meet you at the MTR hotel at 11 a.m. Then we shall all go to the cinema at the Galaxy.

The process of changing a message using code words is called *encoding*. The process of getting the actual message out of the coded message is known as *decoding*.

Example 13.3

A word in English is encoded as follows: Every word is replaced by the next letter in the alphabet. Z is replaced by A.

For example, JAMES BOND would become KBNFT CPOE.

(i) Decode the following words:
DBU, EPH, MJPO, UJHFT.

(ii) Give a mathematical definition of this encoding and decoding procedure.

SOLUTION

(i) To decode a word, every letter has to be replaced by the letter occurring before it in the alphabet. Thus, we get the words as:
CAT, DOG, LION, TIGES.
Now, from a purely mathematical point of view, the solution is complete. But from a practical point of view, you may believe that there is an error and you may encode UJHFT as TIGER. This is an example of error correction.

(ii) Consider the function $f : \{A, B, C, ..., Z\} \longrightarrow \{A, B, C, ..., Z\}$ defined as

$$\begin{aligned} f(A) &= B \\ f(B) &= C \\ \cdots & \quad \cdots \\ f(Y) &= Z \\ f(Z) &= A \end{aligned}$$

Now, if the English word is $x_1x_2...x_k$, then it is encoded as $f(x_1)f(x_2)...f(x_k)$. Let us note that f is one-to-one and onto. Its inverse g is given by

$$\begin{aligned} g(A) &= Z \\ g(B) &= A \\ \cdots & \quad \cdots \\ g(Y) &= X \\ g(Z) &= Y \end{aligned}$$

Now, the encoded word $y_1y_2...y_k$ is decoded as $g(y_1)g(y_2)...g(y_k)$. ■

Maintaining secrecy or preventing someone from knowing the content of the message is not the only purpose of coding. For example, we know that language characters and numerals are processed in the computer using zeros and ones. Every letter is represented by a string of zeros and ones.

The Idea of Redundancy: Consider the following two sentences written in English, Hindi and Tamil:

English	A man came. A woman came.
Hindi	Ek aadmi aayaa. Ek ourath aayee.
Tamil	Oru aan vandaan. Oru pen vandaal.

We note that, in Tamil and Hindi the gender of the subject has an effect on the verb. In Hindi, it is *aayaa* for male and *aayee* for female. In Tamil, it is *vandaan* for male and *vandaal* for female.

In English, the gender of the subject does not have any effect on the verb. It is *came* for both male and female.

After learning English, we may feel that the effect of the gender on the verb is not really necessary. The subject itself informs us about the gender, and why should it be emphasized in the verb again?

Let us now imagine a situation where someone utters these sentences to you in a noisy locality and you could not hear the first part of the sentence. In English, when you hear only the second half of the sentence, you would not know the gender of the individual being talked about. However, in Tamil or Hindi, the second part of the sentence gives information regarding the gender of the person. Knowing the gender of the person does not tell you much. Still, it conveys some information.

A seemingly unnecessary grammatical rule is useful when we do not have the full sentence. This is the idea of redundancy.

Let us consider one more example. Imagine Ram asking Rahim, 'Shall we go to *Galaxy* theater today?' If Rahim knows that *Galaxy* is a theater, then it is sufficient for Ram to simply ask, 'Shall we go to *Galaxy* today?' The word 'theater' is an example of redundancy.

Let us suppose now that we would like to transmit 0111001 by a communication channel. If the channel is noisy, then some zero may become 1 and vice versa. For example, the actual received message may be 1001001. In this communication, three errors have occurred.

Example 13.4

We shall now describe an example of adding redundancy, known as parity check. In this method of encoding, an extra symbol (redundancy) is added to the code word. This extra symbol is 1 if the original code word contains an odd number of ones, and the extra symbol is 0 if the original message contains an even number of ones.

For example, 001 would be transmitted as 0011, 101 would be transmitted as 1010, etc. Suppose that we receive the message 0111, then we can immediately see that the received message is not a correct one. Note that the message part is 011, which contains an even number of ones. So, the last digit should have been 0.

It is possible that when 1001 is sent, it is received as 0000. Here, we cannot detect the error. ■

In this chapter, we shall learn about words, which are sequences of zeros and ones. If the total number of the digits is m, then such a word is an element of B^m where $B = \{0, 1\}$.

We write an unknown element of B^m using commas to separate the components. But a known element like $(1, 0, 0, 1, 1) \in B^5$ is simply written as 10011. $(0, 0, 0, ..., 0)$ is denoted by 0.

Example 13.5

$B^3 = \{000, 001, 010, 100, 110, 101, 011, 111\}$. ■

PRACTICE 13.1

Using the ideas from Chapter 4, show that B^m contains 2^m elements.

Definition 13.1: An encoding e is a one-to-one function $e : B^m \longrightarrow B^n$.

If e is not one-to-one, then we would have two words b_1 and b_2 such that $x = e(b_1) = e(b_2)$ and so we cannot decide which one of the words is represented by the code word y. This is the reason we demand that e is one-to-one.

Most often, in coding, we consider a word $b \in B^m$ and add extra redundant digits to detect errors. The resulting new word x is called the 'code word' representing the word b. If the code word contains n digits, with $n > m$, then it is an element of B^n.

Definition 13.2: The weight $|x|$ of an element $x \in B^n$ is defined as the number of ones in x.

Example 13.6

Consider B^5. Let $x = 11010$, $y = 10101$, $z = 00000$, $u = 11110$. Then $|x| = |y| = 3$, $|z| = 0$ and $|u| = 4$. ■

Example 13.7

The formal definition of the encoding function used in Example 13.4 is as follows: $e : B^m \longrightarrow B^{m+1}$ is defined as follows:

$e(x) = e(x_1, x_2, ..., x_m) = (x_1, x_2, ..., x_m, a)$

where $a = 0$ if $|x|$ is even, and $a = 1$ if $|x|$ is odd.

Let us recall the group $(Z_2, \oplus_2)$. If we denote [0] by 0 and [1] by 1 and define $1 + 1 = [1] \oplus_2 [1] = [0] = 0$, etc., then $(B, +)$ is isomorphic to Z_2. By Example 13.2 of Chapter 8, B^m is a group under the binary operation defined as

$$(x_1, x_2, ..., x_m) \oplus (y_1, y_2, ..., y_m) = (x_1 + y_1, x_2 + y_2, ..., x_m + y_m)$$

We shall see that this binary operation is useful in coding.

Let $*$ denote the binary operation of multiplication (mod 2) on $B = \{0, 1\}$. In other words, $*$ is given by the table

*	0	1
0	0	0
1	0	1

We define an $m \times n$ Boolean matrix as an $m \times n$ matrix whose entries are zeros and ones. The notation $\mathbf{H} = (h_{i,j})$ indicates that $h_{i,j}$ is the (i, j)th entry of $\mathbf{H}$. If $\mathbf{H} = (h_{i,j})$, $\mathbf{K} = (k_{i,j})$ are $m \times n$ Boolean matrices, then $\mathbf{H} + \mathbf{K}$ is defined as $\mathbf{H} + \mathbf{K} = (h_{i,j} + k_{i,j})$. Naturally, here $+$ stands for addition (mod 2) and $\mathbf{H} + \mathbf{K}$ is a Boolean matrix. Now, let $\mathbf{H}$ be an $m \times n$ Boolean matrix and $\mathbf{K}$ be an $n \times k$ matrix. Then $\mathbf{H} * \mathbf{K}$ is the $m \times k$ matrix whose (i, j)th $c_{i,j}$ entry is given by

$$c_{i,j} = a_{i,1} * b_{1,j} + a_{i,2} * b2, j + \cdots + a_{i,n} * b_{n,j}$$ ■

13.2 ERROR DETECTION

Example 13.8

Imagine someone sending a telegram containing the message:

Raja's wife is arriving on 12th

It is likely that this message gets distorted into one of the following wrong messages:

Raja is arriving on 12th

Raja's is arriving on 12th

Although the first wrong message is not the intended message, the receiver cannot detect the error. It is a meaningful sentence and instead of his wife, Raja himself could very well make the trip. But the receiver can easily see that the second wrong message is not the correct one.

In the coding theory, messages containing m symbols are converted into code words that contain n symbols with $n > m$. Now, only certain words made up of n symbols are legitimate code words. *When a word is received, which is not a code word, an error is detected.* Formally speaking, if $e : B^m \longrightarrow B^n$ is an encoding function, the range of e, that is, $e(B^m) \subseteq B^n$ is the set of all code words. If b is the word to be communicated, then $x = e(b)$ is communicated. If y is the received word and $y \in e(B^m)$, then we accept y even though it may not be equal to x. But if $y \notin e(B^m)$, we detect an error. ■

Definition 13.3 defines the notion of distance between code words, which is useful in error detection.

Definition 13.3: The Hamming distance between x and y in B^n is defined as $D(x, y) = |x + y|$.

Example 13.9

Consider B^6. Let $x = 110010$ and $y = 000111$. Then $|x + y| = |110101| = 4$.

Let $x = (a_1, a_2, ..., a_n)$ and $y = (b_1, b_2, ..., b_n)$. Now, if $c = (c_1, c_2, ..., c_n) = x + y$, then $c_i = 0$ if $a_i = b_i$ and $c_i = 1$ if $a_i \neq b_i$. Hence, $|x + y|$ is nothing but *the number of places in which the components of x and y differ.*

Thus, if x is a code word that is sent, and y is the received code word, then $|x + y|$ is the number of errors occurred.

We omit the proof of the following theorem, which can easily be proved by the reader. ■

Theorem 13.1

The Hamming distance has the following properties:

(a) $D(x, y) = D(y, x)$.

(b) $D(x, y) \geq 0$.

(c) $D(x, y) = 0$ if and only if $x = y$.
(d) $D(x, y) \leq D(x, z) + D(z, y)$.

Definition 13.4: Let $e : B^m \longrightarrow B^n$ be a code. The minimum distance of this code is defined as

$$\min\{d(e(x), e(y)) : x, y \in B^m, e(x) \neq e(y)\}$$

Example 13.10

Find the minimum distance of the following code: $e : B^2 \longrightarrow B^3$

$$\begin{aligned} e(00) &= 000 \\ e(01) &= 011 \\ e(10) &= 101 \\ e(11) &= 110 \end{aligned}$$

$d(000, 011) = 2$, $d(000, 101) = 2$, $d(000, 110) = 2$, $d(011, 101) = 2$, $d(011, 110) = 2$, $d(101, 110) = 2$. Hence, the minimum distance of this code is 2. ■

Example 13.11

Define $e : B^2 \longrightarrow B^4$ as $e(x_1, x_2) = (x_1, x_2, x_1, x_2)$. Find the minimum distance of this code.

SOLUTION

Let us first explicitly write the coding function:

$$\begin{aligned} e(00) &= 0000 \\ e(01) &= 0101 \\ e(11) &= 1111 \\ e(10) &= 1010 \end{aligned}$$

In the following table, $d(x, y)$ is given in the box where the column for x and the row for y meet.

d	0000	0101	1111	1010
0000	0	2	4	2
0101	2	0	2	4
1111	4	2	0	2
1010	2	4	2	0

Hence, the minimum distance is 2. ■

Example 13.12

Define $e : B^3 \longrightarrow B^4$ as $e(x_1, x_2, x_3) = (x_1, x_2, x_3, x_1 + x_2)$. Find the minimum distance of this code.

SOLUTION

Let us first explicitly write the coding function:

$$
\begin{aligned}
e(000) &= 0000 \\
e(010) &= 0101 \\
e(001) &= 0010 \\
e(011) &= 0111 \\
e(100) &= 1001 \\
e(110) &= 1110 \\
e(101) &= 1011 \\
e(111) &= 1111
\end{aligned}
$$

We note that $d(0000, 0010) = 1$ and hence the minimum distance is 1 for this code. ■

PRACTICE 13.2

Define $e : B^3 \longrightarrow B^5$ as $e(x_1, x_2, x_3) = (x_1, x_2, x_3, x_1 + x_2, x_2 + x_3)$. Find the minimum distance of this code.

It is not possible to design a code that would detect any number of errors. But we can design a code that can detect when a wrong string has been received, whenever the number of errors is less than or equal to a pre-fixed number k. Such a code has the following property: If $e : B^m \longrightarrow B^n$ is the encoding function, b is the word to be communicated, $x = e(b)$ is the transmitted code word and y is the received string, then whenever $d(x, y) \leq k$, $y \notin e(B^m)$. Theorem 13.2 illustrates the use of minimum distance in the design of such codes.

Definition 13.5: An encoding function $e : B^m \longrightarrow B^n$ is said to be capable of correcting k or fewer errors, if the following happens: whenever the received string y is such that $d(e(b), y) \leq k$, then $y \notin e(B^m)$. That is y is not a code word.

Theorem 13.2

An encoding function $B^m \longrightarrow B^n$ can detect k or fewer errors if and only if

$$\min \left\{ d(e(b), e(c)) : b, c \in B^m, e(b) \neq e(c) \right\} \geq k + 1 \qquad (13.1)$$

Proof

Let equation (13.1) hold. Let $b \in B^m$ $x = e(b)$ be the code word representing b. If

y is the received word, $y \neq x$ and y is a code word, then $d(x, y) \geq k+1$. So, x has been transmitted with $k+1$ or more errors. If x has been transmitted with k or less errors, then y cannot be a code word. That is, we have detected an error.

Now, we assume that equation (13.1) is false. We shall show that k or fewer errors could not be detected. Let

$$l = \min\left\{d(e(b), e(c)) : b, c \in B^m e(b) \neq e(c)\right\} \leq k$$

Choose code words x and y such that $d(x, y) = r$. If x had been transmitted and received as y, then $l \leq k$ errors have been committed. But y being a code word, the error could not have been detected. ■

Example 13.13

Find how many errors can be detected by the code $e : B^2 \longrightarrow B^4$ defined as

$$\begin{aligned} e(00) &= 00000 \\ e(10) &= 10011 \\ e(10) &= 01101 \\ e(10) &= 11110 \end{aligned}$$

SOLUTION

$d(0000, 10011) = 3$, $d(0000, 001101) = 3$, $d(0000, 11110) = 4$, $d(10011, 01101) = w(11110) = 4$, $d(10011, 11110) = w(01101) = 3$, $d(01101, 1110) = w(10011) = 3$. Hence the minimum weight is 3. Thus, this code can detect two or fewer errors by Theorem 13.2. ■

PRACTICE 13.3

Find how many errors can be corrected by the code for $e : B^3 \longrightarrow B^6$ defined as:

$$\begin{aligned} e(000) &= 000000 \\ e(100) &= 100110 \\ e(010) &= 011010 \\ e(001) &= 111100 \\ e(110) &= 110001 \\ e(101) &= 001111 \\ e(011) &= 010011 \\ e(111) &= 101101 \end{aligned}$$

13.2.1 Group Codes

We have already noticed that B^m is a group. We use this observation in the following definition.

Definition 13.6: A group code is an encoding function $e : B^m \longrightarrow B^n$ such that $e(B^m)$ is a subgroup of B^m.

Example 13.14

Let $e : B^2 \longrightarrow B^5$ be defined as:

$$\begin{aligned} e(00) &= 00000 \\ e(01) &= 01110 \\ e(10) &= 10101 \\ e(11) &= 11011 \end{aligned}$$

Show that e is a group code.

SOLUTION

Let us prepare the table of addition.

+	00000	01110	10101	11011
00000	00000	01110	10101	11011
01110	01110	00000	11011	10101
10101	10101	11011	00000	01110
11011	11011	10101	01110	00000

We find that $e(B^2)$ is closed. The identity element $00000 \in e(B^2)$. Further, every element of $e(B^2)$ is its own inverse. Hence $e(B^2)$ is a subgroup of B^5 and hence e is a group code. ■

Example 13.15

Show that the code of Example 13.11 is a group code.

SOLUTION

Let $u, v \in e(B^2)$. Then $u = (a, b, a, b)$ and $v = (c, d, c, d)$ for some $(a, b), (c, d) \in B^2$, $u + v = (a + c, b + d, a + c, b + d) = e(a + c, b + d) \in e(B^2)$.

Now, $u + (-v) = (a, b, a, b) + (-c, -d, -c, -d) = (a - c, b - d, a - c, b - d) = e(a - c, b - d) \in e(B^2)$ Thus, by Proposition 8.8 of Section 8.10, $e(B^2)$ is a group.

SOLUTION

Let $x = (x_1, x_2)$ and $y = (y_1, y_2)$ be elements of B^2. We have

$$\begin{aligned} e(x + y) &= e((x_1, x_2) + (y_1, y_2)) \\ &= e(x_1 + y_1, x_2 + y_2) \end{aligned}$$

$$\begin{aligned}
&= (x_{+}y_1, x_2 + y_2, x_1 + x_2, y_1 + y_2) \\
&= (x_1, x_2, x_1, x_2) + (y_1, y_2, y_1, y_2) \\
&= e((x_1, x_2)) + e((y_1, y_2)) \\
&= e(x) + e(y)
\end{aligned}$$

Hence, e is a group homomorphism and by Proposition 8.10 of Section 8.10, $e(B^2)$ is a subgroup of B^4. ■

Theorem 13.3

$\min\{|e(x) + e(y)| : x \neq yx, y \in B^n\} = \min\{|e(z)| : z \neq 0\}.$

Proof

$|e(x)|$ is the distance between $e(x)$ and the zero code word. If the code is a group code, then $d(x, y) = |e(x) + e(y)| = |e(z)|$ for some $z \in B^n$. Here, we do not claim that $z = x + y$. Since $e(B^m)$ is a group, $e(x) + e(y) \in e(B^m)$ and hence, there exists $z \in B^m$ such that $e(x)+e(y) = e(z)$. ■

If $e : B^m \longrightarrow B^n$ is a code, there are 2^n code words. To find the minimum distance, we need to calculate the distance between $\frac{2^n \times (2^n - 1)}{2} = 2^{n-1} \times (2^n - 1)$ pairs. But if the code is a group code, we need to calculate only the weights of 2^{n-1} code words. If e is a group homomorphism, then e is a group code. But for e to be a group code, it is not necessary that e should be a homomorphism.

PRACTICE 13.4

Show that $e; B^4 \longrightarrow B^5$ defined as follows is a group code:

$$\begin{aligned}
e(0000) &= 11000 \\
e(1000) &= 00000 \\
e(0100) &= 10100 \\
e(0010) &= 10010 \\
e(0001) &= 10001 \\
e(1001) &= 01100 \\
e(0101) &= 01010 \\
e(0011) &= 01001 \\
e(1100) &= 00110 \\
e(1010) &= 00101 \\
e(0110) &= 00011
\end{aligned}$$

$$
\begin{aligned}
e(1101) &= 11110 \\
e(1011) &= 11101 \\
e(0111) &= 11011 \\
e(1110) &= 10111 \\
e(1111) &= 01111
\end{aligned}
$$

Find the minimum weight and also the number of errors that can be detected.

PRACTICE 13.5

Show that e is not a group homomorphism in Practice 13.4.

PRACTICE 13.6

Show that $e : B3 \longrightarrow B7$ defined as

$$
\begin{aligned}
e(000) &= 0000000 \\
e(001) &= 0010110 \\
e(010) &= 0101000 \\
e(011) &= 0111110 \\
e(100) &= 1000101 \\
e(101) &= 1010011 \\
e(110) &= 1101101 \\
e(111) &= 1111011
\end{aligned}
$$

is a group code.

Designing of Group Codes: Let $m < n$ and $r = n - m$. An $n \times r$ Boolean matrix

$$
\mathbf{H} = \begin{bmatrix}
h_{11} & h_{12} & \ldots & h_{1r} \\
h_{21} & h_{22} & \ldots & h_{2r} \\
\cdot & \cdot & \ldots & \cdot \\
\cdot & \cdot & \ldots & \cdot \\
\cdot & \cdot & \ldots & \cdot \\
h_{m1} & h_{m2} & \ldots & h_{2r} \\
1 & 0 & \ldots & 0 \\
0 & 1 & \ldots & 0 \\
\cdot & \cdot & \ldots & \cdot \\
\cdot & \cdot & \ldots & \cdot \\
\cdot & \cdot & \ldots & \cdot \\
0 & 0 & \ldots & 1
\end{bmatrix}
$$

such that the last r rows form the $r \times r$ identity matrix is called a parity-check matrix. Let $\mathbf{N}$ be the $m \times r$ sub-matrix of $\mathbf{H}$ defined as

$$\mathbf{N} = \begin{bmatrix} h_{11} & h_{12} & \dots & h_{1r} \\ h_{21} & h_{22} & \dots & h_{2r} \\ \cdot & \cdot & \dots & \cdot \\ \cdot & \cdot & \dots & \cdot \\ \cdot & \cdot & \dots & \cdot \\ h_{m1} & h_{m2} & \dots & h_{2r} \end{bmatrix}$$

Theorem 13.4

Define $e : B^m \longrightarrow B^n$ as follows: $e(b_1, b_2, \dots, b_m) = (b_1, b_2, \dots, b_m, x_1, x_2, \dots, x_r)$ where $(x_1, x_2, \dots, x_r) = (b_1, b_2, \dots, b_m) * \mathbf{N}$. Then, e is a group code.

Proof

We show that the image $e(B^m)$ is the set

$$\{x \in B^n : x * \mathbf{H} = \mathbf{0}\}$$

Let $x \in e(B^m)$. Then $x = e(b)$ for some $b = (b_1, b_2, \dots, b_m) \in B^m$. So, $x = (b_1, b_2, \dots, b_m, x_1, x_2, \dots, x_r)$, where $(x_1, x_2, \dots, x_r) = b * \mathbf{N}$. Now,

$$\begin{aligned} x * \mathbf{H} &= b * \mathbf{N} + (x_1, x_2, \dots, x_r) \\ &= (x_1, x_2, \dots, x_r) + (x_1, x_2, \dots, x_r) \\ &= 0 \end{aligned}$$

Next, let $x \in B^m$ be such that $x * H = \mathbf{0}$. We must show that $x = e(b)$ for some $b \in B^m$. Let us write $x = (b_1, b_2, \dots, b_m, x_1, x_2, \dots, x_r)$. We have

$$\begin{aligned} \mathbf{0} &= x * \mathbf{H} \\ &= (b_1, b_2, \dots, b_m) * \mathbf{N} + (x_1, x_2, \dots, x_r) \end{aligned}$$

Letting $(b_1, b_2, \dots, b_m) * \mathbf{N} = (y_1, y_2, \dots, y_r)$ we find that $(x_1 + y_1, x_2 - y_2, \dots, x_r + y_r) = \mathbf{0}$. Now in the group $(B, +)$ every element is its own inverse. Hence $x_1 = y_1$, $x_2 = y_2, \dots, x_r = y_r$. Thus, $(x_1, x_2, \dots, x_r) = (b_1, b_2, \dots, b_m) * \mathbf{N}$ and the proof is complete by taking $b = (b_1, b_2, \dots, b_m)$.

Now, we define $f : B^m \longrightarrow B^n$. ■

Example 13.16

Compute the code $e_H : B^4 \longrightarrow B^7$ corresponding to the parity-check matrix

$$H = \begin{bmatrix} 1 & 1 & 0 \\ 1 & 1 & 1 \\ 0 & 1 & 0 \\ 0 & 0 & 1 \\ 1 & 0 & 0 \\ 0 & 1 & 0 \\ 0 & 0 & 1 \end{bmatrix}$$

SOLUTION

$$N = \begin{bmatrix} 1 & 1 & 0 \\ 1 & 1 & 1 \\ 0 & 1 & 0 \\ 0 & 0 & 1 \end{bmatrix}.$$

For $b \in B^4$, $e(b)$ is obtained by appending $b * N$ to b on the right. It is clear that if $b = 1000$, then $b * N$ is the first row of N. Similarly, for $b = 0100$, $b * N$ is the second row of N. If $b = 0101$, then $b * N$ is the sum of the second and fourth rows of N. If $b = 1101$, then $b * N$ is the sum of the first, second and fourth rows of N. Using these observations, we calculate $b * N$ and append it to b to calculate $e(B)$

$$0000 * \begin{bmatrix} 1 & 1 & 0 \\ 1 & 1 & 1 \\ 0 & 1 & 0 \\ 0 & 0 & 1 \end{bmatrix} = 000; \qquad 1000 * \begin{bmatrix} 1 & 1 & 0 \\ 1 & 1 & 1 \\ 0 & 1 & 0 \\ 0 & 0 & 1 \end{bmatrix} = 110$$

$$0100 * \begin{bmatrix} 1 & 1 & 0 \\ 1 & 1 & 1 \\ 0 & 1 & 0 \\ 0 & 0 & 1 \end{bmatrix} = 111; \qquad 0010 * \begin{bmatrix} 1 & 1 & 0 \\ 1 & 1 & 1 \\ 0 & 1 & 0 \\ 0 & 0 & 1 \end{bmatrix} = 010$$

$$0001 * \begin{bmatrix} 1 & 1 & 0 \\ 1 & 1 & 1 \\ 0 & 1 & 0 \\ 0 & 0 & 1 \end{bmatrix} = 001; \qquad 1100 * \begin{bmatrix} 1 & 1 & 0 \\ 1 & 1 & 1 \\ 0 & 1 & 0 \\ 0 & 0 & 1 \end{bmatrix} = 001$$

$$1010 * \begin{bmatrix} 1 & 1 & 0 \\ 1 & 1 & 1 \\ 0 & 1 & 0 \\ 0 & 0 & 1 \end{bmatrix} = 100; \qquad 1001 * \begin{bmatrix} 1 & 1 & 0 \\ 1 & 1 & 1 \\ 0 & 1 & 0 \\ 0 & 0 & 1 \end{bmatrix} = 111$$

$$0110 * \begin{bmatrix} 1 & 1 & 0 \\ 1 & 1 & 1 \\ 0 & 1 & 0 \\ 0 & 0 & 1 \end{bmatrix} = 101; \qquad 0101 * \begin{bmatrix} 1 & 1 & 0 \\ 1 & 1 & 1 \\ 0 & 1 & 0 \\ 0 & 0 & 1 \end{bmatrix} = 110$$

$$0011 * \begin{bmatrix} 1 & 1 & 0 \\ 1 & 1 & 1 \\ 0 & 1 & 0 \\ 0 & 0 & 1 \end{bmatrix} = 011; \qquad 0111 * \begin{bmatrix} 1 & 1 & 0 \\ 1 & 1 & 1 \\ 0 & 1 & 0 \\ 0 & 0 & 1 \end{bmatrix} = 100$$

$$1011 * \begin{bmatrix} 1 & 1 & 0 \\ 1 & 1 & 1 \\ 0 & 1 & 0 \\ 0 & 0 & 1 \end{bmatrix} = 101; \qquad 1101 * \begin{bmatrix} 1 & 1 & 0 \\ 1 & 1 & 1 \\ 0 & 1 & 0 \\ 0 & 0 & 1 \end{bmatrix} = 001$$

$$1110 * \begin{bmatrix} 1 & 1 & 0 \\ 1 & 1 & 1 \\ 0 & 1 & 0 \\ 0 & 0 & 1 \end{bmatrix} = 011; \qquad 1111 * \begin{bmatrix} 1 & 1 & 0 \\ 1 & 1 & 1 \\ 0 & 1 & 0 \\ 0 & 0 & 1 \end{bmatrix} = 010$$

Now we tabulate the code

$$\begin{aligned} e(0000) &= 0000000 \\ e(1000) &= 1000110 \\ e(0100) &= 0100111 \\ e(0010) &= 0010010 \\ e(0001) &= 0001001 \\ e(1100) &= 1100001 \\ e(1010) &= 1010100 \\ e(1001) &= 1001111 \\ e(0110) &= 0110101 \\ e(0011) &= 0011011 \\ e(0111) &= 0111100 \\ e(1111) &= 1111010 \end{aligned}$$

■

PRACTICE 13.7

Find the minimum distance of the code generated by

$$H = \begin{pmatrix} 1 & 0 & 1 \\ 1 & 1 & 0 \\ 0 & 0 & 1 \\ 1 & 0 & 0 \\ 0 & 1 & 0 \\ 0 & 0 & 1 \end{pmatrix}$$

13.3 DECODING...

Example 13.17

Let $S = \{00011, 11010, 10000, 01010, 11011, 01111, 11100\}$ and let $y = 00100$. Let us calculate the distance between y and each of the elements of S

$$\begin{aligned} d(00100, 00011) &= 3 \\ d(00100, 11010) &= 4 \end{aligned}$$

$$d(00100, 10000) = 2$$
$$d(00100, 01010) = 3$$
$$d(00100, 11011) = 5$$
$$d(00100, 01111) = 3$$
$$d(00100, 11100) = 2$$

We see that 2 is the smallest distance in the above list. Thus, among all the strings of S, the strings 10000 and 11100 are the closest to the element 00100. This motivates the following definition. ■

Definition 13.7: Let $S \subseteq B^n$ and $y \in B^n$. We say that $z \in S$ is the nearest element to y in S if

$$d(z, y) \leq d(x, y) \quad \text{for all} \quad x \in S$$

Remark 13.1

We cannot talk about *the nearest* element as there may be more than one such element. In the previous example, both 10000 and 11100 are nearest elements to 00100 in S.

Let $e : B^m \longrightarrow B^n$ be a code. Let $x \in B^m$ be a message and let $e(x)$ be the corresponding code word that has been transmitted. Now, let $y \in B^n$ be the received word. The set of all code words is $e(B^m)$. If $y \in e(B^n)$, then there is a unique $w \in B^m$ such that $e(w) = y$, and we decode the message as w. Though w may not be equal x, this is the best possible thing we can do.

If there had been no error in transmission, then $y = e(w)$ and $w = x$. We have the exact decoding.

Now, how to decode if $y \notin e(B^n)$? We choose a nearest element z to y in $e(B^n)$. If $z = e(w)$, then w is the decoded message. As there may be more than one nearest element, to get a unique decoded message, we arrange the code words in some arbitrary order and choose the first one among the nearest code words. The following example illustrates this procedure.

Example 13.18

Consider the following encoding function $e : B^3 \longrightarrow B^4$:

$$e(000) = 1111$$
$$e(100) = 0110$$
$$e(010) = 1010$$
$$e(001) = 1100$$
$$e(110) = 0011$$

$$e(101) = 0101$$
$$e(011) = 1001$$
$$e(111) = 0000$$

Decode the following received words 1110, 1000 and 1100, where we have arranged the code words in the order given above.

SOLUTION

Note that 1110 is not a code word. We calculate the distance between the received word and the code words.

$d(1110, 1111) = 1$. Now, since a positive weight is greater than or equal to 1, 1111 is one of the nearest elements to 1110. As per our arrangement rule, we take 1111 as the corrected code word and decode it as 000 since $e(000) = 1111$.

Consider 1000. Again this is not a code word. We calculate the distance between 1000 and the code words.

$d(1000, 1111) = 3$. $d(1000, 0110) = 3$, $d(1000, 1010) = 1$. Again, we shall choose as before, 1010 as the nearest element to 1000. It is decoded as 0101, since 1100 is a code word and $e(001)$ is decoded as 001. ■

The process of obtaining the original message from a received string is called decoding. Just as encoding is achieved by a function, decoding is achieved by another function. It is natural that if $e : B^m \longrightarrow b^n$ is the encoding function, then the decoding function d will have the domain as B^n and the co-domain as B^m.

Definition 13.8: An onto function $d : B^n \longrightarrow B^m$ is called a decoding function associated with a code $e : B^m \longrightarrow B^n$ if $d(x) = b$ is such that when there is no error during transmission, and $e(b) = x$.

Theorem 13.2 gives the condition for a code to *detect* k or fewer errors. Theorem 13.5 gives the condition for a code to *correct* k or fewer errors. The relevant definition is as follows:

Definition 13.9: The pair of encoding and decoding functions $e : B^m \longrightarrow B^n$, $d : B^n \longrightarrow B^m$ is said to correct k or fewer errors if the following happens: Whenever $x = e(b)$, y is the received word, $|x + y| \leq k$, then $d(y) = b$.

Theorem 13.5

Suppose that $e : B^m \longrightarrow B^n$ is a coding function and $d : B^n \longrightarrow B^m$ is a maximum likelihood decoding function associated with e. Then the pair (e, d) can correct k or fewer errors if and only if the minimum distance of the code e is at least $2k - 1$.

Proof

Assume that the minimum distance of e is at least $2k+1$. Let $b \in B^m$ and $x = e(b) \in B^n$. Let x be transmitted with k or fewer errors, and let y be the received word. We have $d(x, y) \leq k$. If z is any other code word, then

$$2k + 1 \leq d(x, z) \leq d(x, y) + d(y, z) \leq k + d(y, z)$$

Therefore,

$$k + 1 \leq D(y, z)$$

This implies that x is the unique code word that is closest to y, and $d(y) = b$. Hence k or fewer errors can be corrected. ■

We have already noted that finding the minimum distance for a general encoding function e is a tedious process. If the code is a group code, then the minimum weight gives the minimum distance. Thus, when we want codes detecting or correcting a pre-determined number of errors, we have to find subgroups of B^n having a given minimum weight. First, let us consider how one may obtain various subgroups of B^n.

Proposition 13.1

Show that the set of all elements of even weight is a subgroup of B^n.

Proof

In B^n every element is its own inverse. Thus, any closed subset of B^n containing 0 becomes a subgroup. Now if x and y have even weights, then $x + y$ also has an even weight. Thus, the set of all elements of B^n having an even weight is closed and it is a subgroup. ■

Example 13.19

Design a code $e : B^2 \longrightarrow B^n$ that would

(i) detect one error,

(ii) correct one error.

SOLUTION

For a code to correct one error, as per Theorem 13.2, its minimum weight is $1 + 1 = 2$. Now, since e has to be one-one, $e(B^2)$ must have four elements. The minimum weight has to be 2. B^2 itself has four elements but its minimum weight is 1. Thus, we must look for a subgroup of B^3 having four elements and with minimum weight 2. Let us try out the simplest subgroup of B^3 that we know. It is the set of elements having an even weight. These elements are 000, 110, 101 011. We are lucky that both our requirements are satisfied. Now, we define

$$\begin{aligned} e(00) &= 000 \\ e(01) &= 011 \\ e(10) &= 101 \\ e(11) &= 110 \end{aligned}$$

It is clear that a single error would produce an element with odd weight, which is not a code word, and hence it is detected.

Next, let us consider the problem of designing a code that would detect and correct one error. Now, as per Theorem 13.5, the minimum weight is at least $2 \times 1 + 1 = 3$.

Now, we need a subgroup of some B^n containing four elements with minimum weight 3. It is obvious that the only subgroup of B^3 with minimum weight 3 is $\{000, 111\}$. Thus, we will have to go to higher dimensions. The question that confronts us now is,

"Can we find a subgroup of B^4 containing 4 elements with minimum weight 3?"

The answer is 'no'. The reader is asked to prove this in Practice 13.9.

Denoting $H = \{000, 111\}$, we consider the cartesian product $H \times H = \{000000, 000111, 111000, 111111\}$, which is a subgroup of B^6, satisfying all our requirements. Now, we can define $e : B^2 \longrightarrow B^6$ as

$$\begin{aligned} e(00) &= 000000 \\ e(01) &= 000111 \\ e(10) &= 111000 \\ e(11) &= 111111 \end{aligned}$$

Let us verify if this code actually detects and corrects one error. Let $b = 01$ be the message to be conveyed. $e(b) = 000111$ is transmitted. For example, $y = 000110$ is a case of one error taking place. It is not a code word. Hence the error has been detected. Now, let us use the maximum likelihood procedure to correct the error. To do that, we need to find the distances between the code words and y:

$$\begin{aligned} d(000000, 000110) &= 2 \\ d(000111, 000110) &= 1 \\ d(111000, 000110) &= 2 \\ d(111111, 000110) &= 4 \end{aligned}$$

This shows that 000111 is the unique code word nearest to y and hence no matter how we order the code words, y will be corrected as 000111 to be decoded into 01, which is precisely the message that had to be conveyed.

If we want more errors to be corrected and detected, then we may have to choose a bigger value of n. ■

PRACTICE 13.8

Show that there is no subgroup of order 4 in B^4 having minimum weight 3 by using the fact that such a subgroup must be contained in the subset $\{0000, 1110, 1101, 1011, 1111\}$.

PRACTICE 13.9

Find a subgroup of B^5 containing four elements with minimum weight 3, and design a code $e : B^2 \longrightarrow B^5$ capable of detecting and correcting one error. In what way is this code better than the one discussed in (ii) of Example 13.19?

Example 13.20

Design a code $e : B^4 \longrightarrow B^n$ that would detect and correct three or fewer errors by appropriately choosing n.

SOLUTION

A proper design of a code must address the question of efficiency. For example, it is advantageous to choose as small a value for n as possible. But in this example, we are only interested in showing that a code, which satisfies the given requirements, exists.

We are looking for a group containing $2^4 = 16$ elements with minimum weight 3. We take $H \times H \times H \times H \subseteq B^{12}$, where $H = \{000, 111\}$. A part of the definition of $e : B^4 \longrightarrow B^{12}$ is given below:

$$\begin{aligned} e(0000) &= 000000000000 \\ e(1000) &= 111000000000 \\ e(1010) &= 111000111000 \end{aligned}$$

PRACTICE 13.10

In Example 13.20, take a message in B^4, find $x = e(b)$ and consider a y with $0 < |x + y| \leq 3$. Decode y and show that b is obtained.

The procedure for maximum likelihood decoding can be simplified if e is a group code. We now have that $e(B^m) = H$ is a subgroup of B^n. If y is the received string and $y \notin e(B^n)$, then we have to find $z \in H$ such that

$$D(y, z) \leq D(y, u)$$

for all $u \in N$. That is, to find $y \in H$ such that

$$|y + z| \leq |y + u| \quad \text{for all} \quad u \in H \tag{13.2}$$

Equation (13.2) means that $y + z$ is an element of minimum weight in the coset $y + H$. So, if z is a required nearest element, then $v = z + y$ is an element of the coset $y + H$ and is of minimum weight in $y + H$. Further, $v = y + z$ implies that $z = v + y$. To decode y, we have the following procedure:

(i) Find $v \in y + H$, which has the minimum weight.
(ii) Define $z = v + y$, which is a code word.
(iii) Find the unique $b \in B^m$ such that $z = e(b)$.
(iv) b is required for the decoding of y.

In the following example, we list the elements of all possible cosets in a systematic manner and make the decoding easy.

Example 13.21

Consider the group code defined as

$$\begin{aligned} e(00) &= 0000 \\ e(10) &= 1010 \\ e(01) &= 0101 \\ e(11) &= 1111 \end{aligned}$$

Decode 1110 for this code.

SOLUTION

$$\begin{aligned} B^2 &= \{00, 01, 10, 11\} \\ B^4 &= \{0000, 1000, 0100, 0010, 0001, 0011, 0110, 1100, \\ &\quad 1001, 1010, 0101, 1110, 1011, 0111, 1101, 1111\} \end{aligned}$$

$H = e(B^2)\{0000, 0101, 1010, 1111\}$. For each $x \in H$, the coset xH is nothing but H. Consider $x = 1000 \in B^4$, which is not in H. Let us find its coset:

$$\{1000, 1101, 0010, 0111\}$$

Next, we consider 0100, which is not in either of the above cosets. Its coset is given by

$$\{0100, 0001, 1110, 1011\}$$

Finally, we consider 0011 and its coset

$$\{0011, 0110, 1001, 1100\}$$

Now we tabulate the cosets in the following fashion. Each row represents one of the cosets and every element of the coset is in the corresponding column. The first element in each row is an element of the coset, which has the least weight. If more than two elements have the least weight, the first element is chosen from them in an arbitrary manner.

0000	1010	0101	1111
1000	1101	0010	0111
0100	0001	1110	1011
0011	0110	1001	1100

Let us decode $y = 1110$. We find that it is in the third coset. The first element in the coset, which is one of the elements of minimum weight, is $v = 0100$. Let $z = v + y = 0100 + 1110 = 1010$, which is a code word such that $e(10) = 1010$. Thus, the required message is 01. ■

The above procedure can be further simplified for the case of a code generated by a parity-check matrix. Every code word $x \in H = e(B^m)$ and every element of B^n, which is not a code word, belongs to some coset of H. Instead of trying to locate y in one of the cosets by inspection, we use Theorem 13.6 to decode y.

Theorem 13.6

$x \in y + N$ if and only if $x * H = y * H$.

Proof

$x * H$ is called the syndrome of x. By Theorem 13.4, $x \in N$ if and only if $x * H = 0$. Therefore,

$$\begin{aligned} x \in y + N & \\ & \leftrightarrow \quad x + y \in N \\ & \leftrightarrow \quad (x + y) * H = 0 \\ & \leftrightarrow \quad x * H = y * H \end{aligned}$$

■

Example 13.22

Find the encoding function $e : B^2 \longrightarrow B^4$ given by the parity-check matrix

$$\begin{pmatrix} 1 & 1 \\ 1 & 0 \\ 1 & 0 \\ 0 & 1 \end{pmatrix}$$

Decode 1111, 0101 and 1001

SOLUTION

Here $\mathbf{N} = \begin{pmatrix} 1 & 1 \\ 1 & 0 \end{pmatrix}$. We calculate $b * \mathbf{N}$ for each of the messages as follows:

$00 * \begin{pmatrix} 1 & 1 \\ 1 & 0 \end{pmatrix} = 00$. Hence $e(00) = 0000$.

$01 * \begin{pmatrix} 1 & 1 \\ 1 & 0 \end{pmatrix} = 10$. Hence $e(01) = 0110$.

$10 * \begin{pmatrix} 1 & 1 \\ 1 & 0 \end{pmatrix} = 11$. Hence $e(11) = 1011$.

$11 * \begin{pmatrix} 1 & 1 \\ 1 & 0 \end{pmatrix} = 01$. Hence $e(11) = 1101$.

Next, we calculate the cosets. N is just the set of code words. First, take any element of B^4 not in N and find the corresponding coset. Afterwards, choose an element not in any of these two former sets and so on.

$$\begin{aligned} N &= \{0000, 0110, 1011, 1101\} \\ 1000 + N &= \{1000, 1110, 0011, 0101\} \\ 1100 + N &= \{1100, 0111, 1010, 0001\} \\ 0100 + N &= \{0100, 1111, 0010, 1001\} \end{aligned}$$

From each of these cosets, choose an element of least weight and declare it as the coset leader. Calculate $x * H$ for each coset leader x and prepare the table of syndromes.

$0 * H = 00$. We show the calculation for one coset leader:

$$1000 * \begin{pmatrix} 1 & 1 \\ 1 & 0 \\ 1 & 0 \\ 0 & 1 \end{pmatrix} = 11$$

The table of syndromes:

Coset Leader	*Syndrome*
0000	00
1000	11
0001	01
0100	10

Now, to decode the given words: $1111 * \begin{pmatrix} 1 & 1 \\ 1 & 0 \\ 1 & 0 \\ 0 & 1 \end{pmatrix} = 10$. Since the syndrome of the code word 0100 is 10, 1111 is corrected as $0100 + 1111 = 1011$ and it is decoded as 10. ■

PRACTICE 13.11

Complete Exercise 13.22 by decoding 0101 and 10011.

SUMMARY

- A message b is a finite string of zeros and ones.
- A code word x is a string with more bits than the message to facilitate decoding when errors may occur during transmission.
- A one-one function $e : B^m \longrightarrow B^n$ is called an encoding function.
- A group code is given by an encoding function $e : B^m \longrightarrow B^n$ such that $e(B^m)$ is a subgroup $(B^{\oplus})$.
- $e : B^m \longrightarrow B^n$ is said to detect k or fewer errors whenever x is transmitted with k or fewer errors resulting in the receiving of y. Then y is not a code word.
- The weight of $x \in B^n$ is the number of ones in it.
- The distance between two elements x, y of B^m is the weight of $|x + y|$
- The minimum distance of a code is the smallest of all the positive distances between the code words.
- For a code e to be capable of detecting k or fewer errors, its minimum distance should be at least $k + 1$.
- For a group code, the minimum distance is given by its minimum weight.
- When the received string is not a code word, it is decoded after correcting it as one of its closest elements in the set of code words.
- For a code to be capable of detecting and correcting k errors, its minimum distance should be at least $2k + 1$.
- Parity-check matrices give rise to group codes.

PROBLEMS

1. The following strings are encoded names of some of the famous characters from the plays of Shakespeare. After making some guesses, find the names and also find the encoding and decoding procedures. HSJGSQ, AYCQCP, PMKCM, HSJGCR, AJCMNYRY.
2. Find the weights of the following: 0011, 11000011, 01110, 11111.
3. Find the Hamming distance between the following pairs:
   (i) 1110, 0111.
   (ii) 111101, 010101.
   (ii) 00101, 10001.
4. Find the nearest elements to 1001100 (with respect to the Hamming distance) from the set $\{0000000, 0110011, 1000000\}$.
5. Find the number of elements in B^4 whose weight is 3.
6. In B^4, find all the elements that are at a distance of 3 from 1111.
7. Find the nearest elements to 11111 from the set of all elements of B^5 with an even weight.
8. Explain why $e : B^3 \longrightarrow B^4$ defined as $e(x_1, x_2, x_3) = (x_1, x_2, x_1, x_2)$ is not an encoding function.
9. Find the minimum distance of the code $e : B^2 \longrightarrow B^3$ defined as $e(x_1, x_2) = (x_2, x_1, x_1 * x_2)$. Here $x_1 * x_2$ is mod 2 multiplication.
10. Define $e : B^3 \longrightarrow B^4$ as $e(x_1, x_2, x_3) = (x_1, x_2, x_3, 1)$. Is it a group code?
11. Is the set of all elements of B^n with odd weight a subgroup?
12. Consider the encoding function $e : B^3 \longrightarrow B^4$ defined as $e(000) = 0000)$, $e(100) = 0011$, $e(010) = 0100$, $e(001) = 1000$, $e(110) = 0111$, $e(101) = 1011$, $e(011) = 1100$, $e(111) = 1111$. Find how many errors e can detect.
13. Show that $e : B^2 \longrightarrow B^3$ defined as $e(000) = 1100$, $e(100) = 1001$, $e(010) = 1010$, $e(001) = 0011$, $e(110) = 1111$, $e(101) = 0110$, $e(011) = 0101$, $e(111) = 0000$ is a group code. Find how many errors e can detect.
14. Show that for any k and m there exists an encoding function $e : B^m \longrightarrow B^n$, which detects and corrects k errors, where the choice of n depends on k and m.
15. Let
$$H = \begin{pmatrix} 0 & 1 & 1 \\ 0 & 0 & 1 \\ 1 & 0 & 0 \\ 1 & 0 & 0 \\ 0 & 1 & 0 \\ 0 & 0 & 1 \end{pmatrix}$$
be a parity-check matrix. Decode the following words: 110011, 111000, 001001.
16. Write an algorithm for the procedure of decoding using syndrome.

FURTHER READING

Kolman, B., Busby, R. C. and Ross. S. (2001), *Discrete Mathematical Structures*, Prentice Hall of India, New Delhi.

CHAPTER 14

Discrete Probability

Learning Objectives

On completing this chapter, you should be able to:

- write down the sample space for simple random experiments
- define the notions of an event, mutually exclusive events and compliment of an event
- define the notion of a random variable
- state the axioms of probability
- state and prove the addition law for probabilities
- compute the probability of events in simple random experiments
- identify events associated with a given random variable
- define conditional probability
- define the notions of independent and mutually exclusive events
- state the multiplicative law for probabilities
- state and prove total probability and Bayes theorems
- solve problems based on total probability and Bayes theorems
- recall the notion of probability mass function
- define the quantities of mean, variance and standard deviation
- calculate mean, variance and standard deviation of a random variable whose probability mass function is given
- recognize binomial, geometric, negative binomial and Poisson random variables
- solve problems based on binomial, geometric, negative binomial and Poisson random variables.

14.1 SAMPLE SPACE

Consider putting your hand inside a pot containing water. Your hand would get wet. If we think of putting a hand inside the pot as an experiment, we know what is going to happen. Such experiments are called *deterministic*. If we toss a coin, we do not know *in advance*, whether the head would come up or the tail. This is a classic example of a *random experiment*.

Now, let us take a look at the experiment of switching on a bulb. If the power supply is good and the bulb is in good condition, the bulb shall glow. Otherwise, it may not glow. Hence, *switching on a bulb* can also be regarded as a random experiment.

In each of these random experiments, two outcomes are possible. The possible results of tossing a coin are denoted by H (head) and T (tail). Let us denote the outcomes of the bulb experiment as L (light) and D (dark). We shall come back to these examples later.

Communication channels: It is well known that computers work with strings of *zeros* and *ones*. For example, an e-mail message is converted into a string of zeros and ones using the ASCII code. A communication channel is a device through which strings of bits can be transmitted and received. Consider a communication channel that transmits strings of bits. For the purpose of illustration, we shall assume that when a string of bits is transmitted, each bit is transmitted one after another.

Example 14.1

Let us assume that a three-bit string is communicated. An example of such a three-bit string is 011. No real-life communication channel is perfect. So it is likely that the bit 1 gets transmitted as 0 and vice versa. Therefore, receiving a string becomes a random experiment. Now when the bit 011 is sent, any one of the eight strings 011, 111, 001, 010, 101, 000, 110, 100 could be received. Thus, this random experiment has eight possible outcomes. ■

Example 14.2

An e-mail service demands that a password must contain exactly five characters made up of the digits $0, 1, 2, \ldots, 9$ and the letters $a, b, c, d \ldots$. Repetition of a character is allowed. Though an e-mail account holder may have some logic in the formulation of his or her password, as far as the service provider is concerned, a password is a random sequence of five characters. If we think of obtaining a password from an account holder as a random experiment, some of the outcomes are 77733, *a*35*dge*, *zzzzz*, 5*uio*3, etc. By using the ideas of Section 4.1, the total number of outcomes is $36^5 = 60466176$.

Imagine a hacker trying to open someone's e-mail account illegally. If the hacker randomly types out a sequence of five characters, what is the chance that it would turn out to be the correct one?

The chances are 1 in 60466176. ■

Example 14.3

Hotel Reservation and Cancelation. Let us imagine a small exclusive hotel located in Mamallapuram. It caters only to Americans visiting Mamallapuram and it has only five double rooms. Rooms are booked well in advance. As the visiting Americans may suddenly change their travel plans, there are bound to be some cancelations. If only five rooms are booked and some cancelations take place, there would be some loss. Taking this into account, the manager allows a maximum of seven bookings even though there are only five rooms. Now, let us list some possible situations as below:

(i) Four rooms are booked and three are canceled later.

(ii) Seven rooms are booked and one is canceled later.

(iii) Six rooms are booked and two are canceled later.

The situation (ii) is potentially bad for business because a customer who has booked a room in advance would be denied a room. In situation (i), though no customer would be denied a room, there will not be much profit. The situation (iii) is good for both the customer and the manager.

If we agree that the ordered pair (a, b) denotes the situation where a rooms are booked and b rooms are canceled, then the list of all possible situations is given by

$$\begin{array}{r}
(0,0)\\
(1,0),(1,1)\\
(2,0),(2,1),(2,2)\\
(3,0),(3,1),(3,2),(3,3)\\
(4,0),(4,1),(4,2),(4,3),(4,4)\\
(5,0),(5,1),(5,2),(5,3),(5,4),(5,5)\\
(6,0),(6,1),(6,2),(6,3),(6,4),(6,5),(6,6)\\
(7,0),(7,1),(7,2),(7,3),(7,4),(7,5),(7,6),(7,7)
\end{array}$$

Having looked at some random experiments and their possible outcomes, we shall now define the idea of a sample space. ■

Definition 14.1: The set of all possible outcomes S of a random experiment is called the *sample space*.

Example 14.4

From the introductory examples we can say

(a) The sample space for the experiment of tossing a coin is $\{H, T\}$.

(b) The sample space for the 'bulb experiment' is $\{L, D\}$.

(c) The sample space in Example 14.1 is

$$\{011, 111, 001, 010, 101, 000, 110, 100\}$$

(d) The sample space for Example 14.3, the hotel reservation and cancelation experiment, is

$$\{(0,0)(1,0),(1,1)(2,0),(2,1),(2,2)(3,0),(3,1),(3,2),$$
$$(3,3)(4,0),(4,1),(4,2),(4,3),(4,4)(5,0),(5,1),(5,2)$$
$$(5,3),(5,4),(5,5)(6,0),(6,1),(6,2),(6,3),(6,4),(6,5),$$
$$(6,6)(7,0),(7,1),(7,2),(7,3),(7,4),(7,5),(7,6),(7,7)\}$$

(e) The sample space of the passwords of Example 14.2 contains the elements 77733, *a*35*dge*, *zzzzz*, 5*uio*3, etc.
As it has 60466176 elements, we cannot list all its elements. ■

We now consider further random experiments and their sample spaces

Example 14.5

The sample space for the experiment of tossing two coins simultaneously is $\{HH, HT, TH, TT\}$. Let us note that this very set can also be thought of as the sample space for tossing a single coin twice. ■

Example 14.6

The sample space for the experiment of throwing a die is $\{1,2,3,4,5,6\}$. The sample space for the experiment of throwing a pair of dice is

$$\{(1,1),(1,2),(1,3),(1,4),(1,5),(1,6)$$
$$(2,1),(2,2),(2,3),(2,4),(2,5),(2,6)$$
$$(3,1),(3,2),(3,3),(3,4),(3,5),(3,6)$$
$$(4,1),(4,2),(4,3),(4,4),(4,5),(4,6)$$
$$(5,1),(5,2),(5,3),(5,4),(5,5),(5,6)$$
$$(6,1),(6,2),(6,3),(6,4),(6,5),(6,6)\}$$ ■

Example 14.7

The sample space for the experiment of throwing a die and tossing a coin simultaneously is

$$\{(1,H),(1,T),(2,H),(2,T),(3,H)(3,T),$$
$$(4,H),(4,T),(5,H),(5,T),(6,H)(6,T)\}$$ ■

Example 14.8

Let S_1 and S_2 be the sample spaces for two different random experiments. If these two experiments are carried out simultaneously, then the sample space for this experiment is just the cartesian product $S_1 \times S_2$. ■

Example 14.9

In a communication channel, after many observations of transmission and reception, it has been found that during the transmission of strings of bits, a maximum of two errors occur. Write the sample space for the possible strings received when

(a) 10 is transmitted

(b) 011 is transmitted

SOLUTION

Consider the case when 10 is transmitted. With no error during the transmission, 10 would be received. If one error occurs, either 00 or 11 would be received. If two errors occur, then 01 would be received. Hence the required sample space is $\{10, 00, 11, 01\}$.

Consider the case when 011 is transmitted. Since three errors cannot occur, 100 would not be received. But any other string of three bits can be received. Thus the sample space is

$$\{011, 111, 001, 010, 101, 000, 110\}$$ ■

PRACTICE 14.1

Consider the transmission of 0100 in a communication channel. Write down the sample space of the received strings if during the transmission

(a) any number of errors can occur,

(b) at least one error occurs,

(c) at most three errors occur,

(d) no error occurs.

PRACTICE 14.2

(a) If three coins are tossed simultaneously, what is the resulting sample space?

(b) If a pair of dice and a coin are tossed simultaneously, what is the resulting sample space?

Example 14.10

Consider throwing a die. If the die shows up an odd number, a coin is tossed. What is the sample space?

SOLUTION

The sample space is $\{2, 4, 6, (1, H), (1, T), (3, H), (3, T), (5, H), (5, T)\}$. ■

Example 14.11

A coin is tossed. If it results in a tail, a die is thrown. If the die shows up an even number, the coin is tossed again. Write down the sample space.

SOLUTION

$\{H, (T, 1), (T, 3), (T, 5), (T, 2, T), (T, 2, H), (T, 4, T), (T, 4, H)(T, 6, T), (T, 6, H)\}$. ■

PRACTICE 14.3

A coin is tossed twice. If the second throw results in a head, a die is thrown; otherwise a coin is tossed. What is the sample space?

Example 14.12

There are five red balls, three blue balls and one green ball in an urn. Two balls are drawn randomly one after another without replacing the first ball. Describe the sample space.

SOLUTION

$S = \{RR, RB, RG, BR, BB, BG, GR, GB\}$. ■

Example 14.13

There are five red balls and one blue ball in an urn. A ball is drawn and after examining its colour it is replaced in the urn. Then again a ball is drawn from the urn at random. Describe the sample space.

SOLUTION

$S = \{RR, RB, BR, BB\}$. ■

Example 14.14

There are three red balls, two blue balls and a green ball in an urn. Two balls are simultaneously drawn from the urn. Describe the sample space.

SOLUTION

In this experiment, there is nothing like a first ball or a second ball. We are interested only in the number of the balls of a particular colour. Hence we write the sample space as $\{2R, (1R, 1B)(1R, 1G), 2B, (1B, 1G)\}$. ■

Example 14.15

There are two machines A and B in a factory doing the same job. The process starts with A. After a fixed time interval A is stopped and B is used. Again after a fixed time interval B is stopped and A is started. The process is stopped when either A or B fails. Write down the sample space.

SOLUTION

Let E_A be the outcome that A functions and $\overline{E}_A$ be the outcome that A fails. We similarly define the outcomes E_B and $\overline{E}_B$.

The sample space is

$$\{\overline{E}_A, E_A E_B \overline{E}_A, \ldots$$
$$E_A E_B E_A \overline{E}_B, E_A E_B E_A E_B E_A \overline{E}_B \ldots\}$$

The first row corresponds to outcomes where the process is stopped by the failure of A, whereas the second row corresponds to outcomes where the process is stopped by the failure of B. ■

PRACTICE 14.4

Let us consider a bag containing two red balls, one blue ball and one green ball. Suppose we draw three balls randomly from the bag. Write the sample space.

Example 14.16

Muthu has created his password by inserting two distinct digits between the letters of his name. The two digits are not consecutive. In the experiment of guessing his password, find the size of the sample space.

SOLUTION

It is clear that the possible password can be formed by placing two digits in any of the four blank places shown below

M		U		T		H		U

First we must choose two empty places that can be done in ${}^4C_2$ ways. Then from among the ten digits 0, 1, ..., 9, two has to be arranged in these two places, which can be done in ${}^{10}P_2$ ways. Thus the required number of elements in the sample is ${}^4C_2 \times {}^{10}P_2$. ■

PRACTICE 14.5

Your friend Ram's password consists of five characters, which include the three letters r, a, m and two distinct digits. In the experiment of guessing his password, find the size of the sample space.

Example 14.17

In a small computational lab, there are two computers and two printers. Let us denote them by C_1, C_2, P_1, P_2. Both computers are connected to both the printers. Assuming that a printer is busy only when a computer is busy and a computer can print in only one printer at a given time, we have the following possibilities:

(a) C_1 is busy. C_2, P_1, P_2 are idle.
(b) C_1 is busy and P_1 is busy by being connected to C_1. C_2 and P_2 are idle.
(c) C_1 is busy and P_2 is busy by being connected to C_1. C_2 and P_1 are idle.
(d) C_2 is busy. C_1, P_1, p_2 are idle.
(e) C_2 is busy and P_1 is busy by being connected to C_2. C_1 and P_2 are idle.
(f) C_2 is busy and P_2 is busy by being connected to C_2. C_2 and P_2 are idle.
(g) Both C_1 and C_2 are busy, and P_1 and P_2 are idle.
(h) Both $C_1$and C_2 are busy, and P_1 is busy by being connected to C_1.
(i) Both $C_1$and C_2 are busy, and P_2 is busy by being connected to C_1.
(j) Both $C_1$and C_2 are busy, and P_1 is busy by being connected to C_2.
(k) Both $C_1$and C_2 are busy, and P_2 is busy by being connected to C_2.
(l) Both $C_1$and C_2 are busy, P_1 is busy by being connected to C_1, and P_2 is busy by being connected to C_2.
(m) Both $C_1$and C_2 are busy, P_1 is busy by being connected to C_2, and P_2 is busy by being connected to C_1.

All these situations constitute the sample space of the experiment of trying to get a specific PC and a specific printer to work within the lab.

If the computers and printers are all equally good, then a user need not worry about precisely which computer or a printer is busy. He will only be concerned about the availability of a computer and a printer. In this circumstance, we can list the possibilities as:

(a) neither the computers nor the printers are busy,
(b) only one computer is busy and no printer is busy,
(c) both the computers and only one printer is busy,
(d) both the computers and both the printers are busy.

The purpose of this example is to illustrate the fact that depending on our requirement, we may design different sample spaces for the same circumstance. ■

14.2 EVENTS

In this section, we shall formally define the concept of an event associated with a random experiment. As usual, before giving the mathematical definition, let us look at some examples.

Consider Example 14.3, the hotel reservation and cancelation experiment of Section 14.1. Depending on the number of reservations and cancelations made, there are many possible situations. One possible situation is that every one who has a reservation gets a room. This happens under the following conditions:

(a) when a maximum of five reservations are made,
(b) when there are six reservations and one or more cancelations,
(c) when there are seven reservations and two or more cancelations.

One other possibility is that someone who has made a reservation does not get a room. This situation occurs when the reservation/cancelation combinations are given by $(6, 0)$, $(7, 0)$, $(7, 1)$. These three elements of the sample space represent the event that at least one customer is denied a room.

Now, consider the event of at least one room being unoccupied. This happens if 4 or less rooms are occupied. In the situations $(7, 3)$ and $(6, 2)$, etc., four rooms are occupied. In the situation $(7, 4)$, $(6, 3)$, etc. three rooms are occupied and so on. Hence, the event that at least one room is unoccupied is represented by the following elements $(7, 3), (6, 2), (5, 1), (4, 0), (7, 4), (6, 3), (5, 2), (3, 0), (7, 5), (6, 4), (5, 3), (4, 2), (3, 1), (2, 0), (7, 6), (6, 5), (5, 4), (4, 3), (3, 2), (2, 1), (1, 0)$.

In the experiment of throwing a die, the event that an even number comes up is represented by the elements 2, 4 and 6.

From these examples, it becomes clear that in a random experiment *an event* is represented by *some elements* of the sample space. Mathematically, this event is identified by the subset of the sample space containing all the elements representing the event. Further, any arbitrary subset of the sample space is considered as an event.

Definition 14.2: Any subset E of the sample space is called an event. The empty set ϕ is called the impossible event, and the sample space S is called the sure event.

Example 14.18

In the experiment of throwing a die, the event that an even number comes up is simply the subset $\{2, 4, 6\}$. ■

Example 14.19

In the hotel booking and cancelation experiment, the event that at least one customer is denied a room is the subset $\{(7, 0), (7, 1), (6, 0)\}$. ■

PRACTICE 14.6

In the hotel booking and cancelation experiment, give a verbal description of the event $\{(7, 2), (6, 1), (5, 0)\}$.

Example 14.20

Consider the experiment of throwing a pair of dice. Identify the event that the sum of the outcomes is 4.

SOLUTION

The desired event is $\{(1, 3), (2, 2), (3, 1)\}$. ■

PRACTICE 14.7

In the experiment of Example 14.20, identify the event that the sum of the outcomes is a prime.

Example 14.21

Consider an urn containing four red balls, two blue balls and five green balls. Three balls at random are drawn from the urn. Identify the event that the number of green balls exceeds the number of red balls.

SOLUTION

The sample space for this experiment is

$$\{3R, 2R1B, 2R1G, 2B1R, 2B1G, 1B2G, 3G, 2G1R, 2G1B, 1G1B1R\}$$

The desired event is the subset $\{2B1G, 1B2G, 3G, 2G1R\}$. ■

Let E_1 and E_2 be events. As these are subsets of sample space S, the sets $E_1 \cap E_2$ and $E_1 \cup E_2$ are subsets of S and hence they are also events. $E_1 \cup E_2$ is the event that *at least one of the events* E_1, E_2 occurs and $E_1 \cap E_2$ is the event that *both* the events occur. In the theory of probability, the event $E_1 \cap E_2$ is also denoted by E_1E_2.

Example 14.22

Imagine that you are walking into a residential institution on a holiday. Let F be the event that the first person you see on the campus is a female, let T be the event that the first person you see on the campus is a teaching staff and let S be the event that the first person you see on the campus is a security personnel. It is clear that the events F and T can occur together, F and S can occur together but T and S cannot occur together. Here, we are under the assumption that no teaching member is a part of the security team. The pair of events T and S is an example of a pair of mutually exclusive events. A pair of events E_1 and E_2 are mutually exclusive if the occurrence of one of them prevents the occurrence of the other. The precise set theoretic definition is given below.

Definition 14.3: The events E_1 and E_2 are mutually exclusive if $E_1 \cap E_2 = \phi$, the empty set.

Since we are using the language of sets, Venn diagrams can be used in the context of events. As all the events are subsets of the sample space S, S is regarded as the *universal set*. We illustrate the sample space and an event in Figure 14.1.

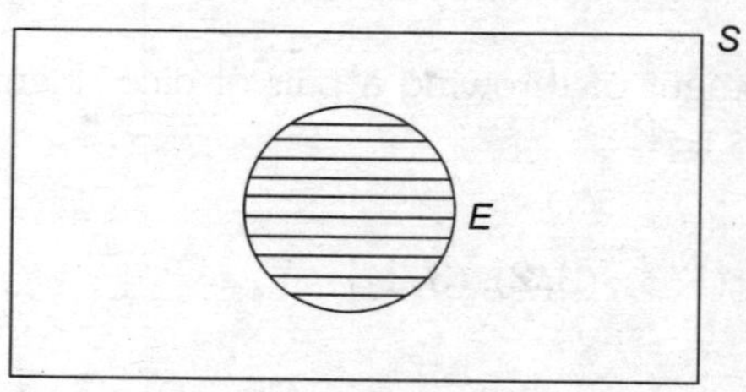

Figure 14.1

Next, we illustrate mutually exclusive events in Figure 14.2.

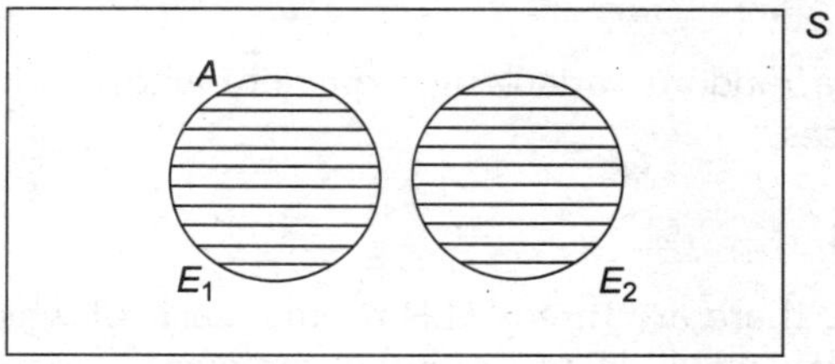

Figure 14.2

Now the subset $S - E$ or E^c is the compliment of the set E. It is nothing but the event that E *does not* occur. It is also denoted by $\overline{E}$.

PRACTICE 14.8

Illustrate E^c using a Venn diagram.

Example 14.23

Consider the experiment of throwing a die and the following events:

(i) A is the event of an odd number coming up.
(ii) B is the event of an even number coming up.
(iii) C is the event of a prime number coming up.
(iv) D is the event of either 2 or 4 coming up.

We know that the occurrence of a prime number does not prevent the occurrence of an even number, but the occurrence of an odd number prevents the occurrence of an even number. These and further observations are described as follows:

(a) A and B are mutually exclusive.
(b) $A^c = B$.
(c) A and D are mutually exclusive but $A^c \neq D$.
(d) B and C are not mutually exclusive.

The above observations can be easily made if we identify these events as subsets of the sample space $\{1, 2, 3, 4, 5, 6\}$ as follows:

(i) $A = \{1, 3, 5\}$,
(ii) $B = \{2, 4, 6\}$,
(iii) $C = \{2, 3, 5\}$,
(iv) $D = \{2, 4\}$,

and so,

(a) $AB = \phi$.
(b) $A^c = \{2, 4, 6\} = B$.
(c) $AD = \phi$ but $A^c = \{2, 4, 6\} \neq D$.
(d) $BC = \{2\}$. ■

14.3 RANDOM VARIABLE

Roughly speaking, a random variable is a quantity whose value depends on some random circumstances.

Example 14.24

In a guest house, there are three V.I.P rooms, each of which is fixed with a 40 Watts tubelight. We would like to estimate the average power consumption due to these fluorescent lights in a day. The following assumptions are reasonable

(a) Between 6 a.m to 6 p.m, no power is consumed.
(b) Between 6 p.m to 10 p.m, all the three tubelights will be switched on, and hence during this time, the power consumption is $3 \times 40 = 120$ Watts.

Now, we consider the time between 10 p.m and 6 a.m. Depending on who is staying in the rooms, some lights would be switched on and some would be switched off. We do not know which lights would be on and which would be off. This is an example of a random phenomenon. The power consumed during this period is a quantity whose value cannot be predicted in advance. This is an example of a random variable.

Let us now list all possible situations and the corresponding quantity of power consumed.

Room 1	Room 2	Room 3	Number of bulbs switched on	Power consumption
On	On	On	3	120 Watts
On	On	Off	2	80 Watts
On	Off	On	2	80 Watts
On	Off	Off	1	40 Watts
Off	On	On	2	80 Watts
Off	On	Off	1	40 Watts
Off	Off	On	1	40 Watts
Off	Off	Off	0	0 Watts

All possible situations of this random phenomenon can now be listed as

$$\{(\text{On, On, On}), (\text{On, On, Off}), (\text{On, Off, On}),$$
$$(\text{On, Off, Off}), (\text{Off, On, On}), (\text{Off, On, Off}),$$
$$(\text{Off, Off, On}), (\text{Off, Off, Off})\}$$

This is just the sample space. Now, the number of bulbs switched on is a function defined on this set. Similarly, the power consumed is also a function defined on this set. These are examples of a random variable. Roughly, a random variable is a quantity that depends on the outcome of a random experiment. The precise definition is given below. ■

Definition 14.4: A random variable is a function $X : S \longrightarrow \mathbb{R}$.

A single element of S, that is, an outcome of the random experiment is denoted by the symbol ω.

Example 14.25

In Example 14.24, let X be the number of bulbs switched on between 6 p.m and 8 p.m. Then X(On, Off, Off) = 1, X(On, Off, On) = 2, etc. Let Y be the power consumption. Then, for ω = (Off, Off, On), $Y(\omega) = 40$. For ω = (On, On, On), $Y(\omega) = 120$. In general,

$$Y(\omega) = 40 \times X(\omega)$$

■

Example 14.26

Let us consider a bag containing two red balls, one blue ball and one green ball. Suppose we draw three balls randomly from the bag. Now, let X_R, X_B and X_G denote the number of red, blue and green balls, respectively, in an outcome. These are examples of random variables. We can explicitly tabulate the values of the functions as follows:

	X_R	X_B	X_G
(1R, 1G, 1B)	1	1	1
(2R, 1B)	2	1	0
(2R, 1G)	2	0	1

■

Example 14.27

Let us consider a bag containing two red balls, one blue ball and one green ball. Suppose that we draw three balls *one by one* without putting the drawn ball back into the bag. Write down the sample space. Let X_R be the number of red balls drawn. Describe X_R explicitly.

SOLUTION

The sample space for this experiment is as follows:

$$\{RBG, RGB, GRB, GBR, BRG, BGR, RRB, RRG, GRR, BRR, RGR, RBR\}$$

$$\begin{aligned}
X_R(RBG) &= 1\\
X_R(RGB) &= 1\\
X_R(GRB) &= 1\\
X_R(GBR) &= 1\\
X_R(BRG) &= 1\\
X_R(BGR) &= 1
\end{aligned}$$

$$\begin{aligned} X_R(RRB) &= 2 \\ X_R(RRG) &= 2 \\ X_R(GRR) &= 2 \\ X_R(BRR) &= 2 \\ X_R(RGR) &= 2 \\ X_R(RBR) &= 2 \end{aligned}$$

■

PRACTICE 14.9

Let us consider a bag containing two identical red balls, one blue ball and one green ball. Suppose that we draw three balls *one by one* and that *we replace* the drawn ball back into the bag before drawing the next one. Write down the sample space for this experiment and give an explicit description of the random variable X_G, the number of green balls.
Hint: There are 27 possible outcomes.

Example 14.28

A pizza restaurant in a small town has employed Amar, Akbar and Antony as pizza makers and Mani and Pandian as delivery boys. Amar is paid Rs 20.00 for making each pizza, Akbar is paid Rs 23.00 and Antony is paid Rs 19.00. For the delivery of each pizza Mani is paid Rs 5.00 and Pandian is paid Rs 7.00. Every time an order is placed, the pizza maker and the delivery boy are chosen randomly. The cost incurred by the firm for each pizza, in addition to the above charges, is Rs 30.00. Each pizza is sold for Rs 100.00. Clearly, the profit per each pizza is a random variable. Describe the sample space and the random variable.

SOLUTION

Clearly, the sample space is

$$\{(\text{Amar, Mani}), (\text{Akbar, Mani}), (\text{Antony, Mani})(\text{Amar, Pandian}), (\text{Akbar, Pandian}), (\text{Antony, Pandian})\}$$

Let x be the amount paid to the pizza maker and y be the amount paid to the delivery boy. The profit for a pizza is $100 - (x + y + 30)$ and hence the random variable 'Profit' is described as follows:

$$\begin{aligned} X((Antony, Pandian)) &= 100 - (19 + 7 + 30) = 44 \\ X((Amar, Mani)) &= 100 - (20 + 5 + 30) = 45 \end{aligned}$$

and so on.

■

PRACTICE 14.10

In Example 14.28, calculate the value of the random variable for the remaining elements of the sample space.

Example 14.29

Consider the sample space of Example 14.9. If we transmit 011 and receive 011 no error has taken place. If 111 has been received, then only the first bit has changed and hence the number of errors is 1. It is clear that the number of errors X is a random variable. The values of X is tabulated as follows:

$$X(011) = 0$$
$$X(111) = 1$$
$$X(001) = 1$$
$$X(010) = 1$$
$$X(101) = 2$$
$$X(000) = 2$$
$$X(110) = 2$$
$$X(100) = 3$$

■

PRACTICE 14.11

In Example 14.17, the number of busy printers is a random variable. Tabulate the values of X.

Example 14.30

George is a manager in a company. He has a mobile phone that would cost him 20 paise per minute when he makes a call to his wife. His official landline would cost him Re 1.00 per minute. During every working day, he makes one call to his wife. If he is in his office, he uses the landline or else he uses his mobile. His presence in the office, the time at which he calls his wife and the length of the call with his wife are all random. His call will last a maximum of 3 minutes. Assume that even a fraction of a minute would be charged its full amount. The amount he spends for this call is a random variable. Describe the sample space and calculate the values of the random variable.

SOLUTION

Let us denote the event of George using his mobile by M and the event of him using the landline by L. The numbers 1, 2 and 3 represent the length of the call. For example, if he talks for 1 minute and 5 seconds, he would be charged for 2 minutes and so this event is represented by 2. Now the sample space here is given by

$$\{(M, 1), (M, 2), (M, 3), (L, 1), (L, 2), (L, 3)\}$$

If X is the amount he spends on his call, it is given by

$$\begin{aligned} X(M, 1) &= 20 \text{ paise} \\ X(M, 2) &= 40 \text{ paise} \\ X(M, 3) &= 60 \text{ paise} \\ X(L, 1) &= 1 \text{ Re} \\ X(L, 2) &= 2 \text{ Rs} \\ X(L, 3) &= 3 \text{ Rs} \end{aligned}$$

■

PRACTICE 14.12

In a gambling game, three coins are tossed by Arjun simultaneously. For every head, he gets 2 rupees and for every tail he looses 3 rupees to his opponent. Write down the sample space. Using negative numbers to denote the loss, let X be the amount due to Arjun and let Y be the amount due to his opponent. Calculate the values of X and Y. Write down the event that Arjun's opponent gets a profit.

14.4 BASIC PROBABILITY

The most important aspect of a random experiment is the probability of various events. The probability of an event is a measure of its likelihood.

A biased coin is one whose weight is distributed more on one side than the other. When such a coin is tossed, the heavier side is more likely to face the ground and the other side would face upwards. If the coin is not biased, when it is tossed many number of times, we believe that the head would come up nearly as many times as the tail would. Such a coin is called an unbiased coin. Imagine that we toss an unbiased coin 1,000 times. The head would come up nearly 500 times and the tail would come up nearly 500 times.

Now, consider the experiment of 'switching a bulb on'. There are again two possible outcomes. We have indicated these outcomes by the symbols L and D. To make the discussion simple, let us assume that the bulb does not fuse. If the power supply is not bad, the outcome L is bound to occur more often than D. Perhaps, if we switch on the bulb 1,000 times, it may glow 850 times, and the remaining 150 times it may not glow.

We summarize the (albeit imaginary) observations of these examples as follows:

In the coin toss experiment, the probability of the outcome **H** is $500/1{,}000 = 1/2$ and the probability of the outcome **T** is $1/2$. In the bulb experiment, the probability of **L** is $850/1{,}000$ and the probability of the outcome of **D** is $150/1{,}000$.

In the following examples, we calculate the probability of an event using common sense.

Example 14.31

A coin is tossed twice. The sample space is $\{TT, TH, HT, HH\}$. The event that the second outcome is head is the subset $\{TH, HH\}$. The probability of this event is $\frac{1}{2}$. ■

Example 14.32

A die is thrown. The event that a prime number turns up is $\{2, 3, 5\}$. The probability of this event is $\frac{3}{6}$. ■

Example 14.33

A pair of dice are thrown. The event that the sum of the outcomes is greater than or equal to 11 is $\{(5, 6), (6, 5), (6, 6)\}$. The probability of this event is $\frac{3}{36}$. ■

Example 14.34

Imagine an unknown random experiment. Let us say that there are n possible outcomes. Let us call these outcomes as $\omega_1, \omega_2, \ldots, \omega_n$. Hence, the sample space is $S = \{\omega_1, \omega_2, \ldots, \omega_n\}$. Clearly, $E = \{\omega_1, \omega_2, \ldots, \omega_m\}$ is an example of an event. What is the probability of the event E? A beginner is tempted to assign the probability $\frac{m}{n}$ for this event. This is wrong. The beginner is misled because he is confronted with standard examples of random experiments in which all possible outcomes are *equally likely*. In the previous examples, all the outcomes were equally likely and the probabilities as we calculated were correct.

But consider the bulb experiment whose sample space is $\{L, D\}$. Here the outcomes are not equally likely. Consider $E = \{D\}$. With reference to Example 14.34, $n = 2$ and $m = 1$. But we noted that $P(\{L\}) > P(\{D\})$ and hence $P(E) \neq \frac{1}{2}$. But we should have $P(L) + P(D) = 1$.

For every event $E \subseteq S$, a number $P(E)$ is assigned and is called the probability of E. Naturally, we would like to know the probabilities of the events. Further, we would like to express the probabilities of $E_1 \cup E_2$ and E_1E_2 in terms of the probabilities $P(E_1)$ and $P(E_2)$. This is not always an easy thing to do. But the following properties must be satisfied. ■

Axioms for Probability:

(a) $0 \leq P(E) \leq 1$.

(b) If E_1, E_2, $E_3, \ldots,$ are events such that $E_i \cap E_j = \phi$, $i \neq j$, then

$$P\left(\bigcup_i^{\infty} E_i\right) = \sum_{i=1}^{\infty} P(E_i)$$

(c) $P(\phi) = 0$.
(d) $P(S) = 1$.

In the following examples, we use the following procedure: If there is a total of n *equally likely* outcomes and m of them are favorable outcomes for an event, then the probability of that event is $\frac{m}{n}$.

Example 14.35

A bag contains five red balls, six blue balls and three black balls. A ball is drawn at random. Find the probability that it is

(i) Red.
(ii) Not blue.
(iii) Blue or black.

SOLUTION

(i) $\frac{5}{14}$.
(ii) $\frac{8}{14}$.
(iii) $\frac{6}{14} + \frac{3}{14}$. ■

Example 14.36

From a standard pack of 52 cards, a card is drawn at random. Find the probability that

(a) It is a 7.
(b) It is a club.
(c) It is a 7 or a club.

SOLUTION

(a) There are four 7's in the deck. Hence the probability of picking a 7 is $\frac{4}{52}$.
(b) There are thirteen clubs in the deck. Hence the probability of picking a club is $\frac{13}{52}$.
(c) There are four 7's and thirteen clubs. But there is a 7 club. Hence the favorable case for us is $4 + 13 - 1 = 16$. Hence the required probability is $\frac{16}{52}$. ■

Motivated by the previous example, we have the following proposition.

Proposition 14.1

Let E_1 and E_2 be events. Then

$$P(E_1 \cup E_2) = P(E_1) + P(E_2) - P(E_1E_2)$$

Proof

Using ideas from set theory we note the following:

(i) $E_1 = E_1E_2^c \cup E_1E_2$.

(ii) $E_2 = E_1^c E_2 \cup E_1 E_2$.

(iii) $E_1 \cup E_2 = E_1 E_2^c \cup E_1 E_2 \cup E_1^c E_2$.

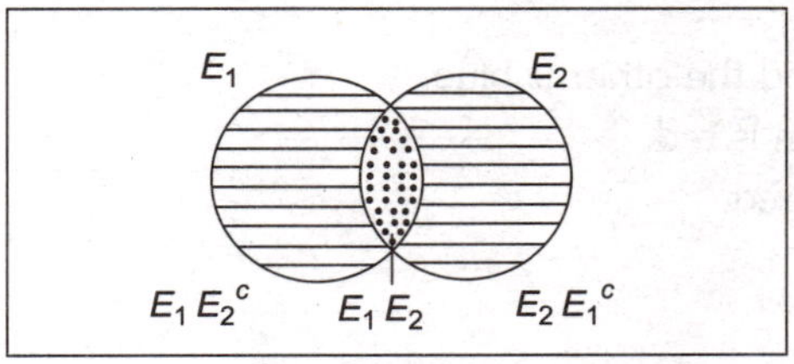

Figure 14.3

Since, $E_1E_2^c$, $E_1^cE_2$ and E_1E_2 are mutually exclusive, we have $P(E_1) = P(E_1E_2^c) + P(E_1E_2)$.

Similarly, $P(E_2) = P(E_1^cE_2) + P(E_1E_2)$ and

$$\begin{aligned} P(E_1 \cup E_2) &= P(E_1E_2^c) + P(E_1^cE_2) + P(E_1E_2) \\ &= P(E_1) - P(E_1E_2) + P(E_2) - P(E_1E_2) + P(E_1E_2) \\ &= P(E_1) + P(E_2) - P(E_1E_2) \end{aligned}$$

Example 14.37

Show that $P(E^c) = 1 - P(E)$. ■

Proof

$S = E \cup E^c$ and further $E \cap E^c = \phi$. So, $1 = P(S) = P(E) + P(E^c)$ and the result follows. ■

Example 14.38

Two cards are drawn from a standard deck of 52 cards.

(i) What is the size of the sample space?

(ii) What is the probability that both the cards are queens?

(iii) What is the probability that both are clubs?

(iv) What is the probability that neither is a club?

(v) What is the probability that at least one is a club?

SOLUTION

(i) ${}^{52}C_2$.

(ii) There are four queens. So, the required answer is $\frac{{}^4C_2}{{}^{52}C_2}$.

(iii) There are 13 clubs. So, the required answer is $\frac{{}^{13}C_2}{{}^{52}C_2}$.

(iv) There are $52 - 13 = 39$ cards, which are not clubs. Hence the required probability is $\frac{{}^{39}C_2}{{}^{52}C_2}$.

(v) Note that the event (iv) is the compliment of the present event. Hence, the required probability is $1 - \frac{{}^{39}C_2}{{}^{52}C_2}$. ■

Example 14.39

An urn contains three red balls, five black balls and seven blue balls. Two balls are selected at random. What is the probability that

(a) Both are red.
(b) One is black and the other is blue.
(c) Neither of them is red.
(d) At least one is red.

SOLUTION

(a) $\frac{{}^3C_2}{{}^{15}C_2}$.

(b) $\frac{{}^5C_1 \times {}^7C_1}{{}^{15}C_2}$.

(c) $\frac{{}^{12}C_2}{{}^{15}C_2}$.

(d) The event *at least one is red* is the compliment of *neither is red* and hence the answer is $1 - \frac{{}^{12}C_2}{{}^{15}C_2}$. ■

Example 14.40

An urn contains three red balls, five black balls and seven blue balls. Nine balls are selected at random. What is the probability for the following events:

(a) Two red balls, three black balls, four blue balls.
(b) None of them is red.
(c) One of them is red.
(d) At least one of them is red.

SOLUTION

(a) The number of ways of choosing two red balls, three black balls, four red balls out of the given collection is ${}^3C_2 \times {}^5C_3 \times {}^7C_4$. The total number of ways of choosing nine balls out of 15 balls is ${}^{15}C_9$ and hence the required probability is $\frac{{}^3C_2 \times {}^5C_3 \times {}^7C_4}{{}^{15}C_9}$.

(b) $\frac{{}^{12}C_9}{{}^{15}C_9}$ is the answer.

Another method for (b):

Black	Blue	Probability
5	4	$\frac{{}^5C_5 \times {}^7C_4}{{}^{15}C_9}$
4	5	$\frac{{}^5C_4 \times {}^7C_5}{{}^{15}C_9}$
3	6	$\frac{{}^5C_3 \times {}^7C_6}{{}^{15}C_9}$
2	7	$\frac{{}^5C_2 \times {}^7C_7}{{}^{15}C_9}$

The required probability is the sum of all entries in the last column.

(c) One of them must be red. Thus, the problem reduces to that of picking up one red ball out of three, and eight balls out of the black and blue balls, and hence the required answer is $\frac{{}^{3}C_{1} \times {}^{12}C_{8}}{{}^{15}C_{9}}$.

(d) Find the probabilities that one of the balls is red, two of the balls are red and three of the balls are red. Add the answers. Or, the event of *at least one red* is the compliment of *none is red*. The required answer is $1 - \frac{{}^{12}C_{9}}{{}^{15}C_{9}}$. ■

Example 14.41

Three dice are rolled. What is the probability that the same number appears on exactly two of the three dice?

SOLUTION

Name the dice as A, B and C. The number of outcomes in which A and B have the same number, and C has a different number is 6×5. Similarly, the number of outcomes in which B and C have the number and not A is 6×5, and the number of outcomes in which A and C have the number and not B is 6×5. Thus, the total number of favorable outcomes is 3×30. The total number of outcomes is $6^3 = 108$. Hence the required probability is $\frac{90}{108}$. ■

PRACTICE 14.13

In a hand of 13 cards drawn randomly from a pack of 52 cards, find the probability of no court cards. Ace, King, Queen and Jack are called court cards.

Example 14.42

Four cards are drawn from a pack of 52 cards without replacement. What is the probability that they are all of different suits?

SOLUTION

We have to choose four out of 52 cards, which can be done in ${}^{52}C_4$ ways. We must have one card of each suit. There are 13 ways of choosing one card from each suit. Hence the required answer is $\frac{13^4}{{}^{52}C_4}$. ■

PRACTICE 14.14

Three cards are drawn from a deck of 52 cards. Find the probability that

(a) Two are jacks and one is a king.
(b) All of them are of different suits.
(c) All of them are of the same suit.

Example 14.43

From a collection of 38 battery cells containing 16 defective ones, 4 cells are selected randomly for quality check. Find the probability that half of them would be defective.

SOLUTION

There must be two good cells and two defective ones. The two good ones can be selected in ${}^{22}C_2$ ways and the two bad defective ones can be selected in ${}^{16}C_2$ ways. Hence the required probability is

$$\frac{{}^{22}C_2 \times {}^{16}C_2}{{}^{38}C_4}.$$

Example 14.44

A bag contains tokens numbered from 1 to 50. What is the probability that a randomly drawn token is a multiple of either 6 or 4?

SOLUTION

Let A be the event that a randomly drawn token is a multiple of 6 and let B be the event that a randomly drawn token is a multiple of 4. There are 8 multiples of 6 among the numbers from 1 to 50. There are 12 multiples of 4 among the numbers from 1 to 50. Hence $P(A) = \frac{8}{50}$ and $P(B) = \frac{12}{50}$. Now, we need $P(A \cup B) = P(A) + P(B) - P(A \cap B)$. Now, the common multiples of 6 and 4 are nothing but multiples of their $l.c.m$, which is 12. There are 4 multiples of 12 among the numbers from 1 to 50. So, $P(A \cap B) = \frac{4}{50}$. Hence the required probability is

$$\frac{8}{50} + \frac{12}{50} - \frac{4}{50}$$

PRACTICE 14.15

A bag contains tokens numbered from 1 to 20. What is the probability that a randomly drawn token is a multiple of either 5 or 7?

Example 14.45

In a class of 25 persons, what is the probability that two of them would have the same birthday?

SOLUTION

All possible number of ways in which the 25 persons would have their birthdays is 365^{25}. The number of ways in which two of them would have the same birthday is ${}^{365}P_{25}$. Hence, the probability that no two of them would have the same birthday is $\frac{{}^{365}P_{25}}{365^{25}}$. Thus the probability that two of them would have the same birthday is

$$1 - \frac{{}^{365}P_{25}}{365^{25}}$$

Example 14.46

Suppose you take a random sample of 10 tickets from a box containing 20 red tickets and 30 blue tickets. What is the chance of getting exactly four red tickets?

SOLUTION

The number of ways of choosing 10 tickets out of 50 tickets is ${}^{50}C_{10}$. Now, we have to pick four red tickets and six blue tickets. As there are 20 red tickets and 30 blue tickets, the number ways of picking four red tickets and six blue tickets is ${}^{20}C_4 \times {}^{30}C_6$. Thus, the required probability is

$$\frac{{}^{20}C_4 \times {}^{30}C_6}{{}^{50}C_{10}}$$

■

14.5 CONDITIONAL PROBABILITY AND BAYES THEOREM

Example 14.47

Let us think of two old women Mrs Ranganayaki Guptha in Chennai and Mrs Neelima Ramanathan in Mumbai. Both of them would like to watch either BBC or Aaj Tak. Being old women, they get sick often. Let us now assign symbols for various events as follows:
W_1 is the event that Mrs Ranganayaki Guptha watches Aaj Tak.
W_2 is the event that Mrs Neelima Ramanathan watches Aaj Tak.
Let F_1 be the event that Mrs Ranganayaki Guptha gets sick.
Let F_2 be the event that Mrs Neelima Ramanathan gets sick.
Consider the following questions:

(a) What is the probability that Mrs Ranganayki Guptha watches Aaj Tak *and* Mrs Neelima Ramanathan gets sick?

(b) What is the probability that Mrs Ranganayki Guptha watches Aaj Tak *and* she gets sick?

Intuitively, we feel that the answers to the above two questions are as follows:

(a) $P(W_1F_2) = P(W_1)P(F_2)$.
(b) $P(W_1F_1) = P(W_1)P(F_1)$.

Now, let R_1 be the event that it would rain in Chennai and let R_2 be the event that it would rain in Mumbai. Consider the following questions:

(c) What is the probability that it would rain in Chennai and Mrs Neelima Ramanathan would get sick?

(d) What is the probability that it would rain in Chennai and Mrs Ranganayaki Guptha would get sick?

The answer to (c) is $P(R_1)P(F_2)$, because we feel that the rain in Chennai is not going to affect the event of someone in Mumbai getting sick.

But is $P(R_1)P(F_1)$ the answer to (d)? We intuitively feel that it would not be so. For, the rain in Chennai may increase the probability that Mrs Ranganayaki

Guptha would get sick. The probability that Mrs Ranganayaki Guptha would get sick given that it is raining in Chennai is denoted by $P(F_1|R_1)$.

Under this notation, we have $P(F_1R_1) = P(F_1)P(F_1|R_1)$. ■

$P(F_1|R_1)$ is an example of conditional probability whose definition is given below.

Definition 14.5: The probability of the event B occurring given that A has occurred is called the conditional probability B, given A; and is denoted by $P(B|A)$.

Proposition 14.2

Let A and B be any two events. Then, $P(AB) = P(A)P(B|A)$.

Proposition 14.2 is called the multiplicative law of probability.

Under what conditions may we have $P(AB) = P(A)P(B)$? If the occurrence of A is not going to affect the occurrence of B, we say that they are independent events. In this case, $P(B|A)$ is going to be equal to $P(B)$.

With reference to Example 14.47, R_1 the event of raining in Chennai and F_2 the event of Mrs Neelima Ramanathan of Mumbai becoming sick are independent events. The formal definition is given below.

Definition 14.6: We say that two events A and B are independent if $P(AB) = P(A)P(B)$

Example 14.48

After transmitting and receiving a bit via a communication channel 1,000 times, it has been found that 965 times the correct bit was received. Based on this observation the probability of correct reception is estimated as 0.965. Let T_1 be the event that 1 is transmitted, let T_0 be the event that 0 is transmitted, let R_1 be the event that 1 is received and let R_0 be the event that 0 is received. Find the conditional probabilities $P(R_1|T_1)$, $P(R_0|T_0)$, $P(R_1|T_0)$, $P(R_0|T_1)$.

SOLUTION

$P(R_1|T_1)$ is the conditional probability that 1 is received given that 1 is transmitted. As this is the case of a correct transmission, $P(R_1|T_1) = 0.965$. Similarly, we have

$$\begin{aligned} P(R_0|T_0) &= 0.965 \\ P(R_1|T_0) &= 1 - 0.965 = .035 \\ P(R_0|T_1) &= 0.035 \end{aligned}$$

■

Example 14.49

Let there be three black balls and five red balls in an urn. Consider the following two different experiments:

(a) We first draw a ball at random from the urn, put it back and then draw another ball at random from the urn.

(b) We first draw a ball at random from the urn, keep it out and then draw another ball at random from the urn.

SOLUTION

Now, we are interested in calculating the probability that the first ball is black and the second ball is red in these two different circumstances. The answers are:

(a) $\frac{3}{8} \times \frac{5}{8}$.

(b) $\frac{3}{8} \times \frac{5}{7}$.

Let A be the event that the first ball is black and let B be the event that the second ball is red.

In the first case, the events A and B are independent. The fact that the first ball is black is not going to affect the probability of the second ball being red as the first ball drawn is placed in the urn again.

In the second case, the events A and B are not independent. After a black ball has been drawn from the urn, the probability of the second ball being red is $\frac{3}{7}$. But if a red ball has been drawn during the first draw, then the probability of the second ball being red would be $\frac{4}{7}$. So, $P(AB) \neq P(A)P(B)$ but it is $P(A)P(B|A)$.■

Example 14.50

A survey has found that in a certain district, only 5% of the womenfolk are colour blind whereas 10% of the menfolk are colour blind. At a village located in this district, there live 1025 women and 1300 men. Find the probability that a randomly selected person from this village is a colour blind woman.

SOLUTION

The essential step in the solution of this problem is the appropriate interpretation of the statement

5% of the women are colour blind.

Let W be the event that a randomly selected person is a woman, let M be the event that a randomly selected person is a man and let C be the event that a randomly selected person is colour blind. Now, $P(W) = \frac{1025}{2325}$, $P(M) = \frac{1300}{2325}$. Now, 5% of the women are colour blind. That is, if a randomly selected person is a woman, then there is a chance of $\frac{5}{100}$ that she is colour blind. So, this means that the conditional probability of a person being colour blind, given that the person is a woman, is $\frac{5}{100}$ Thus, $P(C|W) = \frac{5}{100}$.

Now, we require to find the probability that the randomly selected person is a woman and colour blind. Hence we need to find $P(WC) = P(W)P(C|W)$. Therefore $P(WC) = \frac{1025}{2325} \times \frac{5}{100}$. ■

Example 14.51

Among a group of 17 persons, everyone is a fan of Ilayaraja or R. D. Burman. Ten of them are fans of Ilayaraja and 9 of them are fans of of R .D. Burman. A person is selected at random from this group. Let A be the event that a person is a fan of Ilayaraja and let B be the event that the person is a fan of R. D. Burman. Examine whether these events are independent.

SOLUTION

We need to verify whether $P(AB) = P(A)P(B)$.

$$
\begin{aligned}
P(A \cup B) &= P(A) + P(B) - P(AB) \\
1 &= \frac{10}{17} + \frac{9}{17} - P(AB) \\
P(AB) &= \frac{19}{17} - 1 = \frac{2}{17}
\end{aligned}
$$

Now, $P(A)P(B) = \frac{90}{289} \neq \frac{2}{17}$. Hence, A and B are not independent. ■

Example 14.52

Ram and Lakshman have spotted a deer in the forest and each aims at the deer simultaneously. The probability that Ram would hit the deer is 0.99 and the probability that Lakshman would hit it is 0.98. Find the probability that the deer would escape.

SOLUTION

Method I: Let A be the event that Ram would hit the deer target and let B be the event that Lakshman would hit it. $P(A) = 0.99$ and $P(B) = 0.98$. The probability that the deer would escape is the probability that both of them would miss their target. We can assume that the performances of each of them are independent of the other. Hence the required probability is $P(A^cB^c) = P(A^c)P(B^c) = (1 - 0.99)(1 - 0.98)$.

Method II: The probability that at least one of them would hit the target is

$$
\begin{aligned}
P(A \cup B) &= P(A) + P(B) - P(AB) = P(A) + P(B) - P(A)P(B) \\
&= 0.99 + 0.98 - (0.99)(0.98)
\end{aligned}
$$

Now, the probability that the deer would escape is $1 - P(A \cup B)$. ■

In the above example, we have used the fact that if A and B are independent then A^c and B^c are also independent. We shall prove this in the next proposition.

Proposition 14.3

If A and B are independent show that

(a) A^c and B^c are independent.

(b) A and B^c are also independent.

To show that A^c and B^c are independent, we have to show that $P(A^cB^c) = P(A^c)P(B^c)$. Now,

$$
\begin{aligned}
P(A^c)P(B^c) &= (1 - P(A))(1 - P(B)) \\
&= 1 - P(A) - P(B) + P(A)P(B) \\
&= 1 - [P(A) + P(B) - P(A)P(B)]
\end{aligned}
$$

Since A and B are independent, $P(A)P(B) = P(A \cap B)$. Thus,

$$\begin{aligned}P(A^c)P(B^c) &= 1-[P(A)+P(B)-P(A\cap B]\\ &= 1-P(A\cup B)\\ &= P[(A\cup B)^c]\\ &= P[A^cB^c]\end{aligned}$$

Now, $P(A^c)P(B) = (1-P(A))P(B) = P(B)-P(A)P(B) = P(B)-P(A\cap B)$. Recall that $B = (A\cap B)\cup(A^c\cap B)$, and that $A\cap B$ and $A^c\cap B$ are mutually exclusive. So, $P(B) = P(A\cap B)+P(A^c\cap B)$. Thus, $P(A^c\cap B) = P(B)-P(A\cap B)$ and the result follows.

PRACTICE 14.16

A light bulb company has factories in two cities. The factory in city A produces two-thirds of the company's light bulbs. The remainder are produced in city B and of these, 1% are defective. What proportion is not defective and made in city B?

Example 14.53

There are two machines A and B in a factory doing the same job. The process starts with A and after a certain fixed time interval A is stopped and B is used. Again after a certain fixed time interval B is stopped and A starts. The probability that A functions during the time of its operation is p_A and the probability that B functions during the time of its operation is p_B. The process is stopped when either A or B fails. Find the probability that the process is stopped.

SOLUTION

If the process is stopped on account of the failure of A, the possibilities are as follows:

(i) A fails.
(ii) A functions, B functions, A fails.
(iii) A functions, B functions, A functions, B functions, A fails.
......

and so on. The corresponding probability is

$$\begin{aligned}&(1-p_A)+p_Ap_B(1-p_A)+p_Ap_Bp_Ap_B(1-p_A)+\ldots\\ &\qquad\qquad = (1-p_A)\sum_{i=0}^{\infty}(p_Ap_B)^i\end{aligned}$$

If the process is stopped on account of the failure of B, the possibilities are as follows:

(i) A functions, B fails.
(ii) A functions, B functions, A functions, B fails.
(iii) A functions, B functions, A functions, B functions, A functions, B fails.
....

and so on. The corresponding probability is

$$p_A(1-p_B) + p_A p_B p_A(1-p_B) + p_A p_B p_A p_B p_A(1-p_B).....$$

$$= p_A(1-p_B)\sum_{i=0}^{\infty}(p_A p_B)^i$$

The required probability is the sum of the above probabilities. ■

Example 14.54

If A, B and C are mutually independent, show that $A \cup B$ and C are independent.

SOLUTION

We have $P(AB) = P(A)P(B)$, $P(BC) = P(B)P(C)$, $P(AC) = P(A)P(C)$ and $P(ABC) = P(A)P(B)P(C)$

$$\begin{aligned}
P(A \cup B|C) &= \frac{P((A \cup B) \cap C)}{P(C)} \\
&= \frac{P[(A \cap C) \cup (B \cap C)}{P(C)} \\
&= \frac{P(AC) + P(BC) - P(ABC)}{P(C)} \\
&= \frac{P(A)P(C) + P(B)P(C) - P(A)P(B)P(C)}{P(C)} \\
&= \frac{P(C)[P(A) + P(B) - P(AB)]}{P(C)} \\
&= P(A \cup B)
\end{aligned}$$

and the result follows. ■

Example 14.55

If A and B are mutually exclusive and $P(A \cup B) \neq 0$, show that

$$P(A|A \cup B) = \frac{P(A)}{P(A) + P(B)}$$

SOLUTION

Since A and B are mutually exclusive, $P(A \cap B) = 0$. Now,

$$\begin{aligned}
P(A|A \cup B) &= \frac{P(A \cap (A \cup B)}{P(A \cup B)} \\
&= \frac{P(A)}{P(A) + P(B) - P(A \cap B)} \quad \text{since } A \cap (A \cup B) = A
\end{aligned}$$

$$= \frac{P(A)}{P(A)+P(B)} \qquad \text{since } P(A \cap B) = 0$$

■

Example 14.56

Consider Example 14.24. Assume the probability that an individual would switch on the tubelight between 10 p.m and 6 p.m is 0.3. Find the probability of the configuration (Off, Off, On).

SOLUTION

Let A, B and C be the events that the individual staying, respectively, in the first, second and third room switches on the tubelight. The required event is A^cB^cC. If we assume that the guests do not interact with each other and that their actions are independent of each other, the events A, B and C are independent. Thus, by Proposition 14.3, the required probability is $P(A^c)P(B^c)P(C) = 0.7 \times \times 0.7 \times 0.3$. ■

14.5.1 Total Probability Theorem

Example 14.57

An unbiased die is thrown. If the outcome is less than or equal to 4, an unbiased coin is chosen and is tossed. If the outcome is strictly greater than 4, a biased coin with $P(H) = \frac{7}{12}$ is chosen and is tossed. Find the probability that the outcome of the second toss is a tail.

Let E_1 be the event that the outcome is less than or equal to 4 when the die is thrown. Let E_2 be the event that the outcome is strictly greater than 4. We have $P(E_1) = \frac{2}{3}$ and $P(E_2) = \frac{1}{3}$. Now, if E_1 happens we would get tail with a probability of $\frac{1}{2}$. Hence, $P(T|E_1) = \frac{1}{2}$. Similarly, $P(T|E_2) = \frac{5}{12}$. Now, we must calculate $P(T)$. Using $T = TE_1 \cup TE_2$ and the fact that TE_1 and TE_2 are mutually exclusive, we get

$$\begin{aligned} P(T) &= P(TE_1) + P(TE_2) \\ &= P(E_1)P(T|E_1) + P(E_2)P(T|E_2) \\ &= \frac{2}{3}\frac{1}{2} + \frac{1}{3}\frac{5}{12} \end{aligned}$$

■

Example 14.58

Let us consider Example 14.50 again. Now, how can we calculate $P(C)$, that is, the probability that a randomly selected person is colour blind? Proceeding as in Example 14.57, we can calculate it as

$$P(C) = P(C|M)P(M) + P(C|W)P(W)$$

In the next theorem, we generalize this result. ■

Total Probability Theorem: Let E_i, $i = 1.2...k$ be mutually exclusive such that $\bigcup_{i=1}^{k} E_i = S$, where S is the sample space. Let A be any event. Then,

$$P(A) = \sum_{i=1}^{k} P(A|E_i)P(E_i)$$

Proof
Consider

$$\begin{aligned}
\bigcup_{i=1}^{k} AE_i &= \bigcup_{i=1}^{k} (A \cap E_i) \\
&= A \cap \left(\bigcup_{i=1}^{k} E_i \right) \quad \text{Distributivity} \\
&= A \cap S \\
&= A
\end{aligned}$$

Now, $E_i \cap E_j = \phi$ for $i \neq j$. But $(A \cap E_i) \cap (A \cap E_j) \subseteq E_i \cap E_j$. Hence $(A \cap E_i) \cap (A \cap E_j) = \phi$. Thus the events $AE_i = A \cup E_i$ are mutually exclusive. Hence

$$\begin{aligned}
P(A) &= P\left(\bigcup_{i=1}^{k} AE_i \right) \\
&= \sum_{i=1}^{k} P(AE_i) \\
&= \sum_{i=1}^{k} P(A|E_i)P(E_i)
\end{aligned}$$ ■

Remark 14.1

If $\bigcup_{i=1}^{k} E_i = S$, we say that the events $E_1, E_2, \ldots, E_k$ are exhaustive.

Example 14.59

Let there be three red balls, five blue balls and four green balls in an urn. A ball is drawn at random from the urn and it is kept outside. Then a second ball is drawn. Find the probability that the second ball is blue.

SOLUTION

Let R, B and G be the events that the first ball drawn is, respectively, red, blue and green. Let A be the event that the second ball drawn is blue.

By the total probability theorem,

$$\begin{aligned}
P(A) &= P(R)P(A|R) + P(B)P(A|B) + P(G)P(A|G) \\
&= \frac{3}{12}\frac{5}{11} + \frac{5}{12}\frac{4}{11} + \frac{4}{12}\frac{5}{11}
\end{aligned}$$ ■

14.5.2 Bayes Theorem

Example 14.60

This example is related to guessing the correct bit that is transmitted after receiving a probably wrong bit. Consider a communication channel transmitting bits. Let the probability of correct reception be p. Since errors occur during transmission, when a 1 is received, we cannot be sure that 1 was transmitted. Here we would like to know the probability that 1 was transmitted, when a 1 is received.

Let us write down the relevant events. Let T_1 be the event that 1 is transmitted, let T_0 be the event that 0 is transmitted, let R_1 be the event that 1 is received and let R_0 be the event that 0 is received. The probability of correct reception is p and the probability or wrong reception is $1-p$. So, as in Example 14.48, we have

$$\begin{aligned} P(R_1|T_1) = P(R_0|T_0) &= p \\ P(R_0|T_1) = P(R_1|T_0) &= 1-p \end{aligned}$$

We are interested in the probability that 1 was transmitted *given that 1 has been received.* In other words we are interested in the conditional probability $P(T_1|R_1)$.

We shall solve this problem after proving Bayes theorem.

Example 14.61

In a general election, surveys predict that the party KKP would win with a probability of 0.56 and that the party NKP would win with a probability of 0.44. The probability that KKP would pass the women's reservation bill is 0.70 provided that it comes to power. If NKP comes to power, it would pass the women's reservation bill with a probability of 0.8. Now let us imagine someone who is abroad and too busy to follow the election results. He does not know which party has won the election. Someone tells him that the women's reservation bill has been passed. What is the probability that it was NKP which had won?

The solution to these types of problems is given by the following theorem known as Bayes theorem:

Bayes theorem: Let E_i $i = 1, 2 \ldots k$ be mutually exclusive and be such that $\bigcup_{i=1}^{k} E_i = S$ where S is the sample space. Let A be any event. Then,

$$P(E_i|A) = \frac{P(A|E_i)P(E_i)}{\sum_{i=1}^{k} P(A|E_i)P(E_i)}$$

Proof

By the total probability theorem we know that $P(A) = \sum_{i=1}^{k} P(A|E_i)P(E_i)$. Now,

$$P(AE_i) = P(E_i|A)P(A) = P(A|E_i)P(E_i)$$

and hence

$$P(E_i|A) = \frac{P(A|E_i)P(E_i)}{P(A)} = \frac{P(A|E_i)P(E_i)}{\sum_{i=1}^{k} P(A|E_i)P(E_i)}$$ ■

Solution to Example 14.60: By Bayes theorem

$$P(T_1|R_1) = \frac{P(R_1|T_1)P(T_1)}{P(R_1|T_1)P(T_1) + P(R_1|T_0)P(T_0)}$$

Now from the point of the receiver, both 0 and 1 are equally likely to have been transmitted. So, we have $P(T_0) = P(T_1) = \frac{1}{2}$. Thus,

$$P(T_1|R_1) = \frac{p \times \frac{1}{2}}{p \times \frac{1}{2} + (1-p) \times \frac{1}{2}}$$ ■

PRACTICE 14.17

Solve Example 14.61.

Example 14.62

An electrician bought 250 multi-pins; 128 from Kumar Electronics, 52 from Muthu Electricals and 70 from Guptha Suppliers. It is known that 10% of the multi-pins sold by Kumar Electronics are defective. The percentage of defective multi-pins from the other two shops are, respectively, 21% and 12%.

(i) What is the probability that a randomly selected multi-pin from the collection of 250 is defective?

(ii) If a randomly selected multi-pin turns out to be defective, what is the probability that it was bought from Guptha suppliers?

(iii) If a randomly selected multi-pin turns out to be good, what is the probability that it was bought from Muthu Electricals?

SOLUTION

Let E_1, E_2 and E_3, respectively, be the events that a randomly selected multi-pin is from Kumar Electronics, Muthu Electricals and Guptha Suppliers. Let A be the event that such a multi-pin is defective. Then we have

$$P(E_1) = \tfrac{128}{250}, \qquad P(A|E_1) = \tfrac{10}{100}$$

$$P(E_2) = \tfrac{52}{250}, \qquad P(A|E_2) = \tfrac{21}{100}$$

$$P(E_3) = \tfrac{70}{250}, \qquad P(A|E_3) = \tfrac{12}{100}$$

(i) By total probability theorem,

$$\begin{aligned} P(A) &= P(E_1)P(A|E_1) + P(E_2)P(A|E_2) + P(E_3)P(A|E_3) \\ &= \frac{128}{250} \times \frac{10}{100} + \frac{52}{250} \times \frac{21}{100} + \frac{70}{250} \times \frac{12}{100} \end{aligned}$$

(ii) By Bayes theorem,

$$\begin{aligned} P(E_3|A) &= \frac{P(E_3)P(A|E_3)}{P(E_1)P(A|E_1) + P(E_2)P(A|E_2) + P(E_3)P(A|E_3)} \\ &= \frac{\frac{70}{250}\frac{12}{100}}{\frac{128}{250} \times \frac{10}{100} + \frac{52}{250} \times \frac{21}{100} + \frac{70}{250} \times \frac{12}{100}} \end{aligned}$$

(iii) We need to calculate $P(E_2|A^c)$. Now, $P(A^c|E_1) = 1 - P(A|E_1) = \frac{90}{100}$. Similarly, we can obtain $P(A^c|E_2)$ and $P(A^c|E_3)$:

$$\begin{aligned} &P(E_2|A^c) \\ &= \frac{P(E_2)P(A^c|E_2)}{P(E_1)P(A^c|E_1) + P(E_2)P(A^c|E_2) + P(E_3)P(A^c|E_3)} \\ &= \frac{\frac{52}{250}\frac{79}{100}}{\frac{128}{250} \times \frac{90}{100} + \frac{52}{250} \times \frac{79}{100} + \frac{70}{250} \times \frac{88}{100}} \end{aligned}$$

■

Example 14.63

Kannan's waiting list number is 2 in a railway reservation chart. The probability of no ticket getting canceled is 0.08. The probability that one ticket gets canceled is 0.65 The probability that the passenger whose waiting list is number 1 cancels her ticket if she gets a confirmed ticket is 0.01. Find the probability that Kannan gets a confirmed ticket.

SOLUTION

Let E_0 be the event that no ticket gets canceled, let E_1 be the event that one ticket gets canceled and let E_2 be the event that *two or more* tickets get canceled. Let C be the event that Kannan gets a confirmed ticket. We have $P(E_0) = 0.08$, $P(E_1) = 0.65$ and $P(E_2) = 1 - (0.08 + 0.65) = 0.27$.

Further, we have the following conditional probabilities:

$$P(C|E_0) = 0, \quad P(C|(E_1) = 0.01 \quad \text{and} \quad P(C|E_2) = 1$$

By the total probability theorem, we get

$$\begin{aligned} P(C) &= P(C|E_0)P(E_0) + P(C|E_1)P(E_1) + P(C|E_2)P(E_2) \\ &= 0 \times 0.08 + 0.01 \times 0.65 + 1 \times 0.27 \\ &= 0.2765 \end{aligned}$$

■

PRACTICE 14.18

In Example 14.63, assuming that Kannan has got a confirmed ticket, find the probability that exactly one ticket had been canceled.

PRACTICE 14.19

Mary is a final year B.E. student. In her college, a student is allowed to attend one campus interview at most. She is likely to be called for the campus interview by company A with a probability of 0.45 and by company B with a probability of 0.40. If she attends the interview for company A she has a 90% chance of getting selected and if she attends the interview for company B she has a 85% chance of getting selected.

(i) What is the probability that she gets selected for a job?

(ii) If she has a job now, what is the probability that she was called by the company B?

(iii) If she did not get a job, what is the probability that she was not called by any one of the companies?

PRACTICE 14.20

If A, A_1, A_2 are events such that $A = A_1 \cup A_2$ and $A_1 \cap A_2 = \phi$; prove that for any event E,

$$P(E|A) = P(E|A_1)P(A_1|A) + P(E|A_2)P(A_2|A)$$

14.6 RANDOM VARIABLES, THEIR MEANS AND PROBABILITY MASS FUNCTIONS

Example 14.64

Imagine that Prof Subramnayam needs a driver for 30 days. Let us say that one Mr Kannan works for him as a driver. He charges Rs 200 per day. He works for 24 days after which he falls ill. He sends his friend Mr Kumar to drive Prof Sunbramanyam's car for the remaining 4 days. Mr Kumar demands Rs 250 to be paid and his demand is accepted. Now, we are interested in calculating the average expenditure per day incurred by Prof Subramnayam.

It is very clear that $\frac{250+200}{2}$ is *not* the average. The correct value is calculated as follows

$$\frac{24 \times 200 + 6 \times 250}{30}$$

This is an example of a weighted average. Let us rewrite this quantity in a different manner.

$$\frac{24}{30} \times 200 + \frac{6}{30} \times 250$$

There is nothing probabilistic about this story of Prof Subramanyam. Now, let us change this story a bit. In the new version of the story, Prof Subramanyam needs a driver not just for 30 days but throughout his life. Mr Kannan now becomes a regular driver for him. Well, he is not quite that regular. He takes leave from time to time. He turns up for duty only on 4/5 of the days, but he dutifully sends Mr Kumar on the days he is on leave. Now, the amount of money paid by Prof Subramnayam becomes a *random variable*. The probability that Prof Subramanyam pays Rs 200 is the probability of Kannan's driving, which is $\frac{4}{5}$. The probability that he pays Rs 250 is the probability of Kumar's driving, which is $\frac{1}{5}$.

Now, generalizing from the deterministic story above, the average can be calculated as $200 \times \frac{4}{5} + 250 \times \frac{1}{5}$. This example tells us how to define the average value of a general random variable X.

Let X be a random variable. Let us assume that the list of possible values of X are $x_1, x_2, x_3 ...$, etc. For the random variable of Example 14.64, this list of values is given by 200 and 250. We noted that the probability for Rs 200 was $\frac{4}{5}$ and the probability for 250 was $\frac{1}{5}$. Similarly, for a general random variable X, there is a probability that X assumes the value x_i. If this probability is denoted by p_i, then the mean of the random variable can be computed using the following definition: ■

Definition 14.7: Let us assume that a random variable assumes the values $x_1, x_2, x_3, ..., x_n$. In other words, let the set $\{x_1, x_2, ..., x_n\}$ be the range of $X : S \longrightarrow R$. Further, denote the probability that X assumes the value x_i by p_i. Its *average* is defined as

$$E(X) = \sum_{i}^{n} x_i p_i$$

Further, the average is also known as *the expected value* or *the mean*.

Now, a given random variable $X : S \longrightarrow R$, how do we calculate the probability that $X = x_i$? We shall understand this with the help of subsequent examples. Before the examples, let us also define the concept of probability mass function of a random variable.

Definition 14.8: Let X be a random variable and let its range be $\{x_1, x_2, ...\}$. The probability mass function of X is the function whose domain is the range of X defined as $p(x_i) = P(X = x_i)$.

Example 14.65

Consider Example 14.25. The range of X is the set $\{0, 1, 2, 3\}$. The range of Y is $\{0, 40, 80, 120\}$. ■

Example 14.66

In an experiment of tossing five coins together, if X is the number of heads, then possible values assumed by X are 0,1,2, 3, 4 and 5. That is, its range is the finite set $\{0, 1, 2, 3, 4, 5\}$. ■

PRACTICE 14.21

A biased coin is such that the probability of getting the head is thrice the probability of getting the tail. This coin is simultaneously tossed with an unbiased coin. Describe the sample space and the probabilities of each point in the sample space. Let X be the random variable giving the number of tails. Compute the probability mass function for this random variable.

Example 14.67

From a practical point of view, a random variable is a quantity whose values cannot be predicted in advance. For example, *the maximum temperature recorded in a day during November* at Chennai is a random variable. Here the sample space is the set of all days. If ω is a day, then $T(\omega)$ is the maximum temperature recorded on ω at Chennai. The following data give the values of $T(\omega)$ for the days during the month of November, 2005.

Date	1	2	3	4	5	6	7	8	9	10
Max. Temp.	28	28	27	26	25	23	24	26	24	27

Date	11	12	13	14	15	16	17	18	19	20
Max. Temp.	26	26	25	25	26	25	25	25	25	26

Date	21	22	23	24	25	26	27	28	29	30
Max. Temp.	25	26	26	26	26	27	27	26	26	26

The set of values assumed by this random variable is $\{23, 24, 25, 26, 27, 28\}$.

The following table gives the number of days on which the maximum temperature was a given value.

Temp.	23	24	25	26	27	28
No. of days	1	2	8	13	4	2

Now, out of 30 days, the maximum temperature at Chennai was 27° C on 4 days. Thus, we can assign the probability of $\frac{4}{30}$ for the temperature to be 27° C.

Assuming that the month November of this year 2005 is representative of all November months, we may say that the probability of the maximum temperature being 27° C at Chennai during the month of November is $\frac{4}{30}$. (To be precise, we will have to take the temperature data for the month of November over many years.) Now, probabilities for other maximum temperatures can be tabulated as follows:

Temp.	23	24	25	26	27	28
Probability	1/30	2/30	8/30	13/30	4/30	2/30

Using the notation, we can write $P(X = 24) = \frac{2}{30}$.

The above example is empirical. That is, we estimate the probability using experimental data. In the following examples, we find the probabilities theoretically. ■

Example 14.68

Consider an urn containing 10 red balls and three blue balls. Two balls are drawn one after another randomly from the urn without replacement. Let X be the number of red balls drawn. Find the probabilities for $X = 0$, $X = 1$ and $X = 2$.

SOLUTION

Now, the sample space is $\{BR, BB, RR, RB\}$.

The outcome BB corresponds to $X = 0$. Now, the probability for BB is $\frac{3}{13} \times \frac{2}{12}$. Hence $P(X = 0) = \frac{6}{156}$.

The outcomes BR and RB correspond to $X = 1$. Now, the probability for BR is $\frac{3}{13} \times \frac{10}{12}$. The probability for RB is $\frac{10}{13} \times \frac{3}{12}$ Hence

$$P(X = 1) = \frac{3}{13} \times \frac{10}{12} + \frac{10}{13} \times \frac{3}{12}$$

The outcome RR corresponds to $X = 2$. Hence $P(X = 2) = \frac{10}{13} \times \frac{9}{12} = \frac{90}{156}$. ■

Example 14.69

The probability of correct reception of a bit in a communication channel is p. Assume that the string 10 is transmitted and let X be the number of errors in the received string. Noting that X takes the values 0, 1 and 2, find the probabilities $P(X = 0)$, $P(X = 1)$, $P(X = 2)$.

SOLUTION

The sample is the set of all possible strings of length two, which is

$$\{10, 11, 00, 01\}$$

If the string 10 is received, then both bits have been received correctly. Hence there is no error and $X = 0$. The probability of the correct reception of the first bit is p and the probability of the reception of the second bit is p. These two events are independent and hence by the multiplicative law, the probability of the correct reception 10 is p^2.

The strings 11 and 01 would be received when one of the bits is correctly transmitted and the other is wrongly transmitted. Hence the probability of reception of each of these two strings is $p(1 - p)$. As these two strings correspond to $X = 1$, $P(X = 1) = p(1 - p) + p(1 - p)$.

Similar considerations imply that $P(X = 2) = (1 - p)^2$. Thus, we obtain the following table:

x	0	1	2
$P(X = x)$	p^2	$2p(1-p)$	$(1-p)^2$

■

Example 14.70

Consider Example 14.25. Here, X that is the power consumed between 6 p.m and 10 p.m takes the values 0, 40, 80 and 120 Watts. What is the probability that it is, say, 80 Watts? We get 80 Watts when two of the tubelights are on and this happens for the configurations (On, On, Off), (On, Off, On), (Off, On, On). We know their probabilities to be $(0.3)(0.3)(0.7)$, $(0.3)(0.7)(0.3)$ and $(0.7)(0.3)(0.3)$. Thus, the probability that $X = 80$ is $3 \times (0.3)^2(0.7)$.

Now $\omega =$ (On, On, Off) is an element of the sample space such that $X(\omega) = 80$. $E = \{$(On, On, Off), (On, Off, On), (Off, On, On)$\}$ is nothing but the set of all elements ω of S such that $X(\omega) = 80$. Using the language of sets,

$$E = \{\omega \in S : X(\omega) = 80\}$$

Let us recollect the idea of an inverse image from Section 2.5. It is easy to see that the above set is nothing but the inverse image of $\{80\}$.

Example 14.70 illustrates the following precise statement:
$P(X = x)$, that is, the probability that $X = x$ is nothing but the probability of the event $X^{-1}(x) \subseteq S$.

Let us recall that a random variable is a function $X : S \longrightarrow R$. In Definition 14.8, we have explicitly assumed that the range of X is *a finite set*. But in general, this need not be the case. Consider a lake of depth 10 metres. Theoretically speaking, depending on the rainfall, the water level of this lake can assume any value between 0 and 10 metres. In other words, the range of this random variable is the interval $[0, 10]$. This set is uncountable. See Section 2.8. For such random variables, the concept of probability mass function is not appropriate. The ideas of *continuous random variables* and their *probability density functions* are topics of an advanced course in probability. In this book, we deal only with what are known as *discrete* random variables. Now, we give the proper definition of a discrete random variable. ■

Definition 14.9: A discrete random variable is defined as a random variable $X : S \longrightarrow R$ such that its range is a countable set.

In view of Definition 2.8.5, a random variable whose range is a finite set is a discrete random variable. Examples 14.67 and 14.70 are of this type. Next, we consider an example of a random variable whose range is *countably infinite*.

Example 14.71

Let us consider tossing a coin till the tail appears. If the first trial gives us a tail, we stop. If the first outcome is a head, we toss again. Now, if we get a tail we stop and so on. Hence, the sample space for this experiment is

$$\{T, HT, HHT, HHHT, \ldots\}$$

Now, let X be the number of heads we would obtain before getting a tail. That is $X(T) = 0$, $X(HT) = 1$ and so on. Thus X can assume the values 0, 1, 2, etc. In other words, its range is $\{0, 1, 2, ...\} = \{0\} \cup \mathbb{N}$ and hence it is countable. ■

Remark 14.2

In the theory of probability, when a condition like $X = x_i$ is written within a pair of brackets, it denotes the corresponding events.

Specifically,

$$\begin{aligned}(X = a) &= X^{-1}(a) = \{\omega : X(\omega) = x\} \\ (a < X < b) &= X^{-1}((a, b)) = \{\omega : a < X(\omega) < b\} \\ (a \leq X < b) &= X^{-1}([a, b)) = \{\omega : a \leq X(\omega) < b\}\end{aligned}$$

Example 14.72

Let us consider a bag containing two red balls, one blue ball and one green ball. Suppose we randomly draw three balls one by one without replacement from the bag.

Now, let X_R, X_B and X_G denote the number of red, blue and green balls, respectively, in an outcome. These are examples of random variables. The probability of the various outcomes can be determined as shown:

The sample space can be written as

$$\{RBG, RGB, GRB, GBR, BRG, BGR,$$

$$RRB, RRG, GRR, BRR, RGR, RBR\}$$

It is clear that X_R assumes the values of 1 and 2 only. X_B takes the values of $0, 1$. X_G also takes the values 0 and 1.

Let us calculate the probability of RGR. The probability of the first ball being red is $\frac{2}{4}$. The probability of the second ball being green is $\frac{1}{3}$. The probability of the third ball being red is $\frac{1}{2}$. Thus, the probability of the outcome RGR is $\frac{2}{4} \times \frac{1}{3} \times \frac{1}{2} = \frac{1}{12}$.

Let us now calculate and tabulate the probabilities for the number of red balls X_R.

$X_R = 1$ for the outcomes RBG, RGB, GRB, GBR, BRG, BGR. Their probabilities can be calculated, respectively, as

$\frac{2}{4} \times \frac{1}{3} \times \frac{1}{2}, \frac{2}{4} \times \frac{1}{3} \times \frac{1}{2}, \frac{1}{4} \times \frac{2}{3} \times \frac{1}{2}, \frac{1}{4} \times \frac{1}{3} \times \frac{2}{2}, \frac{1}{4} \times \frac{2}{3} \times \frac{1}{2}, \frac{1}{4} \times \frac{1}{3} \times \frac{2}{2}$

Note that all the above values are equal to $\frac{1}{12}$. Hence $P(X_R = 1) = 6 \times \frac{1}{12} = \frac{1}{2}$.

Similarly, we find that $P(X_R = 2) = \frac{1}{2}$. ■

Hint: Find the probabilities for each of RRB, RBR, RRG, RGR, GRR, BRR and add.

We similarly tabulate the values for X_G and X_B:

X_G	0	1
$P(X_G)$	$\frac{3}{12}$	$\frac{9}{12}$

X_B	0	1
$P(X_B)$	$\frac{3}{12}$	$\frac{9}{12}$

PRACTICE 14.22

Let us consider a bag containing two red balls, one blue ball and one green ball. Suppose we randomly draw three balls one by one *with replacement* from the bag. Let X_B denote the number of blue balls in an outcome. Calculate $P(X_B = x)$ for various values of x.

PRACTICE 14.23

Let a pair of dice be thrown. Let X be the sum of the outcomes. Find the probability $P(X = 7)$.

Example 14.73

If it rains, an umbrella salesman can earn Rs 150 per day. If it does not rain, he can lose Rs 10 per day. What is his average profit if the probability of rain is 0.3?

SOLUTION

In this example, profit is the random variable. Let us denote it by X. The event that the salesman gets a profit of Rs 150 is the event that it rains. Hence $P(X = 150) = 0.3$. $P(X = -10) = 0.7$. Hence the average or the expected profit is $150 \times 0.3 + (-10)(0.7) = 38$. ■

PRACTICE 14.24

A company manufactures electric fuses.The company gets a profit of Rs 7 per fuse. The probability that a fuse lasts for more than 200 hours is 0.81. If a fuse does not last for 200 hours, then the company has to replace the fuse, which costs the company Rs 46 per fuse. Find the average profit per fuse.

Example 14.74

Consider a car rental agency. How will the manager decide how many cars the company needs to buy? If he has only a few cars, he may have to turn down some customers without renting them any. If he has too many cars, then some cars would have to be kept idle. Let X be the number of demands for rental on a day.

X is a random variable. If he keeps n cars with the agency, then the probability that at least one car would be idle is $P(X < n)$ and the probability that at least one customer is turned down is $P(X \geq n)$. ■

Example 14.75

Mr Kumar has a taxi rental service in a certain township and he is urging his friend Mr Kannan to start his own taxi service in a similar township. The following data for a whole year have been collected by Mr Kannan from Mr Kumar. There were 40 days which no one demanded a taxi, there was one demand on 176 days, two demands on 86 days and more than three demands on 63 days. At present, Mr Kannan can afford to buy only one car. When someone engages his taxi on a day he would get a profit of Rs 150.00. The day his taxi is not engaged, the maintenance cost works out to be Rs 160. Calculate Mr Kumar's expected profit

SOLUTION

No. of demands for a taxi	0	1	2	More than 3
Probability	40/365	176/365	86/365	63/365

The 'profit' X on any day is a random variable, which takes the values 150 and -160.

If Y is the number of demands, the event that his taxi would be engaged is $(Y \geq 1)$. Hence $P(X = 150) = P(Y \geq 1) = \frac{325}{365}$. His taxi would be idle with a probability of $P(Y = 0) = \frac{40}{365}$. Hence $P(X = -160) = \frac{40}{365}$.

The expected profit is $150 \times P(X = 150) + (-160) \times P(X = -160)$. ■

PRACTICE 14.25

Find the average of the random variables of Examples 14.25 and 14.72.

PRACTICE 14.26

A merchant can buy an item for Rs 25 and sell it for Rs 36. The probabilities for a demand of 0, 1, 2, 3, 4, 5 or more items are, respectively, 0.05, 0.15, 0.30, 0.25, 0.15 and 0.10. Calculate the expected profit resulting from stocking 0, 1, 2, 3, 4 and 5 items and determine how many items the merchant should keep so as to maximize the expected profit.

Hint: If he stocks 0 item, then no matter what the demand may be, he gets a profit of 0. Hence $E(X) = 0$. Consider the case where he has just 1 item for selling.

If there is no demand he would lose Rs 25. If there are one or more demands, then he would get a profit of Rs 11. The probability for the second is event $1 - 0.05$. We record these observations in the following table

Profit	−25	11
Prob.	0.05	0.95

$E(X) = -25 \times 0.05 + 11 \times 0.95$.
If he keeps two items for sale, then if the demand is 0, he loses Rs 50. If there is one demand he loses $25 - 11$. If there are two or more demands, he gains Rs 22.

Profit	−50	−14	22
Prob.	0.05	0.15	0.80

Consider the other cases and complete the problem.

We have seen that the probabilities $P(X = x_i)$ are important for calculating the average value of X. Practice 14.26 tells us that probabilities like $P(X \geq a)$, etc. are important in many decision-making problems. The following examples tell us how to calculate these probabilities in some situations.

Example 14.76

Consider the experiment of tossing three coins simultaneously. Let X be the number of tails in the outcome.

Identify the following events and their probabilities:

(i) $(X = 3)$.
(ii) $(1.5 \leq X < 4)$.
(iii) $(0.5 < X < 3)$.
(iv) $(X > 1)$.
(v) $(X \leq 2)$.

SOLUTION

In this example, X takes the values 0, 1, 2 or 3. Now, we can simplify the above events as follows:

(ii) $(1.5 \leq X < 4) = (X = 2) \cup (X = 3)$.
(iii) $(0.5 < X < 3) = (X = 1) \cup (X = 2)$.
(iv) $(X > 1) = (X = 2) \cup (X = 3)$.

Next, we identify the events as subsets of the sample space

$$S = \{TTT, TTH, THT, THH, HTT, HTH, HHT, HHH\}$$

(i) $(X = 3) = \{TTT\}$.
(ii) $(1.5 \leq X < 4) = \{TTH, THT, HTT, TTT\}$.
(iii) $(0.5 < X < 3) = \{TTH, THT, THH, HTT, HTH, HHT\}$.
(iv) $(X > 1) = \{TTH, THT, HTT, TTT\}$.

(v) $(X \leq 2) = \{TTH, THT, HTT, THH, HTH, HHT\}$.

The coin being unbiased, the probability of each single outcome is $\left(\frac{1}{2}\right)^3$. Hence, we can calculate the probabilities of these in the following manner: $P(0.5 < X < 3) = P((X = 1) \cup (X = 2)) = 6/8$. ■

PRACTICE 14.27

In Example 14.69, find the probability that the power consumption is more than 100 Watts.

PRACTICE 14.28

The probability of it raining in a city where Ms Selvi lives is 0.20. If it rains on a day, Ms Selvi gets Rs 3.00 from her father and if it does not rain, she gets Re 1.00. Let X be the amount she gets on a day. Describe the sample space and find the probability mass function of X.

PRACTICE 14.29

A bundle of sarees consists of 10 good sarees, three with minor defects and five with major defects. Three sarees are chosen at random without replacement. Let X be the number of sarees with minor defects. Find $E(X)$.

PRACTICE 14.30

The probability mass function of a random variable is given by $P(X = r) = kr^3$, $r = 1, 2, 3, 4$. Find the value of k. Find $P(\frac{1}{2} < X < \frac{5}{2})$ and $P(\frac{3}{2} < X < \frac{7}{2} | X > 3)$.

PRACTICE 14.31

A communication channel is used to transmit strings of five bits. For the transmission of a single bit, the probability of correct transmission is 0.8. Let the string 11110 be transmitted and let X be the number of errors in the received string. Find the probability that the number of errors is 3.

Example 14.77

The mass function of a random variable is given by

X	−1	0	1	2	3
P(X)	0.1	0.2	0.3	0.2	0.2

Find the probability mass functions of $Y = X^2$ and $Z = 2X + 1$.

SOLUTION

It is clear that Y takes the values 1, 0, 4 and 9, and Z takes the values $-1, 1, 3, 5$ and 7.

Recalling that events are subsets of a sample space, we note the following $(Y = 1) = (X = -1) \cup (X = 1)$. Further, the events $(X = 1)$ and $(X = -1)$ are mutually exclusive. Hence

$$\begin{aligned} P(Y = 1) &= P(X = -1) + P(X = 1) \\ &= 0.1 + 0.3 \\ &= 0.4 \end{aligned}$$

It is easy to see that $P(Y = 4) = P(X = 2)$, $P(Z = 3) = P(X = 1)$, etc. Thus, we obtain the tables for the probability mass functions of Y and Z as

Y	0	1	4	9
P(Y)	0.2	0.4	0.2	0.2

Z	−1	1	3	5	7
P(Z)	0.1	0.2	0.3	0.2	0.2

■

PRACTICE 14.32

The mass function of a random variable is given by

X	0	2	5	7	14
P(X)	0.2	0.1	0.2	0.3	0.2

Find the mass function of $Y = 3X + 1$.

PRACTICE 14.33

The mass function of a random variable is given by

X	−3	−2	0	1	2
P(X)	0.2	0.1	0.2	0.3	0.2

Find the mass function of $Y = 3X^2 + 1$.

14.7 VARIANCE, STANDARD DEVIATION AND MOMENTS

Example 14.78

The following two tables give the distribution of marks obtained by the students of Section A and Section B of second Year B.E (CS) of Kathavarayan Deemed University.

Marks	20–30	30–40	40–50	50–60	60–70	70–80	80–90	90–100
Frequency	2	8	10	15	7	8	7	3

Marks	20–30	30–40	40–50	50–60	60–70	70–80	80–90	90–100
Frequency	13	4	6	7	5	9	10	6

The average marks for Section A can be calculated as

$$25\times\frac{2}{60}+35\times\frac{8}{60}+45\times\frac{10}{60}+55\times\frac{15}{60}+65\times\frac{7}{60}+75\times\frac{8}{60}+85\times\frac{7}{60}+95\times\frac{3}{60}$$

Similarly, the average mark for Section B also can be calculated. We shall find that the average marks do not differ too much from each other. But a glance at the tables tells us that the number of students with marks on the higher side is more for Section B. In Section A, many students have scored in the range of 55–65. But in Section B many students have scored well above 'average' or well below 'average'. In other words, there is a lot of variation in the performance of the students of Section B.

Variance is the quantity that measures this variation. For example, if the performance of the students is uniform, the variance would be small. If x is the individual mark of someone, and $\overline{x}$ is the average, then $x-\overline{x}$ is the deviation from the average. This deviation could be positive or negative. The average of the non-negative quantities $(x-\overline{x})^2$ is called *variance*. Its formal definition is given as follows. ■

Definition 14.10: For a discrete random variable X, variance is defined as

$$V(X) = E[X - E(X)]^2$$

The positive square root of the variance is called the *standard deviation* and is denoted by σ. $\sigma = \sqrt{V(X)}$.

In general, the kth *moment* of X about a number a is defined as $E[(X-a)^k]$. Let us note that variance is just *the second moment about the mean*. In the following sections, we shall calculate these quantities for some special distributions.

14.8 BINOMIAL DISTRIBUTION

Example 14.79

Let us consider a floppy manufacturing company that sells boxes of floppies containing 10 floppies each. The company would replace the full box, if at least two floppies are found defective. Naturally, the company would like to know the probability of this event. This probability depends on the probability that a single floppy is defective. Suppose that approximately 25 floppies are found to be defective for every thousand floppies manufactured. Then the probability of a single floppy being defective is 0.025. Ten floppies are chosen randomly to be put in a box. The number of defective floppies X in this box is a random variable. $P(X = 1)$ is the probability that exactly one of these 10 is defective. This would happen in one of the following situations: the first one is defective and the other nine are not, the second one is defective and the other nine are not, etc. There are 10 such possibilities. The probability of each of the above events is $(0.025)(0.975)^9$ and hence $P(X = 1) = 10 \times (0.025)(0.975)^9$.

$X = 2$ if any two are defective and the other eight are not. Let us say that the 3rd floppy and the 6th floppy are defective, and the rest are not. The probability of this event is $(0.025)^2(0.975)^8$. Now, there are ${}^{10}C_2$ pairs among the 10 floppies and hence,

$$P(X = 2) = 10C_2(0.025)^2(0.975)^8$$

It is clear that for $x \leq 10$, $P(X = x) = 10C_x(0.025)^x(0.975)^{10-x}$.

X is an example of a binomial random variable.

Coming back to the original problem of finding the probability that at least two floppies are defective, we have $P(X \geq 2) = \sum_{x=2}^{10} \binom{10}{x} (0.025)^x(0.975)^{10-x}$. ■

Example 14.80

Consider the experiment of tossing 10 unbiased coins together. If X is the number of heads, then proceeding as in the previous example, we can calculate

$$P(X = x) = \binom{10}{x} (0.5)^x(0.5)^{10-x}$$

Now, let us generalize the above two examples. Consider a single random experiment, which has two outcomes. We shall call one of the outcome as 'success' and the other as 'failure'. Let the probability of 'success' be p and hence the probability of failure be q, where $p + q = 1$. Now, we imagine that this experiment is repeated n times one after another, or that n identical experiments are performed simultaneously. Let X be the number of successes amongst the n outcomes. We have

$$P(X = x) = nC_x p^x q^{n-x}$$ ■

Definition 14.11: A random variable taking values 0, 1, 2, ..., n is called a *binomial random variable* if its mass function is given by

$$P(X = x) = nC_x p^x q^{n-x}$$

Example 14.81

Suppose that 15% of the population is left-handed. Find the probability that in a group of 50 individuals, there will be (a) at most 10 left-handers, (b) at least five left-handers, (c) between three and six left-handers inclusive and (c) exactly five left-handers.

SOLUTION

Here $n = 50$, $p = 0.15$ and $q = 0.85$. The random variable X is the number of left-handers.

(a) Here we have to find the probability of $(X \leq 10)$

$$\begin{aligned} P(X \leq 10) &= \sum_{x=0}^{10} P(X = x) \\ &= \sum_{x=0}^{10} \binom{50}{x} (0.15)^x (0.85)^{50-x} \end{aligned}$$

(b)

$$\begin{aligned} P(X \geq 5) &= 1 - P(X < 5) \\ &= 1 - \left(\sum_{x=0}^{4} P(X = x) \right) \\ &= 1 - \left(\sum_{x=0}^{4} \binom{50}{x} (0.15)^x (0.85)^{50-x} \right) \end{aligned}$$

(c)

$$\begin{aligned} P(3 \leq X \leq 6) &= \sum_{x=3}^{6} P(X = x) \\ &= \sum_{x=3}^{6} \binom{50}{x} (0.15)^x (0.85)^{50-x} \end{aligned}$$

(d)

$$P(X = 5) = \binom{50}{5} (0.15)^5 (0.85)^{45}$$

■

PRACTICE 14.34

A bag contains one red and seven white marbles. A marble is drawn from the bag and its colour is observed. Then the marble is put back into the bag and the contents are thoroughly mixed. Find the probability that in eight such drawings a red marble is selected exactly three times.

PRACTICE 14.35

Ten percent of the bolts that a machine produces are defective. Find the probability that in a random sample of 400 bolts produced by the machine (a) at most 30 (b) between 30 and 50 or (c) more that 55 bolts are defective.

Example 14.82

Consider a communication channel that transmits bits. The probability that a bit is received correctly is 0.90. Find the *average* number of errors when a byte is received.

SOLUTION

The number of errors X in the transmission of a byte is a random variable. It assumes the values $0, 1, 2, \ldots, 8$. Further, X follows a binomial distribution with $n = 8$, $p = 0.10$ and $q = 0.90$. Since we are interested in the number of errors, here the word 'success' denotes the reception of a wrong bit. Thus, the probability mass function of the number of errors is

$$P(X) = {}^8C_x \left(\frac{1}{10}\right)^x \left(\frac{9}{10}\right)^{9-x}, \qquad x = 0, 1, 2, \ldots, 8$$

Thus, the average value of X is given by

$$E(X) = \sum_{x=0}^{8} x\, {}^8C_x \left(\frac{1}{10}\right)^x \left(\frac{9}{10}\right)^{9-x}$$

Instead of calculating $E(X)$, in this special case we proceed to calculate the mean of the general binomial random variable. We shall also calculate its variance, which is defined in Section 14.7. ■

Proposition 14.4

The **mean** μ **and variance** σ **of a binomial variable** are given by

$$\mu = np, \qquad \sigma = npq$$

Proof

$$\begin{aligned} E(X) &= \sum xP(X = x) \\ &= \sum_{x=0}^{n} x\ ({}^nC_x)\ p^x q^{n-x} \end{aligned}$$

$$
\begin{aligned}
&= \sum_{x=0}^{n} x({}^nC_x)\, p^x q^{n-x} \\
&= \sum_{x=0}^{n} x\frac{n!}{(n-x)!x!} p^x q^{n-x} \\
&= \sum_{x=1}^{n} \frac{n(n-1)!}{(n-x)!(x-1)!} p \times p^{x-1} q^{(n-1)-(x-1)} \\
&= \sum_{x=1}^{n} np\frac{(n-1)!}{((n-1)-(x-1))!(x-1)!} \times p^{x-1} q^{(n-1)-(x-1)} \\
&= np\sum_{x=1}^{n} \binom{n-1}{x-1} \times p^{x-1} q^{(n-1)-(x-1)} \\
&= np\sum_{x=0}^{n-1} \binom{n-1}{x} \times p^{x} q^{(n-1)-x} \\
&= np(p+q)^{n-1} \\
&= np \\
E(X^2) &= \sum x^2 P(X=x) \\
&= \sum_{x=0}^{n} x^2\, ({}^nC_x)\, p^x q^{n-x}
\end{aligned}
$$

Since $x! = 1.2...(x-1)(x)$, we write x^2 as $x^2 = (x-1)(x) + x$, so that we can conveniently cancel some common terms. Now,

$$
\begin{aligned}
E(X^2) &= \sum_{x=0}^{n} \frac{n!(x(x-1)-x)}{(n-x)!(x)} \times p^x q^{(n-x)} \\
&= \sum_{x=0}^{n} \frac{n!x(x-1)}{(n-x)!(x)!} p^2 \times p^{x-2} q^{(n-2)-(x-2)} \\
&\quad + \sum_{x=0}^{n} \frac{n!(x)}{(n-x)!(x)!} p \times p^{x-1} q^{(n-1)-(x-1)} \\
&= \sum_{x=2}^{n} n(n-1)p^2 \frac{(n-2)!}{(n-x)!(x-2)!} p \times p^{x-2} q^{(n-2)-(x-2)} \\
&\quad + \sum_{x=1}^{n} np\frac{(n-1)!}{(n-x)!(x-1)!} p \times p^{x-1} q^{(n-1)-(x-1)}
\end{aligned}
$$

$$
\begin{aligned}
&= n(n-1)p^2 \sum_{x=2}^{n} \frac{(n-2)!}{((n-2)-(x-2))!(x-2)!} p \\
&\quad \times p^x q^{(n-2)-(x-2)} + np \sum_{x=1}^{n} \frac{(n-1)!}{((n-1)-(x-1))!(x-1)!} p \\
&\quad \times p^x q^{(n-1)-(x-1)} \\
&= n(n-1)p^2 \sum_{x=0}^{n} \frac{(n-2)!}{((n-2)-(x))!(x)!} p \times p^x q^{(n-2)-(x)} \\
&\quad + np \sum_{x=0}^{n} \frac{(n-1)!}{((n-1)-(x))!(x)!} p \times p^x q^{(n-1)-(x)} \\
&= n(n-1)p^2(p+q)^{n-2} + np(p+q)^{n-1} \\
&= n(n-1)p^2 + np \\
\mathrm{Var}(X) &= E(X^2) - [E(X)]^2 \\
&= n(n-1)p^2 + np - n^2p^2 = -np^2 + np \\
&= np(1-p) = npq
\end{aligned}
$$

■

PRACTICE 14.36

Using Proposition 14.4, find the average number of errors in Example 14.81.

14.9 GEOMETRIC RANDOM VARIABLE

Example 14.83

A doctor wants to know if a particular taluk is infected with malaria or not. She is not interested in the percentage of infected persons but just likes to know if there is at least one infected person. She would keep testing individuals and quit testing when she finds the first malarial infected person. The number of persons she would have to test before finding the infected person is a random variable. If the first person is infected, then $X = 0$. If the second person she tests is the first person to be infected, then $X = 1$ and so on. If N is the population of the taluk, then the values assumed by the random variable are $0, 1, 2, \ldots, N$. For simplicity, we shall assume that $N = \infty$. Let the overall probability that a person is infected with malaria be $\frac{3}{100}$. If $X = 4$ then it means that the fifth person is the first infected person and the first four persons were not infected. The probability of this event is $\left(\frac{97}{100}\right)^4 \left(\frac{3}{100}\right)$.

One can easily see that $P(X = x) = \left(\frac{97}{100}\right)^x \left(\frac{3}{100}\right)$. ■

Example 14.84

Imagine a biased coin such that the probability of getting a head is p. We keep tossing the coin till a head turns up. The probability that we get x of tails before getting the first head is $q^x p$. ■

In general, consider an experiment with two possible outcomes, called 'success' and 'failure' with respective probabilities p and $1 - p$. This experiment is repeated till we get one success. If X is the number of failures before we get the first success, then $P(X = x) = q^x p$. Such a random variable is called a geometric random variable.

Definition 14.12: A random variable that assumes the values $x = 0, 1, 2, \ldots$ with probability given by $P(X = x) = q^x p$, where $p > 0$, $q > 0$ and $p + q = 1$ is called a *geometric random variable.*

Example 14.85

Suppose you pick people at random and find out the month of their birth. Find the probability that you will have to ask nine persons until you find a person who was born in August.

SOLUTION

Assuming that all the 12 months are equally likely to be the month of birth of a random person, the probability that a random person is born in the month of August is $\frac{1}{12}$. Thus, the required answer is $\left(\frac{11}{12}\right)^9 \frac{1}{12}$. ■

PRACTICE 14.37

A die is thrown till a number greater than or equal to 4 appears. Find the probability that seven throws are needed to get the desired outcome.

Proposition 14.5

The mean and the standard deviation of the geometric distribution are given by $E(X) = \frac{q}{p}$ and $\sigma = \sqrt{\frac{q^2}{p^2} + \frac{q}{p}}$, respectively.

Proof

$$\begin{aligned} E(X) &= \sum_{x=0}^{\infty} x q^x p \\ &= p \sum_{x=1}^{\infty} x q^x \\ &= pq \sum_{x=1}^{\infty} x q^{x-1} \end{aligned}$$

$$
\begin{aligned}
&= pq\sum_{x=0}^{\infty}(x+1)q^x \\
&= pq(1-q)^{-2} \quad \text{(By binomial expansion)} \\
&= \frac{q}{p} \\
E(X^2) &= \sum_{x=0}^{\infty} x^2q^xp \\
&= \sum_{x=0}^{\infty}[(x)(x-1)+x]q^xp \\
&= \sum_{x=2}^{\infty} x(x-1)q^x + \sum_{x=0}^{\infty} xq^xp \\
&= 2pq^2(1+3q+6q^2+\ldots.)+\frac{q}{p} \\
&= 2pq^2(1-q)^{-3}+\frac{q}{p} \\
&= \frac{2q^2}{p^2}+\frac{q}{p}
\end{aligned}
$$

So, $Var(X) = E(X^2) - [E(X)]^2 = \frac{2q^2}{p^2} + \frac{q}{p} - \frac{q^2}{p^2} = \frac{q^2}{p^2} + \frac{q}{p}$. Hence $\sigma = \sqrt{\frac{q^2}{p^2} + \frac{q}{p}}$.

The reader can find the results of the infinite summations used in the above proof in Appendix A. ■

14.10 NEGATIVE BINOMIAL DISTRIBUTION

Example 14.86

Imagine that we toss a coin till we get three heads. We stop the experiment at the third head. Some elements of the sample space are HTHH, HHTH, TTTTHTTT-THH, HHH, TTHHTTTTTTTTTH, THTHTH, etc. Let X be the number of tails in this experiment. If $X = x$, the total number of trials is $x + 3$. Among this, the last one is a head. So we shall have $3 - 1$ heads and x tails before the last head. Thus, the probability of $X = x$ is

$$
{}^{x+3-1}C_{3-1}\left(\frac{1}{2}\right)^x\left(\frac{1}{2}\right)^{3-1}\frac{1}{2}
$$

In general, consider an experiment with two possible outcomes called success and failure with respective probabilities p and $1-p$. This experiment is repeated till we get r successes. If X is the number of failures before we get the rth success, then

$$P(X = x) = {}^{x+r-1}C_{r-1}(1-p)^x p^{r-1} p = {}^{x+r-1}C_{r-1}(1-p)^x p^r$$

Such a random variable is called a negative binomial random variable. ■

Definition 14.13: A random variable whose mass function is given by

$$P(X = x) = {}^{x+r-1}C_{r-1}(1-p)^x p^{r-1} p = {}^{x+r-1}C_{r-1}(1-p)^x p^r$$

is called a *negative binomial distribution*.

Example 14.87

In an assembly constituency, the probability that a voter would support the political party $P.Q.R$ is 0.45. Find the probability that the tenth person under survey would be the third supporter of $P.Q.R$.

SOLUTION

The number of non-supporters of the party before three supporters are surveyed follows the negative binomial distribution. Hence the required probability is

$$P(X = 7) = {}^{7+3-1}C_{3-1}(0.55)^7(0.45)^3$$ ■

Example 14.88

Two teams are playing a series of games, each of which is independently won by team A with probability p, and by team B with probability $1-p$. The winner is the first team to win four games. Find the probability that seven games are played.

SOLUTION

We calculate the probabilities that A wins or B wins and add them

$${}^{3+4-1}C_{4-1}(1-p)^3p^4 + {}^{3+4-1}C_{4-1}p^3(1-p)^4$$ ■

14.11 POISSON DISTRIBUTION

Consider the time duration, say, between 6 p.m and 8 p.m at a telephone booth in a town. Let the population be 4 lakhs and let the probability that a resident would use this booth during the above-mentioned time be 0.08. Now, we are interested in the probability that there would be nine persons coming to the booth to make a call. Using binomial distribution this probability can be calculated as

$${}^{400000}C_9(0.08)^9(0.92)^{399991}$$

If X is the number of persons making a phone call, then the range of X is 0,1,2..., 4,00,000, neglecting the persons who visit the city and the persons who may leave the city.

This is an example of a binomial random variable with a large n and a small p. For such a random variable, instead of using the formula

$$P(X = x) = {}^nC_x p^x q^{n-x}$$

we use the approximation

$${}^nC_x p^x q^{n-x} \sim \frac{e^{-np}(np)^x}{x!}$$

This appro ximation provides a motivation for the Poisson random variable, which is defined below. Since the average for a binomial variable is given by np, if λ is the average of the Poisson random variable, in the above formula we replace np by λ.

Definition 14.14: A random variable whose range is $\{0, 1, 2, \ldots\}$ is said to be a *Poisson random variable* with mean λ if its mass function is given by

$$P(X = x) = \frac{e^{-\lambda}\lambda^x}{x!}$$

A large value of n and a small value of p appear to be fairly common and the following are examples of random variables known to follow Poisson distribution:

(a) Number of telephone calls in a booth.

(b) Number of printing errors in books.

(c) Number of accidents in a given spot.

We shall prove that λ is in fact the mean of the Poisson random variable later.

Example 14.89

Assume that the number of customers visiting a departmental store is a Poisson random variable with mean 1225. Find the probability that:

(a) No customer comes on a day.

(b) Exactly 3,000 customers visit on a day.

(c) Strictly more than 2,000 customers visit on a day.

(d) 1,000 or less customers visit on a day.

SOLUTION

(a) $P(X = 0) = \frac{e^{-1225}\lambda^0}{0!} = e^{-1225}$

(b) $P(X = 3000) = \frac{e^{-1225}(1225)^{3000}}{3000!}$

(c) $P(X > 2000) = 1 - P(X \leq 2000) = 1 - \left(\sum_{x=0}^{2000} \frac{e^{-1225}(1225)^x}{x!}\right)$

(d)

$$P(X \leq 1000) = \sum_{x=0}^{1000} \frac{e^{-1225}(1225)^x}{x!}$$ ■

Remark 14.3

The answer to (c) can also be written as

$$\sum_{x=2001}^{\infty} \frac{e^{-1225}(1225)^x}{x!}$$

But this involves a summation of infinitely many terms and hence the previous form of the answer is better from the point of view of calculations.

Example 14.90

The number of accidents over a month on a particular intersection is 18. Assuming that the number of accidents on a particular day is Poisson, find the probability of more than 10 accidents occuring on a day.

SOLUTION

The average number of accidents per day is $\frac{18}{30} = 3/5$. Hence the number of accidents is a Poisson random variable with $\lambda = \frac{3}{5}$. So, the required probability is

$$P(X > 10) = 1 - \left(\sum_{x=0}^{10} \frac{e^{-3/5}(3/5)^x}{x!} \right) \quad \blacksquare$$

Example 14.91

Using the Poisson approximation of binomial distribution, find the probability that among a group of 750 people, nine of them would have the Indian Republic Day as their birthdays.

SOLUTION

The probability that an individual's birthday falls on a specific day is $\frac{1}{365}$. Thus, the number of people among 750, having the Indian Republic Day as their birthday is a binomial random variable with $n = 750$ and $p = \frac{1}{365}$. Hence the Poisson distribution that approximates this distribution has the parameter $\lambda = np = 750\frac{1}{365}$.

Therefore,

$$P(X = 9) = \frac{e^{-750/365}(750/365)^9}{9!} \quad \blacksquare$$

PRACTICE 14.38

Between the hours of 2 p.m and 4 p.m, the average number of phone calls per minute coming into the switchboard of a company is 2.5. Find the probability that during one particular minute there will be

(a) No calls.

(b) One call.

(c) Two calls.

(d) More than five calls.

In Sections 14.6 and 14.7, we discussed the mean and standard deviation of random variables. In the following proposition, we compute these quantities for the Poisson random variable.

Proposition 14.6

The mean and the standard deviation of the Poisson distribution are given by λ and $\sqrt{\lambda}$, respectively.

Proof

$$
\begin{aligned}
E(X) &= \sum_{x=0}^{\infty} x\frac{e^{-\lambda}\lambda^x}{x!} \\
&= e^{-\lambda}\sum_{x=1}^{\infty}\frac{\lambda^x}{(x-1)!} \\
&= \lambda e^{-\lambda}\sum_{x=1}^{\infty}\frac{(\lambda)^{x-1}}{(x-1)!} \\
&= \lambda e^{-\lambda}\sum_{x=0}^{\infty}\frac{\lambda^x}{x!} \\
&= \lambda e^{-\lambda}e^{\lambda} \\
&= \lambda \\
E(X^2) &= e^{-\lambda}\sum_{x=0}^{\infty}x^2\frac{\lambda^x}{x!} \\
&= \left(e^{-\lambda}\sum_{x=0}^{\infty}[x(x-1)+x]\frac{\lambda^x}{x!}\right) \\
&= \left(e^{-\lambda}\sum_{x=0}^{\infty}[x(x-1)]\frac{\lambda^x}{x!}\right)+\left(e^{-\lambda}\sum_{x=0}^{\infty}x\frac{\lambda^x}{x!}\right) \\
&= \left(e^{-\lambda}\sum_{x=2}^{\infty}\frac{\lambda^x}{(x-2)!}\right)+\lambda \\
&= \left(e^{-\lambda}\lambda^2\sum_{x=2}^{\infty}\frac{\lambda^{(x-2)}}{(x-2)!}\right)+\lambda \\
&= \left(e^{-\lambda}\lambda^2\sum_{x=0}^{\infty}\frac{\lambda^x}{x!}\right)+\lambda \\
&= e^{-\lambda}\lambda^2e^{\lambda}+\lambda \\
&= \lambda^2+\lambda \\
Var(X) = E(X^2)-[E(X)]^2 &= \lambda^2+\lambda-\lambda^2 = \lambda and hence \\
\sigma &= \sqrt{\lambda}
\end{aligned}
$$

■

SUMMARY

- The sample space S of a random experiment is the set of all its outcomes.
- An event is a subset of the sample space.
- Two events E_1 and E_2 are said to be mutually exclusive if $E_1 \cap E_2 = \phi$.
- E^c is the event that E does not occur.
- A random variable is a function $X : S \longrightarrow R$.
- $P(S) = 1$ and for event E, $0 \leq P(E) \leq 1$.
- If E_i is a sequence of mutually exclusive events, then $\sum_i P(E_i) = P(\bigcup_i E_i)$.
- $P(E_1 \cup E_2) = P(E_1) + P(E_2) - P(E_1 \cap E_2)$.
- $P(A|B)$ is the conditional probability that A occurs given that B has occurred.
- $P(A \cap B) = P(B)P(A|B)$.
- For independent events A and B, $P(A \cap B) = P(A)P(B)$.
- The total probability theorem asserts that if $\bigcup_i E_i = S$ and $\bigcap_i E_i = \phi S$ then $P(A) = \sum_i P(A|E_i)P(E_i)$.
- Bayes theorem asserts that $P(E_j|A) = \frac{P(A|E_j)P(E_j)}{\sum_j P(A|E_j)P(E_j)}$.
- The probability mass function of X is defined as $p(x) = P(X = x)$.
- The mean of a random variable is given by $E(X) = \sum xp(x)$.
- The variance $var(X)$ of a random variable is given by $Var(X) = E[(X - \overline{X})]^2$.
- The standard deviation of a random variable is given by $\sigma = \sqrt{Var(X)}$.
- A random variable is said to be a geometric random variable if its mass function is given by $P(X = x) = q^x p$.
- A random variable is said to be a negative binomial random variable if its mass function is given by

$$P(X = x) = {}^{x+r-1} C_{r-1}(1-p)^x p^{r-1}$$

$$p = {}^{x+r-1} C_{r-1}(1-p)^x p^r$$

- A random variable is said to be a Poisson random variable if its mass function is given by $P(X = x) = \frac{e^{-\lambda}\lambda^x}{x!}$, $x = 0, 1, 2, \ldots$

PROBLEMS

1. Let us consider a bag containing four red balls, two blue balls and one green ball. Suppose we draw four balls randomly from the bag. Write the sample space.
2. In the previous problem, write down the sample space if four balls are drawn one after another
   (a) with replacement,
   (b) without replacement.
3. From a standard deck of 52 cards, three cards are drawn randomly. Find the probability that there are two kings and a five of some suit.
4. From an urn containing three red balls, two blue balls and one green ball, two balls are drawn together. Find the probability that
   (a) they are of the same colour,
   (b) they are of different colours.
5. From a standard deck of 52 cards, the king of diamonds and the queen of hearts are missing. A card is drawn at random from this pack. If it is a 5, what is the probability that it is a heart?

6. In a university, there are 2525 engineering students and 3634 science students. Thirty percent of engineering students and 60% of science students are poor in mathemtics. An individual student was chosen at random and was found to be good in mathematics. Find the probability that this student is a science student.
7. Over a 10-minute period, a counter records an average of 1.3 gamma particles per millisecond coming from a radioactive substance. To a good approximation, the distribution of the count X of gamma particles during the next millisecond is Poisson. Find the probability that one or more gamma particles will be recorded during the next millisecond.
8. It has been claimed that in 60% of all solar heat installations, the utility bill is reduced by at least one-third. Accordingly, what is the probability that the utility bill will be reduced by at least one-third in
   (i) four of five installations,
   (ii) at least four of five installations.
9. An expert shot hits a target 95% of the time. What is the probability that the expert will miss the target for the first time on the fifteenth shot?
10. In a given city, 6% of all drivers get at least one parking ticket per year. Use the Poisson approximation to the binomial distribution to determine the probability that among 80 drivers:
    (a) four will get at least one parking ticket per year,
    (b) at least three will get at least one parking ticket in any given year,
    (c) anywhere from three to six, inclusive of three and six, will get at least one parking ticket in any given year.
11. A company designs swimming pools and its records show that the probability is 0.20 that one of its new pools will require repair within a year. What is the probability that the sixth pool it builds in a given year will be the first one to require repairs within a year?
12. Assume that the probability of an incorrect transmission through a digital communication channel is 0.1. Assuming that the transmissions are independent events, find the probability that the third error occurs at the 10 bit.
13. If a bank receives, on the average, six bad cheques per day, what is the probability that it will receive
    (a) four bad cheques on any given day;
    (b) 10 bad cheques over any two consecutive days.

    Assume that the distribution is Poisson.
14. A certain large city averages three murders per week and their occurrences follow a Poisson distribution. On an average, how many weeks in a year can this city expect to have no murders?
15. A scientist needs three diseased rabbits for an experiment. He has 20 rabbits available and inoculates them one at a time with a serum, quitting if and when he gets three positive reactions. If the probability is 0.25 that a rabbit can contract the disease from the serum, what is the probability that the scientist is able to get three diseased rabbits from 20?
16. In the inspection of tin plate produced by a continuous electrolytic process, 0.2 imperfection is spotted on an average per minute. Assuming Poisson distribution, find the probability of spotting:
    (a) one imperfection in 3 minutes,
    (b) at least two imperfections in 5 minutes,
    (c) at most one imperfection in 15 minutes.
17. A manufacturer produces IC chips, 1% of which are defective. Find the probability that in a box containing 100 chips, no defectives are found.

18. A student has taken a five-answer multiple choice examination orally. He continues to answer questions until he gets five correct answers. What is the probability that he gets them on or before the twenty-fifth question, if he guesses at each answer?
19. Assume that the arrival of automobiles at a gasoline station is Poisson and occurs at an average rate of 50 per hour. The station has only one gasoline pump. If automobiles are assumed to require 1 minute to obtain fuel, what is the probability that a queue will get lined at the pump?

FURTHER READING

Pitman, J. (1996), *Probability*, Narosa Publishing House, New Delhi.

Ross, S. M. (2000), *Introduction to Probability Models*, Academic Press, San Diego.

Praba, B., Arunachalam, P. S. and Sujatha, R. (2003), *Probability and Statistics*, Scitech Publications, Chennai.

Tirvedi, K. S. (2002), *Probability and Statistics with Reliability, Queuing and Computer Science Applications*, John Wiley and Sons, Inc., Singapore.

APPENDIX A

Mathematical Preliminaries For Special Distributions

The Sigma Notation: The symbol $\sum$, known as sigma, is used to denote the summation of a sequence of numbers. If $a_1, a_2, \ldots$ is a sequence of numbers then, for example, the summation $a_4 + a_5 + \cdots + a_{11}$ is denoted by

$$\sum_{i=4}^{11} a_i$$

As further examples, we have

$$\sum_{k=1}^{\infty} b_k = b_1 + b_2 + \cdots$$

$$\sum_{j=0}^{N} A_j = A_0 + A_1 + \cdots + A_N$$

Binomial Theorem: We have learnt the binomial theorem in Chapter 4, which states that for real numbers x, y and a positive integer n

$$(x+y)^n = \binom{n}{0} x^n + \binom{n}{1} x^{n-1} y + \cdots + \binom{n}{r} x^{n-r} y^r + \cdots + \binom{n}{0} y^n \quad \text{(A.1)}$$

The following formulae that are taught in high schools are also familiar to us:

Finite Geometric Series:

$$a + ar + ar^2 + \cdots + ar^n = \frac{a(r^{n+1} - 1)}{r - 1} \quad \text{(A.2)}$$

Infinite Geometric Series:

$$a + ar + ar^2 + \cdots + ar^{n-1} \ldots \quad \text{(A.3)}$$

If $|r| < 1$, then equation (A.3) is meaningful and its value is given by

$$\sum_{x=0}^{\infty} ar^x = \frac{a}{1-r} \tag{A.4}$$

As a special case, we have

$$\frac{1}{1-r} = 1 + r + r^2 + \cdots \tag{A.5}$$

Equation (A.4) is a special case of the binomial series:

$$(1+s)^n = 1 + ns + \frac{n(n-1)}{2}s^2 + \frac{n(n-1)(n-2)}{3!}s^3 + \cdots \tag{A.6}$$

Here n is not necessarily a positive integer and the above infinite summation is meaningful when $|s| < 1$.

In the special case where n is a positive integer, the product $n(n-1)(n-2)\ldots(n-(x-1)) = 0$ when $x \geq n-1$. This happens because when $x \geq n-1$, $(n-n)$ is a factor of this product. So, the infinite summation (A_4) reduces to that of a finite summation when n is a positive integer.

Let us now consider an important special case of equation (A.6). For $n = -2$, we obtain the following expansion:

$$\begin{aligned}(1-s)^{-2} = [1+(-s)]^{-2} &= 1 + (-2)(-s) + \frac{(-2)(-3)}{2!}(-s)^2 \\ &\quad + \frac{(-2)(-3)(-4)}{3!}(-s)^3 + \ldots \\ &= 1 + 2s + 3s^2 + 4s^3 \ldots\end{aligned}$$

Example A.1

Expand as a power series

$$\frac{1}{as+b}$$

SOLUTION

$$\begin{aligned}\frac{1}{as+b} &= \frac{1}{b\left(\frac{as}{b}+1\right)} \\ &= \frac{1}{b}\frac{1}{\left(\frac{as}{b}+1\right)} \\ &= \frac{1}{b}\frac{1}{\left(1-\left(\frac{-as}{b}\right)\right)} \\ &= \frac{1}{b}\left(1 - \frac{as}{b} + \left(\frac{as}{b}\right)^2 - \left(\frac{as}{b}\right)^3 \cdots\right)\end{aligned}$$

Finally, the reader can easily show that

$$(1-q)^{-3} = (1 + 3q + 6q^2 + \cdots) \quad \blacksquare$$

APPENDIX B

Trigonometry

Consider any right-angled triangle with one of the acute angles being θ. We define the trigonometric functions as follows:

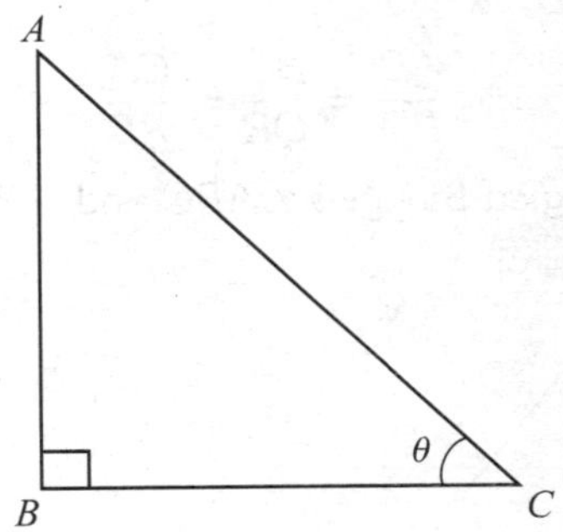

Figure B.1

$$\begin{aligned}
\sin\theta &= \frac{\text{Opposite side}}{\text{Hypotenuse}} = \frac{AB}{AC} \\
\cos\theta &= \frac{\text{Adjacent side}}{\text{Hypotenuse}} = \frac{BC}{AC} \\
\tan\theta &= \frac{\text{Opposite side}}{\text{Adjacent side}} = \frac{AB}{BC} \\
\csc\theta &= \frac{1}{\sin\theta} \\
\sec\theta &= \frac{1}{\cos\theta} \\
\cot\theta &= \frac{1}{\tan\theta}
\end{aligned}$$

Do the values of the trigonometric functions depend on the specific triangle we consider? No. We shall now explain this in detail. Let us recollect the idea of similarity of triangles, which we learnt in high school. If $\triangle ABC$ and $\triangle PQR$ are such that each angle of $\triangle ABC$ is equal to an angle of $\triangle PQR$, then these two triangles are said to be similar. For similar triangles, the three pairs of sides are in the same ratio. To be specific, let us assume that $\angle A = \angle P$, $\angle B = \angle Q$ and $\angle C = \angle R$.

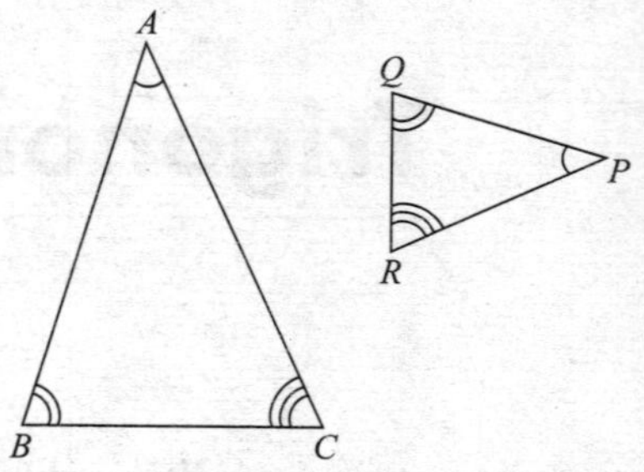

Figure B.2

Then we have

$$\frac{AB}{PQ} = \frac{BC}{QR} = \frac{CA}{RP}$$

Now consider two right-angled triangles $\triangle ABC$ and $\triangle PQR$ with $\angle B = \angle Q = 90°$. Let $\angle C = \angle R$.

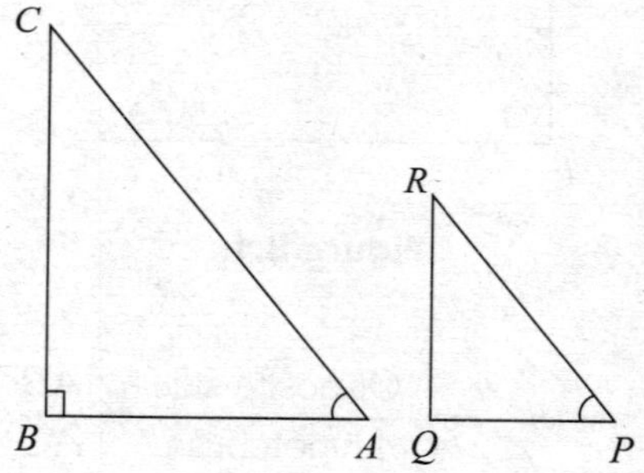

Figure B.3

Under these conditions, we obtain $\angle A = \angle P$, and hence the two triangles are similar. So, we have

$$\frac{AB}{AC} = \frac{PQ}{PR}$$

$$\frac{BC}{AC} = \frac{QR}{PR}$$

$$\frac{AB}{BC} = \frac{PQ}{QR}$$

These equalities show that the values of trigonometric functions do not depend on the specific triangle under consideration.

Let us observe that the above definitions cannot be used to define sin $120°$ as there cannot be any right-angled triangle with one angle having the measure $120°$. We first show that there is a need to define the values of trigonometric functions for angles greater than $90°$. Next, we actually define them.

Projection Formula: The idea of *Projection* is motivated by shadows of an object falling on a surface. When sunlight is falling on Earth vertically, if we hold a stick horizontally then its shadow on the Earth will have the same length as that of the stick.

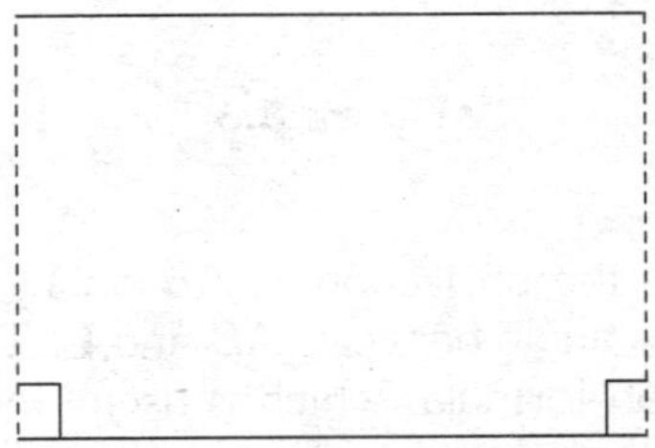

Figure B.4

Now, if we hold the stick at an angle, then the length of the shadow will no more be equal to that of the stick. So, we ask ourselves what would be the length of the shadow. Trigonometry gives us the answer.

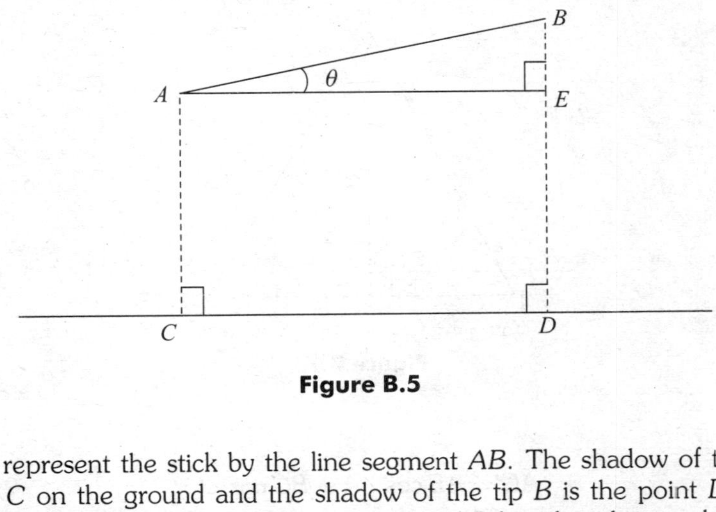

Figure B.5

Let us represent the stick by the line segment AB. The shadow of the tip A is the point C on the ground and the shadow of the tip B is the point D. Now we want to compute the length CD. Let us join B and D by a line that would be vertical and draw a horizontal line from A to meet BD at E. Since the figure $AEDC$ is a rectangle, $AE = DC$. If we denote the angle $\angle BAE$ by θ, then $\cos\theta = \frac{AE}{AB}$ and hence $CD = AE = AB\cos\theta$.

The projection of a line segment AB on a line $\mathbf{L}$ is defined as follows:

Let C and D be the feet of perpendiculars drawn, respectively, from A and B to the line $\mathbf{L}$.

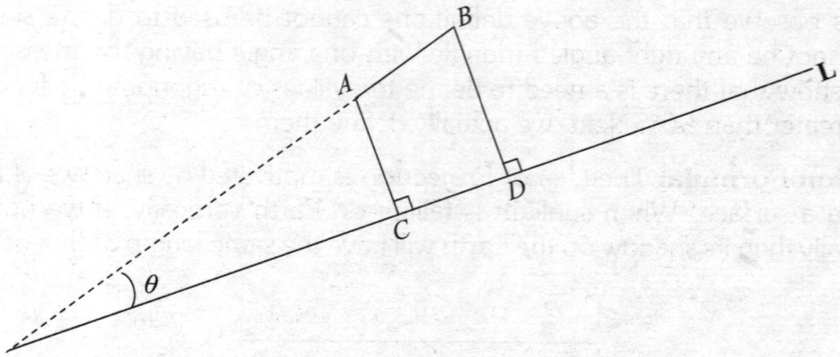

Figure B.6

The line segment CD is the projection of AB on **L**. The length of CD is given by $AB\cos\theta$, where θ is the angle between AB and **L**. The following theorem is an application of the projection formula, which is useful in computing the length of a side of a triangle if two sides and two angles are given. The reader can easily prove it using the idea of the projection and Figure B.7.

Theorem B.1

Let $\triangle ABC$ be an acute-angled triangle.

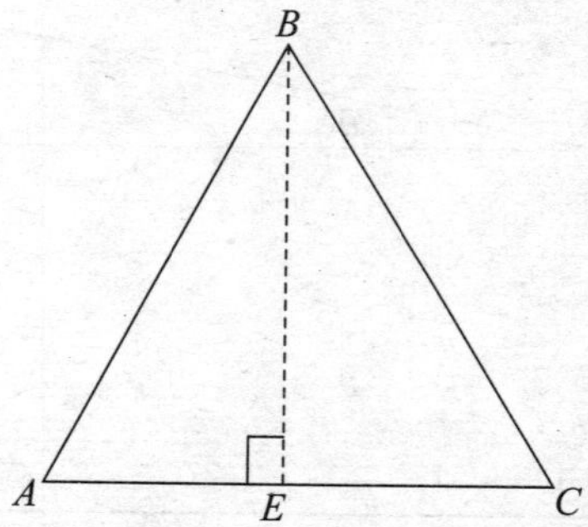

Figure B.7

Then

$$AC = AB\cos\angle A + BC\cos\angle C \tag{B.1}$$

Let us now consider a triangle $\triangle ABC$ with $\angle A$ being an obtuse angle. In this case, equation (B.1) is not meaningful since we have not yet defined $\cos\angle A$ when $\angle A > 90°$. It is not difficult to get the following formula:

$$AC = -AB\cos(180 - A) + BC\cos\angle C \tag{B.2}$$

Equation (B.2) follows from Figure B.8 and the fact that $AC = EC - EA$.

A comparison of equations (B.1) and (B.2) suggests that when $\angle A$ is obtuse, if we define $\cos A = -\cos(180 - A)$, then we do not need the two different formulae (B.1) and (B.2). Equation (B.1) covers both the cases.

Before we define the trigonometric functions for any angle θ, which is not necessarily acute, we recall the radian measure of an angle. Consider a circle of radius 1.

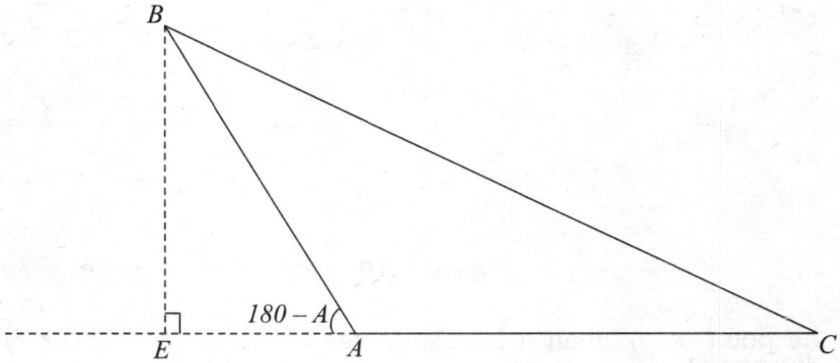

Figure B.8

The length of the arc subtending an angle $\theta°$ is given by $\frac{\theta}{360}2\pi$.

In particular, the length of the semicircle that subtends an angle 180° at the center is π. The radian unit for an angle is defined in such a way that the radian measure of an angle is numerically equal to the corresponding arc length in a circle of radius 1. In other words,

$$180 \text{ degrees} = \pi \text{ radians} \tag{B.3}$$

Thus, we have $\pi/2$ radians $= 90°$, $\frac{2\pi}{3}$ radians $= 120°$, etc.

From now on, it is understood that we use only radian as the unit for measuring angles and that we explicitly omit mentioning the word radian.

We make use of the xy-plane to generalize the definitions of trigonometric functions. Consider a line segment OP of unit length, one of whose end is fixed at the origin O. Initially, it is along the positive half of the x-axis. Hence, P coincides with the point $(1, 0)$.

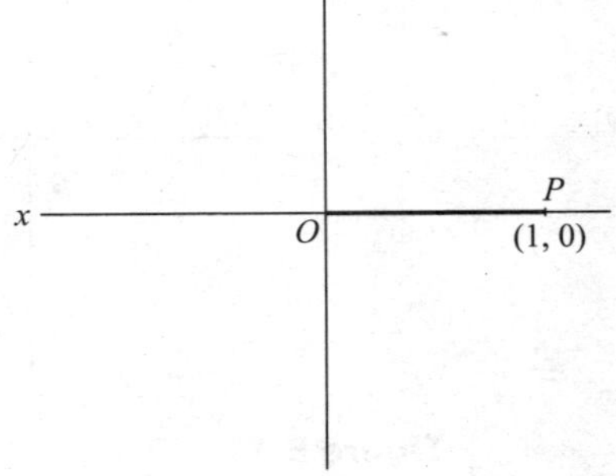

Figure B.9

Imagine that it is rotating about the origin in an anti-clockwise manner. Consider the instant when it makes an angle of $\pi/6$ with the x-axis.

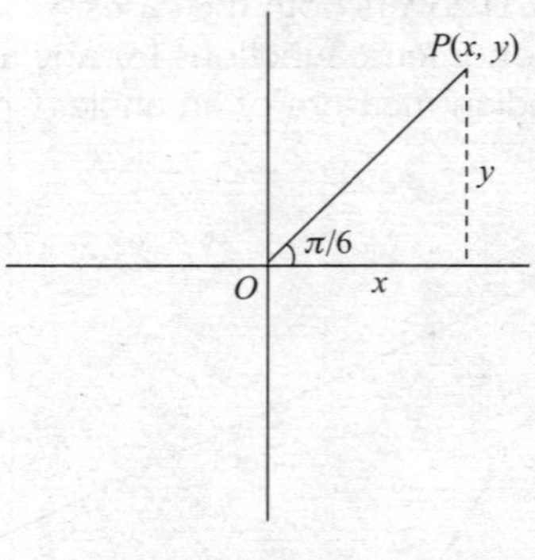

Figure B.10

If P is the point (x, y), then it is easy to see that $x = \frac{\sqrt{3}}{2} = \cos\pi/6$ and $y = \sin\pi/6$. This follows because $OP = 1$ and x is nothing but the projection of OP on the x-axis and hence, $x = OP\cos\pi/6$. $y = \sin\pi/6$ follows from the use of the Pythagoras theorem.

In fact, if $0 < \theta < \pi/2$ and if OP makes an angle θ with the x-axis, then $x = \cos\theta$ and $y = \sin\theta$.

Now, we do not restrict ourselves to the case $0 < \theta < \pi/2$ anymore.

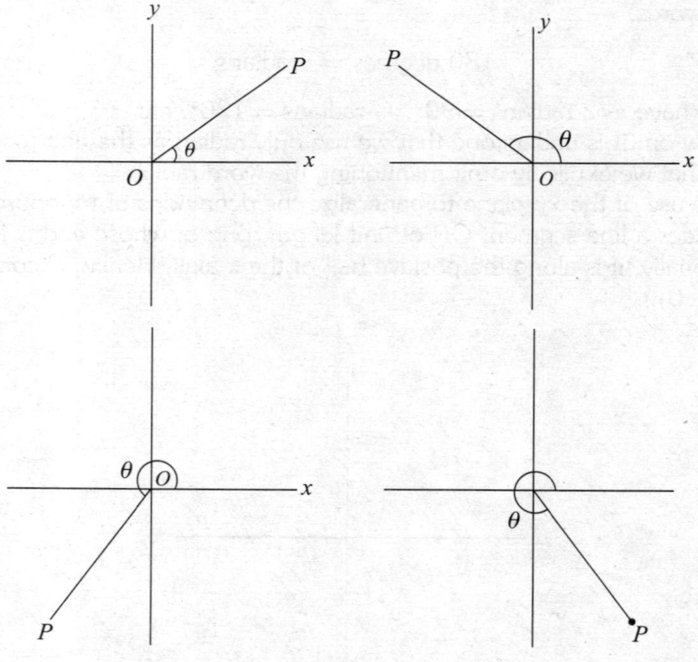

Figure B.11

Let x and y be the co-ordinates of P when OP makes an angle θ with the x-axis. We now *define* $\sin\theta = y$ and $\cos\theta = x$.

First, let is find the values $\sin\theta$ and $\cos\theta$ for $\theta = 0, \pi/2, \pi, 3\pi/2$ and 2π.

As we have already seen, when OP makes an angle 0 with the x-axis, P coincides with the point $(1, 0)$.

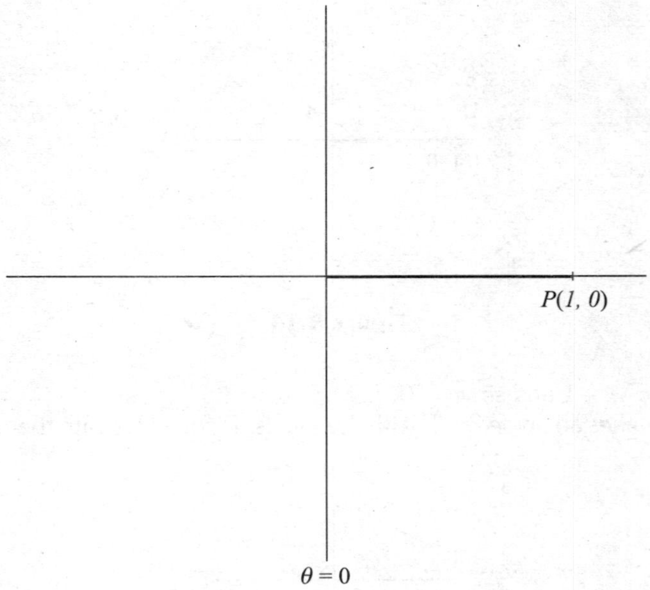

Figure B.12

Hence, $\cos 0 = 1$ and $\sin 0 = 0$. When OP makes an angle $\pi/2$ with the x-axis, P coincides with the point $(0, 1)$.

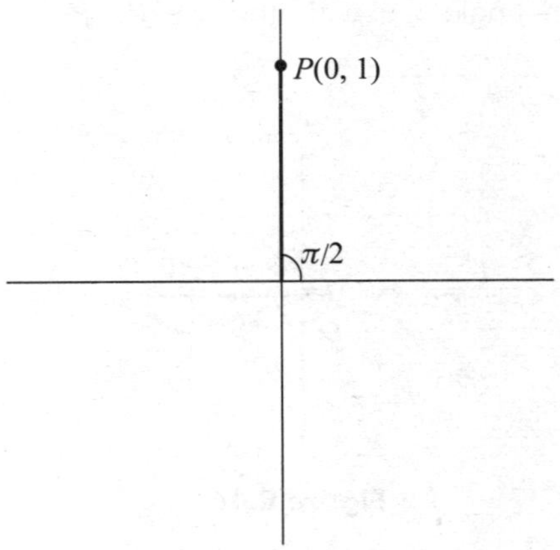

Figure B.13

Hence, $\cos \pi/2 = 0$ and $\sin \pi/2 = 1$. In high school, we learn that $\sin \pi/2 = 1$, since in a right-angled triangle, the hypotenuse is the opposite side of the right angle and hence opposite side/hypotenuse= 1. The explanation for $\cos \pi/2 = 1$ is usually avoided at the high school level. But there is no confusion under the new definition.

When OP makes an angle π with the x-axis, P coincides with the point $(-1, 0)$.

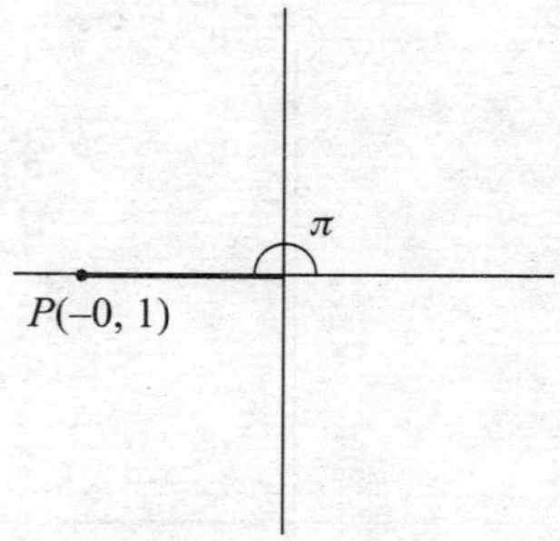

Figure B.14

Hence $\cos \pi = -1$ and $\sin \pi = 0$.

When OP makes an angle $3\pi/2$ with the x-axis, P coincides with the point $(0, -1)$.

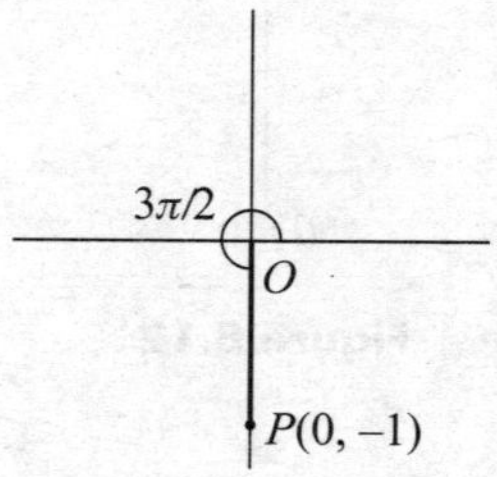

Figure B.15

Hence $\cos 3\pi/2 = 0$ and $\sin 3\pi/2 = -1$.

When OP makes an angle 2π with the x-axis, P *again* coincides with the point $(1, 0)$.

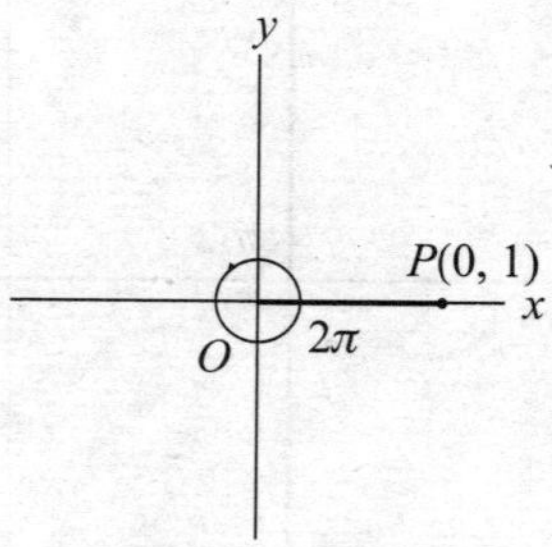

Figure B.16

Hence $\cos 2\pi = 1$ and $\sin 2\pi = 0$.

Now that the rotating line segment OP has come back to its original position, should we stop it? Not if we want to define $\sin\theta$ and $\cos\theta$ for values of θ greater than 2π. If $\theta > 2\pi$, it means that OP has completed at least one full revolution.

Let OP be a line segment of unit length, which makes an angle θ with the x-axis and let OQ be a line segment of unit length, which makes an angle $6 + 2\pi$ with the x-axis. It is clear that OP and OQ coincide with each other. Hence, P and Q have the same co-ordinates. Thus, we have

$$\sin(\theta + 2\pi) = \sin\theta$$

and

$$\cos(\theta + 2\pi) = \cos\theta$$

Since $OP = 1 = x^2 + y^2$, it follows that $\sin^2\theta + \cos^2\theta = 1$.

Let us recall that the x- and y-axes divide the plane into four quadrants.

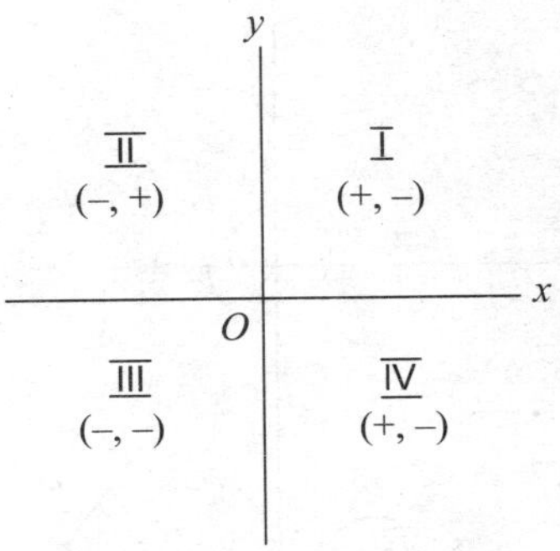

Figure B.17

The first quadrant is the set of all points whose x and y co-ordinates are non-negative. The second quadrant is the set of all points whose x co-ordinates are non-positive and y co-ordinates are non-negative. The third quadrant is the set of all points whose x and y co-ordinates are non-positive

The fourth quadrant is the set of all points whose x co-ordinates are non-negative and y co-ordinates are non-positive.

If θ is the angle made by a line segment OP with the x-axis, the signs of the co-ordinates of P are tabulated as follows:

	x	y
$0 \le \theta \le \pi/2$	+	+
$\pi/2 \le \theta \le \pi$	−	+
$\pi \le \theta \le 3\pi/2$	−	−
$3\pi/2 \le 2\pi$	+	−

So far, we have considered only non-negative values of θ. A negative angle is produced by a line segment moving in the *clockwise* direction. For example, the angle $-\pi/4$ is shown in the following figure:

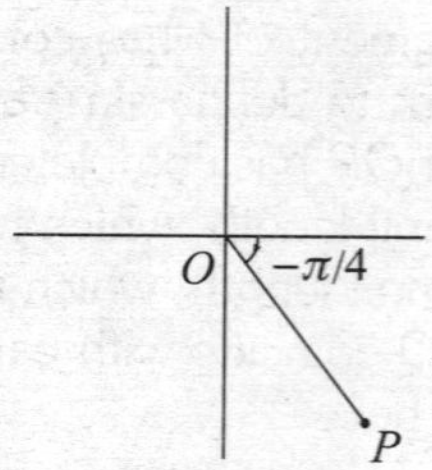

Figure B.18

It is clear that a line segment, which has moved clockwise by θ, coincides with a line segment that has moved $(2\pi - \theta)$ in the anti-clockwise direction.

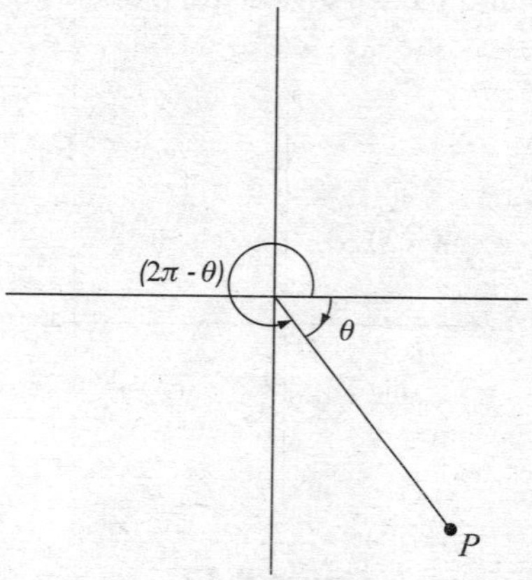

Figure B.19

Let θ be a negative number and consider the sequence of numbers $\theta + 2\pi$, $\theta + 3\pi$,..., etc. It is clear that this sequence will produce a positive number $\theta + 2k\pi$ for a sufficiently large positive integer k. Now, we define $\sin\theta = \sin(\theta + 2k\pi)$ and $\cos\theta = \cos(\theta + 2k\pi)$. This completes the definitions of sin and cos for any real number θ.

If OP is a line segment of unit length and $P = (x, y)$, then by the distance formula, we have $x^2 + y^2 = 1$. Hence, $|x| \leq 1$ and $|y| \leq 1$. Thus, we have the important inequalities

$$|\cos\theta| \leq 1$$
$$|\sin\theta| \leq 1$$

These inequalities can also be written as

$$-1 \leq \cos\theta \leq 1$$
$$-1 \leq \sin\theta \leq 1$$

Let n be an integer. Since a line segment making an angle $n\pi$ with the positive direction of the x-axis lies along the y-axis, $\sin n\pi = 0$. Further, if θ is such that

$\sin\theta = 0$ and OP is a line segment making an angle θ with the x-axis, then the y co-ordinate of P is zero and hence OP must lie along the x-axis. Hence $\theta = n\pi$ for some integer n. We have proved that

$$\sin\theta = 0 \leftrightarrow \theta = n\pi \qquad \text{for some integer } n \tag{B.4}$$

Let us note that a line segment making an angle, which is an odd multiple of $\pi/2$, with the x-axis lies along the y-axis.

This shows that

$$\cos\theta = 0 \leftrightarrow \theta = (2n+1)\pi/2 \qquad \text{for some integer } n \tag{B.5}$$

Noting

$$\begin{aligned} \cos\pi &= -1 \\ \cos 2\pi &= 1 \\ \cos 3\pi &= -1 \\ \cos 4\pi &= 1 \end{aligned}$$

we can easily conclude that

$$\cos n\pi = (-1)^n$$

Similarly, the reader can prove that

$$\sin(2n+1)\pi/2 = (-1)^n$$

Next, we prove

$$\sin\left(\frac{\pi}{2}+\theta\right) = \cos\theta \tag{B.6}$$

$$\cos\left(\frac{\pi}{2}+\theta\right) = -\sin\theta \tag{B.7}$$

Any arbitrary θ can be written as $\theta = \phi + 2k\pi$, where $0 \le \phi \le 2\pi$, and k is an integer. So, it is enough to prove equation (B.7) for $0 \le \theta \le 2\pi$.

Let OP and OQ be line segments, respectively, making an angle of θ and $\frac{\pi}{2}+\theta$ with the x-axis.

Let P be (x, y) and Q be (α, β). We have

$$\begin{aligned} \cos\theta &= x \\ \sin\theta &= y \\ \cos\left(\theta+\frac{\pi}{2}\right) &= \alpha \\ \sin\left(\theta+\frac{\pi}{2}\right) &= \beta \end{aligned}$$

Our aim now is to express α and β in terms of x and y. From Figure B.20, it is clear that $|y| = |\alpha|$ and $|x| = |\beta|$.

Since Q is in the second quadrant, $\alpha \le 0$ and $\beta \ge 0$. Thus, we must have $\alpha = -y$ and $\beta = x$. We have proved equations (B.6) and (B.7) for $0 \le \theta \le \pi/2$. The other

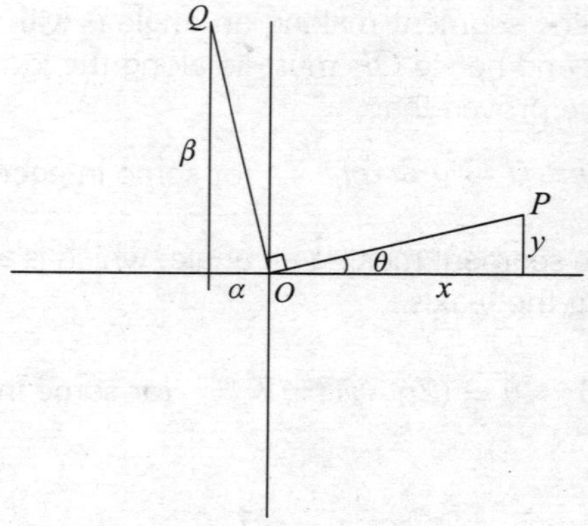

Figure B.20

three cases of θ being in the remaining three quadrants can easily be proved by the reader on the same lines.

The following figure gives us the proof of $\cos(-\theta) = \cos\theta$ and $\sin(-\theta) = -\sin\theta$ for $0 < \theta < \pi/2$ and for $-\pi/2 < \theta < 0$. Let us note that

$$0 < \theta < \pi/2 \leftrightarrow -\pi/2 < -\theta < 0$$

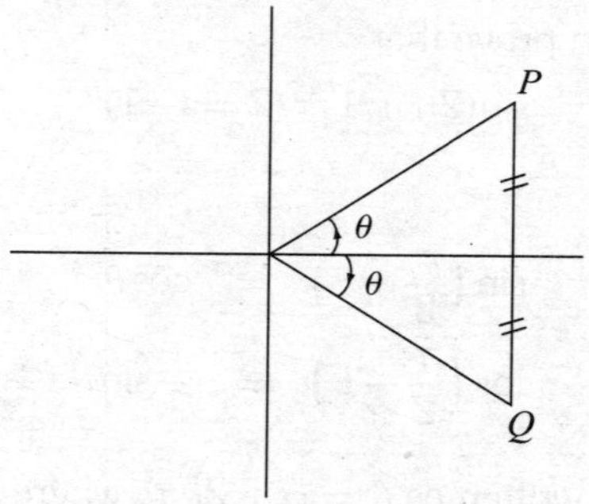

Figure B.21

Now $\sin(\pi + \theta) = \sin(\pi/2 + \pi/2 + \theta) = \cos(\pi/2 + \theta) = -\sin\theta$.

The reader can easily prove the following identities:

(i) $\sin(3\pi/2 + \theta) = -\cos\theta \cos(3\pi/2 + \theta) = \sin\theta$.

Next, we shall indicate the proof of the identities

(i) $\cos(A - B) = \cos A \cos B + \sin A \sin B$.

(ii) $\cos(A + B) = \cos A \cos B - \sin A \sin B$.

(iii) $\sin(A - B) = \sin A \cos B - \cos A \sin B$.

(iv) $\sin(A + B) = \sin A \cos B + \cos A \sin B$.

It is enough to prove (i), and the others can be proved with the help of the results such as $\sin(\pi/2 - \theta) = \cos\theta$ and by writing $A + B = A - (-B)$.

Since $\cos(B - A) = -\cos(A - B)$, without loss of generality, let us assume that the measure of B is smaller than the measure of A.

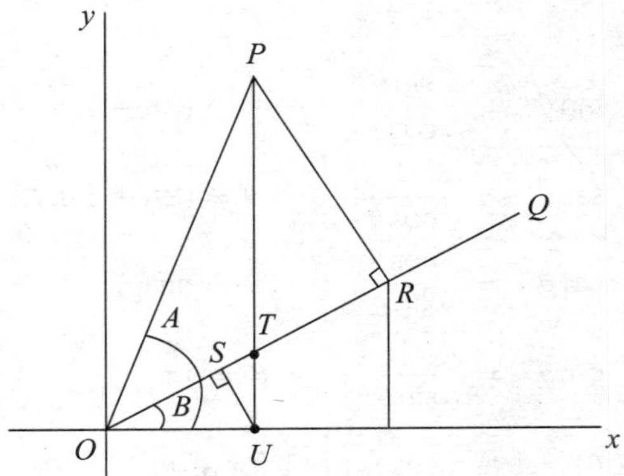

Figure B.22

In Figure B.22, OP and OQ are line segments of length one unit, respectively, making angles of A and B with the x-axis. We drop perpendiculars PR to OQ and PU to the x-axis. It is clear that $PU = \sin A$ and $OU = \cos A$.

OR is the projection of OP on OQ and the angle between OP and OQ is $A - B$. Hence

$$OR = OP\cos(B - A) = \cos(B - A)$$

OS is the projection of OU on OQ and hence

$$OS = OU\cos B = \cos A\cos B$$

$\angle OTU = (\pi/2) - B$ and ST is the projection of TU on OQ and hence

$$ST = TU\cos(\pi/2 - B) = TU\sin B$$

Similarly,

$$TR = PT\sin B$$

Now,

$$\begin{aligned}\cos(A - B) &= OR = OS + ST + TR\\ &= \cos A\cos B + TU\sin B + PT\sin B\\ &= \cos A\cos B + (TU + PT)\sin B\\ &= \cos A\cos B + PU\sin B\\ &= \cos A\cos B + \sin A\sin B\end{aligned}$$

We have considered the case when both A and B are acute angles and the measure of B is smaller than the measure of A. Other cases can be handled by writing $A = n\pi/2 + \theta$ and $B = m\pi/2 + \phi$ for some integers m, n and non-negative angles θ and ϕ whose measure is less than or equal to $\pi/2$.

Other trigonometric functions tan, cot, sec and csc are defined as follows:

$$
\begin{aligned}
\tan\theta &= \frac{\sin\theta}{\cos\theta}, & \theta &\neq (2n+1)\pi/2 \\
\sec\theta &= \frac{1}{\cos\theta}, & \theta &\neq (2n+1)\pi/2 \\
\cot\theta &= \frac{\cos\theta}{\sin\theta}, & \theta &\neq n\pi \\
\csc\theta &= \frac{1}{\sin\theta}, & \theta &\neq n\pi
\end{aligned}
$$

It is clear that $|\sec\theta| \geq 1$ and $|\csc\theta| \geq 1$ for all θ.

The reader can now easily see that we cannot find angles θ for which $\sin\theta = -2$ or $\cos\theta = 3$ or $\sec\theta = 0.7$ or $\csc\theta = -0.1$, etc.

For every real number r, we can find an angle θ (whose measure is a real number) such that $\tan\theta = r$. In fact, infinitely many values of θ exist. A similar statement is valid for $\cot\theta$.

APPENDIX

C

Matrices

Informally, a matrix is a rectangular array of numbers such as

$$\begin{pmatrix} 3 & 4 \\ 5 & -1 \\ 0 & 9 \end{pmatrix} \quad \begin{pmatrix} 13 & 0.74 & 0 \\ 5 & -1 & 7 \end{pmatrix} \quad \begin{pmatrix} 2.5 & 7 & 7 \\ 5 & -1 & 5 \\ 8.1 & \pi & 1 \end{pmatrix}$$

Every horizontal sequence of numbers in the matrix is known as a row, and every vertical sequence of the matrix is known as a column. For example, $(2.5\ 7\ 7)$ is a row of the third matrix above and $\begin{pmatrix} 3 \\ 5 \\ 0 \end{pmatrix}$ is a column of the first matrix. Let us note that the second matrix has two rows and three columns. It is an example of 2×3 matrix. An $m \times n$ matrix is a rectangular array of numbers with m rows and n columns. The entry in the ith row and the jth column is denoted by double-subscripted symbols like a_{ij}. The notation $A = (a_{ij})$ denotes the fact that the entry in the ith row and the jth coloumn of the matrix A is a_{ij}. Now, let us look at some examples of special types of matrices. $\begin{pmatrix} 8 \\ 7 \\ 9 \\ -5 \end{pmatrix}, \begin{pmatrix} 3 \\ 5 \\ 0 \end{pmatrix}$ are examples of column matrices and $(\ 6 \quad -3 \quad 0.7 \quad 2 \quad 9\ )$ is an example of a row matrix. $\begin{pmatrix} 7 & 9 & 0 & 1 \\ 6 & 6 & -4 & 4 \\ 4 & 2 & 4 & 1 \\ 0 & 2 & 2 & 0 \end{pmatrix}$ is a square matrix.

Definition C.1: An $m \times 1$ matrix is said to be column matrix and a $1 \times n$ matrix is called a row matrix. A column matrix is also called a column vector and similarly a row matrix is called a row vector. An $n \times n$ matrix is called a square matrix.

If A is a square matrix, then the entries a_{11}, a_{22},...,a_{nn} are called the diagonal entries. A matrix is said to be a diagonal matrix if $a_{ij} = 0$ for $i \neq j$. In other words, every non-diagonal entry of a diagonal matrix is zero. Some examples of diagonal matrices are given below:

$$\begin{pmatrix} 6.4 & 0 \\ 0 & -5 \end{pmatrix} \quad \begin{pmatrix} 15 & 0 & 0 \\ 0 & 8 & 0 \\ 0 & 0 & 47 \end{pmatrix} \quad \begin{pmatrix} 0 & 0 & 0 \\ 0 & 26 & 0 \\ 0 & 0 & 50 \end{pmatrix} \quad \begin{pmatrix} a & 0 & 0 & 0 \\ 0 & a & 0 & 0 \\ 0 & 0 & a & 0 \\ 0 & 0 & 0 & a \end{pmatrix}$$

The last example above is a diagonal matrix whose diagonal entries are all equal to each other. Such a matrix is called a scalar matrix.

Addition of Matrices: When we want to add two matrices A and B, they must have the same order. We do not define addition for matrices having different orders.

We can add $A = \begin{pmatrix} 7 & 9 & 0 & 1 \\ 6 & 6 & -4 & 4 \\ 4 & 2 & 4 & 1 \end{pmatrix}$ and $B = \begin{pmatrix} 0 & -3 & 7 & 11 \\ 0.7 & \pi & 3 & 0 \\ 0 & 0 & 0 & 0 \end{pmatrix}$.

But we cannot add $A = \begin{pmatrix} 7 & 9 & 0 & 1 \\ 6 & 6 & -4 & 4 \\ 4 & 2 & 4 & 1 \end{pmatrix}$ and $C = \begin{pmatrix} 7 & 9 & 0 & 1 \\ 6 & 6 & -4 & 4 \end{pmatrix}$.

Now,

$$A+B = \begin{pmatrix} 7+0 & 9+(-3) & 0+7 & 1+11 \\ 6+0.7 & 6+\pi & -4+3 & 4+0 \\ 4+0 & 2+0 & 4+0 & 1+0 \end{pmatrix}$$

$$= \begin{pmatrix} 7 & 6 & 7 & 12 \\ 6.7 & 6+\pi & -1 & 4 \\ 4 & 2 & 4 & 1 \end{pmatrix}$$

Thus, if $A = a_{ij}$ and $B = (b_{ij})$ are both $m \times n$ matrices, then the $m \times n$ matrix $C = A + B$ is defined as

$$C = A + B = (a_{ij} + bij) \tag{C.1}$$

Subtraction of Matrices: The reader can easily modify equation (C.1) as

$$A - B = (a_{i,j} - bij)$$

We can also define subtraction in an equivalent but different procedure. First, we define $-A$ for a matrix A called the negative of A.

If

$$A = \begin{pmatrix} 7 & 9 & 0 & 1 \\ 69 & 6 & -4 & 14 \\ 2.4 & -2 & 4 & -1 \end{pmatrix}$$

$$-A = \begin{pmatrix} -7 & -9 & 0 & -1 \\ -69 & -6 & 4 & -14 \\ -2.4 & 2 & -4 & 1 \end{pmatrix}$$

Definition C.2: If $A = (a_{ij})$, then $-A = (-a_{ij})$. If A and B are matrices of the same order, then $A - B$ is defined as $A + (-B)$.

The reader can easily check that $A + (-A)$ is a matrix having the same order as that of A and all of whose entries are zeros.

After learning the definition of matrix addition, it is natural to define matrix multiplication for matrices having the same order by taking entry-wise product. For example, if $A = \begin{pmatrix} 6 & 9 \\ 1 & 10 \end{pmatrix}$ and $B = \begin{pmatrix} 1 & 2 \\ 5 & 0 \end{pmatrix}$ then one may define AB as

$$\begin{pmatrix} 6 \times 1 & 9 \times 2 \\ 1 \times 5 & 10 \times 0 \end{pmatrix} = \begin{pmatrix} 6 & 18 \\ 5 & 0 \end{pmatrix}$$

But this definition has not been found to be of much use to mathematicians, and matrix multiplication is defined in a manner that appears to be peculiar, when one learns it for the first time. However, it is useful.

Definition C.3: Let $A = (a_{ij})$ be a matrix of order $m \times n$ and $B = (b_{jk})$ be a matrix of order $n \times l$. Then AB is a matrix of order $m \times l$, which is defined as follows:

$$AB = \left(\sum_{j=1}^{n} a_{ij} b_{jk} \right)$$

Consider the system of equations

$$\begin{aligned} ax + by &= p \\ cx + dy &= r \end{aligned} \tag{C.2}$$

Now this system can be written in the matrix form as follows:

Let $A = \begin{pmatrix} a & b \\ c & d \end{pmatrix}$, $X = \begin{pmatrix} x \\ y \end{pmatrix}$, $B = \begin{pmatrix} p \\ r \end{pmatrix}$. The system (C.2) now takes the form

$$AX = B \tag{C.3}$$

Consider the system of m equations in n unknowns

$$\begin{aligned} a_{11}x_1 + a_{12}x_2 + \ldots a_{1n}x_n &= b_1 \\ a_{21}x_1 + a_{22}x_2 + \ldots a_{2n}x_n &= b_2 \\ \ldots &= \\ b_{m1}x_1 + a_{m2}x_2 + \ldots a_{mn}x_n &= b_m \end{aligned} \tag{C.4}$$

With $A = (a_{ij})$, $B = \begin{pmatrix} b_1 \\ b_2 \\ \cdots \\ b_m \end{pmatrix}$ and $X = \begin{pmatrix} x_1 \\ x_2 \\ \cdots \\ x_n \end{pmatrix}$, the system (C.4) takes the form $AX = B$.

Let us note that even if AB is defined, BA need not be defined. For example, if A is a 2×3 matrix and B is a 3×4 matrix, AB is defined but BA is not defined. Further, even when both AB and BA are defined, they may not have the same order. For example, if A is a 1×3 matrix and B is a 3×1 matrix, then AB is a 1×1 matrix but BA is a 3×3 matrix. Consider two square matrices A and B of the same order. Now both AB and BA are defined and have the same order. But even here it is not necessary that $AB = BA$. The reader can easily find examples of 2×2 matrices A and B such that $AB \neq BA$.

Two $n \times n$ matrices A and B are said to commute if $AB = BA$.

C.1 Determinants

Consider the system of equations

$$\begin{aligned} ax + by &= p \\ cx + dy &= q \end{aligned} \tag{C.5}$$

Using the method of elimination learnt in high school, we can easily show that the above system has the solution

$$x = \frac{pd - qb}{ad - bc}$$
$$y = \frac{aq - cp}{ad - bc}$$

Here, we must naturally assume that $ad - bc \neq 0$. The quantity $ad - bc$ is called the determinant of the matrix $\begin{pmatrix} a & b \\ c & d \end{pmatrix}$. Determinants are denoted by writing the entries of the corresponding matrix between two vertical bars. Specifically, if A=$\begin{pmatrix} a & b \\ c & d \end{pmatrix}$, then

$$|A| = \left|\begin{pmatrix} a & b \\ c & d \end{pmatrix}\right| = \begin{vmatrix} a & b \\ c & d \end{vmatrix} = ad - bc$$

Using the notation for the determinants, we can write down the solutions of the system (C.5) in the following manner:

$$x = \frac{\begin{vmatrix} p & b \\ q & d \end{vmatrix}}{\begin{vmatrix} a & b \\ c & d \end{vmatrix}}, \quad y = \frac{\begin{vmatrix} a & p \\ c & q \end{vmatrix}}{\begin{vmatrix} a & b \\ c & d \end{vmatrix}}$$

We define the determinant of a 3×3 matrix as follows: Let $A = \begin{pmatrix} a & b & c \\ p & q & r \\ l & m & n \end{pmatrix}$.

Then the determinant of A is defined as

$$|A| = \begin{vmatrix} a & b & c \\ p & q & r \\ l & m & n \end{vmatrix} = a\begin{vmatrix} q & r \\ m & n \end{vmatrix} - b\begin{vmatrix} p & r \\ l & n \end{vmatrix} + c\begin{vmatrix} p & q \\ l & m \end{vmatrix}$$

Let us now describe how the determinant of an $n \times n$ matrix is defined. D_j is the determinant of the $(n-1) \times (n-1)$ matrix obtained by omitting the jth row and the jth column of A. Then

$$|A| = \sum +j = 1^n (-1)^{j+1} a_{1j} D_j$$

We have already seen the importance of determinants in the solution of a system of equations. The real significance of the concept of a determinant is related to the idea of inverse of a matrix, which is described below.

We can easily check that

$$\begin{pmatrix} 1 & 0 \\ 0 & 1 \end{pmatrix}\begin{pmatrix} a & b \\ c & d \end{pmatrix} = \begin{pmatrix} a & b \\ c & d \end{pmatrix}\begin{pmatrix} 1 & 0 \\ 0 & 1 \end{pmatrix} = \begin{pmatrix} a & b \\ c & d \end{pmatrix}$$

$$\begin{pmatrix} 1 & 0 & 0 \\ 0 & 1 & 0 \\ 0 & 0 & 1 \end{pmatrix}\begin{pmatrix} a & b & c \\ p & q & r \\ l & m & n \end{pmatrix} = \begin{pmatrix} a & b & c \\ p & q & r \\ l & m & n \end{pmatrix}\begin{pmatrix} 1 & 0 & 0 \\ 0 & 1 & 0 \\ 0 & 0 & 1 \end{pmatrix}$$
$$= \begin{pmatrix} a & b & c \\ p & q & r \\ l & m & n \end{pmatrix}$$

In fact, if I_n is the $n \times n$ matrix whose diagonal entries are 1 and the rest of the entries are zero, then for any $n \times n$ matrix,

$$I_n A = A I_n = A$$

I_n is called the identity matrix of order n.

The role of I_n in the matrix multiplication is similar to that of 1 in the multiplication of real numbers.

Naturally, we would like to know if there is a notion analogous to the reciprocal of a real number.

Let A be a square matrix. A square matrix B of the same order is said to be the inverse of A if

$$AB = I = BA$$

where I is the identity matrix of the same order. A square matrix A with $|A| \neq 0$ is called a non-singular matrix. A non-singular matrix is also called an invertible matrix in view of the following state important results:

(i) A square matrix A has an inverse if and only if $|A| \neq 0$.

(ii) Let A be an $n \times n$ matrix and B be an $n \times 1$ column vector. The system of equations $Ax = b$ has a unique solution for every B if and only if $|A| \neq 0$. The solution is given by $x = A^{-1}b$, where A^{-1} is the inverse of A.

Informally, we may think of A^{-1} as the reciprocal of A. Just as the real number 0 does not have a reciprocal, square matrices A with $|A| = 0$ do not have inverses. Now, we shall describe the procedure for finding A^{-1} when $|A| \neq 0$.

Let A be an $n \times n$ matrix. C_{ij} is the $(n-1) \times (n-1)$ matrix, which is obtained by omitting the ith row and the jth column of A. We define $c_{ij} = \frac{(-1)^{i+j}|C_{ji}|}{|A|}$. Note the change in the order of occurrence of i and j. A^{-1} is the matrix whose entries are given by c_{ij}.

For a 2×2 non-singular matrix, we can easily write down the inverse. If $A = \begin{pmatrix} a & b \\ c & d \end{pmatrix}$ with $ad - bc \neq 0$,

$$A^{-1} = \begin{pmatrix} \frac{d}{ad-bc} & \frac{-b}{ad-bc} \\ \frac{-c}{ad-bc} & \frac{a}{ad-bc} \end{pmatrix}$$

The set of all $n \times n$ non-singular matrices whose entries are real numbers is denoted by $GL_n(R)$. Among the members of $GL_n(R)$, we have matrices with $|A| = 1$. The set of all such matrices is denoted by $SL_n(R)$.

If A and B are non-singular, is AB non-singular? In other words, if $A, B \in GL_n(R)$, is $AB \in GL_n(R)$? The answer is yes, and it follows from the following result on determinants:

$$|AB| = |A||B|$$

If $A, B \in GL_n(R)$, then $|A| \neq 0$ and $|B| \neq 0$ and hence $|AB| = |A|B| \neq 0$. So, $AB \in GL_n(R)$.

The reader can easily prove that if $A, B \in SL_n(R)$, then $AB \in SL_n(R)$.

List of Symbols

Chapter 1

$\rightarrow$	Implies
$\leftrightarrow$	Implies and implied by
$<$	Less than
$\leq$	Less than or equal to
$>$	Greater than
$\geq$	Greater than or equal to

Chapter 2

C	The set of complex numbers
$\mathbb{R}$	The set of real numbers
Q	The set of rational numbers
Z	The set of integers
$\mathbb{N}$	The set of natural numbers
$\in$	Is an element of
$\subseteq$	Is a subset of
$\cup$	Union
$\cap$	Intersection
$[a, b]$	$\{x \in \mathbb{R} : a \leq x \leq b\}$
(a, b)	$\{x \in \mathbb{R} : a < x < b\}$
$(a, b]$	$\{x \in \mathbb{R} : a < x \leq b\}$
$[a, b)$	$\{x \in \mathbb{R} : a \leq x < b\}$
(r, ∞)	$\{x \in \mathbb{R} : r < x\}$
$[r, \infty)$	$\{x \in \mathbb{R} : r \leq x\}$
$(-\infty, r)$	$\{x \in \mathbb{R} : x < r\}$
$(-\infty, r]$	$\{x \in \mathbb{R} : x \leq r\}$
$P(A)$	The power set of A = The set of all subsets of A
$a R b$	a is related to b under the relation R
$[x]$	The equivalence class of x under the equivalence relation R

Chapter 3

$\vee$	OR
$\wedge$	AND
$\Rightarrow$	Implies (tautological)
$\Leftrightarrow$	Implies and implied by (tautological)

$\forall$ For all
$\exists$ There exists

Chapter 4

$nC_r = \binom{n}{r}$	Combination of r objects out of n objects
nPr	Permutation of r objects out of n objects

Chapter 5

$\lfloor x \rfloor$ The greatest integer less than or equal to x

Chapter 7

$a \cong b$ (mod m) $a - b$ is a multiple of m

Chapter 8

$\otimes$, $\odot$	Symbols denoting binary operations
$\oplus_n$	Addition modulo n
$\oplus_n$	Multiplication modulo m
$GLn(\mathbb{R})$	The set of all non-singular $n \times n$ matrices with real entries
$SLn(\mathbb{R})$	The set of all $n \times n$ matrices with a determinant value as 1

Chapter 9

$G = (V, E, \varphi)$	A graph whose vertex set is V, edge set is E and φ is the function defined on E giving the end vertices of the edges
$deg(v)$	The degree of a vertex v
$n(G)$	The number of connected components of a graph G

Chapter 10

$\leq$	A partial order
$a \vee b = a \oplus b$	$\sup\{a, b\}$
$a \wedge b = a \odot b$	$\inf\{a, b\}$

Chapter 14

$P(A)$	The probability of the event A
$A^c = \bar{A}$	The event that A does not occur
$A \cup B$	The event that either A or B occurs
$AB = A \cap B$	The event that both A and B occur

Index